86332
WITHD
FRO
UNIVERS
AT
MEDWAY
LIBRARY
D1759199
DWAY RY

ANIMAL FEEDING AND NUTRITION

A Series of Monographs and Treatises

Tony J. Cunha, Editor

Distinguished Service Professor Emeritus
University of Florida
Gainesville, Florida

and

Dean Emeritus, School of Agriculture
California State Polytechnic University
Pomona, California

Tony J. Cunha, SWINE FEEDING AND NUTRITION, 1977

W. J. Miller, DAIRY CATTLE FEEDING AND NUTRITION, 1979

Tilden Wayne Perry, BEEF CATTLE FEEDING AND NUTRITION, 1980

Tony J. Cunha, HORSE FEEDING AND NUTRITION, 1980

Charles T. Robbins, WILDLIFE FEEDING AND NUTRITION, 1983

Tilden Wayne Perry, ANIMAL LIFE-CYCLE FEEDING AND NUTRITION, 1984

Lee Russell McDowell, NUTRITION OF GRAZING RUMINANTS IN WARM CLIMATES, 1985

Ray L. Shirley, NITROGEN AND ENERGY NUTRITION OF RUMINANTS, 1986

Peter R. Cheeke, RABBIT FEEDING AND NUTRITION, 1987

Lee Russell McDowell, VITAMINS IN ANIMAL NUTRITION: COMPARATIVE ASPECTS TO HUMAN NUTRITION, 1989

Tony J. Cunha, HORSE FEEDING AND NUTRITION, 2E, 1990

Dennis J. Minson, FORAGE IN RUMINANT NUTRITION, 1990

FORAGE IN RUMINANT NUTRITION

FORAGE IN RUMINANT NUTRITION

Dennis J. Minson
Division of Tropical Crops and Pastures
Commonwealth Scientific and Industrial Research Organisation
St. Lucia, Queensland, Australia

UNIVERSITY OF GREENWICH LIBRARY
636.2085 MIN

ACADEMIC PRESS, INC.
Harcourt Brace Jovanovich, Publishers
San Diego New York Boston
London Sydney Tokyo Toronto

Cover photograph: Belmont Reds at Narayen Research Station in Queensland, Australia. Courtesy of CSIRO, Division of Tropical Crops and Pastures.

This book is printed on acid-free paper. ∞

Copyright © 1990 by Academic Press, Inc.
All Rights Reserved.
No part of this publication may be reproduced or transmitted in any form or by any means, electronic or mechanical, including photocopy, recording, or any information storage and retrieval system, without permission in writing from the publisher.

Academic Press, Inc.
San Diego, California 92101

United Kingdom Edition published by
Academic Press Limited
24–28 Oval Road, London NW1 7DX

Library of Congress Cataloging-in-Publication Data

Minson, Dennis J.
Forage in ruminant nutrition / Dennis J. Minson.
p. cm.
Includes bibliographical references and indexes.
ISBN 0-12-498310-3 (alk. paper)
1. Ruminants--Feeding and feeds. 2. Forage plants. 3. Ruminants--Nutrition. I. Title.
SF95.M6585 1990
636.2'084--dc20 90-778
CIP

b9212681

Printed in the United States of America
90 91 92 93 9 8 7 6 5 4 3 2 1

Contents

10 Sodium

11 Copper

12 Iodine

13 Zinc

14 Manganese

15 Selenium

16 Cobalt

Foreword

This is the twelfth in a series of books about animal feeding and nutrition. The books in this series are designed to keep the reader abreast of the rapid developments that have occurred in this field. As the volume of scientific literature expands, interpretation becomes more complex, and a continuing need exists for summation and for up-to-date books.

D. J. Minson is a distinguished scientist in the use of forages in feeding ruminant animals. He has spent 35 years studying the environmental, chemical, and physical factors that control the nutritive value of grazed and conserved temperate and tropical forages. He graduated in Agricultural Chemistry from Reading University, worked on various aspects of forage quality at the Grassland Research Institute at Hurley, the Animal Research Institute at Ottawa, and the Ruakura Animal Research Station in New Zealand. In 1963 he joined the Australian Commonwealth Scientific and Industrial Research Organisation. He is currently a Chief Research Scientist with the Division of Tropical Crops and Pastures and Officer in Charge of the Cunningham Laboratory. He was awarded a D.Sc. from Reading University in recognition of the importance of his distinguished research career.

Many scientists feel there is a need to double animal protein production in the next 20 to 25 years in order to improve the protein intake status of the world's rapidly growing human population. Every $2\frac{1}{2}$ to 3 years, the world's population increases by approximately 230 million people, and about 87 percent of this population growth takes place in countries that are the least able to feed themselves. Many third-world countries have increased food production, but the increase is not keeping pace with rapid population growth. It is estimated that over 1 billion people now suffer from chronic malnutrition. The developing countries have about 60 percent of the world's animals but produce only 20 percent of the world's meat, milk, and eggs. Better feeding and nutrition programs would increase their production of animal foods for human consumption.

The total population of ruminants that make some contribution of food for humans approaches 2 billion head. Cattle and sheep are the most numerous: each group contains more than 1 billion head. There are approximately 400 million goats. About 40 percent of the cattle and 50 percent of the sheep are in the developed countries and more than 90 percent of the goats are in the developing countries. It is apparent, therefore, that ruminant animals are important throughout the world. It is estimated that ruminants produce two-thirds of the world's meat and 80 percent of the world's milk in the developed countries. Approximately 90 percent of the feed available to ruminants throughout the world consists of forage-grass, browse, legumes, hay, and straw. Therefore, the information in *Forage in Ruminant Nutrition* is of considerable value for increasing production and efficiency.

Tony J. Cunha

Preface

Domestic ruminants make a major contribution to human welfare. They provide 70% of the total animal protein eaten and 10% of the natural fiber used by humans. In tropical countries they are often a source of draft power. This is achieved without seriously reducing the quantity of food available for direct human consumption. Ruminants, with their symbiotic population of rumen microbes and ability to chew partially digested food, have the capacity to utilize forages that contain too much fiber for human consumption.

The efficiency of grazing ruminants, as producers of meat, milk, and fiber, depends on many factors. Many books have been written on plant physiology, nutrition, and composition. At the other end of the spectrum are books on biochemistry, rumen microbiology, and nutrient requirements of ruminants. Between these two extremes is forage nutritive value and ruminant production, the subject of this book.

As a source of nutrients, forage has the most variable composition of any feedstuff, being affected by forage species, soil fertility, stage of growth, and management practices. Efficient use of forage can only be achieved by understanding these factors and the way they can be manipulated by humans.

Forage in ruminant nutrition is a very broad area and it has been necessary to place limits on the scope of this review. I have achieved this by considering only those nutrients likely to be deficient in grazed and conserved forage, how these deficiencies may be identified, and what management practices can be adopted to improve production.

This book is a hybrid, drawing on information from both the plant and animal sciences. I hope this will stimulate the interest of students, lecturers, research workers, and extension officers in the more efficient use of forage and thus make some small contribution to feeding an expanding world population.

In preparing this book I have obtained numerous helpful suggestions from eminent forage and ruminant scientists. I am especially grateful to the following: A. C. Field, R. J. W. Gartner, J. B. Hacker, D. P. Henry, D. P. Poppi, N. L. Suttle, and R. J. Williams.

I am particularly grateful to Alison Minson for her many useful suggestions and untiring assistance in the editing of the entire book. Likewise, I wish to acknowledge with thanks and appreciation the skill and care of Kathy Mitchell and Joanne Lawton for typing and Peter Tuckett and Judy Thompson for valuable assistance. Also, I am indebted to R. J. Clements, Chief of the Division of Tropical Crops and Pastures of the Commonwealth Scientific and Industrial Organisation, St. Lucia, Brisbane, for encouraging me to complete this book.

Dennis J. Minson

Terminology and Symbols Used

AD	=	Apparent digestibility
ADF	=	Acid detergent fiber
AIA	=	Acid-insoluble ash
ARDOM	=	Apparent rumen degradable organic matter
°C	=	Degree centigrade
Ca	=	Calcium
cc	=	Cubic centimeters
CF	=	Crude fiber
Co	=	Cobalt
CP	=	Crude protein
Cu	=	Copper
d	=	Day
DCP	=	Digestible crude protein
DE	=	Digestible energy
DM	=	Dry matter
DMD	=	Dry-matter digestibility coefficient
DOM	=	Digestible organic matter
DOMD	=	Digestible organic matter in dry matter
D-value	=	DOMD in 100 g DM
ED	=	Energy digestibility coefficient
F	=	Feces, fecal loss
g	=	Grams
GE	=	Gross energy
ha	=	Hectare
H	=	Heat increment
I	=	Iodine
in sacco	=	In bag
in vitro	=	In glass
in vivo	=	In animal

IRD = Intraruminal slow-release device
K = Potassium
k_f = Efficiency of utilization of ME for growth and fattening
kg = Kilograms
kg W = Body weight in kilograms
LW = Liveweight
M = Metabolic secretions
m = Meters
MADF = Modified acid detergent fiber
ME = Milliequivalent
mEq = Metabolizable energy
Mg = Magnesium
mg = Milligram (10^{-3}g)
ml = Milliliters
mm = Millimeters
Mn = Manganese
Mo = Molybdenum
N = Nitrogen
Na = Sodium
NDF = Neutral detergent fiber
NE = Net energy
ng = Nanogram (10^{-9}g)
NIR = Near-infrared radiation, also near infrared reflectance
NMD = Nutritional muscular dystrophy
NPN = Nonprotein nitrogen
OM = Organic matter
OMD = Organic matter digestibility coefficient
P = Phosphorus
PD = Potential Digestibility
RSD = Residual standard deviation
S = Sulfur
Se = Selenium
TD = True digestibility
TDN = Total digestible nutrients
VFA = Volatile fatty acids

U = Urine
μg = Microgram (10^{-6}g)
VI = Voluntary intake
yr = Year
Zn = Zinc

FORAGE IN RUMINANT NUTRITION

1

Ruminant Production and Forage Nutrients

I. INTRODUCTION

Domestic ruminants are kept by humans to produce milk, meat, and wool from plant material, which, for the most part, is unsuitable for direct human consumption. In some cultures ruminants are also an important source of power (Copland, 1985), are utilized as wealth or status, and are ceremonial. Potential production of these animals has been raised during centuries of breeding and selection by humans, while losses in developed countries caused by disease, toxic plants, and bad husbandry practices have been reduced to low levels. Maximum production of meat, milk, or wool will be achieved only if animals are supplied with sufficient quantities of the raw materials required for the synthesis of those products. This can occur when housed ruminants are fed grain-based diets supplemented with protein, minerals, and vitamins, but when forage is the sole source of nutrients, production is invariably much lower than the genetic potential of the animal.

There are many available ways to improve the quality of forage-based diets and increase profit. Identification of the optimum forage strategies for use on an individual property or in a region requires a knowledge of the different nutrients required for production, the ability of the forage to supply these nutrients, how to identify which aspect of forage quality is failing, and the ways this deficiency may be prevented. Any potential solution must obviously take into account the relevant local economic, environmental, and social factors.

The digestive tract of all herbivores contains bacteria, protozoa, and fungi capable of hydrolyzing cellulose, hemicellulose, and other substances resistant to digestion by enzymes secreted by the host animal. Microbial hydrolysis of forage is a slow process and the digestive tracts of herbivores are modified in various ways to increase the quantity of forage retained and hence the time it is exposed to the microflora. In ruminants, the adaptation takes the form of an enlargement of the forestomach to form the reticulorumen and the ability to regurgitate and chew forage that has been partially digested and softened in the reticulorumen.

The quantity of each nutrient absorbed depends on (1) the quantity of forage dry matter eaten each day, and (2) the concentration and availability of that particular nutrient in each kilogram (kg) of forage dry matter (DM). The factors controlling voluntary food intake will be considered in Chapters 2 and 3, while the remaining chapters will consider the different nutrients that may limit ruminant production from forage.

This chapter will consider the composition of ruminant products, the nutrients to be found in forage, and those nutrients that could possibly limit ruminant production from forage and hence warrant further consideration in this volume.

II. COMPOSITION OF RUMINANT PRODUCTS

A. Wool

The fleece of the sheep comprises three fractions: the actual wool fiber, suint (the secretion of the sweat glands), and fat (the secretion of the sebaceous glands). The relative proportions of these three fractions vary between breeds and managements. For British breeds of sheep a typical fleece contains 80% wool, 12% fat, and 8% suint (ARC, 1980). The wool fibers consist almost entirely of the protein keratin, which is characterized by a high content of cysteine, a sulfur-containing amino acid. The chemical composition of clean dry fleece of British breeds is shown in Table 1.1.

B. Milk

The composition of milk produced by ruminants varies with species, breed, age, stage of lactation, and nutrition (ARC, 1980). The main constituent of milk is water, ranging from 83.6 to 87.3% for sheep and temperate cattle, respectively (Armsby and Moulton, 1925). Solids in the milk of temperate breeds of cattle contain approximately equal quantities of protein, fat, and carbohydrate, in the form of lactose (Table 1.1). The remaining solid consists of a range of mineral elements and vitamins (Table 1.1).

C. Body Tissue

The body of ruminants is mainly composed of protein, fat, and water with a smaller quantity of mineral matter. With increasing maturity there

TABLE 1.1

Mean Chemical Composition of Fleece, Milk, and Tissue Gain Produced by Ruminant Compared with the Nutrients in Forage (g/kg DM)

Components	Fleece[a] (British breeds)	Milk[a] (temperate cattle)	Body tissue[a] (cattle)	Forage[b] (mean)
Energy constituents				
Protein	809	266	230	142
Fat	125	289	707	54
Carbohydrate	—	388	—	657
Lignin	—	—	—	41
Mineral constituents				
Total ash	—	57	63	106
Calcium	1.4	9	22	9.0
Phosphorus	0.3	7	13	2.9
Magnesium	0.3	1	1	2.8
Potassium[c]	17.0	11	3	2.7
Sodium[c]	1.1	4	2	2.2
Chlorine[c]	—	8	2	4.2

[a]Fleece (ARC, 1980); milk (Armsby and Moulton, 1925; ARC, 1980). Growth composition at 400 kg empty body weight (ARC, 1980).

[b]To be discussed in following chapters.

[c]Usually no field response to feeding these elements as supplements.

is a decrease in the proportion of protein and an increase in the proportion of fat. For example, calves with an empty weight of 50 kg contain four times as much protein as fat, but by the time they reach 500 kg the body contains twice as much fat as protein (ARC, 1980). This change is due to the high fat content of new growth, particularly in mature animals (Fig. 1.1). At the final stages of cattle fattening, 86% of the energy which contributes to increased weight is stored as fat and only 14% as protein. Similar changes in body composition have been found in sheep (ARC, 1980). A small part of the gain in weight is in the form of bone. The main elements involved are calcium and phosphorus with smaller quantities of magnesium and sodium (Table 1.1).

III. NUTRIENTS REQUIRED FOR MAINTENANCE AND REPRODUCTION

All animals require nutrients in order to maintain body processes and for reproduction. Energy is required for the muscular work of circulation and respiration and in mammals for maintaining body temperature.

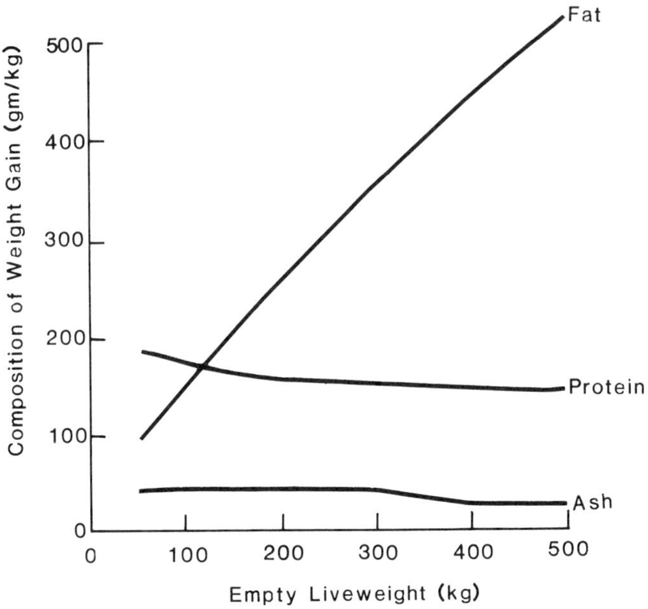

Fig. 1.1. Effect of maturity on body composition of cattle. Adapted from Armsby and Moulton (1925) and ARC (1980).

Muscles and enzymes have to be restored, and minerals unavoidably lost in the urine, feces, or sweat must be replaced. Trace elements and vitamins are also required (Table 1.2). The functions of these different nutrients are adequately reviewed in most standard texts on animal and human nutrition and only their availability to ruminants from forage will be considered in this volume.

IV. NUTRIENTS IN FORAGE

A. Water

The water required by ruminants is derived from three sources: water in the forage, water formed within the body as a result of oxidation of the nutrients absorbed from the forage, and water drunk. Lactating cows grazing forage in a temperate environment and provided with a water supply drank on average 40 kg/day, although the quantity was influenced by rainfall, maximum air temperature, and dry-matter content of the herbage (Castle, 1972). The water requirement of lactating cows is positively correlated with the quantity of milk produced (Little *et al.*, 1978a). Dry cattle

TABLE 1.2

Trace Elements and Vitamins Required by Ruminants[a]

Trace elements	Vitamins
Iron[b]	Fat-soluble
Copper	A retinol
Iodine	D_2 ergocalciferol
Zinc	D_3 cholecalciferol
Manganese	E tocopherol
Selenium	K phylloquinone[c]
Cobalt	
Molybdenum[b]	Water-soluble
	B complex
Cadmium[d]	B_1 thiamin[c]
Lithium[d]	B_2 riboflavin[c]
Nickel[d]	nicotinamide[c]
	B_6 pyridoxine[c]
	panthothenic acid[c]
	biotin[c]
	folacin[c]
	choline[c]
	B_{12} cyanocobalamin[c]
	C ascorbic acid[b]

[a]Data from Nielsen (1984) and McDonald *et al.* (1988).

[b]Usually present in all forages in quantities exceeding ruminant requirements.

[c]Synthesized by microorganisms in the digestive tract.

[d]Ultratrace elements. No evidence of deficiencies in ruminants fed forage.

require less drinking water than do lactating cattle, and sheep and goats probably require proportionally less water than do cattle, due to the higher dry-matter concentration of their feces. Other aspects of water requirements of ruminants are considered in a publication by the ARC (1980).

The water content of forages varies with weather conditions, species, and stage of maturity. Published values for water content range from 920 g/kg for grazed forage (Davies, 1962) to less than 100 g/kg for dried grass. It is considered good practice for grazing animals to have access to drinking water at all times because forage usually cannot provide all the water needed by ruminants. To meet these requirements the water content of forage would have to be raised to such a level as to have an adverse effect on voluntary intake (Chapter 2). Forage as a source of water for ruminants will not be considered in this volume. Readers interested in the provision of drinking water are referred to reviews by ARC (1980) and by Shirley (1985).

B. Protein

Protein is a major component of all ruminant products (Table 1.1) and is also required for maintenance and reproduction. Proteins are complex organic compounds of high molecular weight containing 22 amino acids in varying proportions. In common with carbohydrates and fats, amino acids contain carbon, hydrogen, and oxygen, but additionally all amino acids contain nitrogen (N). Three amino acids, cystine, cysteine, and methionine, also contain sulfur (S). Sulfur and nitrogen are closely associated in all proteins and there appears to be no advantage to considering them as independent nutrients and reviewing their role in separate chapters.

For lactation and for the growth of young ruminants, protein is often the main nutrient that limits production. This is illustrated by the improved production achieved by providing additional protein. Other forms of production require less protein, and excess protein in the forage is converted into glucose which is used for the synthesis of fat, lactose, etc. (Blaxter, 1962; Preston and Leng, 1987).

The protein content of forage is very variable and undergoes large changes in the rumen before being absorbed in the small intestine. These and other aspects of protein supply from forages will be reviewed in Chapter 5.

C. Energy

The gross energy (GE) of forage is relatively constant but there are large differences in the availability of this energy to the animal. The simplest measure of available energy is digestible energy (DE). This takes account of the energy lost in the feces (F).

$$DE = GE - F$$

The DE of forage is closely correlated with the proportion of forage dry matter and organic matter (OM) digested. Since these parameters are less expensive to measure than DE, they have been the main method of expressing the energy availability of forage. This information will be presented in Chapter 4.

Energy requirements of ruminants are now quoted in terms of metabolizable energy (ARC, 1980). The metabolizable energy (ME) value of a forage takes account of the energy lost in the urine (U) and as methane (M), in addition to the fecal loss (F).

$$ME = GE - F - U - M$$

The ME content of forage is closely correlated with the more readily measured DM and OM digestion coefficient of the forage and for many purposes ME can be estimated with sufficient accuracy from either of these parameters. Relatively few ME values have been determined for forage and most of those published in feedstuffs tables are predicted from DM digestibility or chemical composition. No attempt has been made to review this information since any conclusions would be the same as those drawn from the results of the digestibility studies considered in Chapter 4.

The true energy value of a forage is the quantity of energy that can be retained in a product or used to spare the catabolism of the body reserves for maintenance purposes. This is the net energy (NE) value of a feed and is determined by subtracting the additional heat produced when a forage is eaten (heat increment, H) from the ME value.

$$NE = ME - H$$

Forages with a high ME value generally have a high NE value but there are major exceptions of practical importance. The ME in spring forages is utilized more efficiently than autumn forage and pelleted forage more efficiently than chopped material. It is also possible to increase the efficiency of utilization of ME. These and other aspects of the efficiency of conversion of ME to NE will be reviewed in Chapter 5.

D. Mineral Elements

All the mineral elements known to be required by ruminants are shown in Tables 1.1 and 1.2. Seven of these elements, potassium, chlorine, iron, molybdenum, cadmium, lithium, and nickel, usually appear to be present in sufficient quantities in all forages to meet ruminant requirements. For these elements there appear to be only a few reports of an improvement in production when they are fed as a supplement to grazing ruminants (McDowell, 1985), so they will not be considered in this volume.

E. Vitamins

The vitamin requirements of ruminants and their availability in forages have been thoroughly reviewed by the ARC (1980) and by McDowell

(1985). There is, however, one vitamin, B_{12}, that will be considered. Vitamin B_{12} is absent from forage but is synthesized by microbes in the rumen. The quantity of vitamin B_{12} synthesized by rumen microorganisms is determined primarily by the cobalt concentration in the diet, and symptoms of cobalt deficiency and vitamin B_{12} deficiency are identical. The concentration of cobalt in forage will be considered in Chapter 16. There are few situations where grazing ruminants have responded to feeding vitamins other than B_{12}, so they will not be considered in this volume.

V. SOURCES OF NUTRIENTS

The main feature of domestic ruminants is their ability to survive and produce on fibrous diets unsuitable for pigs, poultry, and humans. The most important sources of fibrous nutrients for ruminants are forage grasses and legumes which are grazed or eaten after conservation as hay or silage. In this volume the factors that control the voluntary intake, digestibility, and chemical composition of forage will be reviewed, with the aim of establishing principles that can be applied to all types of forage. No attempt will be made to list data for the different forage species, as this information has already been collated (Schneider, 1947; Gohl, 1975; see also various publications of the International Network of Feed Information Centres). These publications also contain details on the chemical composition and digestibility of native browse shrubs and agricultural byproducts, so these will not be considered in this volume except where they illustrate a principle that could apply to all forages.

Cereal straws are another source of nutrients for ruminants and the upgrading of these low-quality materials has received considerable attention in the last decade. No attempt will be made to consider this work because it has been the subject of many recent reviews (Pearce, 1983; Sundstol and Owen, 1984; Devendra, 1988). However, reference will be made to some of the upgrading processes used for straw where these have a potential for improving the nutritive value of hay.

Domestic ruminants can also use low-fiber diets based on cereal grains and, where forage diets are of poor quality, supplements of cereal grains can be used to achieve higher levels of production. The effect of grain supplements on voluntary intake, digestibility, and net energy will be reviewed but the value of different types of grain as a source of nutrient for ruminants is beyond the scope of this volume.

2

Intake of Forage by Housed Ruminants

I. INTRODUCTION

The quantity of forage dry matter eaten is the most important factor controlling ruminant production from forages (see Chapter 1). Voluntary intake (VI) of forage may be defined as the quantity of dry matter eaten each day when animals are offered excess feed. By measuring the VI of forage with housed animals two major objectives can be achieved:

1. The effect of forage species, varieties, plant parts, processing, and the influence of soils, as well as climate in which forages are grown, can be accurately measured.
2. Plant attributes that can be used to predict VI from physical or chemical analysis of small samples of the forage, or from a knowledge of the way the forage was grown and processed, may be identified.

This chapter will consider the factors that control the VI of forage by housed ruminants. These factors also apply to grazing animals but there are additional factors which are specific to grazed forage. These will be considered in Chapter 3. Also included in this section will be discussion of studies necessarily carried out indoors but which are more strictly related to the outdoor situation (see Sections I,D,5 and I,D,7).

II. MEASUREMENT OF VOLUNTARY INTAKE

A. Forage Form

Forage may be cut and fed daily or fed after freezing, drying, ensiling, or pelleting. When forage is field dried, ensiled, or pelleted, there are often large changes in VI compared to fresh forage (Section III,H) and these conservation methods are not used in studies where the results are to be applied to grazed forage.

The method of cutting and previous history of the sward can affect VI.

Cutting with a flail forage harvester depressed the VI of forage from clean swards by between 2 and 7%, but when the forage had been previously grazed the presence of very small quantities of feces depressed VI by 25 and 43% for chopped and lacerated forage, respectively (Tayler and Rudman, 1965).

Equipment has been developed for rapidly cutting, elevating, and transporting large quantities of fresh forage from the field with minimal chopping, bruising, and laceration (Hutton *et al.,* 1975). Chopping fresh *Festuca arundinacea* (2- to 4-cm pieces) increased the VI of sheep by 6 to 9% (Chenost and Demarquilly, 1982), but with a grass/legume hay chopping had no effects on VI (Webb *et al.,* 1957). Fresh forage is usually fed twice each day, with weighing and discarding of refused forage each morning, prior to feeding (Chenost and Demarquilly, 1982).

B. Feeding Method

1. Level of Excess Forage

In order to measure VI, forage must be available at all times and more forage must be offered than can be eaten. For example, cattle ate 25% less hay when access time was reduced to 5 hr each day (Freer and Campling, 1963), and larger depressions have been recorded with silage (Harb and Campling, 1983). Voluntary intake is not altered by the number of times forage is given each day provided it is available more than 18 hr each day (Blaxter *et al.,* 1961; Ronning and Dobie, 1967).

Voluntary intake of forage depends on the quantity of excess feed offered (Fig. 2.1). This increase in intake is caused by two factors. Appetite varies from day to day, and sufficient forage must be available on days when appetite is high if intake is not to be depressed by a shortage of forage. The second reason for variation in intake is the increased opportunity for the animal to select the more desirable and usually less fibrous parts of the forage when excess forage is made available. With *Andropogon gayanus, Pennisetum purpureum,* and several tropical legumes, the leaf fraction was preferred and the diet selected was very different from that offered (Butterworth, 1965; Zemmelink *et al.,* 1972; Zemmelink, 1980). In one study sheep were offered *P. purpureum* at three stages of maturity but the quantity offered was so large that the sheep were able to select a diet of similar composition from these contrasting forages and VI was hence unaffected (Fig. 2.2). When measuring VI a reasonable compromise was achieved by limiting the excess of forage offered to 5–10% (Wilson and McCarrick, 1967), 10–15% (Blaxter *et al.,* 1961; Heaney

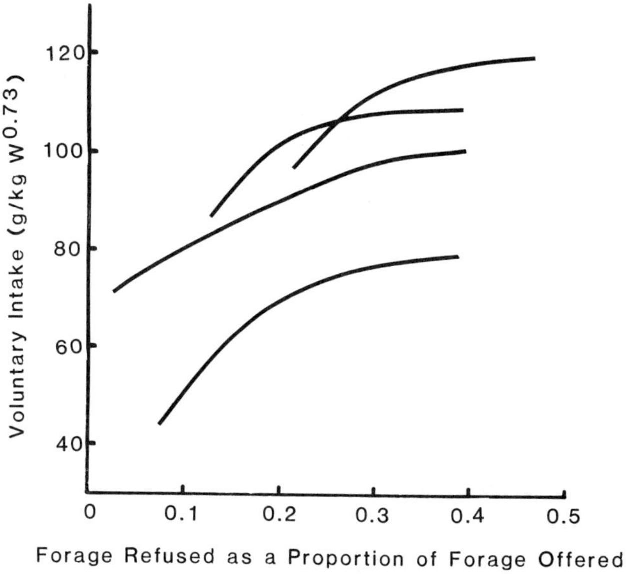

Fig. 2.1. The voluntary intake of cattle offered different quantities of four forages. Data from Tayler and Rudman (1965).

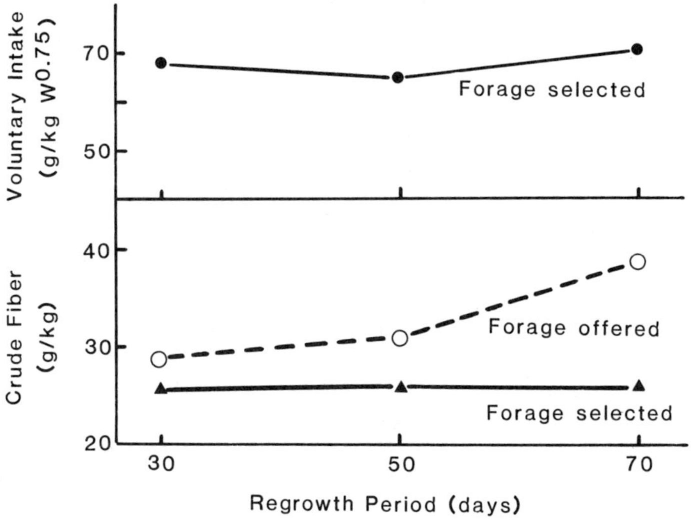

Fig. 2.2. The constant voluntary intake and fiber content of diets selected by sheep from forage at three stages of maturity. Data from Butterworth (1965).

et al., 1969; Cammell, 1977; Chenost and Demarquilly, 1982), or between 5 and 20% (Heaney, 1973). When dry forages are fed it is possible to estimate VI with minimal selection by topping up the manger each day and collecting the excess feed at the end of the measurement period.

2. Preliminary and Measurement Periods

Preliminary feeding periods have varied from 0 to 30 days but 7 days is generally considered sufficient time provided the animals are accustomed to the feeding facilities and there has been no major change of diet. Voluntary intake is influenced by the previous levels of feeding (Pienaar *et al.*, 1983), especially where these cause large differences in body condition (Section II,D,2). The measurement period has varied from 5 days to 1 yr, but for most purposes 10–15 days is sufficient (Blaxter *et al.*, 1961).

Recommended lengths of the preliminary and measurement periods are based on the assumptions that the composition of the forage does not change during the study and the appetite of the animal is neither depressed by a toxic component in the forage nor is the forage deficient in an essential nutrient.

C. Number of Animals

Since the coefficient of variation of VI of sheep of the same size and condition is between 7 and 16% (Crampton *et al.*, 1960; Blaxter *et al.*, 1961; Minson *et al.*, 1964; Chenost and Demarquilly, 1982), the number of animals used to measure VI depends on the accuracy required (Heaney *et al.*, 1968; Chenost and Demarquilly, 1982). Low variation has been found when feeding frozen forage (c.v. 10.3%), and high variation has been found with silage (c.v. 21.0%) (Heaney *et al.*, 1968). Sheep kept in metabolism pens are less variable (c.v. 11.5%) than are sheep kept in individual stalls within an exercise yard (c.v. 25.9%) (Heaney *et al.*, 1968).

Six sheep are often used (Demarquilly and Weiss, 1970; Heaney, 1979), but where the results are to be used for developing laboratory methods for predicting VI, 8–10 sheep are recommended. Voluntary intake is subject to unexpected week-to-week differences (Heaney *et al.*, 1968), so all forages being compared are usually fed at the same time in the same animal house and this requires large numbers of animals. Labor-efficient facilities suitable for measuring the VI of forage with 76 sheep have been described (Minson and Milford, 1968b).

D. Animal Species and Condition

1. Differences between Species

Voluntary intake of forages varies between animal species, a difference that is mainly due to body size (Anderson *et al.*, 1977; Ternouth *et al.*, 1979). Differences in VI between sheep are related to metabolic size, calculated as body weight (kg) raised usually to the power of 0.75 (Crampton *et al.*, 1960). However, in a number of trials cattle ate more forage than sheep even when VI was expressed on the basis of the 0.75 power of body weight (Playne, 1970b; Ternouth *et al.*, 1979; Chenost and Martin-Rosset, 1985), but when the 0.9 power of body weight was used VI was similar for the two species (Table 2.1).

There are also differences in VI between ruminant species of similar size. The VI of *Bos taurus* is higher than that of *Bos indicus* (Ledger *et al.*, 1970; Moran *et al.*, 1979), *Bubalus bubalis* (Grant *et al.*, 1974; Moran *et al.*, 1979), and *Bos banteng* (Moran *et al.*, 1979). Sheep and goats had similar VI when compared on the basis of the 0.75 power of body weight (Gihad, 1976; Watson and Norton, 1982), but there was considerable variation.

2. Compensatory Intake

When animals of the same age but different body condition are fed *ad libitum* the lighter animals eat more forage and grow faster, an effect described as compensatory growth (Wilson and Osbourn, 1960; Allden, 1970; O'Donovan, 1984). Thin cows ate 8.2 kg/day of *Lolium perenne* hay compared with 6.3 kg eaten by fat cows and had 17% more digesta in the

TABLE 2.1

Voluntary Intake of Eight Tropical Forages by Cattle and Sheep When Expressed in Different Ways[a]

Measurement	Cattle	Sheep	Significance of difference
Mean weight (kg)	481	48	
Voluntary intake			
kg DM/day	6.40	0.84	*
g DM/$W^{0.75}$	62.3	46.0	*
g DM/$W^{0.90}$	24.7	25.8	NS
g DM/$W^{1.0}$	13.3	17.5	*

[a]Data from Ternouth *et al.* (1979) and Poppi (1979).
*$p < 0.01$.

rumen (Bines *et al.*, 1969). When allowance was made for differences in body size, VI of the thin cows was almost twice that of the fat cows (90 versus 51 g/kg $W^{0.75}$). Sheep also exhibit this compensatory effect on VI, thin and fat ewes eating 1.9 and 1.4 kg/day of a dried grass, respectively (Foot, 1972). Compensatory intakes have also been reported for grazing cattle (O'Donovan *et al.*, 1972) and sheep (Allden and Young, 1964; Arnold and Birrell, 1977).

3. Pregnancy

Appetite increases during early and mid-pregnancy but generally declines in the few weeks prior to parturition (Forbes, 1970a). This fall in appetite has been found with both silage and hay, the extent of the reduction being inversely related to the subsequent birth weight of lambs (Forbes, 1970b).

4. Lactation

Following parturition, VI of both cattle and sheep increases (Forbes, 1970a) (Table 2.2). These increases in VI are found with all forages; they are in response to the additional energy required for milk production (Bines, 1976) and lead to an increase in the size of the rumen (Tulloh *et al.*, 1965; Weston and Cantle, 1982).

5. Shearing

When sheep are kept in a cold environment, shearing increases the heat production by about 19% (Davey and Holmes, 1977) and leads to a 20–51% rise in VI (Wheeler *et al.*, 1963; Wodzicka-Tomaszewska, 1963; Love *et al.*, 1978). The increase is related to both ambient temperature and quality of the forage; the largest increases occur at the lowest temperature and with the poorest quality forage (Minson and Ternouth, 1971). Increases in VI have also been reported for grazing sheep (Wheeler *et al.*, 1963; Arnold and Birrell, 1977) but only when the temperature is low. Shearing had no effect on VI when mean ambient temperature exceeded 10°C (Arnold and Birrell, 1977).

6. Wool Production

The energy required for wool growth is small and VI of high-wool-producing sheep fed a diet containing *Medicago sativa* was only 5% higher than that of sheep of a low-wool-producing line (Williams and Miller, 1965; Wodzicka-Tomaszewska, 1966). However there was no dif-

TABLE 2.2

Effect of Lactation on the Voluntary Intake of Forages by Cattle and Sheep

Forage	Increase in voluntary intake (%)	Reference
Cattle		
Avena sativa/Medicago sativa	16	Hunter and Siebert (1986a)
Temperate pasture	52	Hutton (1963)
Temperate pasture (grazed)	31	Dijkstra (1971)
Temperate hay	28	Campling (1966)
Veld (grazed)	28	Elliott and Fokkema (1961)
Mean for cattle	30	
Sheep		
Avena sativa	26	Owen *et al.* (1980)
Phalaris aquatica/Trifolium subterraneum (grazed)	38	Arnold and Dudzinski (1967)
	31	Arnold (1975)
Medicago sativa mixture	25	Weston and Cantle (1982)
Range forage (grazed)	28	Cook *et al.* (1961)
Temperate pasture (grazed)	35	Doney *et al.* (1981)
Mean for sheep	32	

ference in VI between lines when a low-protein diet was fed (Williams and Miller, 1965).

7. Exercise

Walking increases the energy requirements of sheep (Osuji, 1974), but when exercised on a treadmill sheep ate 3 to 10% *less* hay (Welch *et al.*, 1982). The sheep had been previously conditioned to the treadmill, so the depression in intake is difficult to explain.

8. Internal Parasites

Internal parasites are mainly a problem of grazing animals, but most studies of their effect on VI have been conducted with housed animals (see review by Sykes and Coop, 1976). On mixed diets, VI was depressed by high levels of roundworms, particularly *Trichostrongyle* species, which inhabit the small intestine (Sykes and Coop, 1976). With ewes *Ostertagia circumcinata* larvae reduced VI during lactation but had no effect on pregnant ewes (Leyva *et al.*, 1982).

E. Temperature

1. Cold Stress

When ambient temperature is below the critical temperature of the animal, there is an increased requirement for energy and a higher VI would be expected (Weston, 1970). With shorn sheep, changing room temperature from 21–26°C to 6–7°C increased VI by 10% (Weston, 1970). When the temperature was dropped to −5°C, VI and rumination time were increased by 27 and 23%, respectively (Kennedy, 1985).

2. Heat Stress

High ambient temperatures cause a depression in VI (Johnson and Yeck, 1964). With cattle fed *Dactylis glomerata, F. arundinacea,* or *M. sativa,* VI was depressed by 4% when ambient temperature was increased from 18 to 32°C (Warren *et al.*, 1974) and by 16% for a mixed forage (Vohnout and Bateman, 1972). The depression is likely to be greater for immature forages eaten in large quantities but no data appears to have been published on this point.

F. Units of Measurement

The water content of forages is very variable and values for VI are normally expressed on a moisture-free basis. Voluntary intake has been expressed as grams of dry matter per day but, to overcome differences in the size of the animals used to measure VI, results are usually expressed as g/kg body weight *(W)* or g/kg $W^{0.75}$ (Section I,D). The latter term was first used by Heaney *et al.* (1963) and the 0.75 power was adopted the following year by the Third Symposium on Energy Metabolism. Prior to this decision the 0.73 power of body weight had been used. With this exponent, values of VI are 7–14% higher than with the 0.75 exponent. Results have also been expressed on the basis of intake relative to that of *M. sativa* (Crampton *et al.*, 1960), but this method appears to have no advantage over the simpler g/kg $W^{0.75}$.

Care must be exercised when comparing the VI of similar forages measured in different studies due to differences caused by animal species, their physiological state, climatic stress, and internal parasites (Sections II,D and II,E). This problem can be reduced by including a standard forage in all studies (Abrams *et al.*, 1987), but the logistics of producing, storing, and distributing a standard forage throughout the world are daunting.

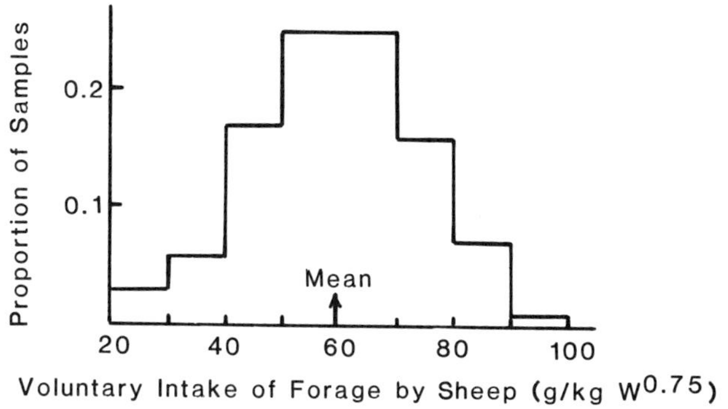

Fig. 2.3. Voluntary intake by sheep of 1215 different forages. Data from world literature.

III. FORAGE FACTORS CONTROLLING VOLUNTARY INTAKE

A. Introduction

The VI of forage is affected by forage species, cultivar, stage of growth, soil fertility, climatic condition, and conservation process. This wide variation is illustrated in Fig. 2.3. For 1215 published values for cut forages fed to sheep the mean VI was 59 g/kg $W^{0.75}$ with a standard deviation of ± 13 g/kg $W^{0.75}$. The forage factors contributing to this large variation will now be considered.

B. Species Differences

1. Legumes versus Grasses

The VI of forage depends on its resistance to breakdown (Balch and Campling, 1962; Troelsen and Campbell, 1968). Digesta particles must be reduced to a size where they will pass through a 1-mm screen before they can readily flow from the rumen (Poppi *et al.*, 1980). The critical size for passage of particles from the rumen is almost the same for grasses and legumes (Poppi *et al.*, 1980) and for sheep and cattle (Poppi *et al.*, 1985). Chewing during eating and ruminating is the most important way in which forage particles are reduced in size (Balch and Campling, 1962; Ulyatt *et al.*, 1986; McLeod and Minson, 1988).

Temperate legumes are usually eaten in greater quantities than grasses. This is illustrated in Fig. 2.4, which shows the frequency distribution for the VI of the 96 temperate legumes and 670 temperate grasses.

Temperate legumes are eaten in greater quantities than are grasses because they have a lower resistance to breakdown during eating and ruminating. Sheep spent 6.5 hr chewing each kilogram DM of *Trifolium repens* compared with 8.7 hr for a grass which had a higher DM digestibility (Weston, 1985). With *T. repens* and *L. perenne*, chewing times were 6.4 and 9.1 hr/kg DM, respectively (Aitchison *et al.*, 1986). In a comparison of *Trifolium pratense, Phalaris arundinacea,* and *Bromus inermis,* each

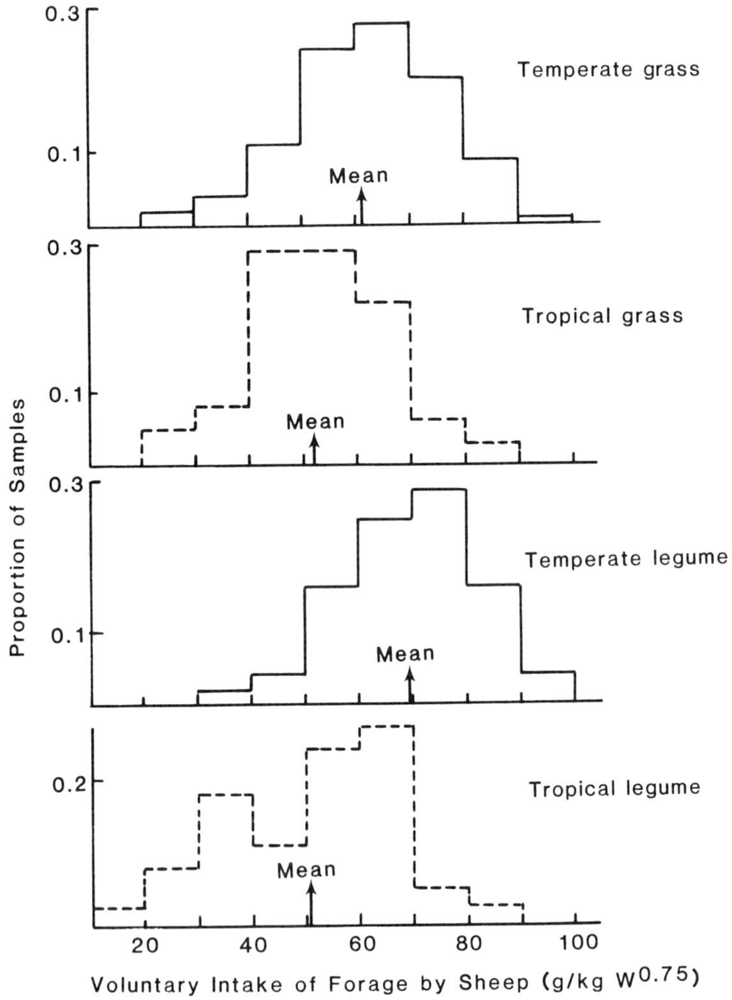

Fig. 2.4. Voluntary intake by sheep of temperate and tropical grasses and legumes. Data from world literature.

gram of legume DM was chewed 42 times compared with 64 and 70 times, respectively, for the two grasses (Chai *et al.*, 1985). The lower resistance of temperate legumes to breakdown by chewing results in the legumes being retained in the rumen for a shorter time than are grasses (Ingalls *et al.*, 1966; Ulyatt, 1970; Thornton and Minson, 1973b). Large legume particles also break down more rapidly than do large grass particles in the rumen (Fig. 2.5).

The lower resistance of legumes to chewing is probably due to the smaller quantity of cell wall constituents (Reid and Jung, 1965b; Ulyatt, 1970; Weston, 1985; Chai *et al.*, 1985; Aitchison *et al.*, 1986) and smaller length-to-width (aspect) ratio of the fibers (Troelsen and Campbell, 1968). Less energy is required to grind legumes than grasses (Weston, 1985) even though legumes contain more lignin (Reid and Jung, 1965b; Ulyatt, 1970; Aitchison *et al.*, 1986).

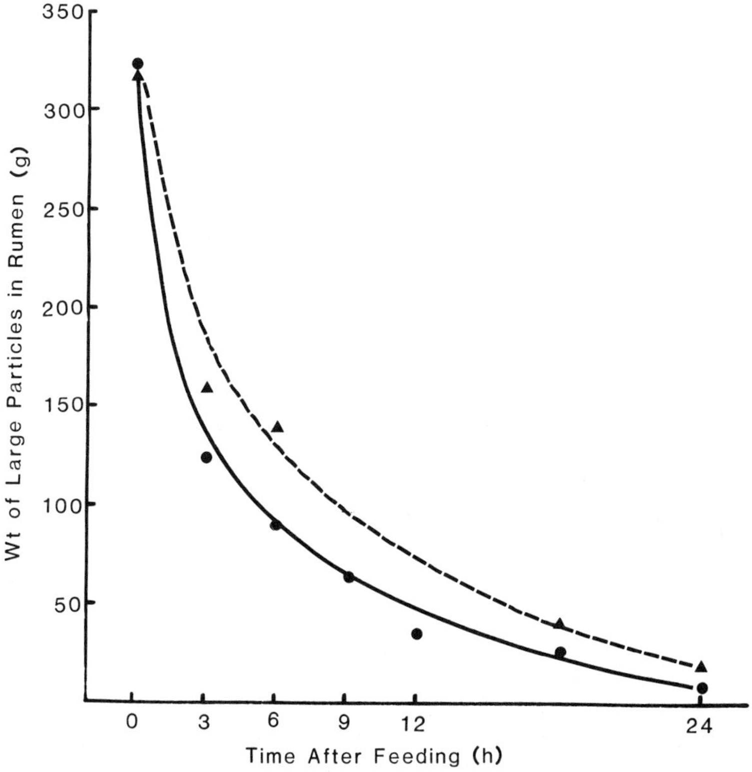

Fig. 2.5. Disappearance of large forage particles (>1 mm) from the rumen of sheep fed *Lolium perenne* (▲ — — —) and *Trifolium repens* (● ———). Data from Moseley and Jones (1984).

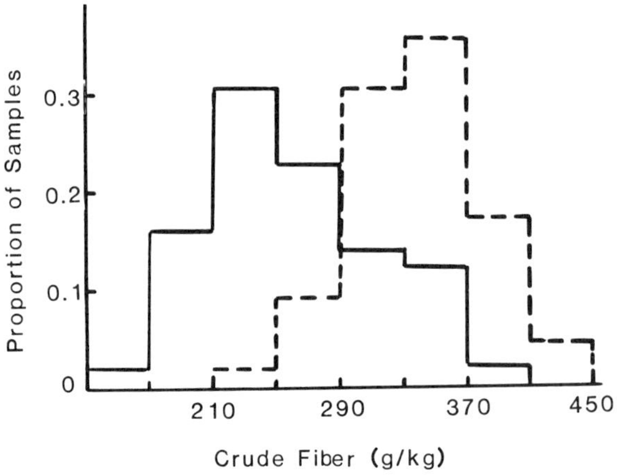

Fig. 2.6. The fiber content of a range of temperate (___) and tropical (— — —) forages. Data from world literature.

2. Temperate versus Tropical Species

Temperate forages generally have a higher VI than tropical forages, a difference that applies to both legumes and grasses (Fig. 2.4). For the temperate forages the mean VI was 61 g/kg $W^{0.75}$ compared with 50 g/kg $W^{0.75}$ for the tropical forages. The higher VI of temperate forages (Fig. 2.6) was associated with a lower level of fiber and higher digestibility of the dry matter (Chapter 4).

3. Mixtures of Grasses and Legumes

Where both the grass and the legume component of a mixture contains sufficient crude protein (CP) and minerals, VI is linearly related to the proportion of legume in the mixture and there is no synergism between the two species (Moseley, 1974) (Fig. 2.7).

Synergism will occur where one species in a mixture is deficient in an essential nutrient and the other species contains a high level of this nutrient. This occurred when *M. sativa* containing 220 g CP/kg DM was fed with *Digitaria decumbens* containing 38 g CP/kg DM. Voluntary intake was 630 g/day when the grass was fed alone but increased to over 900 g/day when the mixture contained 10% legume (Fig. 2.8). By extrapolating the linear regression back to zero legume in the mixture it was possible to show that the synergistic effect of the legume was equal to 50% of the VI of *D. decumbens*. In another experiment the synergistic effect of *M. sativa* was 17% when fed to lambs in combination with *F. arundinacea* containing 69 g CP/kg DM (Hunt *et al.*, 1985).

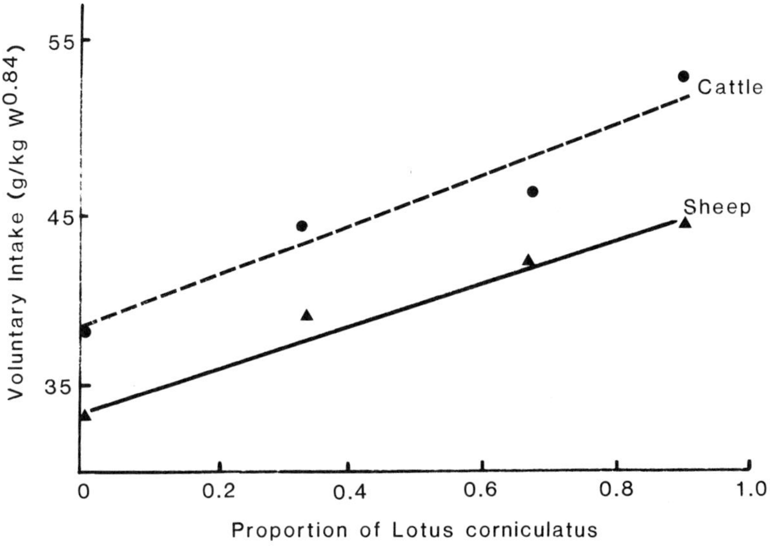

Fig. 2.7. The voluntary intake of mixtures containing different proportions of *Phleum pratense* and *Lotus corniculatus*. Data from Monson and Reid (1968).

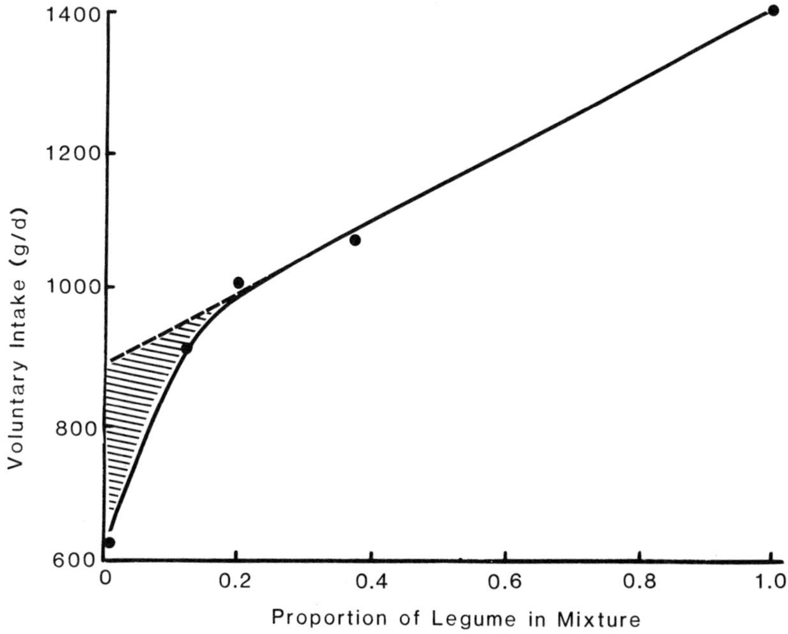

Fig. 2.8. The voluntary intake of sheep of mixtures of nitrogen-deficient *Digitaria decumbens* and *Medicago sativa* showing supplementary effect of the legume. Data from Minson and Milford (1967b).

C. Cultivars and Selections

1. Primary Growth

Differences in VI have been found between cultivars of temperate species when harvested on the same date in spring. Where this occurred it was often associated with a difference in stage of maturity. The problem of the correct interpretation of differences in VI is illustrated by the results of three studies.

The largest reported difference (8 g/kg $W^{0.73}$) between cultivars of *Phleum pratense* was found between S.51 and S.352 (Fig. 2.9). The yield, leafiness, growth-stage index, and content of digestible organic matter of the two grasses were similar (Walters, 1971), so there was a true difference in VI that was attributed to a difference in lignin content (Walters,

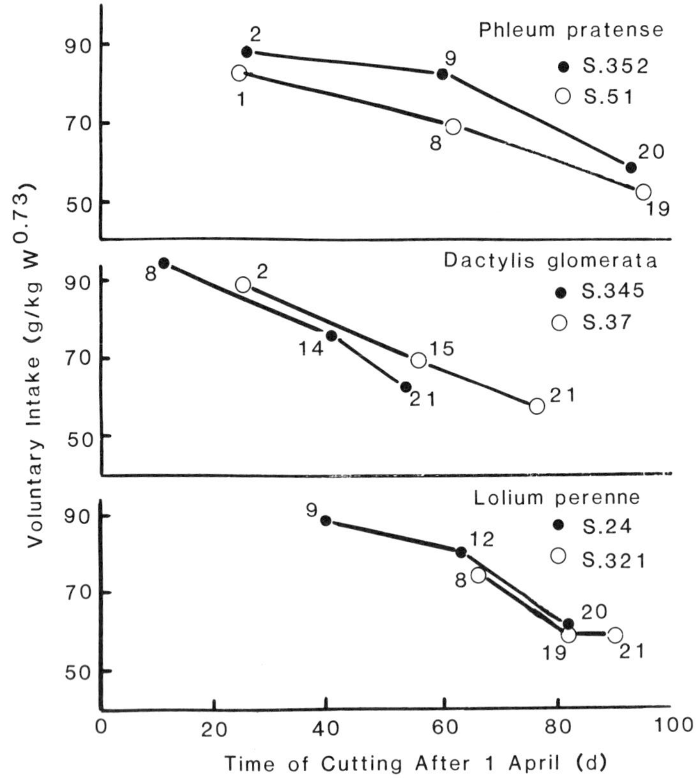

Fig. 2.9. Voluntary intake of three temperate grasses as affected by both cutting date and stage-of-growth index. Data from Walters (1971).

1974). However, in another comparison, the VI of a pasture-type cultivar was only higher than hay-type cultivars when compared in late summer (Heaney *et al.*, 1966a).

A small difference (5 g/kg $W^{0.73}$) was found between two cultivars of *D. glomerata* (Fig. 2.9), but this was associated with the later maturity of the cultivar S.37 (Walters, 1971). Cultivar S.37 was also superior to the cultivar Common, with differences in VI of 5 and 8 g/kg $W^{0.73}$ when cut in late May and early June, respectively (Haenlein *et al.*, 1966). However, the VI of S.37 was lower than that of the cultivar Latar, which matures later than S.37. In another study the VI of early-maturing cultivar Virginia 70 was 18% lower than that of a late-maturing cultivar Jackson cut on the same day (Bryant *et al.*, 1976).

The third example presented in Fig. 2.9 refers to two cultivars of *L. perenne* where there was no difference in VI, a result that was not unexpected in view of the similar growth-stage index of the two grasses (Walters, 1971). In another study diploid and tetraploid cultivars of *Lolium multiflorum* cut at an immature stage of growth had a similar VI, but as the grasses matured the VI of the tetraploid became progressively lower than that of the diploid cultivar (Thomson, 1971), possibly due to a higher proportion of stem.

2. Regrowth

When regrowths are compared at the same age, differences in VI are usually small (Table 2.3). However, dwarf *Pennisetum glaucum* was eaten in greater quantities than a tall cultivar due to a lower proportion of stem (Burton *et al.*, 1969). In *Zea mays* a brown midrib mutant had a lower lignin concentration and higher VI than the normal *Zea mays* (Rook *et al.*, 1977).

D. Plant Parts

The leaf fraction of legumes and grasses of both temperate and tropical forages was eaten in greater quantity than the stem fraction (Table 2.4). In studies of *D. decumbens* and *Chloris gayana* the leaf fraction had a similar dry-matter digestibility to that of the stem fraction but contained less neutral detergent fiber, acid detergent fiber, and lignin (Laredo and Minson, 1973, 1975a,b; Poppi *et al.*, 1981a). The difference in VI was caused by the lower resistance of the leaf fraction to chewing. More than twice as many large particles in the leaf fraction were broken down to small particles (<1 mm) during eating and this reduced the time that forage stayed in the rumen (Table 2.5). The lower resistance of the leaf to physical breakdown was confirmed by the small quantity of energy

TABLE 2.3

Range of Voluntary Intake of Forages Harvested as Regrowths of Several Species

Species and line	Voluntary intake ($g/kg\ W^{0.75}$) Range	Mean	Reference
Cenchrus ciliaris			
Nine cultivars/ecotypes	7 (max)[a]	67	Donaldson and Rootman (1980)
Five lines	7 (max)[a]	52	Minson and Bray (1985)
Five lines	9 (max)	53	Minson and Bray (1986)
Chloris gayana			
Six cultivars	3 (max)[a]	52	Milford and Minson (1968b)
Cynodon dactylon			
Coastcross 1 v. Coastal Bermuda	4 (NS)[b]	84	Lowrey *et al.* (1968)
Tifton 44 v. Coastal Bermuda (pelleted)	0	109	Utley *et al.* (1978)
Dactylis glomerata			
Currie v. Apanui	2 (NS)	75	Michell (1973a)
Digitaria milanjiana			
Six genotypes	9[a]	59	Minson and Hacker (1986)
Digitaria species			
Five species	5[a]	52	Minson (1984)
Lolium multiflorum			
S.22 v. *Tetrone*	0	—	Thomson (1971)
Lolium perenne			
Tasmanian No. 1 v. Ariki	1	79	Michell (1973a)
Panicum coloratum			
Three cultivars	6[a]	56	Minson (1971a)
Panicum maximum			
Hamil v. Coloniao	3[a]	66	Minson (1971a)
Pennisetum glaucum			
Dwarf v. tall line	21% higher		Burton *et al.* (1969)
Phalaris arundinacea			
High- and low-preference lines	0	49	O'Donovan *et al.* (1967)
High- and low-preference lines	2	59	Barnes and Mott (1970)
Zea mays			
Brown midrib v. normal (silage)	10% higher		Rook *et al.* (1977)

[a]Differences significant at $p < 0.05$ for line listed first.
[b]NS, not significant.

required to grind the leaf fraction through a 1-mm screen (Laredo and Minson, 1973, 1975a,b).

E. Stage of Growth

1. Cause of Reduced Voluntary Intake

As forages grow and mature they pass through a succession of growth stages; from a nutritional viewpoint, these may be classed as vegetative,

TABLE 2.4

Voluntary Intake of Separated Leaf and Stem Fractions of Forages

Species	Animal	Voluntary intake ($g/kg\ W^{0.75}$)			Reference
		Leaf	Stem	Difference	
Digitaria decumbens	Sheep	58	40	18	Laredo and Minson (1973)
		44	34	10	Laredo and Minson (1975b)
		51	44	7	Poppi *et al.* (1981a)
	Cattle	68	49	19	Poppi *et al.* (1981a)
Chloris gayana	Sheep	57	45	12	Laredo and Minson (1973)
		36	29	7	Laredo and Minson (1975b)
		49	38	11	Poppi *et al.* (1981a)
	Cattle	76	57	19	Poppi *et al.* (1981a)
Lablab purpureus	Sheep	91	53	38	Hendricksen *et al.* (1981)
	Cattle	91	51	40	Hendricksen *et al.* (1981)
Lolium perenne	Sheep	74	62	12	Laredo and Minson (1975a)
	Cattle	74	60	14	McLeod and Minson (1988)
Medicago sativa	Cattle	84	42	42	McLeod and Minson (1988)
Panicum maximum	Sheep	64	47	17	Laredo and Minson (1973)
Pennisetum clandestinum	Sheep	50	35	15	Laredo and Minson (1973)
Setaria sphacelata	Sheep	59	32	27	Laredo and Minson (1973)
		41	27	14	Laredo and Minson (1975b)
Mean		61	44	17	

TABLE 2.5

Voluntary Intake of Leaf and Stem Fractions of Tropical Grasses, Proportional Breakdown of Large Particles during Eating and Time Forage Is Retained in the Rumen[a]

Measurement	Animal	Leaf	Stem	Difference
Large particle breakdown during eating (%)	Sheep	34	19	15
	Cattle	32	12	20
Time dry matter retained in rumen (hr)	Sheep	27	33	6
	Cattle	35	45	10
Voluntary intake ($g/kg\ W^{0.75}$)	Sheep	50	41	9
	Cattle	72	53	19

[a]Data from Poppi *et al.* (1981a–c).

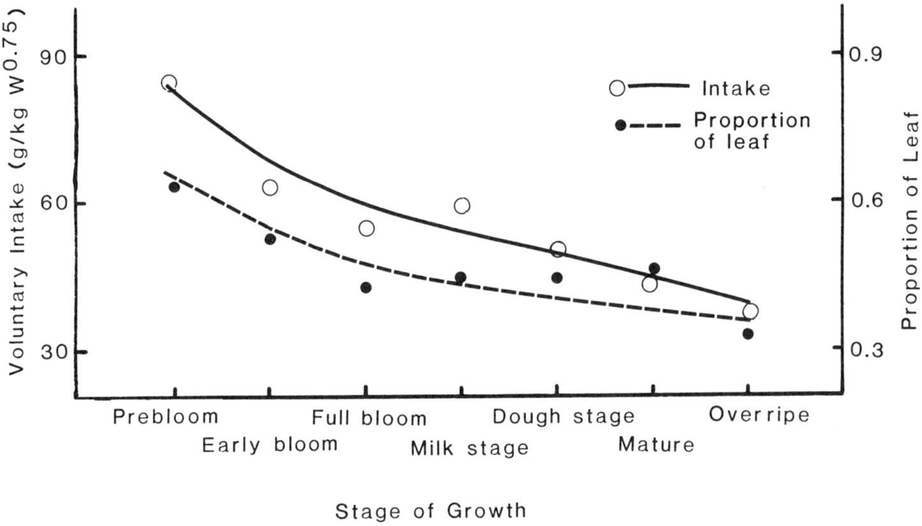

Fig. 2.10. Mean voluntary intake by sheep and proportion of leaf in four grasses cut at different stages of growth. Data from Troelsen and Campbell (1969).

prebloom, early bloom, full bloom, milk stage, dough stage, mature, and overripe. These changes in maturity are accompanied by increases in yield and proportion of stem and flowering head, a decrease in the proportion of leaf, and a fall in VI (Fig. 2.10). The fall in VI is caused by three factors: an increase in the proportion of stem (which is eaten in smaller quantities than leaf), a fall in VI of both leaf and stem fraction, and nutrient deficiencies in the mature forages. These last two mechanisms will now be considered.

The VI of both leaf and stem fractions decreases as the forage matures. In a study of five tropical grasses the VI of leaf fractions declined from 69 to 52 g/kg $W^{0.75}$ over 37 days, while the corresponding decrease in VI of the stem fraction was from 49 to 35 g/kg $W^{0.75}$ (Laredo and Minson, 1973). This decrease in VI was associated with increases in lignin, grinding energy, and the time leaf and stem were retained in the rumen (Table 2.6). Another possible cause of the low VI of mature forage is a nutrient deficiency, most commonly protein. When CP falls below about 70 g/kg DM there is a rapid fall in VI (Milford and Minson, 1965c). This deficiency can be overcome, and VI increased, by applying fertilizer nitrogen (see Section III,F) or feeding a protein supplement (Fig. 2.11).

As forages mature there is a rise in fiber concentration. With *L. perenne*, *P. pratense*, and a mixed temperate sward, VI was negatively correlated with crude fiber content (Wilson and McCarrick, 1967). Regressions have also been published relating VI of a range of forages to the

TABLE 2.6

Composition, Physical Character, Retention Time in the Rumen, and Voluntary Intake by Sheep of Leaf and Stem Fractions of Five Tropical Grasses[a]

	Fraction			
	Leaf		Stem	
Measurement	Immature	Mature	Immature	Mature
Lignin (g/kg)	27	37	35	51
Grinding energy (KJ/kg)	202	250	340	450
Retention time in rumen (hr)	18	21	23	34
Voluntary intake (g/kg $W^{0.75}$)	69	52	49	35

[a]From Laredo and Minson (1973).

concentration of neutral detergent fiber, acid detergent fiber, and lignin in the forage (Section IV,B). Fiber depresses VI through its effect on the resistance of the forage to chewing during eating and ruminating (Table 2.7). Confirmation of the dominant role of chewing in limiting VI is the increase that is achieved by grinding and pelleting, a process which overcomes the need for the animal to break down forage particles to a size that can readily leave the rumen (Section III,G).

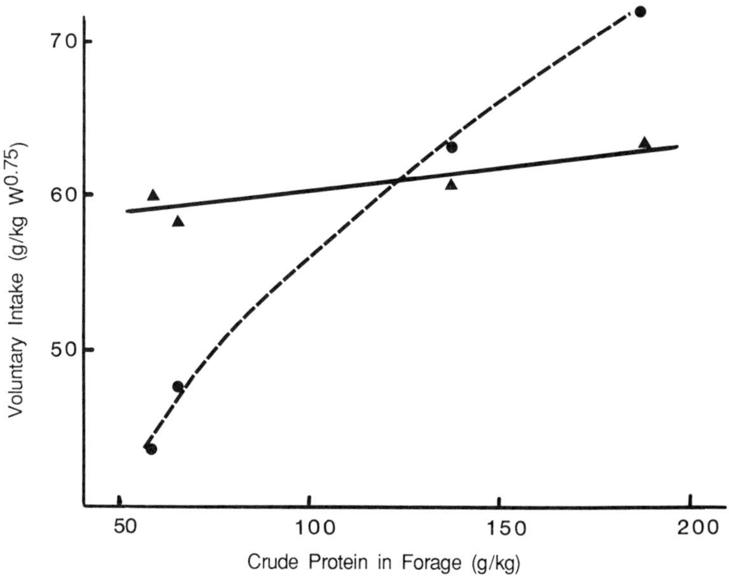

Fig. 2.11. Voluntary intake of *Digitaria decumbens* with and without a protein supplement of 4 g/kg $W^{0.75}$ soybean meal. (● — — —) Forage only; (▲ ———) forage and soybean meal. Data from Ventura *et al.* (1975).

TABLE 2.7

Composition, Resistance to Comminution, and Voluntary Intake of Grasses Grouped on the Basis of Dry-Matter Digestibility[a]

Attribute	Digestibility		
	High	Medium	Low
Cell wall (g/kg)	430	560	670
Grinding energy (KJ/kg)	31	64	117
DM digestibility	0.81	0.72	0.56
Eating time (hr/kg DM)	3.6	5.2	6.8
Ruminating time (hr/kg DM)	5.1	8.4	11.1
Total chewing time (hr/kg DM)	8.7	13.6	16.9
Voluntary intake (g/kg $W^{0.75}$)	75	65	51

[a]From Weston (1985).

The digestibility of the forage organic matter or dry matter is also related to the fiber content (Chapter 4). This leads to *noncausal* correlations between VI and the digestibility of forage energy (Blaxter, 1960; Blaxter *et al.*, 1961), organic matter (Minson *et al.*, 1964; Osbourn *et al.*, 1966; Thomson, 1971), and dry matter (Reid and Jung, 1965b; Wilson and McCarrick, 1967; Troelsen and Campbell, 1969). These regressions will be considered in Section IV,C.

2. Changes in Voluntary Intake of Primary and Subsequent Regrowths

a. Primary Growth. Intake of forages commencing growth in spring or following the first rains falls as the forage matures. The rate of fall in VI of 26 forage species is presented in Table 2.8. These have a mean fall of 0.39 g/kg $W^{0.75}$/day with no evidence of any difference in rate of fall between temperate and tropical grasses and legumes. Large differences in rate of fall in VI of particular species have been measured, but insufficient information has been published to decide whether these are real differences or whether they are due to the portion of the maturity curve used in the study.

b. Regrowths. Very little has been published on the rate of fall in VI of forage regrowths. The mean rate of fall in a number of tropical grasses was 0.17 g/kg $W^{0.75}$/day but the rates were very variable both between and within forage species (Table 2.9). In a study with *D. decumbens* cut at regular intervals, the rate of fall in VI decreased as the interval between harvests increased (Fig. 2.12).

TABLE 2.8

Rate of Fall in Voluntary Intake by Sheep of Primary Growths Cut at Different Stages of Maturity

Species	Rate of fall ($g/kg\ W^{0.75}/day$) Mean	Range	Reference
Agropyron cristatum	0.48	—	Troelsen and Campbell (1969)
Bromus inermis	0.47	—	Troelsen and Campbell (1969)
Cenchrus ciliaris	0.22	0.19–0.27	Minson and Milford (1968a), Playne (1970b)
Chloris gayana	0.18	0.11–0.23	Minson and Milford (1967a), Minson (1972)
Cynodon dactylon	0.32	0–0.63	Golding *et al.* (1976), Abrams *et al.* (1983)
Dactylis glomerata	0.68	0.57–0.98	Minson *et al.* (1964), Reid and Jung (1965b), Haenlein *et al.* (1966), Walters (1971, 1973)
Digitaria decumbens	0.32	0.10–0.52	Minson (1972), Ventura *et al.* (1975), Abrams *et al.* (1983)
Elymus junceus	0.35		Troelsen and Campbell (1969)
Festuca arundinacea	0.49	0.21–0.66	Minson *et al.* (1964), Walters (1971)
Festuca pratensis	0.34	—	Minson *et al.* (1964)
Heteropogon contortus	0.34	0.31–0.36	Playne and Haydock (1972)
Lolium multiflorum	0.62	—	Minson *et al.* (1964)
Lolium perenne	0.56	—	Minson *et al.* (1964), Walters (1971)
Medicago sativa	0.35	0.32–0.38	Troelsen and Campbell (1969)
Panicum coloratum	0.33	0.26–0.40	Minson (1971a)
Panicum maximum	0.28	0.10–0.43	Minson (1971a, 1972)
Paspalum dilatatum	0.22	—	Minson (1972)
Paspalum notatum	0.20	0.50–0.36	Moore *et al.* (1970), Abrams *et al.* (1983)
Pennisetum clandestinum	0.04	—	Minson (1972)
Phalaris arundinacea	0.39	—	Troelsen and Campbell (1969)
Phleum pratense	0.39	0–0.84	Minson *et al.* (1964), Heaney *et al.* (1966a), Walters (1971)
Setaria sphacelata var. *splendida*	0.35	0.32–0.38	Minson and Milford (1968a), Minson (1972)
Sorghum bicolor var. *sudanensis*	0.40	0.34–0.47	Reid *et al.* (1964a)
Stylosanthes humilis	0.84	—	Playne and Haydock (1972)
Urochloa mosambicensis	0.23	—	Playne (1972b)
Vigna unguiculata	0.20	—	Milford and Minson (1968a)
Mean	0.39	0–0.98	

TABLE 2.9

Rate of Reduction in Voluntary Intake by Sheep of Forage Harvested after Different Regrowth Periods

Species	Rate of reduction (g/kg $W^{0.75}$/day) Mean	Range	Reference
Cenchrus ciliaris	0.33	—	Donaldson and Rootman (1977)
Chloris gayana	0.02	−0.02–0.08	Milford and Minson (1968a), Minson (1972)
Digitaria species	0.24	−0.08–0.38	Chenost (1975), Minson (1972, 1984), Minson and Hacker (1986)
Panicum maximum	0.20	—	Minson (1972)
Pennesetum clandestinum	0.04	—	Minson (1972)
Setaria sphacelata var. *splendida*	0.16	—	Minson (1972)
Mean	0.17	−0.08–0.38	

F. Soil Fertility

The effect of soil fertility on VI has been determined by growing forage on soil that has received different quantities of mineral fertilizers. Studies have been conducted with fertilizer nitrogen, phosphorus, calcium, sulfur, and magnesium.

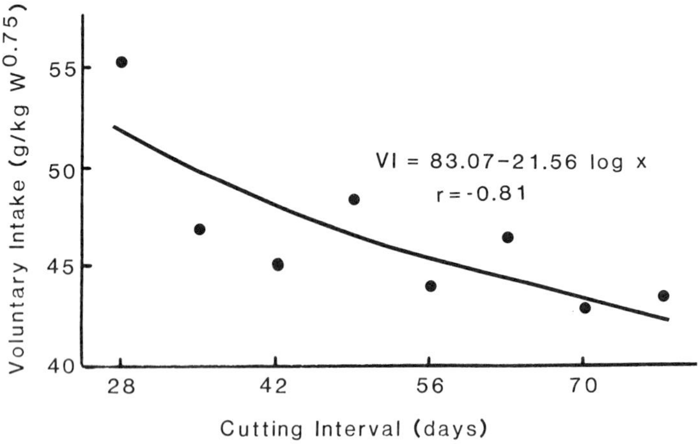

Fig. 2.12. Effect of cutting interval on the voluntary intake of *Digitaria decumbens*. Data from Chenost (1975).

1. Nitrogen Fertilizer

Fertilizer nitrogen (N) has been applied at two or more levels and the forage cut after growing for the same length of time, no allowance being made for differences in yield or stage of growth caused by the different level of N.

Fertilizer N has no consistent effect on the VI of temperate or tropical forages. In some studies VI increased while in other comparisons VI was depressed (Table 2.10). Where fertilizer N changed dry-matter digestibility of tropical grasses, VI changed in the same direction (Minson, 1973).

Although N has no consistent effect on VI in controlled experiments, where all treatments are harvested on the same day it would be wrong to conclude that fertilizer N has no effect on VI in production systems.

TABLE 2.10

Effect of Low and High Levels of Fertilizer Nitrogen on Voluntary Intake of Forage by Sheep

	Voluntary intake (g/kg $W^{0.75}$)		
Species	Low N	High N	Reference
Cenchrus ciliaris	14	15	Minson and Milford (1968a)
	62	62	Donaldson and Rootman (1977)
Chloris gayana	63	66	Minson (1973)
Dactylis glomerata	62	61	Reid *et al.* (1966)
	95	96	Reid *et al.* (1967b)
Digitaria decumbens	36	46	Minson (1967)
	50	53	Chacon *et al.* (1971b)
	61	65	Minson (1973)
Festuca arundinacea	62	64	Reid and Jung (1965a)
	65	63	Reid *et al.* (1967a)
Lolium perenne	67	62	Hight *et al.* (1968)
Reveille, first cycle	74	74	Demarquilly (1970b)
second cycle	73	78	
third cycle	66	64	
Melle, second cycle	74	77	
Panicum coloratum	50	53	Chacon *et al.* (1971a)
Pennesetum clandestinum	62	62	Minson (1973)
Phleum pratense	53	53	Cameron (1967)
Sorghum bicolor var. *sudanensis*	65	62	Reid *et al.* (1964a)
Temperate pasture			
Second cycle	71	65	Demarquilly (1970b)
Third cycle	64	54	
Mean	61	62	

Fertilizer N increases yield and this forage will generally be used for cutting or grazing at an earlier date when VI will be higher (Section III,E). In future studies, VI should possibly be compared when the yield of forage is the same from the low- and high-fertility treatments.

When the N concentration of forage is below about 10 g/kg, VI is often depressed by an N deficiency (Section III,E). By applying a late dressing of fertilizer N, the crude protein content of *D. decumbens* was increased from 39 to 52 g/kg DM and when fed to cattle VI increased by 70% (Chapman and Kretschmer, 1964). A similar response was obtained with sheep fed *D. decumbens* initially containing 37 g CP/kg DM. With very mature *Cenchrus ciliaris* containing 31 g CP/kg DM, fertilizer N failed to increase VI, indicating that the critical CP level probably varies with forage maturity (Minson and Milford, 1968a).

2. PHOSPHORUS FERTILIZER

There is no unequivocal evidence that fertilizer phosphorus (P) can alter VI (Table 2.11). Increases in VI occurred with *Stylosanthes humilis* and *M. sativa,* but P was applied as superphosphate and the effect could have been due to a rise in the S content of the legume.

3. CALCIUM FERTILIZER

Liming increased the VI of *D. glomerata* in the year of application, but in subsequent years it had no effect on VI (Reid *et al.,* 1969). With *D. decumbens,* fertilizer Ca increased VI from 39 to 43 g/kg $W^{0.75}$ and reduced the time forage was retained in the rumen (Rees and Minson, 1976). This small improvement was caused by structural changes in the forage

TABLE 2.11

Effect of Low and High Levels of Fertilizer Phosphorus on Voluntary Intake of Forage by Sheep

Species	Voluntary intake (g/kg $W^{0.75}$) Low P	High P	Reference
Digitaria decumbens	42	43	Rees and Minson (1982)
Festuca arundinacea	62	64	Reid and Jung (1965a), Reid *et al.* (1967a)
Medicago sativa	78	132[a]	Weir *et al.* (1958)
Stylosanthes humilis	75	84[a]	Playne (1972a)
Mean	63	73	

[a]Superphosphate used.

and VI was not increased by feeding sheep the unfertilized grass with a Ca supplement.

4. Sulfur Fertilizer

Applying fertilizer sulfur (S) to *D. decumbens* increased the level of S from 0.9 to 1.5 g/kg and raised VI by sheep from 44 to 64 g/kg $W^{0.75}$ (Rees *et al.*, 1974). This improvement was mainly due to the additional S overcoming an S deficiency in the control grass. The increase (I) in VI (g/kg $W^{0.75}$) varied with the S content of the grass (g/kg DM) according to the formula $I = 20.6 - 11.65\,S$ (Rees and Minson, 1978). When the S content of the control forage was above 1.8 g/kg DM, fertilizer S had no effect on the intake of forage by sheep.

5. Magnesium Fertilizer

Applying kieserite (magnesium sulfate) to a low-magnesium (Mg) soil increased the Mg concentration of *P. pratense* from 0.8 to 1.6 g/kg DM but had no effect on the VI of lambs (Reid *et al.*, 1984).

G. Climate

Seasonal changes in VI have been found in *A. gayanus* but these were associated with differences in N content of the forage ($r = 0.83$) (Haggar and Ahmed, 1970) and are probably more a reflection of differences in stage of growth than a response to climatic variation. When forages are cut and fed to ruminants throughout the season it is difficult to decide whether differences in VI are caused by differences in the forage or seasonal changes in the appetite of the animal (Michell, 1973a). This objection does not apply if all forages are cut after the same regrowth period, conserved, and fed back at the same time. When this method was used there was no evidence of a seasonal variation in VI of monthly regrowths of six temperate grasses (Minson *et al.*, 1964), *D. glomerata* (Ishiguri, 1979, 1983), *M. sativa* (Ishiguri, 1980), or a range of tropical grasses (Milford and Minson, 1968a; Minson, 1971a, 1972, 1984; Minson and Bray, 1985, 1986).

There is evidence that VI is affected by light intensity. When *L. perenne* was grown in the shade, VI was depressed from 69 to 60 g/kg $W^{0.75}$. This difference was associated with a fall in soluble carbohydrate and a rise in lignin concentration (Hight *et al.*, 1968). No studies have been conducted on the effect of either water deficiency or frost on the VI of forage. However low VI values have been published for frosted tropical grasses and legumes (Milford, 1960a, 1967) and *D. glomerata* (Reid *et al.*, 1967b). Drought and frost will depress VI if they cause a loss of leaf, and further reductions would be expected if the frosted forage became infected with microorganisms.

H. Processing

1. Freezing

Freezing and storing at −20°C has been used as a method of conserving forage for use in metabolism studies (Raymond *et al.*, 1953a). Frozen forage usually had the same VI as fresh grass, although the results were variable (Table 2.12). Freezing has been suggested as a method of conserving forage for feeding ruminants in the winter (Milne, 1953), but the capital and running costs of the equipment would be prohibitive.

2. Drying

The water content of forage ranges from 0 in recently dried material to as much as 920 g/kg in fresh forage (Davies, 1962). There is no evidence that the VI of DM is improved by the presence of water in the forage. Conversely, levels of water exceeding about 780 g/kg fresh forage have a detrimental effect on VI (Arnold, 1962; Davies, 1962; John and Ulyatt, 1987). The intake of *L. multiflorum* containing 875 g water/kg fresh forage was increased by wilting but no increase occurred when the forage initially contained 855 or 763 g/water/kg (Wilson, 1978). When the water

TABLE 2.12

Voluntary Intake of Fresh, Frozen, and Dried Forage

Species	Voluntary intake (g/kg $W^{0.75}$)			Reference
	Fresh	Frozen[a]	Dried[b]	
Digitaria decumbens	56	53	54	Minson (1966)
Panicum maximum	62	62	67	Minson (1966)
Sorghum almum	40	40	37	Minson (1966)
Sorghum vulgare var. *sudanense*	58	61	60	Heaney *et al.* (1966b)
Temperate pasture				
First growth	76	82	76	Heaney *et al.* (1966b)
Second growth	92	90	87	
Third growth	94	106	101	
Mean	68	71	69	

[a]Frozen at −20°C.
[b]Dried at 80–90°C.

content of *P. purpureum* was reduced by wilting from 880 to 854 g/kg, VI was increased by 15% (Grant *et al.*, 1974). The most comprehensive study of the effect of water in forage was conducted with lactating cows; VI decreased by 0.337 kg DM for each 10 g/kg rise in water content above 819 g/kg (Vérité and Journet, 1970).

Although it is well established that the VI of forage is depressed by high levels of water the cause of this depression has only recently been elucidated. Introducing large quantities of water into the rumen via a fistula had no effect on VI (Davies, 1962), but spraying water onto forage to increase the moisture content from 779 to 854 g/kg reduced the VI by cattle by 22% (Butris and Phillips, 1987). The cattle fed sprayed forage spent 70% more time ruminating, probably because the wet forage was swallowed before maximum particle breakdown had occurred.

When forage contained less than 780 g water/kg fresh material, VI was not limited by the presence of water and drying had no effect on VI (Duckworth and Shirlaw, 1958; Minson, 1966; Lake *et al.*, 1973; Bull and Tamplin, 1974).

Drying is the most common method of conserving forage. The changes associated with conservation vary depending on the method adopted, but when forage is rapidly dried at temperatures around 100°C the VI of the fresh and dried material is similar (Table 2.12). If a flail-type harvester is used to cut the forage, then VI of the dried material is depressed (Demarquilly, 1970b), as compared with a reciprocating-blade harvester.

Natural drying of forage in the field leads to losses of dry matter by respiration and leaf shatter, resulting in more stemmy material and hence reduced VI. Temperate pasture dried on tetrapods had a VI 24% lower than that dried in a conveyor-type grass drier at 175°C (Wilson and McCarrick, 1966). Field losses can be reduced by baling the forage at a higher moisture content, but any advantage may be offset by molding, which may depress VI by as much as 20% (Cochrane, 1976). The development of mold may be reduced by adding a preservative, but this can have an adverse effect on both VI and the growth of cattle (Johnson and McCormick, 1976).

3. Pelleting

Pellets made from ground forage are usually eaten in larger quantities than is chopped forage, a difference found with both temperate and tropical legumes and grasses fed to both sheep (Table 2.13) and cattle (Table 2.14). The improvement in VI is not constant but varies with species of ruminant, stage of forage maturity, and concentration of nitrogen in the forage. These three factors will be considered before the cause of the improvement in VI is discussed.

TABLE 2.13

Voluntary Intake of Dry Matter of Chopped and Pelleted Forage by Sheep

Species	Voluntary intake (g/kg $W^{0.75}$)		Reference
	Chopped	Pelleted	
Cenchrus ciliaris	35	50	Minson and Milford (1968a)
Chloris gayana	32	67	Laredo and Minson (1975b)
Dactylis glomerata	58	95	Heaney *et al.* (1963)
Digitaria decumbens	41	50	Minson (1967)
	48	71	Minson and Milford (1968a)
	39	70	Laredo and Minson (1975b)
Festuca arundinacea	83	94	Tetlow and Wilkins (1974)
Lolium multiflorum	60	79	Milne and Campling (1972)
Lolium perenne	72	86	Wilkins *et al.* (1972)
	55	85	Greenhalgh and Reid (1973)
	58	81	Greenhalgh and Reid (1974)
	78	88	Tetlow and Wilkins (1974)
	52	83	Tetlow and Wilkins (1977)
Medicago sativa	80	104	Meyer *et al.* (1959a)
	102	135	Meyer *et al.* (1960)
	67	100	Heaney *et al.* (1963)
	54	91	Weston and Hogan (1967)
Pennesetum clandestinum	48	65	Minson and Milford (1968a)
Phleum pratense	60	101	Heaney *et al.* (1963)
	62	75	Miles *et al.* (1969)
Setaria sphacelata	38	64	Minson and Milford (1968a)
Setaria splendida	34	48	Laredo and Minson (1975b)
Triticum aestivum	36	50	Weston (1967)
	34	49[a]	Weston and Hogan (1967)
Mean	65	92	

[a]Ground but not pelleted.

Pelleting of forage increased the mean VI of sheep by 42% (Table 2.13), but with cattle the mean difference between pelleted and chopped forage was only 23% (Table 2.14). This difference in response to pelleting between sheep and cattle occurs at all stages of forage maturity and ages of ruminants (Greenhalgh and Reid, 1973). Calves and yearlings are more affected by the physical form of the forage than are cows (Buchman and Hemken, 1964).

Pelleting causes large increases in the VI of poor-quality forage but has a much smaller effect on high-quality immature forages (Minson, 1963; Greenhalgh and Wainman, 1972). This difference in response is illustrated in Fig. 2.13, which is based on results obtained with *D. glomerata, M.*

TABLE 2.14

Voluntary Intake of Dry Matter of Chopped and Pelleted Forage by Cattle

Species	Voluntary intake (g/kg $W^{0.75}$)		Reference
	Chopped	Pelleted	
Avena sativa	52	66	Campling and Freer (1966)
Chloris gayana	69	105	Holmes *et al.* (1966)
Cynodon dactylon	80	96	Beaty *et al.* (1960)
Dactylis glomerata	57	64	Miles *et al.* (1969)
Lolium perenne	85	87	Campling *et al.* (1963)
	68	77	Miles *et al.* (1969)
	90	98	Wilkins *et al.* (1972)
	79	90	Greenhalgh and Reid (1973)
Medicago sativa	85	122	Webb *et al.* (1957), Webb and Cmarik (1958)
Phleum pratense mixture	84	124	Webb *et al.* (1957), Webb and Cmarik (1958)
	96	102	Klosterman *et al.* (1960)
Setaria sphacelata	75	112	Holmes *et al.* (1966)
Lespedeza cuneata	89	102	Webb *et al.* (1957), Webb and Cmarik (1958)
Temperate pasture	97	95	Campling and Freer (1966)
Mean	79	97	

sativa, and *P. pratense*, each cut at three stages of growth. Where the VI of forage was limited by a low level of crude protein, pelleting had little effect on VI (Minson, 1967). When this protein deficiency was eliminated by supplementing with urea or gluten, or by applying urea to the forage several weeks before it was cut, then pelleting increased VI (Table 2.15).

The increase in VI caused by pelleting is due to the shorter time pelleted forage is retained in the rumen. Lower retention times have been reported for *Avena sativa* (Campling and Freer, 1966), *C. gayana* (Laredo and Minson, 1975b), *D. decumbens* (Minson, 1967; Laredo and Minson, 1975b), *F. arundinacea* (Tetlow and Wilkins, 1974), *L. multiflorum* (Milne and Campling, 1972), *L. perenne* (Tetlow and Wilkins, 1974), and *Setaria sphacelata* var. *splendida* (Laredo and Minson, 1975b). Pelleting also reduces the time spent on rumination (Minson, 1963; Campling and Freer, 1966). It has been suggested that "fine grinding is probably the major factor causing the increased feed consumption of pelleted hay and the pelleting process serves to put a fine dusty feed in a more palatable form" (Meyer *et al.*, 1959b).

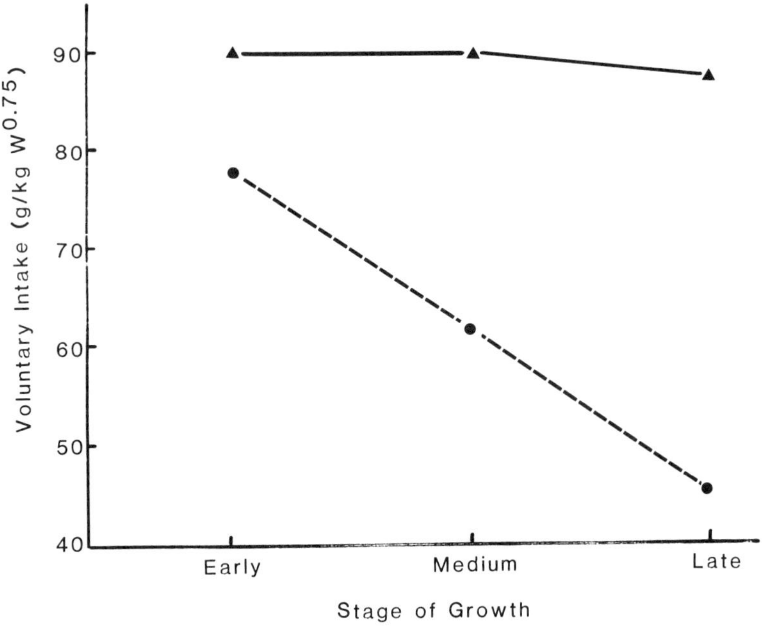

Fig. 2.13. Voluntary intake of pelleted (▲ ———) or chopped (• — — —) temperate forages at three stages of growth. Data from Heaney *et al.* (1963).

4. ENSILING

a. Direct-Cut Silage. It is well established that the VI of silage made from unwilted forage is lower than that of the corresponding fresh, frozen, or dried forage (Table 2.16). The reduction in VI is more for sheep than for cattle. In a study with three temperate grasses, VI of silage was 59 and 20% lower than the original forage when eaten by sheep and cattle, respectively (Demarquilly and Dulphy, 1977). Ensiling caused a larger reduction in VI with lambs than mature sheep, 29 and 14%, respectively (Harris *et al.*, 1966). The reduction in VI of forage occurred within a few days of its being ensiled (Lancaster, 1975).

b. Cause of the Intake Depression. The reasons for the low intake of direct-cut silage are not well understood. The high water content is not the limiting factor; fresh and oven-dried silages are eaten in similar quantities and addition of water to wilted silage has no effect on VI (Thomas *et al.*, 1961). The depression in VI of silage is associated with changes in physical structure, breakdown of protein, a reduction in pH, and production of organic acids (Demarquilly and Dulphy, 1977).

TABLE 2.15

The Effect of Nitrogen on the Voluntary Intake of Ground or Pelleted Forage

		Dietary nitrogen (g/kg)		Voluntary intake (g/kg $W^{0.75}$)		
Species	Treatment	Control	Treatment	Control	Treatment	Reference
Avena sativa	Urea supplement	7	13	66	101	Campling and Freer (1966)
Digitaria decumbens	Urea fertilizer	7	13	39	60	Minson (1967)
Triticum aestivum	Gluten supplement	7	11	50	80	Weston (1967)
Mean		7	11	52	80	

TABLE 2.16

Voluntary Intake by Sheep and Cattle of Direct-Cut Silage Compared with the Original Forage Fed Fresh, Frozen, or Dried

Species	Voluntary intake ($g/kg\ W^{0.75}DM$)		Reference
	Control	Silage	
Dactylis glomerata	74	23	Harris and Raymond (1963)
Festuca pratensis	47	19	Harris and Raymond (1963)
Lolium multiflorum	69	16	Harris and Raymond (1963)
Lolium perenne	48	36	Wernli and Wilkins (1980a)
Phleum pratense	59	34	Harris and Raymond (1963)
Medicago sativa	60	41	Moore *et al.* (1960)
	105	100	Byers (1965)
	100	72[a]	Waldo *et al.* (1966)
	64	52	Goering *et al.* (1976)
Temperate grass	66	42	Demarquilly (1973)
Cattle	80	64	Demarquilly and Dulphy (1977)
Sheep	57	24	Demarquilly and Dulphy (1977)
Temperate forage	65	63	Harris *et al.* (1966)
	90	66	Demarquilly and Dulphy (1977)
Trifolium pratense	82	64	Demarquilly (1973)
Mean	72	48	

[a]Assuming liveweight of 400 kg.

The structure of long silage is a major factor limiting VI. When temperate grasses were ensiled in either the long or the chopped form and fed to cattle and sheep, VI of the chopped silage was 33 and 94% higher, respectively, than that of the long unchopped silage (Demarquilly and Dulphy, 1977). The low VI of long silage appears to be partly caused by a greater resistance to breakdown of silage particles to a size that could leave the rumen. Silage was retained in the rumen for a longer time than hay, was eaten more slowly, required more ruminating time per kilogram of dry matter, and more boluses (15%) were regurgitated (Campling, 1966). Fluoroscopic and cineradiographic studies showed that swallowed silage did not accumulate in the cardiac region of the rumen but was forced into the dorsal sac of the rumen by the contractions of the reticulum and cranial sac (Deswysen and Ehrlein, 1981). This leads to a high proportion of pseudorumination cycles with rumen fluid containing few feed particles, delayed regurgitation of digesta, less efficient rumination, and lower VI (Deswysen *et al.*, 1978).

During silage fermentation a large proportion of the protein is hydrolyzed but the low VI does not appear to be caused by a protein deficiency. Feeding fish meal, groundnut meal, or soybean meal increased the VI of high-moisture silages by only 4 to 9% (Garstang *et al.*, 1979; Gordon, 1979; Kaiser *et al.*, 1982a; Gill and England, 1984), and there was no improvement in the VI of *L. perenne* silage when casein was infused into the abomasum or duodenum of sheep (Hutchinson *et al.*, 1971).

The VI of ruminants is depressed by a reduction in rumen pH (Bhattacharya and Warner, 1968), but partial neutralization of silages with sodium bicarbonate before feeding has little effect on VI by sheep (McLeod *et al.*, 1970; Lancaster and Wilson, 1975; Farhan and Thomas, 1978) or calves (McLeod *et al.*, 1970; Thomas and Wilkinson, 1975). With cows, both positive and negative responses have been found when silage was neutralized (Lancaster and Wilson, 1975; Farhan and Thomas, 1978; Shaver *et al.*, 1985).

Many studies have shown that the depression in VI of silage is correlated with the products of the silage fermentation. When silage effluent was put into the rumen through a small fistula the VI of hay by heifers was immediately depressed by 25% (Thomas *et al.*, 1961). Another possible factor limiting VI is the high osmolality in the rumen when silages are fed (Ternouth, 1967; Phillip *et al.*, 1981).

Aerating forage during silage making has very little effect on the dry-matter content of the silage but increases the pH and ammonia level (Hutton *et al.*, 1971). Despite these adverse effects on fermentation, aerated silage is eaten in greater quantities (Harris *et al.*, 1966; Hutton *et al.*, 1971; Lancaster, 1975), illustrating the complexity of silage intake.

The low VI of direct-cut silage is related in some way to the changes that occur during fermentation. Reducing the extent of fermentation should reduce or prevent this depression.

c. *Wilted Silage.* The most important way of increasing the VI of silage is to reduce fermentation by wilting the forage before ensiling. For example, the VI of heifers was 4.04 and 5.90 kg DM/day when fed *M. sativa* silage containing 240 and 450 g DM/kg, respectively (Moore *et al.*, 1960). Subsequent work showed that VI of silage is linearly related to dry-matter content over the range 180–540 g/kg (Thomas *et al.*, 1961), an effect that applied to all classes of cattle (Fig. 2.14).

Wilting increased the VI of silage made from *L. perenne* and mixtures of temperate species but the improvement was variable (Table 2.17). The optimum dry-matter content required for maximum VI of silage is about 300 to 350 g/kg; wilting beyond this point fails to increase VI (Jackson and Forbes, 1970; Forbes and Jackson, 1971).

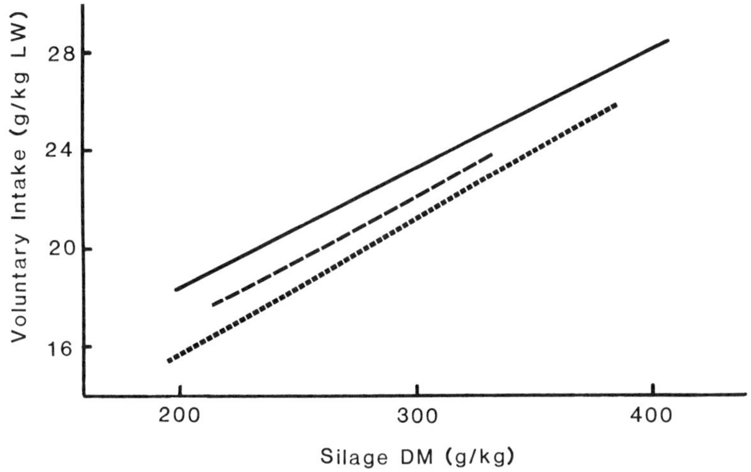

Fig. 2.14. Effect of moisture content of *Sorghum vulgare* silage on voluntary intake by three classes of ruminant: (———) lactating cows; (– – – –) beef steers; (- - - -) beef heifers. Data from Ward *et al.* (1965).

TABLE 2.17

Effect of Wilting on the Voluntary Intake of Silage by Sheep and Cattle

	Silage dry matter (g/kg)		Voluntary intake (g/kg $W^{0.75}$DM)		
Species	Unwilted	Wilted	Unwilted	Wilted	Reference
Lolium perenne	175	365	71	69	Morgan *et al.* (1980)
	186	217	62	68	England and Gill (1983)
	201	261	84	89	Castle and Watson (1984b)[a]
	205	292	78	81	Gordon and Peoples (1986)[a]
Medicago sativa	240	450	45	66	Moore *et al.* (1960)[b]
	240	450	90	113	Moore *et al.* (1960)[a]
	187	265	79	94	Murdoch (1960)[b]
Temperate pastures	148	215	48	56	Harris *et al.* (1966)
	188	235	77	83	McCarrick and Wilson (1966)
	205	318	85	86	Castle and Watson (1970)[b]
	190	323	78	93	Jackson and Forbes (1970)
	187	352	57	86	Forbes and Jackson (1971)
Mean	196	312	71	82	

[a]Protein and energy supplemented.
[b]Assumed liveweight of 400 kg.

TABLE 2.18

Influence of Fine Chopping of Forage before Ensiling on the Voluntary Intake by Sheep and Cattle[a]

Animal	Voluntary intake (g/kg $W^{0.75}$ DM) Long	Chopped	Increase (%)
Sheep	38	54	42
Heifers (1 yr)	69	84	22
Dairy cows	98	110	12

[a]From Demarquilly and Dulphy (1977).

d. Particle Size. The VI of silage made from chopped forages is higher than that for lacerated cut silage (Gordon *et al.*, 1958; Murdoch, 1965), a difference that is larger for sheep than for cattle (Table 2.18). The VI of flail-harvested silage was increased by fine chopping just before feeding (Table 2.19), indicating that physical structure is a major factor limiting the VI of direct-cut silage. The higher intake was due to more efficient regurgitation and less time spent ruminating each kilogram of finely chopped silage dry matter (Demarquilly and Dulphy, 1977; Deswysen *et al.*, 1978; Castle *et al.*, 1979).

The improvement in intake achieved by precision chopping is much less when the silage is supplemented with concentrates. Precision chopping increased the VI of silage by 21% in the absence of concentrates but this fell to 11% when the ration contained 36% concentrates (Murdoch, 1965). In another study, VI of *L. perenne* silage was not improved by fine

TABLE 2.19

Effect of Fine Chopping Just before Feeding on the Voluntary Intake of Flail-Harvested Silage[a]

Animal	Forage species	Voluntary intake (g/kg $W^{0.75}$ DM) Control	Chopped	Increase (%)
Sheep	Grass	27	40	48
	Medicago sativa	43	55	28
Cattle	Grass	68	76	12

[a]From Demarquilly and Dulphy (1977).

TABLE 2.20

The Voluntary Intake of Green Forage and Finely Chopped Silage with and without Addition of Formic Acid[a]

Attribute	Fresh forage	Silage	
		Untreated	Formic acid
Voluntary intake (g/kg $W^{0.75}$)			
Heifers	102	77	94
Sheep	77	56	65
pH	—	4.49	3.88
Ammonia (% of total N)	—	11	6
Lactic acid (g/kg DM)	—	48	60
Acetic acid (g/kg DM)	—	70	20
Propionic acid (g/kg DM)	—	3	Trace
Alcohols (g/kg DM)	—	33	8

[a]Data from Demarquilly and Dulphy (1977).

chopping when the diet contained 46% concentrates (Gordon, 1982). With *Z. mays* silage containing 50% grain, VI by cows was similar for silage chopped to 4, 8, and 13 mm (Stockdale and Beavis, 1988).

e. *Silage Additives.* The extent of silage fermentation may be reduced by the use of chemical additives. Sulfuric and hydrochloric acids were used by Virtanen (1933), but these have generally been replaced, for safety reasons, by weaker acids such as formic acid. In some studies formic acid had no effect on VI (Barry *et al.*, 1978; McIlmoyle and Murdoch, 1979), but in other comparisons VI was improved 10–25% (Castle and Watson, 1970; Lancaster *et al.*, 1977; Barry *et al.*, 1978) (Table 2.20). Another additive, formaldehyde, increased the VI of legume silage by 18–54% (Barry *et al.*, 1973, 1978), a rise attributed to a reduction in protein degradation in the rumen and an increase in the absorption of amino acids (Barry *et al.*, 1973). With *L. perenne* silage fed to cattle, VI was increased by 17% when a formaldehyde–sulfuric acid mixture was added (McIlmoyle and Murdoch, 1979). Excess formaldehyde reduced VI but intake was resorted by feeding a urea supplement (Lonsdale *et al.*, 1977).

IV. PREDICTION OF VOLUNTARY INTAKE

Producers, stock advisers, and plant breeders require rapid methods for predicting VI. Voluntary intake may be predicted from the physical and chemical composition of the forage, from bioassays, and from the use of mathematical models. These different methods of predicting VI will now be considered.

A. Physical Measurements

1. Leaf Proportion

Leaf is eaten in greater quantities than stem of similar chemical composition and dry-matter digestibility (Section III,D). Farmers use visual estimates of the proportion of leaf to estimate relative quality of both growing and conserved forages. In the laboratory the proportion of leaf in dried chopped forage was measured following separation using seed-cleaning equipment (Rao *et al.*, 1987). Voluntary intake was correlated with the proportion of leaf in samples of legumes ($r = 0.87$) and grasses ($r = 0.66$), but the method was sensitive to the way the forage was grown and the characteristics of the leaf separator.

2. Bulk Density

Pelleted forages are eaten in greater quantities than chopped forages and have a higher density (g/ml). Equations have been developed for predicting VI from density (Baile and Pfander, 1967; Laredo and Minson, 1973) and gross packed volume (ml/g). With chopped temperate forages, VI was negatively correlated ($r = -0.97$) with gross packed volume (Seoane *et al.*, 1982).

3. Resistance to Breakdown

Laboratory methods have been developed that simulate the physical resistance of forage to breakdown by chewing. These are based on the energy required to grind dry forage through a 1-mm screen or the extent that particles in fresh forage are reduced in size by an artificial masticator.

a. Grinding Energy. Grinding energy (also called fibrousness index) may be determined by measuring the electrical energy required to pulverize 5 g of oven-dried forage though a 1-mm screen in a laboratory mill (Chenost, 1966). The relation between VI and grinding energy for 25 temperate legume and grass hays is shown in Fig. 2.15. For 30 leaf and stem samples of 5 tropical grasses, there was a negative correlation ($r = -0.81$) between VI and grinding energy (Laredo and Minson, 1973).

b. Artificial Mastication. When the forage is cycled as a slurry through a gear water pump for 10 min and then wet sieved, the final

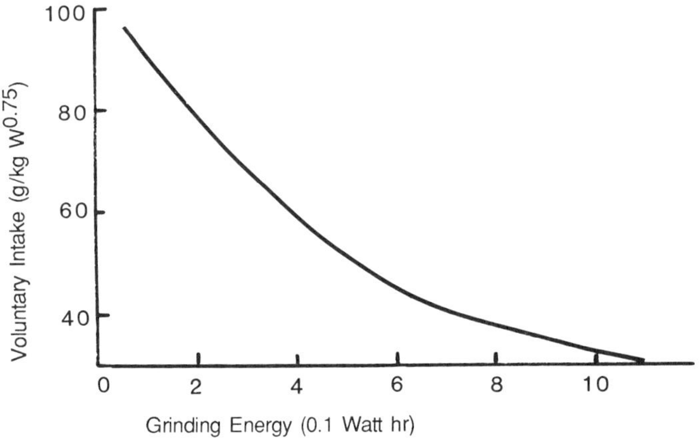

Fig. 2.15. Prediction of voluntary intake of forages by sheep from grinding energy. (y = 91.74 − 55 log x; r = −0.90; RSD ± 6.6 g/kg $W^{0.75}$). Data from Chenost (1966).

particle-size index is correlated with VI (Fig. 2.16). This result is encouraging but the technique has not been used in routine work because it is laborious. Simpler and more rapid methods need to be developed for measuring the resistance of forage to chewing.

c. *Leaf Tensile Strength.* Instruments have been developed for measuring the force required to break individual grass leaves (Evans, 1964). No attempt has been made to relate VI to leaf tensile strength. The value of the technique in plant breeding and selection has been reviewed by Hacker (1982).

4. Near-Infrared Reflectance Spectroscopy

Voluntary intake of forages can be predicted from the near-infrared reflection (NIR) from samples of ground forage. In a study by Norris *et al.* (1976) the residual standard deviation (RSD) of the regression relating VI to reflectance was ±7.8 g/kg $W^{0.75}$. For arid and semiarid rangeland forage grown in New Mexico the RSD of the regression was ±9.6 g/kg $W^{0.75}$ (Ward *et al.*, 1982), and with temperate forages fed to cattle and sheep it was ±7.6 and ±6.3 g/kg $W^{0.75}$, respectively (Redshaw *et al.*, 1986). Similarly, low errors have been found when the method was applied to forage species used in the calibration, but prediction errors were much higher when NIR was used to predict the VI of forage types not included in the calibration (Minson *et al.*, 1983). The method is no longer used to predict VI.

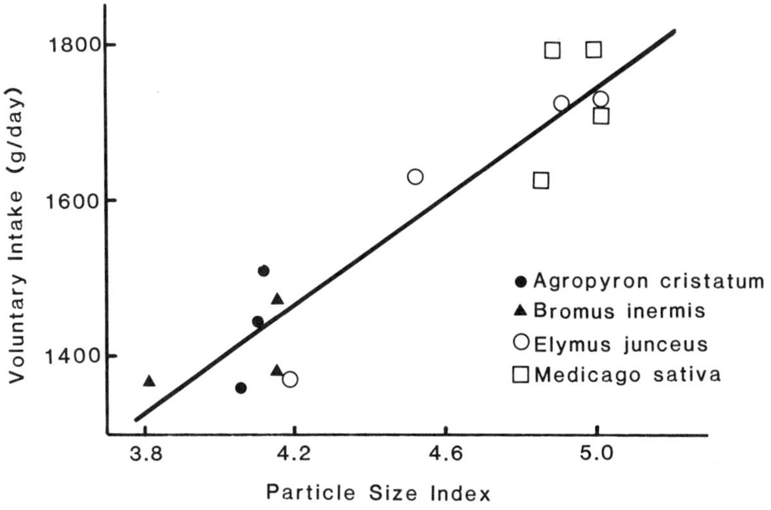

Fig. 2.16. Predicting voluntary intake by sheep from particle-size distribution following artificial mastication. ($y = 353x - 14$; $r = 0.94$.) Data from Troelsen and Bigsby (1964).

B. Chemical Analysis

The VI of forages by sheep is correlated with different chemical fractions and many regressions have been published for predicting VI (Table 2.21). The RSD is the minimum error that should be applied to estimates of VI predicted from these regressions and it varied from 3.7 to 11.8 g/kg $W^{0.75}$. The lowest RSD generally occurred where only a limited range of samples was used to derive the regression. When contrasting forages were included in the regression, the RSD was generally high and it was possible to calculate separate regressions for the different classes or species of forage (Fig. 2.17).

The VI of silage is controlled by both the composition of the original forage and the type of fermentation. Regressions have been published for predicting VI of silage from the concentration of water, protein, neutral detergent fiber, and acid detergent fiber (Jones *et al.*, 1980) and from the Flieg index, lactic acid, acetic acid, total acids, nitrogen, ammonia, and pH. Regressions have also been derived incorporating two or more variables, but different equations were applicable to legumes and grasses (Wilkins *et al.*, 1971). Intake of silage by cattle was predicted from a regression which included liveweight (LW) (measured in kg), DM concentration, pH, and intake of hay and supplements (Richards and Wolton, 1975).

TABLE 2.21

Chemical Methods of Predicting the Voluntary Intake of Forages by Sheep (g/kg $W^{0.75}$/day)

Species	Regression	*r*	RSD	Reference
Solubility in water (g/kg)				
Temperate forage	VI = 13.25 + 0.358*x*	0.94	± 4.9	Seoane *et al.* (1982)
Tropical grass	—	0.75	—	Hawkins *et al.* (1964)
Pepsin-soluble dry matter or organic matter (g/kg)				
Panicum spp.	VI = 24.0 + 0.194*x*	0.67	± 7.7	Minson and Haydock (1971)
Temperate forage	—	0.94	—	Donefer *et al.* (1966)
Temperate grass	—	0.91	± 3.7	Jones and Walters (1975)
Tropical grass	—	0.26	—	Laredo and Minson (1975c)
Crude protein (g/kg)				
Temperate forage	—	0.54	—	Van Soest (1965b)
Temperate forage	VI = 33.5 + 0.22*x*	0.86	± 7.2	Wilson and McCarrick (1967)
	VI = 31.7 + 0.21*x*	0.67	± 9.7	Wilson and McCarrick (1967)
	VI = 36.6 + 0.31*x*	0.90	± 6.2	Seoane *et al.* (1982)
Temperate silage	VI = 17.7 + 0.237*x*	0.57	±14.6	Wilkins *et al.* (1971)
	VI = 35.8 + 0.239*x*	0.92	± 5.4	Laforest *et al.* (1986)
Tropical grass	—	0.47	—	Abrams *et al.* (1983)
Neutral detergent fiber (g/kg)				
Temperate forage	VI = 110.4 − (1716/100 − *x*)	−0.65	—	Van Soest (1965b)
	VI = 95 − 0.073*x*	−0.88	—	Osbourn *et al.* (1974)
	VI = 165.7 − 0.15*x*	−0.90	± 5.9	Seoane *et al.* (1982)
Temperate grass	—	−0.89	± 4.1	Jones and Walters (1975)
Temperate silage	VI = 132.7 − 0.105*x*	−0.87	± 6.8	Laforest *et al.* (1986)
Tropical grass	—	−0.38	—	Abrams *et al.* (1983)
Acid detergent fiber (g/kg)				
Temperate grass	—	−0.90	± 3.9	Jones and Walters (1975)
Temperate forage	—	−0.53	—	Van Soest (1965b)
Tropical grass	—	−0.54	—	Abrams *et al.* (1983)
Modified acid detergent fiber (g/kg)				
Temperate forage	VI = 118.6 − 0.181*x*	−0.82	± 7.0	Clancy and Wilson (1966)
Crude fiber (g/kg)				
Temperate forage	VI = 98.3 − 0.11*x*	−0.62	±11.8	Chenost (1966)
Cut in 1965	VI = 110.0 − 0.17*x*	−0.89	± 6.2	Wilson and McCarrick (1967)
Cut in 1966	VI = 109.8 − 0.18*x*	−0.62	±10.3	Wilson and McCarrick (1967)
Lignin				
Cynodon dactylon	—	−0.61	—	Hawkins *et al.* (1964)
Temperate forage	—	−0.13	—	Van Soest (1965b)
	—	−0.69	± 6.3	Jones and Walters (1975)
Tropical grass	—	−0.45	—	Abrams *et al.* (1983)
Cellulose solubility in cupriethylenediamine (g/kg)				
Temperate forage	VI = 201.1 − 0.372*x*	0.74	—	Johnson *et al.* (1965)
Dry matter soluble in 1 *N* sulfuric acid (g/kg)				
Temperate forage	VI = 9.4 + 0.339*x*	0.76	—	Johnson *et al.* (1965)

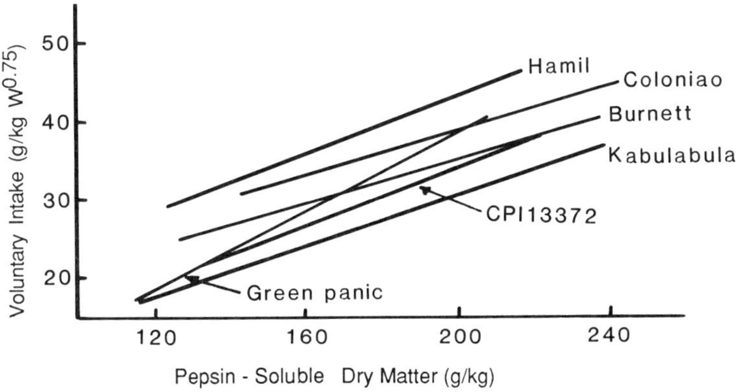

Fig. 2.17. Varietal and species differences in the relation between voluntary intake of *Panicum* spp. by sheep and pepsin-soluble dry matter. Data from Minson and Haydock (1971).

C. Bioassay

The VI of forages is correlated with the *in vivo* digestibility of the dry matter, organic matter and energy (Table 2.22), and dry-matter digestibility (DMD) estimated by many different laboratory methods (Chapter 4). The *in vitro* techniques for estimating DMD have the lowest RSD and provide a useful indirect way of predicting VI. The DMD can be estimated by either the *in vitro* rumen fluid/pepsin technique or the pepsin/cellulase technique and both methods will be described in Chapter 4.

Although VI may be predicted from *in vitro* DMD, a further improvement is to measure the *rate* of breakdown of forage *in vitro*. In 1960 it was shown that cellulose in legumes was digested more rapidly *in vitro* than cellulose in grasses and that this difference was correlated (r = 0.83) with VI (Donefer *et al.*, 1960). With cultivars of *C. gayana* cut at four stages of growth VI was correlated with the disappearance of dry matter *in vitro* after 12, 24, 48, and 72 hr (Minson and Milford, 1967a). However, there was no difference in the rate of digestion of leaf and stem fractions of tropical grasses, although the VI of the leaf fraction was very much higher (Laredo and Minson, 1973). Chewing rather than digestion is the main factor responsible for the breakdown of forage particles to a size that can readily leave the rumen (McLeod and Minson, 1988), so any correlation between VI and rate of digestion is probably not causal but due to an association between rate of digestion and resistance to breakdown by chewing.

In early studies, rate of digestion was measured *in vitro*, but more recent studies have used the "*in sacco*" technique with chopped or coarsely ground forage samples suspended in the rumen in a permeable bag made of indigestible cloth (Chapter 4). The VI of contrasting hays

TABLE 2.22

Prediction of Voluntary Intake by Sheep from the *in Vivo* Digestibility Coefficient of the Dry Matter (DMD),Organic Matter (OMD), or Energy (ED) in Forage

Species	Equation	r	RSD	Reference
Dactylis glomerata	Log VI[a] = 0.45*DMD* + 1.512	0.50	—	Demarquilly (1965)
	Log VI = 2.23*DMD* + 0.364	0.72	—	Demarquilly (1965)
Festuca pratensis	Log VI = 1.03*DMD* + 1.090	0.62	—	Demarquilly (1965)
Lolium multiflorum				
Tetraploid	VI[a] = 182*OMD* − 20.5	—	±1.6	Osbourn *et al.* (1966)
Diploid	VI[a] = 142*OMD* − 53.6	—	±1.8	Osbourn *et al.* (1966)
Lolium perenne	Log VI = 0.77*DMD* + 1.246	0.44	—	Demarquilly (1965)
Medicago sativa	Log VI = 1.01*DMD* + 1.699	0.82	—	Demarquilly (1965)
	VI = 358*DMD* + 170DMD^2 − 85.3	—	±7.3	Troelsen and Campbell (1969)
Temperate forage	VI[a] = 36*OMD* − 25.6	—	—	Minson *et al.* (1964)
	Log VI = 0.21*DMD* + 1.426	0.16	—	Demarquilly (1965)
	VI[a] = 158*ED* − 20.5	—	±6.3	Blaxter *et al.* (1966)
	VI = 151.1*OMD* − 36.16	0.86	±7.8	Chenost (1966)
	VI = 105.7*DMD* − 10.5	0.77	±7.8	Clancy and Wilson (1966)
	VI = 130*DMD* − 11.2	0.79	±7.9	Hovell *et al.* (1986)
	VI = 34.8*DMD* + 100DMD^2 + 2.37	—	±8.5	Troelsen and Campbell (1969)

[a]VI, voluntary intake (g/kg $W^{0.73}$ DM).

was more closely related to the extent of DM disappearance, after 12-, 24-, 48-, and 72-hr digestion, than to *in vivo* digestibility (Hovell *et al.*, 1986). The VI of straws was also closely related to rate of digestion (r = 0.88) (Orskov *et al.*, 1988b). Further work appears warranted on the relative advantages of the *in vitro* and *in sacco* techniques and the way in which rate of digestion limits VI. There is a need to determine how the resistance to chewing of digesta during rumination is related to rate of digestion and whether the statistical relation between VI and rate of digestion has a physiological basis.

D. Models

The VI of forage by ruminants can be predicted from a mathematical model containing all animal, plant, and management factors known to control VI. The error in predicting VI depends on the accuracy of the data used to developed the model, the extent that it is used to predict VI for animals and forages not included in the model, and the validity of the model. The use of models to predict VI and production of ruminants is a special problem beyond the scope of this volume and readers are referred to work by Mertens (1977, 1987), Forbes (1977a,b), Meissner *et al.* (1979), Sibbald *et al.* (1979), Yungblut *et al.* (1981), Bywater (1984), and Jarrige *et al.* (1986).

V. INCREASING VOLUNTARY INTAKE

The VI of forages may be improved by various management strategies, feeding supplements, chemical treatment of the forage before feeding, and plant breeding. These four approaches to increasing VI of forages will now be considered.

A. Management Practices

When sowing new pastures, high-VI varieties or species can be included provided they also have the desired agronomic characteristics. Management strategies adopted should encourage the survival of the high-VI species in existing pastures or natural rangeland.

Maximum VI is achieved by cutting or grazing pastures while the forage is immature and leafy (Section III,E), but a compromise has to be maintained between high VI, forage yield, and survival of the pasture. With more mature swards the proportion of leaf in the diet can be increased by feeding excess forage and allowing animals to select a diet with a higher VI (Fig. 2.1).

The VI will be reduced if the forage is deficient in any essential nutrient and the animal has no other source of that nutrient either as a feed supplement or from body reserves. Most deficiencies can be rectified by applying the appropriate fertilizer to growing forage (Table 2.23). A notable exception is sodium, where the concentration depends on the forage species and cultivar and not on the sodium level in the soil (Chapter 10).

B. Protein and Nonprotein Nitrogen Supplements

The VI of mature grasses may be limited by a deficiency of crude protein when the concentration falls below about 62 g/kg (Section III,E). Feeding supplementary protein will overcome this deficiency, increase the rate of digestion in the rumen (Schlink and Lindsay, 1988), and raise the VI of the forage (Table 2.24). Many forms of protein supplement have been used, but of special interest is the use of forage legumes (Minson and Milford, 1967b; Siebert and Kennedy, 1972), which can be grown as a mixture with, or as a special-purpose pasture adjacent to, protein-

TABLE 2.23

Increases in Voluntary Intake of Nutrient-Deficient Forage Following Application of Appropriate Fertilizer

	Voluntary intake ($g/kg\ W^{0.75}$)			
Deficiency	Control	Fertilized	Increase (%)	Reference
Protein				
Digitaria decumbens	41[a]	69[a]	68	Chapman and Kretschmer (1964)
	36	46	28	Minson (1967)
Phosphorus				
Medicago sativa	78	132	69	Weir *et al.* (1958)
Stylosanthes humilis	75	84	12	Playne (1972a)
Sulfur				
Digitaria decumbens	44	64	45	Rees *et al.* (1974)

[a]Assuming liveweight of 400 kg.

deficient forage. This supplementary effect of a legume is illustrated in Fig. 2.7.

Responses to supplementing with CP are generally restricted to forages containing less than 62 g CP/kg DM. An exception was *Calluna vulgaris*, where an increase in VI occurred although the forage contained 78 g CP/kg DM. This difference was possibly caused by the low availability of protein in this browse plant (Milne and Bagley, 1976).

During silage fermentation, protein is degraded and VI can be limited by the quantity of amino acids absorbed from the small intestine (Section III,H). When lactating cows fed silage containing 128 g CP/kg DM were supplemented with 1 kg fish meal (replacing 1 kg barley) the VI of silage was increased from 8.0 to 8.6 kg DM/day and milk yield increased from 17.7 to 18.9 kg/day (Kassem *et al.*, 1987). A fish-meal supplement increased VI by 7% and doubled the quantity of protein retained by lambs. However, much smaller increases in VI were found when 4-month-old steers fed silage containing 103 g CP/kg DM were supplemented with fish meal (Gill and England, 1984), possibly due to the lower protein requirement of the steers.

Protein supplements are usually expensive and VI may be increased by feeding nonprotein nitrogen, usually in the form of urea (Table 2.25). The VI has sometimes been improved by using only a urea supplement (Campling *et al.*, 1962), but in other studies urea has failed to improve the VI of low-nitrogen *A. sativa, Triticum aestivum* (Minson and Pigden, 1961), and *Heteropogon contortus* (Siebert and Kennedy, 1972). With

TABLE 2.24

Effect of Feeding Protein Supplements on the Voluntary Intake of Protein-Deficient Forages

Species	Forage crude protein (g/kg DM)	Increase in voluntary intake due to supplementary protein (%)	Reference
Calluna vulgaris	78	58	Milne and Bagley (1976)
Digitaria decumbens	36	38	Minson and Milford (1967b)
	32	14	Ventura *et al.* (1975)
	59	27	Ventura *et al.* (1975)
Heteropogon contortus			
Cattle	42	39	Siebert and Kennedy (1972)
Sheep	42	61	Siebert and Kennedy (1972)
Cattle	45	38	Hunter and Siebert (1980)
Sheep	39	77	Playne (1969b)
	25	73	Lindsay *et al.* (1988)
Lolium multiflorum	45	44	Crabtree and Williams (1971b)
Paspalum notatum	66	14	Moore *et al.* (1970)
Paspalum plicatulum	39	20	Kellaway and Leibholz (1983)
Prairie hay	61	27	McCollum and Galyean (1985)
Temperate forage	48	16	Lamb and Eadie (1979)
Tropical hay	46	16	Morris (1958)
	39	44	Hennessy *et al.* (1983)
	27	75	Lee *et al.* (1985)
Mean	45	40	

some diets, maximum increase in VI is only achieved if molasses or sucrose is fed in addition to urea (Hemsley and Moir, 1963). Molasses is a source of the branch-chain volatile acids, isobutyric, *n*-valeric, and isovaleric acid, required by rumen microbes (Hemsley and Moir, 1963). Rumen bacteria also require sulfur and with some diets both sulfur and urea are necessary for maximum VI. This is particularly important when feeding sheep, which are more sensitive to sulfur deficiency than are cattle (Bird, 1974).

Some protein-deficient forages contain sufficient sulfur, readily available carbohydrate, and precursors of branch-chain volatile fatty acids and

TABLE 2.25

Effect of Supplementary Urea on the Voluntary Intake of Protein-Deficient Forages

Species	Forage crude protein (g/kg DM)	Increase in voluntary intake due to supplementary urea (%)	Reference
Avena sativa	32	42	Campling *et al.* (1962)
	31	40	Freer *et al.* (1962)
	44	46	Hemsley and Moir (1963)
	35	21	Mulholland *et al.* (1974)
	52	37	Egan and Doyle (1985)
Digitaria decumbens	49	28	Hunter and Siebert (1985b)
	60	12	Hunter and Siebert (1987)
Heteropogon contortus	45	28	Hunter and Siebert (1980)
	39	8	Hunter and Siebert (1985b)
	22	26	Hunter and Siebert (1987)
Hordeum vulgare	28	58	Tudor and Morris (1971)
	31	24	Mulholland *et al.* (1974)
Paspalum plicatulum	38	20	Leibholz (1981)
	25	85	Lindsay *et al.* (1988)
Triticum vulgare	19	104	Bird (1974)
	29	10	Mulholland *et al.* (1974)
Temperate hay	50	12	Fleck *et al.* (1988)
Tropical hay	29	29	Ernst *et al.* (1975)
	36	25	Hennessy *et al.* (1978)
Mean	37	34	

VI can be increased by supplementing with urea only. However, with other forages maximum response to urea will only be obtained if sulfur and branch-chain volatile fatty acids or their precursors are also supplied. At present it is not possible to identify forage that is deficient in branch-chain acids and therefore molasses and sulfur are included in most urea supplements.

C. Energy Supplements

Energy supplements have a low fiber content and in a choice situation are eaten in greater quantities than forage. When fed in combination with forage total VI is higher than for forage alone, but the increase is less than the quantity of supplement included in the diet. This is because the VI of the forage is depressed by the presence of the supplement, a substitution effect that has practical and economic implications. When the substitution coefficient is 1.0, the VI of forage is depressed by the same amount as the quantity of energy supplement eaten and total VI is unchanged by feeding the supplement. Conversely, if the substitution coefficient is 0 then total intake is increased by the same quantity as the amount of supplement fed.

Substitution coefficients for a range of forages are shown in Table 2.26. These substitution coefficients were correlated with the CP concentration of the unsupplemented forage ($r = 0.51$). Energy supplements have only a marginal effect on total intake of food when the forage is immature and highly digestible, but large increases in VI occur when mature or tropical forages are supplemented.

High substitution coefficients were found when large quantities of *Z. mays* were fed to cattle receiving *Agropyron elongatum* (0.88), *A. sativa* (0.76), and *Phalaris aquatica* (0.89) (Torres and Boelcke, 1978). However, when the diets contained less than one-third *Z. mays,* the substitution coefficients were smaller and varied between 0.03 and 0.66.

The substitution coefficient can be changed by chemical treatment of the forage. With unprocessed temperate hay (78 g CP/kg DM) the substitution coefficient was 0.52 but this increased to 1.15 when the hay was upgraded with ammonia (Wylie and Steen, 1988).

D. Mineral Supplements

Voluntary intake may be depressed if forage is deficient in any essential element. There are many reports of VI being increased by feeding supplementary minerals and these will be considered in later chapters.

E. Physical and Chemical Treatment

The VI of forage is increased by any treatment that reduces the resistance to breakdown during chewing. Grinding and pelleting is the most dramatic illustration of this effect and was considered in Section III,H. The resistance of forage to chewing is also reduced by any chemical treatment that increases the digestibility of the energy (Chapter 4). Many different chemicals are used to increase the digestibility of forage (Sundstol and Owen, 1984), but data on the effect of chemicals on VI is only available for sodium hydroxide, anhydrous ammonia, and ammonia produced *in situ* from urea. These treatments increase the VI of forages from 6 to

TABLE 2.26

Depression in Voluntary Intake of Forage for Each Unit of Grain Supplement

Forage	Substitution coefficient[a]	Forage crude protein (g/kg DM)	Reference
Agropyron elongatum	0.34	66	Torres and Boelcke (1978)
Avena sativa (straw)	0.34	39	Crabtree and Williams (1971a)
	0.46	24	Torres and Boelcke (1978)
	0.25	22	Horton (1978)
	0.01	35	Lamb and Eadie (1979)
Chloris gayana (silage)	0.50	52	Thomas (1977)
Dactylis glomerata	0.53	67	Crabtree and Williams (1971a)
Hordeum vulgare	0.20	39	Horton (1978)
Lolium multiflorum	0.22	45	Crabtree and Williams (1971b)
Lolium perenne	0.91	172	Tayler and Wilkinson (1972)
	0.69	238	Milne *et al.* (1981)
	0.63	116	Orr and Treacher (1984)
	0.30	172	Kassem *et al.* (1987)
Lolium perenne (silage)	0.52	137	McIlmoyle and Murdoch (1977)
	0.69	125	Wernli and Wilkins (1980b)
Medicago sativa (wafer)	0.78	244	Bath *et al.* (1974)
Phalaris aquatica	0.27	46	Torres and Boelcke (1978)
Phleum pratense (straw)	0.14	47	Lamb and Eadie (1979)
Triticum aestivum (straw)	0.11	23	Horton (1978)
Temperate grass	0.83	154	Forbes *et al.* (1966)
	0.38	210	Forbes *et al.* (1967a)
	0.25	90	McCullough (1976)
	0.35	88	Lamb and Eadie (1979)
	0.22	62	
	0.69	91	
	0.56	136	Vadiveloo and Holmes (1979a)
	0.28	103	
	0.14	155	
Temperate grass (silage)	0.0	135	McCarrick (1963)
	0.17	168	McCullough (1972)
	0.58	—	Moisey and Leaver (1982)
	0.44	175	Harb and Campling (1983)
Mean	0.40	108	

[a]When the intake of forage was depressed by the same amount as the quantity of grain fed, the substitution coefficient was 1.0. If the supplementary grain had no effect on the intake of the forage, the substitution coefficient was zero.

TABLE 2.27

Proportional Increase in Voluntary Intake Following Chemical Treatment of Different Forages

Species	Crude protein in original forage (g/kg DM)	Increase in voluntary intake (%)	Reference
Sodium hydroxide			
Hordeum vulgare (straw)	38	34	Owen and Nwadukwe (1980)
Saccharum officinarum	—	19	Losada *et al.* (1979)
Triticum aestivum (straw)	77	50	Alawa and Owen (1984)
	—	53	Davis and Weston (1986)
Anhydrous ammonia			
Avena sativa (straw)	22	41	Horton (1978)
	26	12	Horton and Steacy (1979)
Festuca arundinacea			
Lambs	79	32	Buettner *et al.* (1982)
Cattle	79	52	Buettner *et al.* (1982)
Hordeum vulgare (straw)	39	40	Horton (1978)
	39	21	Horton and Steacy (1979)
	38	34	Owen and Nwadukwe (1980)
	42	24	Jewell and Campling (1986)
	—	14	Orskov *et al.* (1988a)

50% (Table 2.27) but there is insufficient information to determine whether there is any difference between the methods. These improvements in VI are due to a softening of the fiber and reduced resistance to chewing during eating and ruminating (Davis and Weston, 1986) and only occurred when forage was the main component in the diet. When *A. sativa* straw was fed alone, anhydrous ammonia increased VI by 41%, but when the diet contained only one-third straw there was no advantage in treating the straw (Horton, 1978).

F. Plant Breeding and Selection

In the early stages of a forage breeding and selection program many samples must be examined. The most reliable indicator of VI, albeit an indirect one, is digestibility (Section IV,C), so screening forages on the basis of *in vitro* dry-matter digestibility should also increase VI. Differences in dry-matter digestibility exist between lines of *Cynodon dactylon,* and a hybrid, Coastcross 1, has a higher VI (9%) than a low-digestibility line (Lowrey *et al.*, 1968). Large genotypic differences in the *in vitro* digestibility of *Digitaria* spp. were reported (Strickland, 1970; Strickland

and Haydock, 1978) with the digestibility of *Digitaria setivalva* generally being higher than other introductions. When fed to sheep, *D. setivalva* had 8% higher VI than the other genotypes at all stages of growth (Minson, 1984). In *C. ciliaris* lines selected on the basis of high and low *in vitro* digestibility there was a 12% difference in VI (Minson and Bray, 1985).

Cafeteria trials have been used to identify lines that may possess improved VI characteristics. When grown as pure swards, genotypes of *P. arundinacea* selected on the basis of preference were eaten in greater quantities and produced higher liveweight gains in lambs than low-preference lines (O'Donovan *et al.*, 1967). In a subsequent indoor study, high-preference genotypes were found to have higher dry-matter digestibilities but there was little difference in VI (Barnes and Mott, 1970). Cattle grazing a cultivar of *Eragrostis curvula,* selected on the basis of preference in a cafeteria-type study, gained 12% more per head than when grazing an unpalatable line (Voigt *et al.*, 1970). However, with *C. ciliaris* there was no relation between VI and preference rating (Minson and Bray, 1986). In *Paspalum notatum,* there was no relation between liveweight gain and preference rating (Burton, 1974).

VI. CONCLUSION

The quantity of forage eaten is the primary factor limiting ruminant production. The appetite of ruminants is affected by animal size, pregnancy, lactation, growth rate, parasites, climatic stress, and previous nutrition, and, in the case of sheep, wool length. The quantity of forage eaten depends on the method of feeding and the resistance of the forage to breakdown by chewing to a particle size (1 mm) that can pass easily from the rumen.

Voluntary intake is higher for legumes than for grasses, for temperate than for tropical forages, and for immature than for mature forages. Differences in VI occur between forage species and cultivars, and plant parts, but not between fertilized and unfertilized forage except where VI is limited by a nutrient present in the fertilizer. Low light intensity, frost, and water deficiency probably depress VI.

High levels of water (>780 g/kg) depress VI, but artificial drying or haymaking has little effect on VI unless leaf loss occurs. Pelleting increases VI especially with mature forage. Ensiling fresh forage causes a larger depression in VI by sheep than by cattle. The depressing effect of ensiling on VI may be reduced by fine chopping, wilting, addition of formic acid or formaldehyde, or feeding grain supplements.

Voluntary intake of forages has been predicted from leaf proportion, bulk density, grinding energy, resistance to chewing, near-infrared reflectance spectroscopy, *in vitro* and *in sacco* digestion techniques, a range of chemical analyses, and models including all animals, plant, and management factors known to control VI.

Forage intake can be increased by the adoption of management practices that encourage superior forage species and keep pastures leafy. Intake of protein-deficient forage is increased by supplementary protein, nonprotein nitrogen, or sulfur. Energy supplements depress the VI of the forage but increase total intake, especially with mature forages. Intake of mature forages is increased by treatment with sodium hydroxide or ammonia. Plant breeding and selection have been successful in increasing the VI of some tropical and temperate grasses.

3

Intake of Grazed Forage

I. INTRODUCTION

The intake of grazed forage is controlled by many animal, forage, and environmental factors. Most of these are the same as apply to pen-fed animals (Chapter 2) but some factors are unique to grazed forage. These include the opportunity to graze selectively and the influence of availability of desired forage on bite size. Some studies have aimed at confirming under-grazing conditions principles already established with cut forage fed to penned ruminants, but in other studies the work is directed toward an understanding of phenomena specific to the field situation. Many different techniques have been used to measure intake of grazed forage. These will be briefly described, highlighting their advantages and disadvantages before considering the factors influencing the intake of grazed forage.

II. MEASUREMENT OF FORAGE INTAKE

The intake of forage by the grazing animal can be measured in many different ways. Comprehensive reviews have been published on the techniques available for measuring forage intake (Corbett, 1978; Le Du and Penning, 1982), so it is only necessary to highlight the main features of these methods and under what conditions they may be applied. The methods can be classified according to the duration of the measurement period (Table 3.1).

A. Short-Term Changes in Liveweight

The first method used to measure the quantity of forage eaten was to weigh the animals before (W_1) and after grazing (W_2) (Erizian, 1932). It was necessary to correct the change in body weight for a loss in weight

TABLE 3.1

Techniques for Determining the Forage Intake of Grazing Ruminants Classified According to Measurement Period

Minimum time	Technique for measuring forage intake
5 min	Short-term changes in liveweight
15 min	Grazing behavior (bite size, rate, and time)
1 day	Cutting method
1 week	Fecal techniques
1 month	Production and liveweight

caused by the excretion of feces (F), urine (U), and insensible loss (I) and for any gain in weight when the animals drank water (L).

$$\text{Intake} = (W_2 + \text{F} + \text{U} + \text{I}) - W_1 - \text{L}$$

Changes in liveweight can now be determined by weighing animals with an integrated electronic balance (Penning and Hooper, 1985) or continuously by fitting load cells to their feet and monitoring liveweight using radio telemetry. Short-term changes in weight were used to study the rate of forage intake of sheep offered different types of sward (Allden and Whittaker, 1970; Penning and Hooper, 1985). The main source of error is the insensible weight loss, which varies between sheep and is influenced by environmental conditions (Penning and Hooper, 1985). The intake of fresh forage can be converted into intake of dry matter by measuring the moisture content of plucked samples of the forage selected, but this introduces a subjective source of error.

B. Grazing Behavior

Forage intake may be calculated from the time spent grazing (T), the number of bites per unit of time (R), and the average size of each bite (S) (Spedding *et al.*, 1966):

$$\text{Intake} = \text{T} \times \text{R} \times \text{S}$$

This equation can be simplified to:

$$\text{Intake} = \text{N} \times \text{S}$$

where N is the total number of bites (Chacon *et al.*, 1976).

Equipment has been developed for measuring number of harvesting bites (Stobbs and Cowper, 1972) and bite size (Stobbs, 1973). The method is suitable for measuring forage intake for periods of 15–30 min. Bite size varies throughout the day and allowance for these changes in required if the method is used to measure intake over long periods (Chacon *et al.*, 1976).

The number of bites is recorded with a jaw switch and harvesting bites are separated from ruminating bites using a head-position switch (Stobbs and Cowper, 1972). The head-position switch is only effective with short forage; for long forage an accelerometer must be used to separate total bites into harvesting and ruminating bites (Chambers *et al.*, 1981). Electronic methods are now available that determine whether the pattern of jaw movements is associated with grazing or ruminating (Penning, 1983).

C. Cutting Method

The intake of grazed forage has been estimated from the difference between the weight of forage before and after grazing (Woodward, 1936; Linehan, 1952; Cox *et al.*, 1956; Walters and Evans, 1979; Meijs, 1981; Meijs *et al.*, 1982). The accuracy of the estimate of intake depends on three factors: the error in estimating the initial and final yields of forage; the proportion of forage offered that is eaten; and the growth of forage that occurs while the pasture is being grazed, or loss of pasture through senescence or activity of insects, etc. With extensively grazed pastures, intake of cows is overestimated by 8–16 kg DM/head/day, but this bias is reduced to 0.3–1.1 kg DM/head/day when the pastures are strip grazed (Van der Kley, 1956). In-and-out cuts are widely used to measure the intake of forage by strip-grazed animals (Stehr and Kirchgessner, 1976; Trigg and Marsh, 1979; Walters and Evans, 1979; Meijs *et al.*, 1982).

With continuously grazed pasture, forage growth can be measured on protected areas and used to correct intake estimated from the in-and-out cuts (Linehan, 1952; Meijs *et al.*, 1982). This method has been recently used to measure forage intake by beef cattle (Mitchell *et al.*, 1986) and dairy cows (Meijs, 1986; Rearte *et al.*, 1986).

D. Fecal Techniques

In 1934 Garrigus described a simple method of estimating intake (DM) of grazing steers using the quantity of feces (DM) collected in canvas bags attached to the animal and the apparent digestibility coefficient (DMD) of the forage determined by feeding cut forage to cattle in stalls.

$$\text{Intake (DM)} = \frac{\text{Feces output}}{(1 - \text{DMD})}$$

Total collection of feces was laborious and the determination of DMD with cut forage usually underestimates the DMD of grazed forage because animals select a diet higher in DMD (see Section III). Indirect methods of estimating fecal output and DMD have been developed to overcome these problems.

1. Estimating Fecal Production

Total collection of feces has been avoided by the use of an indigestible tracer. A fixed quantity (Q) of an indigestible tracer is fed each day and the concentration of tracer (C) determined in a representative sample of the feces:

$$\text{Daily fecal output} = \frac{Q}{C}$$

Many indigestible tracers have been used in nutrition studies (Kotb and Luckey, 1972), but for estimating the fecal production of grazing animals chromic oxide (Cr_2O_3) is the tracer most often used. The indigestible tracer is administered once or twice daily in gelatine capsules containing 1 or 10 g Cr_2O_3 in an oil base (Raymond and Minson, 1955), paper impregnated with Cr_2O_3 (Corbett *et al.*, 1958; Langlands *et al.*, 1963), cereal pellets (Zoby and Holmes, 1983), or as intraruminal capsules which release Cr_2O_3 at a controlled rate over a period of 3 weeks (Ellis *et al.*, 1981; Laby *et al.*, 1984).

One of the greatest problems in measuring fecal production with indigestible tracers is their uneven excretion in the feces. When fed twice daily there are large diurnal variations in tracer concentration in the feces of grazing beef cattle (Hardison and Reid, 1953), dairy cattle (Smith and Reid, 1955), and sheep (Raymond and Minson, 1955). The magnitude and time characteristics of these excretion curves are related to the level of feeding and forage quality (Raymond and Minson, 1955; Lambourne, 1957) and pose problems in obtaining a representative sample for tracer analysis. Biased estimates of fecal production occur when feces are collected directly from the rectum at fixed times each day (Hardison and Reid, 1953). This is avoided by collecting a random sample of feces deposited on the pasture (Raymond and Minson, 1955), feces from different animals being identified by feeding different-colored polystyrene particles (Minson *et al.*, 1960a). The magnitude of the diurnal variation is reduced when using Cr_2O_3 paper (Langlands *et al.*, 1963) and intraruminal capsules, but diurnal variation can never be entirely eliminated because there are diurnal variations in forage intake and fecal excretion.

A recent development has been to use synthetic *n*-alkanes of even chain length (C_{28} and C_{32}) to measure fecal production (Mayes *et al.*, 1986) and to use naturally occurring forage *n*-alkanes of uneven chain length (C_{27}, C_{29}, C_{31}, C_{33}, and C_{35}) to estimate DMD (Chapter 4). Both artificial and naturally occurring *n*-alkanes are only partially recovered in the feces, but by using both artificial and natural *n*-alkanes of similar chain length any overestimation of fecal production is balanced by an underestimation of DMD, so forage intake can be predicted with a low error (Mayes *et al.*, 1986).

2. Estimating Dry-Matter Digestibility of Grazed Forage

Many methods have been used to determine the DMD of selectively grazed forage. These may be divided into two distinct groups: methods which use regressions previously derived indoors, which relate DMD to the chemical composition of the feces (fecal-index technique) and methods that use a sample of the selectively grazed forage. Fecal-index components used to measure the DMD of forage include chromogen, nitrogen, and fiber (Chapter 4). The main source of error in the technique is that differences in DMD associated with animal species, level of feeding, and endoparasites are not taken into account when deriving the fecal-index regressions (Chapter 4).

Samples of selectively grazed forage may be collected by hand plucking or by fitting animals with esophageal fistulae (Van Dyne and Torrell, 1964). The *in vivo* DMD of these samples can be estimated from their chemical composition or by an *in vitro* technique (Chapter 4). All these methods rely on regressions derived with sheep fed in pens and errors can be introduced when they are applied to grazing studies where the animal species, level of feeding, and level of endoparasites may be very different. This source of error does not apply when DMD is estimated from the concentration of a naturally occurring indigestible tracer present in the forage (X_1) and feces (X_2):

$$\text{DMD} = \frac{X_2 - X_1}{X_2}$$

Many naturally occurring chemical components in forage have been used as internal tracers including silica, iron, lignin, chromogen, potentially indigestible cellulose, acid-insoluble ash, and *n*-alkanes. In some studies these have proved to be ideal tracers with complete recovery in the feces, but incomplete recoveries have also been reported (Chapter 4). Since Cr_2O_3 is not naturally present in forage it cannot act as a tracer to measure the digestibility of grazed forage.

E. Production and Liveweight

Equations have been published that can be used to estimate the quantity of forage eaten by the animal from the average weight of the animal and level of production.

The following equations have been published.

Beef cattle

Minson and McDonald (1987):

$$I = (1.185 + 0.00454W - 0.0000026W^2 + 0.315\,G)^2$$

where I = kg DM/day, W = liveweight (kg), and G = growth (kg/day).

Lactating cattle

Cox *et al.* (1956):

$$I = 0.13FCM + 0.0053W + 0.96G$$

where I = kg DM/day, FCM = 4% fat-corrected milk (kg/day), W = liveweight (kg), and G = growth (kg/day).

Bines *et al.* (1977):

$$I = 0.16M + 0.0113W + 2.45G + 4.25$$

where I = kg DM/day, M = milk yield (kg/day), W = liveweight (kg), and G = growth (kg/day).

Neal *et al.* (1984):

$$I = 0.20M + 0.022W$$

where I = kg DM/day, M = milk yield (kg/day), and W = liveweight (kg).

When concentrates are fed, forage intake can be calculated using the equation of Vadiveloo and Holmes (1979b), as modified by Caird and Holmes (1986):

$$I = 3.476 + 0.404C + 0.013W - 0.129T + 4.120\log T - 0.140M$$

where I = kg DM/day, C = concentrates (kg/day), W = liveweight (kg), T = weeks after calving, and M = milk yield (kg/day).

Other equations have been published that include herbage allowance, DMD, and forage type (Meijs and Hoekstra, 1984; Stockdale, 1985).

TABLE 3.2

Chemical Compounds Correlated with Forage Preference by Ruminants

Species	Chemical compound	Reference
Dactylis glomerata	Soluble carbohydrate	Bland and Dent (1964)
Eragrostis curvula	Organic acids	Arnold (1981)
Phalaris arundinacea	Indole-alkylamine alkaloids	Simons and Marten (1971), Marten *et al.* (1973)

III. SELECTIVE GRAZING

Grazing animals have the opportunity to select their diet and many factors interact to determine what they choose on a particular day (Arnold, 1981). The composition of the diet selected may be very different from that of the pasture on offer. For example, when grazing a semiarid rangeland, the diet selected came from plants that contributed only 1% of the total forage on offer (Leigh and Mulham, 1966a,b).

The forage selected by grazing ruminants usually contains a higher proportion of leaf than the forage offered (Alder and Minson, 1963; Arnold, 1981; Ruiz *et al.*, 1981). In mixed-temperate pasture, sheep and cattle usually prefer legumes (Curll *et al.*, 1985; Clark and Harris, 1985; Sollenberger *et al.*, 1987). These preferences sometimes reflect active discrimination by animals against the taste, smell, or texture of particular species or plant parts (Arnold, 1981). Preference ratings in cafeteria studies can sometimes be related to differences in chemical composition (Table 3.2), but it is unwise to assume that this always controls selective grazing. In many cases the composition of the diet selected simply reflects the species and plant parts present in the horizon of the canopy being grazed (Milne *et al.*, 1982; Sollenberger *et al.*, 1987) or lower resistance to prehension (Hendricksen and Minson, 1980). Where selective grazing increases the proportion of leaf in the diet there is a rise in intake because leaf is eaten in greater quantities than stem, even when there is no difference in DMD (Chapter 2).

IV. PHYSICAL CONSTRAINTS

The quantity of forage eaten each day depends on the time spent grazing, the rate of biting, and the size of each bite (see Section II,B). As a sward is grazed down there are large changes in bite size, but differences

TABLE 3.3

Differences in Grazing Behavior and Forage Intake of Cattle Grazing a *Lablab purpureus* Sward[a]

Attribute	Range of values		
	Low	High	High/low
Grazing time (min/day)	468	685	1.5
Biting rate (bite/min)	55	65	1.2
Bite size (g/OM/bite)	0.09	0.41	4.6
Forage intake (kg/day)	2.8	11.5	4.1

[a]From Hendricksen and Minson (1980).

in grazing time and biting rate are small (Table 3.3). There appears to be an upper limit to the time that animals will spend grazing each day (Hancock, 1952; Hubbard, 1952), and difference in bite size is the primary behavioral component that limits the intake of grazing ruminants. With dairy cows intake is depressed when bite size falls below about 0.3 g OM (Stobbs, 1973). Although the critical bite size has not yet been determined for other ruminants, the concept of a critical bite size provides a useful basis for understanding the way ruminants react to different types of sward. In the following sections the intake, grazing behavior, and production of ruminants will be considered in relation to the forage mass, allowance, and heterogeneity of the sward.

A. Forage Mass

Where the yield of young, desired forage exceeds about 2000 kg DM/ha and grazing is unrestricted, ruminants have no difficulty satisfying their appetite, taking in large quantities of forage with each bite (Allden, 1962; Allden and Whittaker, 1970). Bite size on young uniform swards varies with the physical dimensions of the individual bite and the quantity of forage within the volume encompassed by the teeth (Hodgson, 1985). Biting is concentrated in the upper horizons of the sward. For example, when a *Digitaria decumbens* sward was grazed by lactating cows most of the forage eaten (94%) came from above 20 cm, although this horizon contained only half the dry matter (Ruiz *et al.*, 1981).

When the forage mass is reduced to below 2000 kg DM/ha there is a reduction in bite size. This reduction is only partially offset by an increase in time spent grazing, so total forage intake declines (Fig. 3.1). This fall in intake is probably caused by a reduction in length of the tillers since

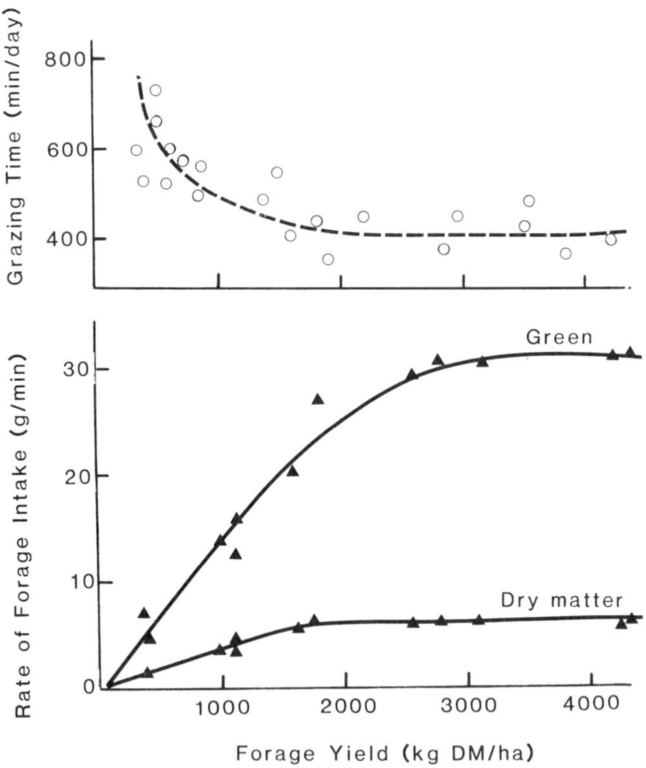

Fig. 3.1. Relation between grazing time by sheep, rate of forage intake, and forage yield. Data from Allden and Whittaker (1970).

bite size of sheep is directly proportional to tiller length over the range 4 to 37 cm (Allden and Whittaker, 1970).

Artificial swards prepared by threading tillers of *Lolium perenne* and *Pennisetum clandestinum* through a board with holes at different spacings have been used to study the factors controlling bite size. Bite size is positively related to both sward height and density but is best described by the mass per unit area effectively covered by one bite, provided forage mass exceeds 1000 kg/ha (Black and Kennedy, 1984). In field studies, bite size, depth of biting, and volume of each bite by sheep were positively related to sward height, r = 0.88, 0.97, and 0.87, respectively (Burlison and Hodgson, 1985). In a comparison of *L. perenne* swards maintained at heights of 3, 6, 9, and 12 cm, intake of ewes was similar for the 6-, 9-, and 12-cm swards but was reduced by about 50% when grazing the 3-cm

TABLE 3.4

Effect of Forage Height on Behavior, Forage Intake, and Production of Lactating Cows[a]

Attribute	Forage height[b] (cm) 4.8	6.4
Forage (kg OM/ha)	1810	2734
Tiller density (1000/m^2)	17	16
Leaf proportion	0.51	0.55
Green forage allowance (kg OM/cow/day)	17	21
Digestibility of selected forage (OM)	0.76	0.77
Grazing time (min/day)	575	565
Grazing bites (thousand/day)	44.7	43.3
Bite size (mg OM/bite)	282	345
Forage intake (kg OM/cow/day)	12.7	15.1
Milk yield (kg FCM/day)	26.3	28.1
Liveweight change (kg/day)	−0.67	+0.15

[a]From Kibon and Holmes (1987).
[b]Determined with a falling plate exerting a pressure of 4.8 kg/m^2.

sward (Penning *et al.*, 1984). With lactating cows offered excess pasture maintained at mean heights of 4.8 and 6.4 cm, intake, bite size, milk production, and liveweight gain were lower for the shorter forage (Table 3.4).

B. Forage Allowance

When a sward is intensively grazed over a period of several days there are large changes in grazing behavior, quality of forage selected, intake, and production. Examples will be presented for lactating cows, beef cattle, and sheep grazing young uniform swards.

1. Lactating Cows

The relation between intake and forage allowance is generally curvilinear (Fig. 3.2). Once the allowance of desired forage is less than twice the maximum intake there is a progressive fall in the quantity of forage eaten (Combellas and Hodgson, 1979; Ernst *et al.*, 1980). This is associated with a reduction in grazing time, rate of biting, number of bites, and bite size (Table 3.5). With rotational grazing of immature temperate forages, maximum intake is achieved when cows leave an 8- to 10-cm-high stubble, but when they are forced to graze down to 5 cm, intake is depressed by 10 to 15% (Ernst *et al.*, 1980).

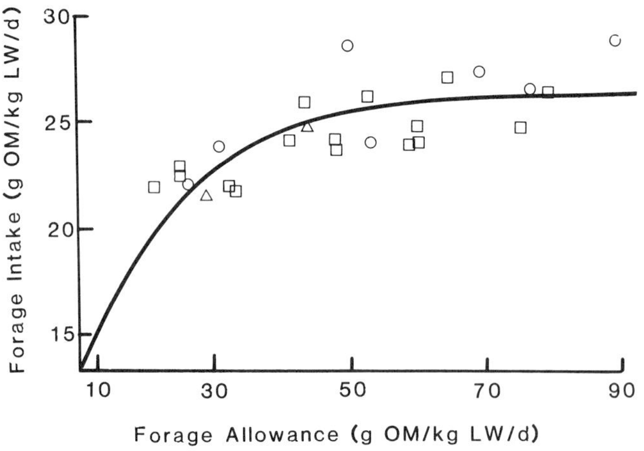

Fig. 3.2. The effect of forage allowance on intake by calves (○), beef (△), and dairy (□) cows. Data from Ernst *et al.* (1980).

TABLE 3.5

Effect of Forage Allowance on the Forage Intake and Production of Lactating Cows[a]

Attribute	Daily forage allowance (g DM/kg liveweight) 30	50	70
Forage intake			
kg OM/day	10.7	13.3	14.1
g OM/kg liveweight	22.3	26.3	28.0
Grazing behavior			
Grazing time (hr)	7.6	8.7	8.8
Biting rate (per min)	62	66	65
Total daily bites (thousand/day)	28	34	34
Bite size (g OM/bite)	0.39	0.40	0.43
Milk production			
Mean production (kg/day)	14.0	17.1	17.7
Peak production (kg/day)	21.1	24.5	23.8
Rate of fall (kg/week)	0.75	0.74	0.63
Liveweight change (kg/day)	−0.01	0.45	0.40

[a]From Le Du *et al.* (1979).

TABLE 3.6

Effect of Forage Allowance on the Intake and Behavior of Calves Grazing *Lolium perenne*[a]

Attribute	Daily forage allowance (g DM/kg liveweight)			
	30	50	70	90
Forage intake (g OM/kg liveweight)	24.1	29.0	27.9	29.6
Digestibility (OM)	0.80	0.82	0.82	0.82
Residual forage				
Mass (kg OM/ha)	1240	1630	2060	2310
Height (cm)	5.4	7.4	10.5	13.6
Green material (%)	50	64	68	71
Grazing behavior				
Grazing time (hr)	7.5	7.9	7.6	7.9
Biting rate (per min)	49	50	49	50
Bite size (g OM/bite)	0.28	0.26	0.29	0.27

[a]From Jamieson and Hodgson (1979).

2. Beef Cattle

The influence of forage allowance on the intake of beef cattle is similar to that reported for lactating cows (Fig. 3.2). The intake of beef calves is depressed approximately 18% when the daily allowance of forage is reduced from 90 to 30 g DM/kg liveweight, and this is associated with a decline in height of the residual forage from 7.4 to 5.4 cm and a lower OM digestibility (Table 3.6). The effect of forage allowance on intake is similar for steers of 5–6 and 15–18 months of age (Trigg and Marsh, 1979). With suckler cows and calves, intake of *L. perenne* is depressed when the allowance falls below 34 g DM/kg liveweight and animals are required to graze forage less than 6 cm in height (Table 3.7).

3. Sheep

Lactating ewes require a relatively high forage allowance to achieve maximum intake. In three studies, intake of lactating ewes grazing *L. perenne* was linearly related to forage allowance (Gibb and Treacher, 1978; Gibb *et al.*, 1981; Penning *et al.*, 1986), with no indication that a forage allowance of 160 kg OM/kg liveweight was sufficient to achieve maximum intake (Fig. 3.3). Ewes at the highest allowance selected a diet of higher OM digestibility and spent less time ruminating (Penning *et al.*, 1986). In contrast to this result from England, a curvilinear trend over a similar

TABLE 3.7

Effect of Allowance of *Lolium perenne* on the Intake, Behavior, and Production of Suckler Cows and Calves[a]

Attribute	Forage allowance (g DM/kg liveweight)		
	17	34	51
Height of sward (cm)			
Pregrazing	23.4	23.8	24.8
Postgrazing	3.6	6.2	8.9
Grazing (hr/day)			
Cows	7.3	8.5	8.8
Calves	4.3	5.2	5.3
Forage intake (kg OM/day)			
Cows	7.0	10.2	11.6
Calves	1.5	1.7	1.9
Milk yield (kg/day)	8.7	9.9	10.4
Liveweight change (kg/day)			
Cows	−0.1	0.5	0.7
Calves	0.9	1.2	1.2

[a]From Baker *et al.* (1981).

range of allowances was reported in New Zealand (Geenty and Sykes, 1986). This difference in response may have been caused by a greater uniformity of the pasture used in the New Zealand study, a subject to be discussed in the following section.

The intake of *L. perenne* by lambs grazing with ewes is linearly related to forage allowance up to at least 120 g OM/kg liveweight (Fig. 3.3), but with older lambs grazing alone, an asymptotic curve describes the relation between intake and forage allowance (Fig. 3.4). With 6-month-old lambs grazing *L. perenne* or *Trifolium repens* intake was increased by 23 and 64% respectively when forage allowance was doubled (Hodgson, 1975). For wethers, grazing *Calluna vulgaris*, intake was similar when they removed 40 or 80% of the current seasons shoots in summer, but in autumn intake was higher at the lower grazing pressure (Milne *et al.*, 1979).

C. Forage Heterogeneity

Swards are rarely uniform and the diversity provides ruminants with the opportunity to graze selectively. Some parts of the sward will be re-

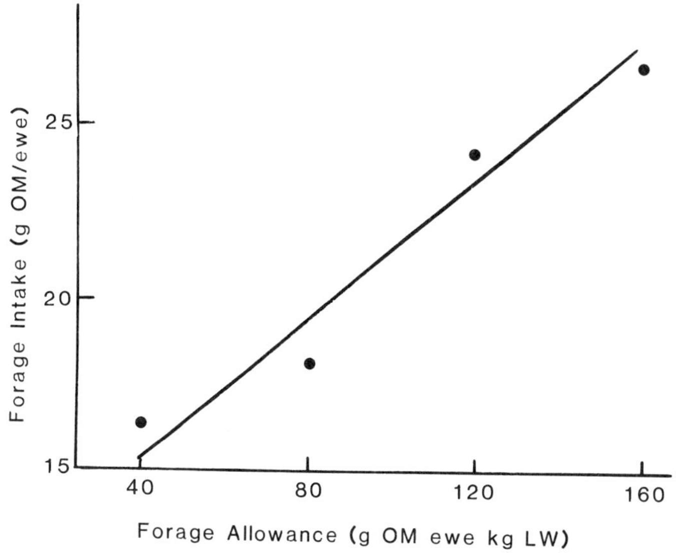

Fig. 3.3. Intake of *Lolium perenne* by lactating ewes at four forage allowances. Data from Penning *et al.* (1986).

jected and this may reach the point where this fraction should not be included when estimating forage allowance. Four forms of sward heterogeneity can affect the relation between intake and forage allowance: leaf versus stem, green versus dead, species differences, and soiled versus unsoiled.

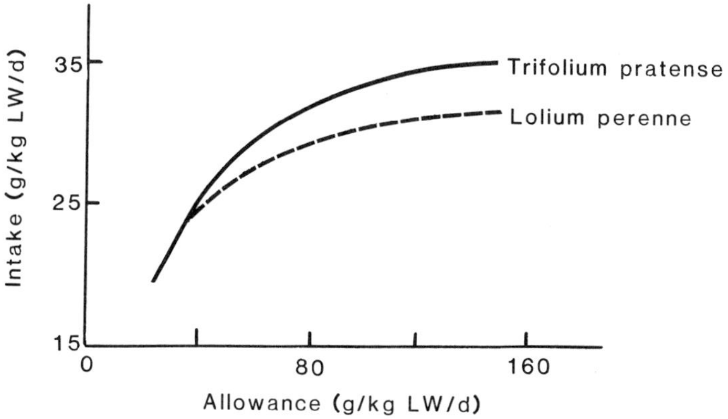

Fig. 3.4. Effect of forage allowance upon the intake of lambs. Data from Gibb and Treacher (1976).

1. Leaf versus Stem

In mature temperate and most tropical swards there are large physical differences between the leaf and the stem fractions. For example, the shear load required to sever the stem of *Lablab purpureus* is up to 10 times that required to harvest the leaf fraction (Hendricksen and Minson, 1980). Ruminants accustomed to eating leaf appear to restrict their diet to leaf, even when very little leaf is present (Chacon and Stobbs, 1976). This effect was demonstrated by the changes that occurred when cattle rotationally grazed *L. purpureus* (Fig. 3.5). On the first day bite size averaged 0.41 g OM, but by the twelfth day it had fallen to 0.09 g OM although large quantities of green stem were still available to the cattle. Very much larger differences exist with browse shrubs, and only the consumed portion of the plant should be used when calculating forage allowance and production of edible DM.

2. Green versus Dead Forage

Most swards contain both green and dead material, particularly toward the end of the grazing season, and liveweight gain of sheep is more closely related to the quantity of green forage DM offered than to the total forage DM present (Willoughby, 1958; Arnold and Duszinski, 1966). Cattle have been found to eat very little dead forage provided some green leaf is available (Mannetje, 1974; Chacon and Stobbs, 1976; Hendricksen and Minson, 1980).

2. Species Differences

Many swards contain more than one forage species and this provides another opportunity for intake to be modified by selective grazing. There will usually be a difference in preference between species, and some tropical legumes are almost completely rejected in spring and summer but readily eaten later in the year. Cattle grazing a *Setaria sphacelata/Macroptilium atropurpureum* sward selected a diet containing 10% legume in summer but this increased to 62–73% in autumn (Stobbs, 1977). Preference for legume depends on soil fertility and can be modified by the application of fertilizer. Applying superphosphate to a native grass sward containing *Stylosanthes* spp. increased the proportion of legume in the diet selected by cattle from 12 to 58%, although the proportion of legume in the sward only doubled (McLean *et al.*, 1981).

4. Soiled versus Clean Forage

It was shown in Chapter 2 that the intake of cut forage is depressed when contaminated with feces. When slurry is spread on a sward, cattle

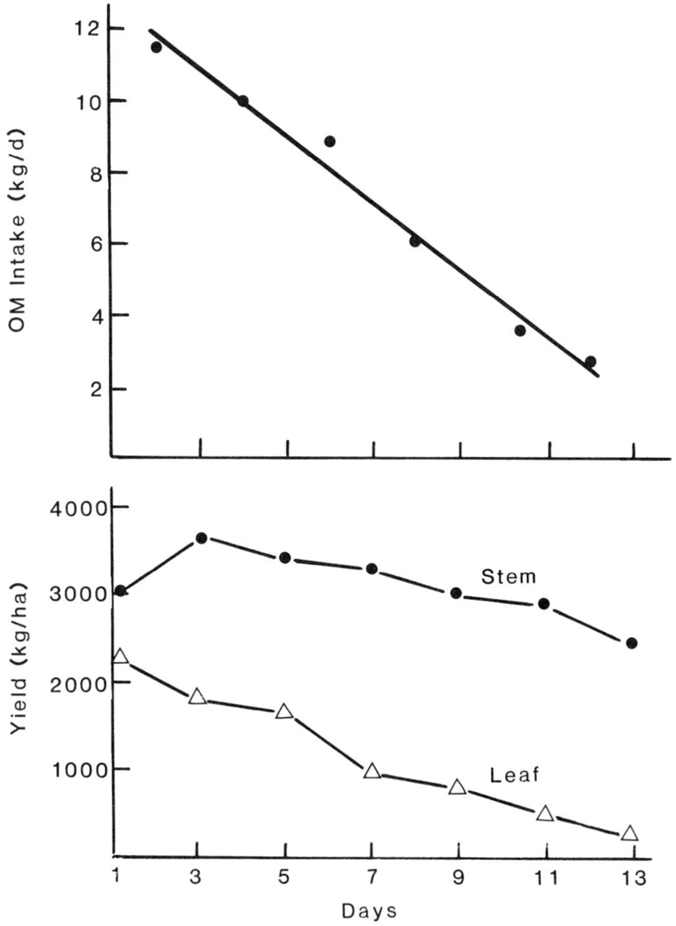

Fig. 3.5. Composition of a sward of *Lablab purpureus* during rotational grazing by cattle and the effect on intake. Data from Hendricksen and Minson (1980).

spend a longer period each day grazing, walk more while grazing, take smaller bites, and remove less forage (Pain and Broom, 1978). As the grazing season advances, there is an increase in the area of forage rejected due to fouling and accumulation of stem and dead material. Bite size of cattle decreases from 0.6 g DM/bite in spring to 0.3 g DM/bite in late summer (Leaver, 1987). The adverse effect of fouling cannot be overcome by "conditioning" the cattle to eat fouled forage (Reid *et al.*, 1972) or by the use odor-masking materials (aniseed) or appetizers (salt and molasses) sprayed onto fouled areas (Garstang and Mudd, 1971).

V. FORAGE DIFFERENCES

A. Legumes versus Grass

Pen studies have shown that legumes are eaten in greater quantities than grasses due to a lower resistance of legumes to breakdown during chewing and the shorter time they are retained in the rumen (Chapter 2). Field studies with both sheep and cattle have also shown higher intake of most temperate legumes when grazed separately (Table 3.8). The mean intake of legumes was 17% higher than that of grasses, but there were occasions when grasses and legumes were eaten in similar quantities (Fels *et al.*, 1959; Hodgson, 1975; Freer and Jones, 1984). There is no information available comparing intakes of tropical grasses and legumes when grazed separately.

B. Species and Cultivars

1. Forage Species

Differences in intake were found between grass species when grazed by both sheep and cattle (Table 3.9). In most studies there is insufficient

TABLE 3.8

Relative Intake of Legumes and Grasses by Grazing Ruminants

Species compared	Animal species	Relative intake (legume/grass)	Reference
Medicago sativa v. *Dactylis glomerata*	Sheep	1.24	Forbes and Garrigus (1950)
	Cattle	1.05	
		1.11	Alder and Minson (1963)
M. sativa v. *Lolium rigidum*	Sheep	1.16	Freer and Jones (1984)
Trifolium pratense v. *Lolium perenne*	Sheep	1.18	Hodgson (1975)
	Sheep	1.09	Gibb and Treacher (1976)
T. pratense v. *Poa pratensis*	Cattle	1.24	Garrigus (1934)
Trifolium repens v. *D. glomerata*	Sheep	1.17	Ulyatt (1969)
	Sheep	1.14	Ulyatt (1981)
Trifolium subterraneum v. *L. rigidum*	Sheep	1.07	Freer and Jones (1984)
T. subterraneum v. Various grasses	Sheep	1.05	Fels *et al.* (1959)
Mean		1.17	

data to rank the different species but there is an indication that *Lolium* spp. are eaten in greater quantities than *Festuca arundinacea*.

2. Forage Cultivars

Different forage cultivars vary in intake when grazed by sheep and cattle (Table 3.10). The extent of these cultivar differences is similar to that found between forage species (Table 3.9), indicating that there is probably as much variation in intake within species as there is between species.

C. Stage of Growth

The intake of cattle set-stocked on mature tropical grass pasture remained constant for many weeks and then rapidly decreased (Fig. 3.6). The cattle initially selected a diet with an OM digestibility of 0.6 and digestible crude protein of 70 g/kg DM. Intake remained constant for 12 weeks before suddenly decreasing when the cattle were obliged to eat a poor-quality diet. In this study the intake of forage by cattle was related to the OM digestibility of the selected forage (Fig. 3.7). Other regressions

TABLE 3.9

Relative Intake of Different Forage Grasses by Grazing Ruminants

Species compared	Animal species	Relative intake[a]	Reference
Bromus inermis v. *Festuca arundinacea*	Sheep	1.27	Forbes and Garrigus (1950)
Dactylis glomerata v. *F. arundinacea*	Sheep	1.31	Forbes and Garrigus (1950)
	Cattle	1.20	
Lolium hybridum v. *Lolium perenne*	Sheep	1.00	McLean *et al.* (1962)
	Sheep	1.05	Ulyatt (1969)
L. perenne v. *D. glomerata*	Cattle	1.13	Alder and Cooper (1967)
Lolium rigidum v. *Phalaris aquatica*	Sheep	1.07	Freer and Jones (1984)
Phleum pratense v. *L. perenne*	Sheep	1.23	McLean *et al.* (1962)
Poa pratensis v. *F. arundinacea*	Cattle	1.03	Forbes and Garrigus (1950)
Mean		1.14	

[a]Intake of grass with higher intake divided by that of grass with lower intake.

TABLE 3.10

Relative Intake of Different Cultivars by Grazing Ruminants

Species compared	Animal species	Relative intake[a]	Reference
Dactylis glomerata			
Cambria v. Sylvan	Sheep	1.20	Walters and Evans (1979)
Lespedeza juncea			
High v. low tannin	Cattle	1.14	Donnelly *et al.* (1971)
Lolium multiflorum			
Tetrone v. Danish	Cattle	1.10	Alder (1968)
L. multiflorum			
Sabalan v. Tetrone	Sheep	1.14	Walters and Evans (1974)
	Cattle	1.13	Walters and Evans (1974)
	Cattle	1.01	Wilkinson *et al.* (1982)
Mean		1.13	

[a]Intake of cultivar with higher intake divided by that of cultivar with lower intake.

have subsequently been published relating intake and digestibility of temperate and tropical forage grazed by sheep and cattle (Hodgson *et al.*, 1977). There are differences between these regressions with an indication that tropical forages are eaten in larger quantities than temperate forages of similar digestibility (Fig. 3.8). This difference is probably caused by the higher proportion of leaf in tropical forage when compared with temperate forages of the same digestibility.

D. Fertilizer

Fertilizer nitrogen increases the yield of forage but has no consistent effect on either intake or DMD when the forage is fed indoors (Chapters 2 and 4). Similar results have been obtained with grazing ruminants. Calves grazing *L. perenne* pastures fertilized with low and high levels of ammonium nitrate consumed 2.20 and 2.14 g OM/kg liveweight, respectively (Hodgson, 1968).

E. Seasonal Changes

Seasonal changes in intake occur even when there are no apparent changes in forage maturity. Lactating cows grazing a temperate pasture consumed 10% less forage in autumn than in spring although there was no difference in OM digestibility (Corbett *et al.*, 1963). The autumn forage

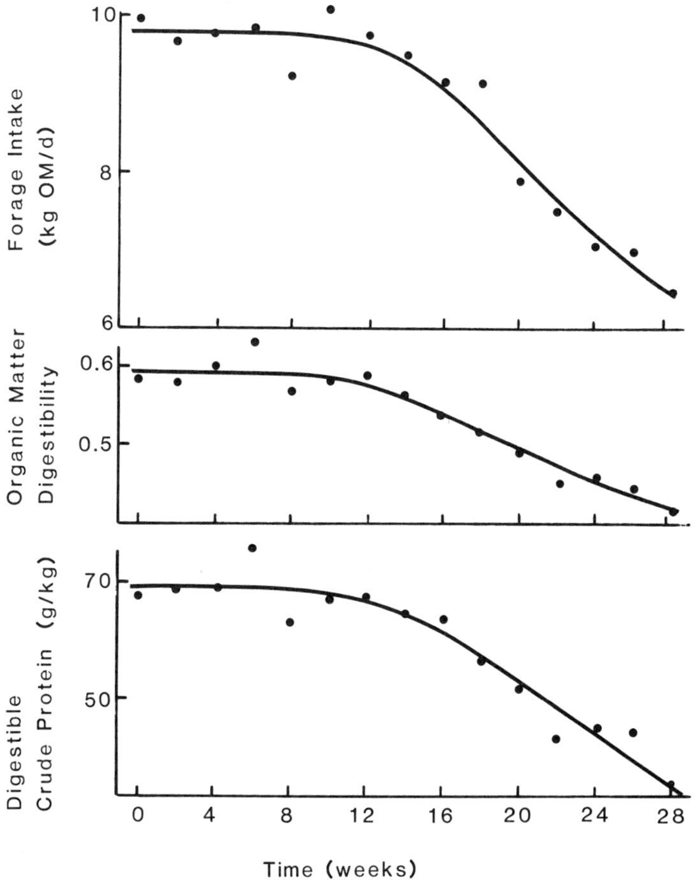

Fig. 3.6. Seasonal changes in forage intake and composition of selected forage. Data from Elliott *et al.* (1961).

had a "lush leafy appearance and entirely lacked inflorescences" and it was suggested "that though the two growths were of similar digestibility they were in fact digested at different rates." Other reasons suggested for the lower intake of autumn forage were presence of excreta voided during grazing periods earlier in the season, fungal infections such as rusts, soil contamination, and excess moisture. Large seasonal differences in intake were also found with cattle grazing *Dactylis glomerata, Medicago sativa,* or mixtures of the two species (Alder and Minson, 1963). In autumn the cattle ate 9% less forage than in summer although the autumn forage was slightly more digestible and the animals were 17% heavier. Calves are

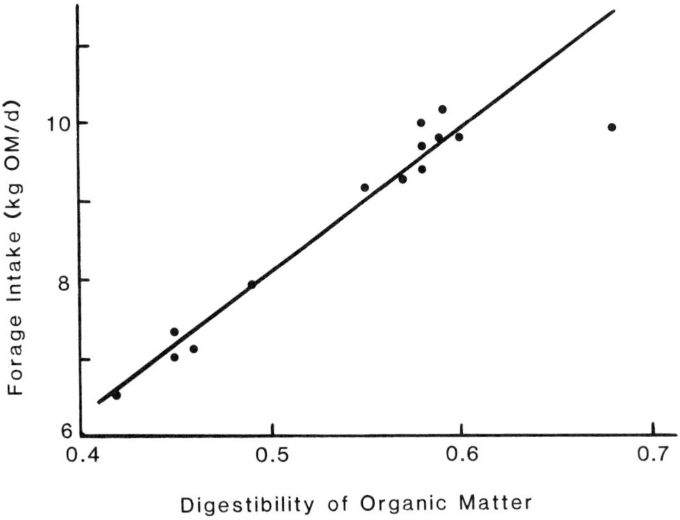

Fig. 3.7. Relation between intake of tropical forage and digestibility of the diet selected. ($y = 18.11x - 0.92$; $r = 0.96$). Data from Elliott *et al.* (1961).

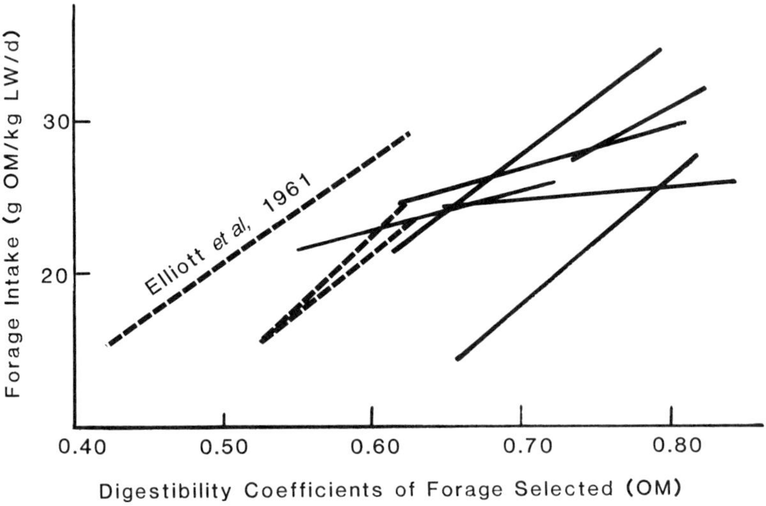

Fig. 3.8. Relation between intake and digestibility of temperate (—) and tropical (— — —) forages by grazing ruminants. Data from Hodgson *et al.* (1977).

more susceptible to these seasonal differences and in autumn ate 25% less *L. perenne* and grew 40% slower than in spring (Hodgson, 1968).

F. Water Content

Indoor studies have shown that forage with a high moisture content (water >780 g/kg) is eaten in smaller quantities than drier forage of the same maturity (Chapter 2). It is probable that the intake of grazed forage is also depressed by a high water content but no relevant data appear to have been published. The intake of *D. glomerata* and *M. sativa* by grazing cattle was not increased by cutting and wilting for a day, probably because the comparison was conducted in the middle of summer, when moisture contents of the unwilted forage would be insufficient to limit intake (Alder and Minson, 1963).

VI. IMPROVING INTAKE

A. Supplementary Feeding

Supplementary feeding provides producers with a means of increasing total food intake when pasture yields are low or quality is poor. Supplements given to grazing ruminants may conveniently be divided into three classes: energy-rich, protein, and mineral supplements. Protein and mineral supplements will be reviewed in later chapters and only energy supplements will now be considered.

1. Supplementation When Forage Is Abundant

When excess forage is available for grazing, feeding an energy supplement increases *total* food intake but the increase is less than the quantity of supplement eaten. This is due to the supplement depressing the intake of forage. If the depression in forage intake (DM) is equal to the quantity of supplement fed (DM), then the substitution coefficient is unity and the supplement will have little effect on production. Conversely, if supplements have no effect on the intake of forage, the substitution coefficient is zero and full benefit of feeding the supplement will be achieved.

With grazed forage, substitution coefficients for energy supplements vary between 0.25 and 1.67, with a mean of 0.69 (Table 3.11), indicating that the average improvement in intake and production is only one-third of the expected increase. The substitution coefficient varies with the type of supplement, the time of feeding, and the quality of the forage but is not affected by species of ruminant or form of production. When a *Zea*

TABLE 3.11

Depression in Intake of Grazed Forage When Energy Supplements Are Fed to Ruminants

Forage grazed	Animal	Substitution coefficient[a]	Reference
Cynodon dactylon	Lactating cows	0.90	Ruiz *et al.* (1981)
Cenchrus ciliaris	Lactating cows	0.53	Combellas *et al.* 1979)
Elymus junceus			
Morning supplement	Steers	1.67	Adams (1985)
Afternoon supplement		0.67	
Festuca pratensis/	Steers	0.90	Forbes *et a.* (1966)
Phleum pratense		0.38	Forbes *et al.* (1967b)
Lolium perenne	Lambs	0.48	Newton and Young (1974)
	Dry cows	0.54	Sarker and Holmes (1974)
	Ewes and lambs	0.79	Milne *et al.* (1979)
		0.85	Young *et al.* (1980)
	Lactating cows	0.50	Meijs and Hoekstra (1984)
Poa pratensis	Lactating cows	0.64	Seath *et al.* (1956)
		0.97	Cole *et al.* (1957)
Range forage	Wethers	0.85	Cook and Harris (1968)
Temperate forage	Lactating cows	0.38	MacLusky (1955)
		0.50	Corbett and Boyne (1958)
	Wethers	0.88	Holder (1962)
		0.67	Allden (1969)
	Steers	0.51	Lake *et al.* (1974)
	Lactating cows	0.25	Rearte *et al.* (1986)
Mean		0.69	

[a] $\frac{\text{Depression in forage intake (g)}}{\text{Quantity of supplement (g)}}$.

mays supplement is fed early in the morning, total food intake by grazing steers is depressed by more than the quantity of supplement fed and there is no improvement in growth. However, when the supplement is fed early in the afternoon, forage intake is only slightly depressed and total food intake and production are improved (Adams, 1985).

The response to energy supplements also depends on the quality of the grazed forage. With both lactating cows and steers, higher substitution coefficients are found in spring than in autumn, when the intake of grazed forage is lower (Corbett and Boyne, 1958; Forbes *et al.*, 1967b). With cows grazing irrigated *Cenchrus ciliaris* the substitution coefficient was 0.64 in the rainy season but 0.42 in the dry season (Combellas *et al.*, 1979). The depression in forage intake caused by energy supplements is similar to that found in pen studies (Chapter 2) and is associated with a

TABLE 3.12

Influence of Energy Supplements on Grazing Behavior of Lactating Cows[a]

	Supplement (kg/day)		
Behavior	0	3	6
Forage intake (kg OM/day)	9.25	7.57	6.42
Total intake (kg OM/day)	9.25	10.20	11.70
Grazing time (hr)	7.7	7.2	6.6
Rate of biting (bites/min)	51.7	50.4	49.9
Total bites (thousand/day)	24.0	22.0	19.9
Bite size (g OM/bite)	0.41	0.36	0.34
Milk yield (kg/day)	8.3	9.2	9.9
Liveweight gain (kg/day)	0.21	0.23	0.27

[a]Data from Combellas *et al.* (1979).

progressive decrease in grazing time, rate of biting, and size of bite as the level of supplements is increased (Table 3.12).

2. Supplementation When Forage Is Sparse

When forage is sparse, grazing animals cannot achieve maximum intake and feeding energy supplements will have little adverse effect on intake of grazed forage. With lactating cows, substitution coefficients decreased from 0.5 to 0.1 when forage allowance was reduced from 24 to 16 kg OM/cow (Meijs and Hoekstra, 1984). Similar changes have been reported for dry sheep; substitution coefficients changed from 0.75 to 0.33 when the quantity of a temperate pasture offered was reduced from 4800 to 760 kg/ha (Langlands, 1969). With lactating ewes grazing *L. perenne* in spring, the substitution coefficient was zero when the sward was less than 4.6 cm high, but when forage intake was not limited by sward height (8 cm) the substitution coefficient was unity. In early winter the sward declined in height from 5 to 3 cm, forage intake halved, and the substitution coefficient changed from 0.56 to 0.30 (Milne and Mayes, 1985).

B. Grazing Management

The aim of good pasture management is to provide animals with sufficient young forage so that intake and production is not limited by a low bite size. This can be achieved if the mass and height of grazed swards are above the critical levels discussed in Section IV. Swards are often leniently grazed in the spring to ensue maximum intake of forage but a high proportion of stem is left and this will depress intake when swards are again grazed. A traditional method of overcoming this problem is to use a leader-and-follower system of grazing. The leading animals have a higher potential for production and require more forage, while the followers are dry stock with a lower requirement. When cows of equal milk potential grazed, as two herds, a sward of *M. sativa*/*D. glomerata*, intake and milk production of the leader group were 18 and 20% higher, respectively, than intake and production of the followers (Bryant *et al.*, 1961). Commercially, high-producing cows are usually included in the leading herd and low-potential cows in the follower group (Mayne *et al.*, 1988). Other aspects of the split-herd grazing system have been reviewed by Fontenot and Blaser (1965).

VII. CONCLUSION

Intake of forage by grazing ruminants can be determined by techniques which measure intake over periods varying from 5 min to 1 month. These methods include short-term changes in liveweight, grazing behavior, cutting methods, fecal techniques, and regressions which predict intake from liveweight and production.

The intake of grazed forage is controlled by the same factors that affect cut forage (Chapter 2) plus two factors that are unique to grazed forage. These are the ability of animals to graze selectively and the depression of intake by any feature of the sward that limits bite size. Bite size is reduced by low forage mass per hectare, low allowance per animal, and presence in the sward of plant material rejected by the animal. Grazing animals generally eat more temperate legume than grass, with differences in intake between forage species and cultivars. Intake is depressed by forage maturity with a tendency for tropical forage grasses to be eaten in larger quantities than temperate forages of the same dry-matter digestibility. Intake of forage is lower in autumn than in spring.

The feeding of energy supplements depresses intake of forage but increases total feed intake, the improvement varying with quality and quantity of forage available. When forage availability is low, feeding an energy supplement has little effect on forage intake and increases total feed intake. In a leader-and-follower system of grazing, intake of forage and production by the leader animals is higher than that of followers.

4

Digestible Energy of Forage

I. DIGESTIBILITY AND ITS MEASUREMENT

A. Plant Structure and Chemical Composition

When examined under the microscope, forage is found to contain five different types of tissues: vascular bundles containing phloem and xylem cells, parenchyma bundle sheath(s) surrounding the vascular tissue, sclerenchyma patches connecting the vascular bundles to the epidermis, mesophyll cells between the vascular bundles and epidermal layers, and, on the exterior, a single layer of epidermal cells covered by a protective cuticle (Akin, 1982). These tissues are digested to varying extents in the rumen. Resistance to digestion increases in the following order: mesophyll and phleom < epidermis and parenchyma sheath < sclerenchyma < lignified vascular tissue. The proportion of these tissues varies among species, plant parts, and stage of growth and is affected by management factors. It is these differences in structure that cause forages to have a wider range of digestibility than any other feed eaten by ruminants.

Forages, in common with all plants, are made up of variously modified cells; these contain two major components: the cell contents and the "membrane" (Jarrige, 1960) or cell-wall constituents (Van Soest, 1965b). The cell-contents fraction contains most of the organic acids, soluble carbohydrates, crude protein, fats, and soluble ash, while the cell-wall fraction includes hemicellulose, cellulose, lignin, cutin, and silica (Fig. 4.1). This simple model links plant anatomy to chemical composition and is the basis for differences in the potential digestibility of the various fractions.

B. Potential Digestibility

1. Chemical Fractions

The concept of potential digestibility of forage was developed by Wilkins (1969), who defined it "as the maximum digestibility attainable when the conditions and duration of fermentation are not limiting factors." The concept was first applied to the cellulose fraction of temperate and

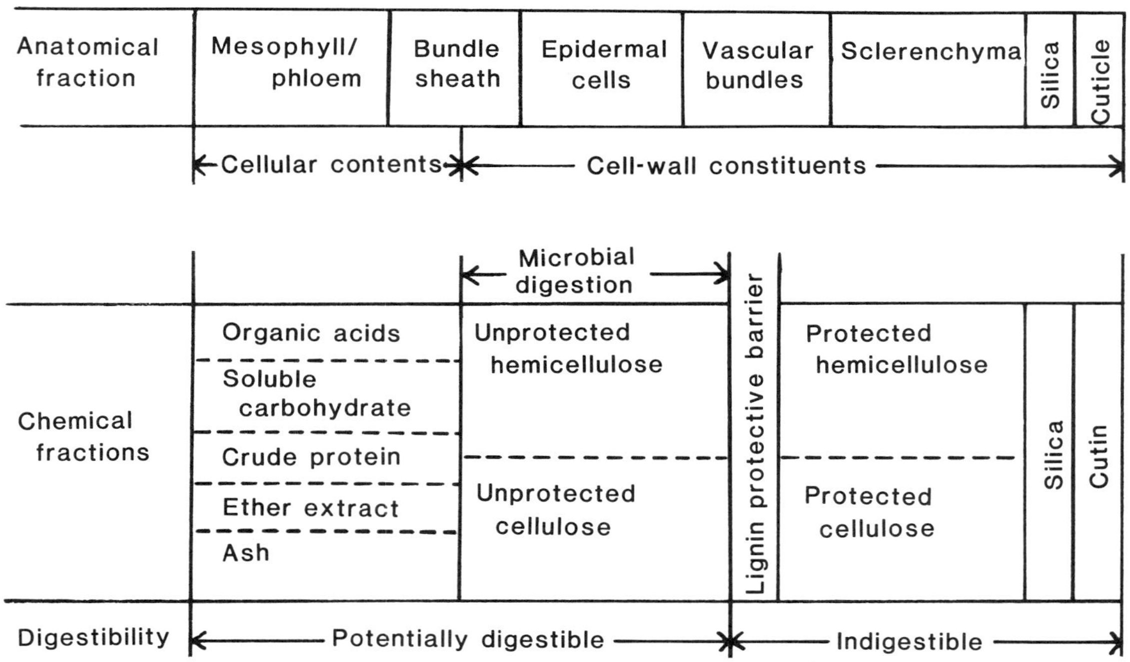

Fig. 4.1. Conceptual model of the relation between plant anatomy and chemical fractions indicating areas of potential digestibility.

tropical forages but can be expanded to include all fractions of the plant. The organic acids and soluble carbohydrates are virtually absent in the feces and thus have a potential digestibility near 1.00. Conversely, nearly all the lignin, silica, and cutin are excreted in the feces and are classified as "completely indigestible."

In forage, much of the cellulose and hemicellulose is protected from the rumen microflora by a layer of indigestible lignin that can only be disrupted by ball milling (Dehority and Johnson, 1961) or chemical treatment (Section IV,D). Thus the cellulose and hemicellulose can be divided into potentially digestible and indigestible fractions separated by a protective layer of lignin (Fig. 4.1). The potential digestibilities of the different fractions are shown in Table 4.1.

For each sample of forage there will be only one potential digestibility (PD) but any number of *in vivo* apparent digestibilities (AD) depending on the quantity of the potentially digestible material which escapes digestion in the animal (E) and the unavoidable metabolic secretions in the feces (M). The relation between these four parameters is given by the equation

$$AD = PD - (E + M)$$

The true digestibility (TD) of a forage has been occasionally used in forage studies and is the apparent digestibility plus the unavoidable metabolic secretions in the feces: TD = AD + M. No advantage has been demonstrated for the use of TD in forage evaluation and apparent digestibility will be used throughout the remainder of this chapter when describing the availability of forage energy to ruminants.

2. Unavoidable Metabolic Secretions

The potential digestibility of the cell contents is very high (Table 4.1), but when forages are fed to animals digestibility is reduced by the unavoidable loss in the feces of protein, fats, and minerals associated with the digestion and passage of forage through the digestive tract. For immature forages these metabolic losses are small relative to the quantity of cell contents in the forage, so the apparent digestibility of the cell contents is also high. However, with tropical forages the proportion of cell contents is often low, so the metabolic secretions are proportionally greater and the cell contents have a low apparent digestibility. This leads to the unexpected situation where the apparent digestibility of the cell contents of tropical forage (and mature temperate forage) is often lower than that of the fibrous cell-wall fraction (French, 1957, 1961).

The quantity of metabolic material unavoidably lost in the feces is di-

TABLE 4.1

Potential Digestibilities of Different Forage Fractions

Fractions				
Anatomical	Chemical	Diet	Digestibility coefficient	Reference
Cell contents	Monosaccharides	Grasses	1.00	Gaillard (1962)
	Disaccharides	Grasses	1.00	Gaillard (1962)
	Fructosan	Grasses	1.00	Gaillard (1962)
	Cytoplasmic carbohydrates	Legumes	0.99	Jarrige (1960)
		Grasses	0.99	Jarrige (1960)
	Soluble carbohydrates	Grasses	0.99	Jarrige and Minson (1964)
		Grasses	1.00	Waite *et al.* (1964)
		Lolium perenne	1.00	Beever *et al.* (1971b)
		Pelleted grasses	1.00	Beever *et al.* (1981)
	Crude protein	Temperate pasture	0.93	Holter and Reid (1959)
		Tropical pasture	0.90	Milford and Minson (1965a)
		Legumes	0.92	Combellas *et al.* (1971)
		Grasses	0.97	Combellas *et al.* (1971)
	Neutral detergent solubles	Temperate pasture	0.98	Van Soest (1967)
		Temperate pasture	0.90	Colburn *et al.* (1968)
		Tropical pastures	0.83	Combellas *et al.* (1971)
		Temperate pastures	0.96	Osbourn *et al.* (1974)
		Mixed pasture	1.01	Deinum (1973)
Cell wall	Cellulose	*Lolium perenne*	0.77	Wilkins (1969)
		Dactylis glomerata	0.73	Wilkins (1969)
		Chloris gayana	0.60	Wilkins (1969)
	Hemicellulose	Straw	0.55	McAnally (1942)
	Lignin	Mixed Pasture	0.02	Crampton and Maynard (1938)
		Medicago sativa	0.09	Gaillard (1962)
		Trifolium pratense	0.01	Gaillard (1962)
		Grasses	0.02	Gaillard (1962)
		Lolium perenne	0	Jarrige and Minson (1964)
		Dactylis glomerata	0	Jarrige and Minson (1964)
	Silica	Mixed Pasture	0.03	Jones and Handreck (1965)
		Medicago sativa	0.01	Jones and Handreck (1965)
		Avena sativa	0	Jones and Handreck (1965)
	Cutin	—	—	No data published

rectly proportional to the quantity of dry matter eaten, irrespective of forage type. For every gram of dry matter eaten, 0.129 g of metabolic material is secreted in the feces and this is chemically indistinguishable from the cell contents of the forage (Van Soest, 1967). Subsequent work confirmed the high level of this unavoidable metabolic loss with values of 0.098, 0.107, and 0.111 g/g forage (Colburn *et al.*, 1968; Combellas *et al.*, 1971; Osbourn *et al.*, 1974).

The cell contents of forage are rapidly fermented by the rumen microflora, so any change in activity of the microflora or length of time the forage is fermented within the rumen does not affect the digestion of cell contents (Deinum, 1973). Pelleting the forage and increasing the level of feed also has no effect on fecal metabolic loss (Osbourn *et al.*, 1976, 1981). Metabolic secretions are, however, increased by high levels of internal parasites. These reduce by 0.014 to 0.016 the proportion of forage organic matter apparently digested and at least one-third of this depression is due to a greater loss of protein (Spedding, 1954, 1955).

3. Loss of Potentially Digestible Cell Walls

The digestion of cell walls by the rumen microflora is a relatively slow process, and maximum digestion of the fibrous fraction will only be achieved when forage is exposed to the action of microflora for many days or weeks (Wilkins, 1969). Any factor that reduces this exposure time or the activity of the cellulolytic microorganisms will lead to a loss of potentially digestible cell wall in the feces and a depression in digestibility of the cell wall.

Rumination increases the surface area available for microbial action and is generally thought to increase digestibility, but it actually reduces the time feed is retained in the rumen and depresses digestibility. By muzzling sheep between meals, rumination of a chaffed diet is reduced and digestibility of the fiber fraction is increased from 0.59 to 0.71 (Pearce and Moir, 1964).

The effect of grinding and pelleting on digestive efficiency is similar to that caused by rumination. When sheep were fed hay ground through screens of different size (12.7, 4.8, 3.1, and 1.0 mm), hay with the smallest particle size was retained for the shortest time in the digestive tract and both organic matter and fiber were less efficiently digested (Alwash and Thomas, 1974).

Increasing the quantity of forage eaten reduces the digestive efficiency of ruminants for both organic matter (Raymond *et al.*, 1959; Alwash and Thomas, 1971) and energy (Blaxter and Wainman, 1964). Most of the depression is caused by a reduction in the extent of fiber digestion and is associated with a reduction in the mean time feed is retained in the digestive tract, a decrease in pH, and a lower rate of digestion (Alwash and Thomas, 1971). The extent of the depression in digestive efficiency caused by increased intake is related to the proportion of grain in the diet (Tyrrell and Moe, 1975). With high-grain diets large decreases in digestive efficiency occur when the level of feeding is increased, but with low-grain diets changes in digestive efficiency are small.

Cattle digest the dry matter in forage more efficiently than do sheep fed diets of similar chemical composition (Cipolloni *et al.*, 1951). Many comparisons have since been made using the same forage (Fig. 4.2). The mean difference in dry-matter digestibility (DMD) in favor of cattle is 0.024, with a tendency for cattle and sheep to digest legumes with similar efficiency. The lower digestive efficiency of sheep is associated with reduced fiber digestion caused by the shorter time forage is retained in the reticulorumen (Rees and Little, 1980; Poppi *et al.*, 1981b). There is no difference between sheep and cattle in crude protein digestibility (Chenost and Martin-Rosset, 1985). Sheep are more sensitive to a sulfur (S) deficiency than cattle (Bird, 1974), and with S-deficient forages very large

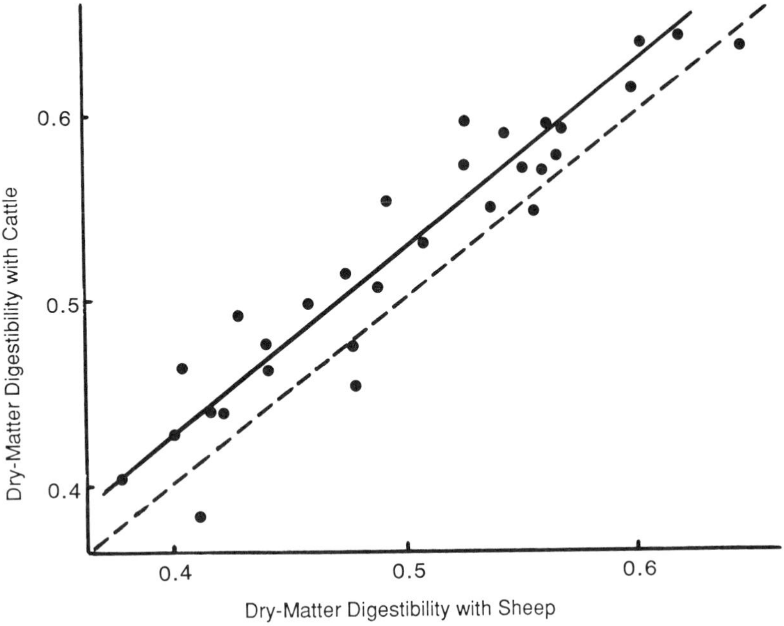

Fig. 4.2. The dry-matter digestibility of forages by sheep and cattle. ($y = 0.039 + 0.970x$; $r = 0.95$.)

differences (0.15) in DMD are found between sheep and cattle (Playne, 1978).

Rumen bacteria require a supply of protein, amino acids or their precursors, nonprotein nitrogen, and sulfur (Hungate, 1966). Urea increases the DMD of low-nitrogen roughage fed to sheep (Harris and Mitchell, 1941; Graham, 1967) and cattle (Campling *et al.*, 1962; Bird, 1974), while S increases the DMD of S-deficient forages eaten by sheep (Bird, 1974; Rees *et al.*, 1974; Rees and Minson, 1978). It appears that this increased loss of potentially digestible material is associated with a reduction in rumen microbial activity rather than a change in the digestive efficiency of the animal per se.

C. Determination of Digestibility *in Vivo*

1. Introduction

The apparent digestibility of forage is the proportional difference between the quantities consumed and excreted in the feces. When measured with ruminants, forage is given "in exact quantities for long periods, in order to ensure that a 'steady state' of faecal excretion is reached, and then to collect the faeces excreted during a measured interval of time" (Blaxter *et al.*, 1956). Forage digestibility may also be predicted by many chemical, physical, and biological methods (see Section III), but *in vivo* digestibility is the standard against which the accuracy of indirect methods is assessed. The equipment, conduct of experiments, and factors affecting digestive efficiency have been reviewed (Grassland Research Institute, 1961; Heaney *et al.*, 1969; Schneider and Flatt, 1975). In the following section only the special problems associated with the measurement of forage digestibility will be considered.

2. Choice of Animal and Conduct of Trial

Forage digestibilities have been measured with golden hamsters (Masuda *et al.*, 1977), rabbits (Crampton *et al.*, 1940), goats (Devendra, 1978), sheep (Homb, 1953), cattle, and buffalo (Moran *et al.*, 1979). Meadow voles have also been used for evaluating diets low in fiber (Keys and Van Soest, 1970). The smaller animals require less forage and cage space, but this advantage is offset by their low digestive efficiency compared with sheep and cattle (Crampton *et al.*, 1940; Masuda *et al.*, 1977). Sheep digest forage less efficiently than cattle (Fig. 4.2) but are used in most studies since they rank different forages in the same order as cattle (Playne, 1970b) and eat less feed.

The number of animals used to measure digestibility depends on the size of the difference to be determined and the level of statistical significance required. Sheep vary in their digestive efficiency, the variation increasing with declining digestibility (Fig. 4.3). Each forage is usually fed to 3 or 4 sheep (Raymond *et al.*, 1953a), but when voluntary food intake is also measured, then 8 or 10 sheep are normally used (Chapter 2). For the purpose of estimating metabolizable energy, it is recommended that digestibility should be determined at the maintenance level of intake (ARC, 1980).

Forage for digestion trials may be cut and fed once or twice daily (Greenhalgh *et al., 1959;* Martz *et al.*, 1960; Hutton, 1961) but the method has three disadvantages: the composition of the forage changes throughout the measurement period, cutting large quantities of forage can be difficult when the ground is waterlogged, and forage grown at different times of the year cannot be compared in the same study. Various methods have been introduced for overcoming these limitations.

Drying at 100°C has been used to conserve forage for digestion studies

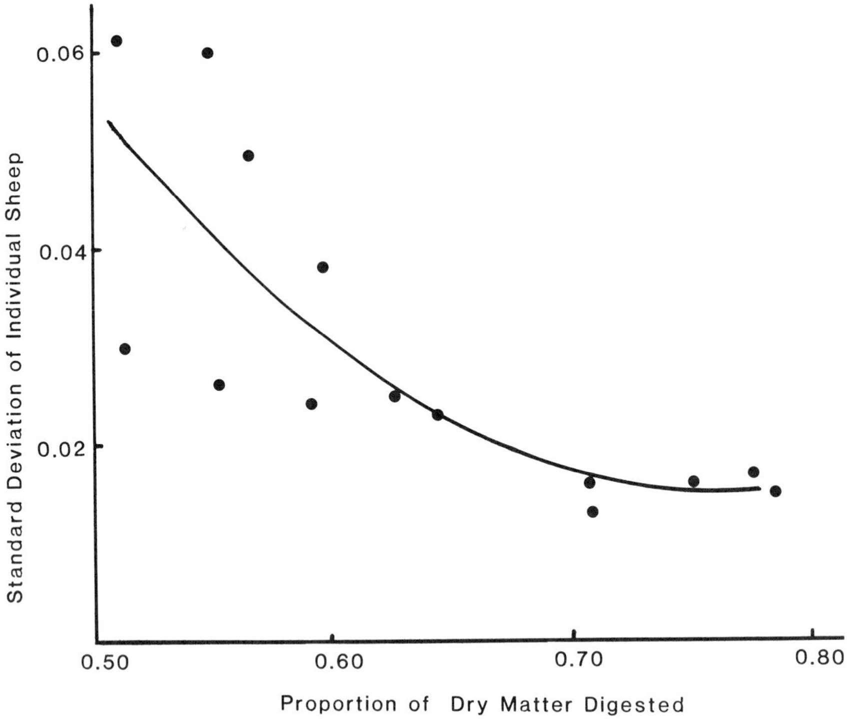

Fig. 4.3. Effect of forage quality on the variation in digestive efficiency of sheep. Adapted from Wilson and McCarrick (1967).

with sheep and cattle. Provided the temperature does not exceed 100°C and the forage is dried rapidly, changes in digestibility are small (Section II,J). Rapid freezing and storage at −15°C is a useful way of preserving forage for digestibility studies (Raymond *et al.*, 1949; Grassland Research Institute, 1961; Heaney *et al.*, 1969; Hutton *et al.*, 1975) and has only a small effect on digestibility (Section II,G). Ensiling is not used as a method of preserving forage for digestibility studies because major changes occur in the composition of the crude protein and carbohydrates during the ensiling process. Of course, ensiled forages are fed in digestibility studies, but then the objective is to study the digestibility of the silage, not the original forage.

D. Dry-Matter Digestibility and Digestible Energy

The energy requirements of ruminants are expressed in terms of metabolizable energy (ME) or net energy (NE) (Chapter 1). To measure ME it is necessary to determine methane production, which requires expensive equipment. Most studies of the energy value of forage have been limited to measuring the loss of forage dry matter in the feces, results being expressed as a dry-matter digestibility coefficient (DMD). In a few studies organic-matter digestibility coefficient (OMD), digestible energy (DE), ME, and NE have also been measured.

The ME and DE of forages are closely correlated and for a range of forages the ratio of ME:DE is 0.80, varying from 0.77 to 0.83 (Table 4.2). The coefficient is not effected by stage of growth or chemical composition of the forage but is increased by high levels of feeding (Armstrong, 1964; MacRae *et al.*, 1985) and by pelleting (Blaxter and Graham, 1956; MacRae *et al.*, 1985).

The DE values (MJ/kg DM) for forages can be estimated from the DMD coefficient. Four equations have been published.

Temperate grasses and legumes:

$$\mathrm{DE} = 19.66\mathrm{DMD} - 0.70,\ r = 0.96$$ (Heaney and Pigden, 1963)

Digitaria decumbens:

$$\mathrm{DE} = 17.99\mathrm{DMD} - 0.48,\ r = 0.99$$ (Minson and Milford, 1966)

Macroptilium atropurpureum:

$$\mathrm{DE} = 18.41\mathrm{DMD} - 0.34,\ r = 0.95$$ (Minson and Milford, 1966)

Sorghum almum:

$$\mathrm{DE} = 16.32\mathrm{DMD} + 0.62,\ r = 0.99$$ (Minson and Milford, 1966)

TABLE 4.2

Metabolizable Energy as a Proportion of the Digested Energy in Forage Fed Near the Maintenance Level

Forage	ME:DE Mean	ME:DE Range	Reference
Dactylis glomerata	0.81	0.81–0.82	Armstrong (1964)
Desmodium uncinatum	0.81	—	Graham (1967)
Digitaria decumbens	0.80	0.77–0.82	Graham (1967), Margan *et al.* (1989)
Festuca pratensis	0.82	0.81–0.83	Blaxter and Wilson (1963)
Lolium multiflorum	0.79	—	Margan *et al.* (1989)
Lolium perenne	0.81	0.78–0.83	Armstrong (1964), Rattray and Joyce (1974), Blaxter *et al.* (1971), Beever *et al.* (1985)
Phleum pratense	0.81	0.81–0.82	Armstrong (1964)
Setaria sphacelata	0.76	—	Margan *et al.* (1989)
Sorghum almum	0.80	—	Graham (1967)
Trifolium repens	0.80	0.78–0.82	Rattray and Joyce (1974), Beever *et al.* (1985)
Trifolium resupinatum	0.80	—	Graham (1969)
Trifolium subterraneum	0.78	—	Graham (1969)
Mixed forages	0.81	0.80–0.82	Joshi (1973), Rattray and Joyce (1974)
Mean	0.80	0.77–0.83	

These equations do not take into account differences in energy content caused by variations in the ash concentration of the forage. This source of error is overcome by expressing results as digestible organic matter (DOMD) in 100 g of forage DM, or D value (Minson *et al.*, 1960b). This parameter is widely used in forage extension work in the United Kingdom and can be used to predict total digestible nutrients (TDN), DE, and ME (MJ/kg DM):

Temperate grasses and legumes:

$$\text{TDN} = 1.018\text{DOMD} - 0.27,\ r = 0.99 \text{ (Heaney and Pigden, 1963)}$$

$$\text{DE} = 0.234\text{DOMD} - 2.34,\ r = 0.97 \text{ (Heaney and Pigden, 1963)}$$

Hay, dried grass, and straw:

$$\text{ME} = 0.15\text{DOMD (MAFF, 1975)}$$

Green grasses and legumes:

$$\text{ME} = 0.16\text{DOMD (MAFF, 1975)}$$

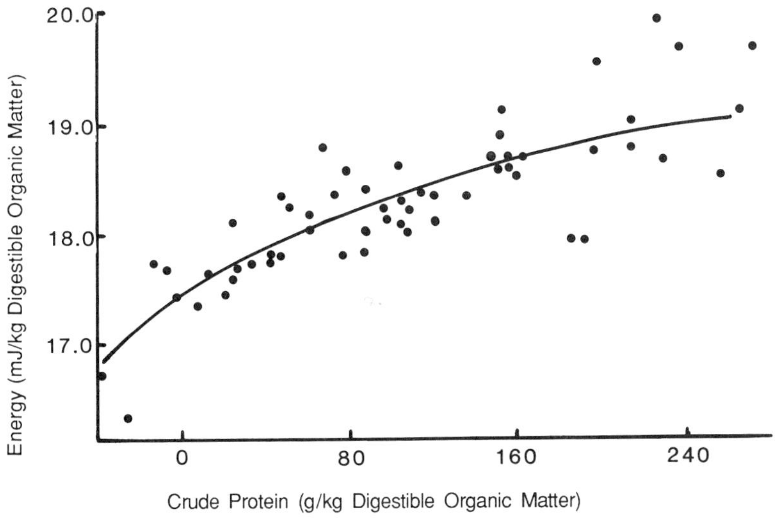

Fig. 4.4. Effect of crude protein on the energy concentration of digested organic matter. Adapted from Minson and Milford (1966).

The energy content of digested organic matter varies with the concentration of digested crude protein (Fig. 4.4). Inclusion of this variable will reduce the error when predicting ME from DOMD. Other factors that can influence the energy value of the DOMD are the proportions of organic acids, soluble carbohydrates, fiber, and fats in the diet. The gross energy values of some of these forage components are listed in Table 4.3. The relation between NE and ME will be considered in Chapter 5.

II. VARIATION IN FORAGE DIGESTIBILITY

A. Species Differences

The DMD of forage is very variable. This is illustrated by the frequency distribution of *in vivo* DMD for over 1500 forage samples grown in different parts of the world (Fig. 4.5). The reasons for this large variation in *in vivo* DMD will be considered in the following sections. *In vitro* DMD of forages will only be considered when no *in vivo* data are available.

1. Temperate versus Tropical Forages

The mean DMD coefficient of temperate grasses is higher by 0.13 than that of tropical grasses (Minson and McLeod, 1970) (Fig. 4.6). In

TABLE 4.3

Gross Energy Value of Components Likely to be Present in Fresh Forage or Silage (MJ/kg DM)[a]

Component	Gross energy (MJ/kg DM)
Oxalic acid	2.80
Succinic acid	12.64
Acetic acid	14.60
n-Butyric acid	24.89
Fumaric acid	11.55
Malic acid	11.55
Lactic acid	15.15
Xylose	15.65
Glucose	15.65
Galactose	15.56
Fructose	15.69
Sucrose	16.48
Raffinose	14.27
Starch	17.48
Cellulose	17.49
Protein	23.58
Fat	39.32

[a]Adapted from Armsby and Moulton (1925), Blaxter (1962), and Weast *et al.* (1965).

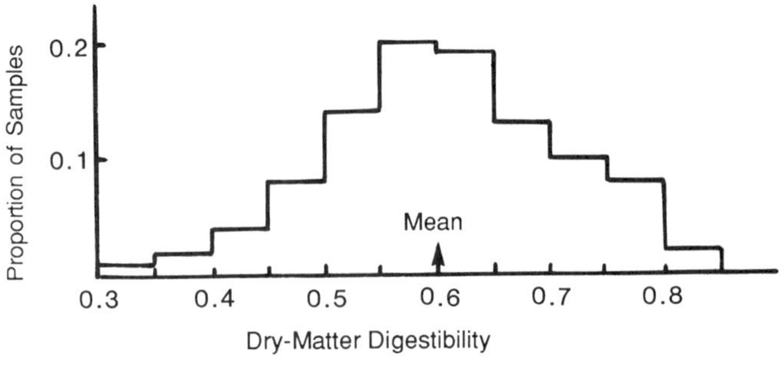

Fig. 4.5. *In vivo* dry-matter digestibility of a wide range of forages (World literature).

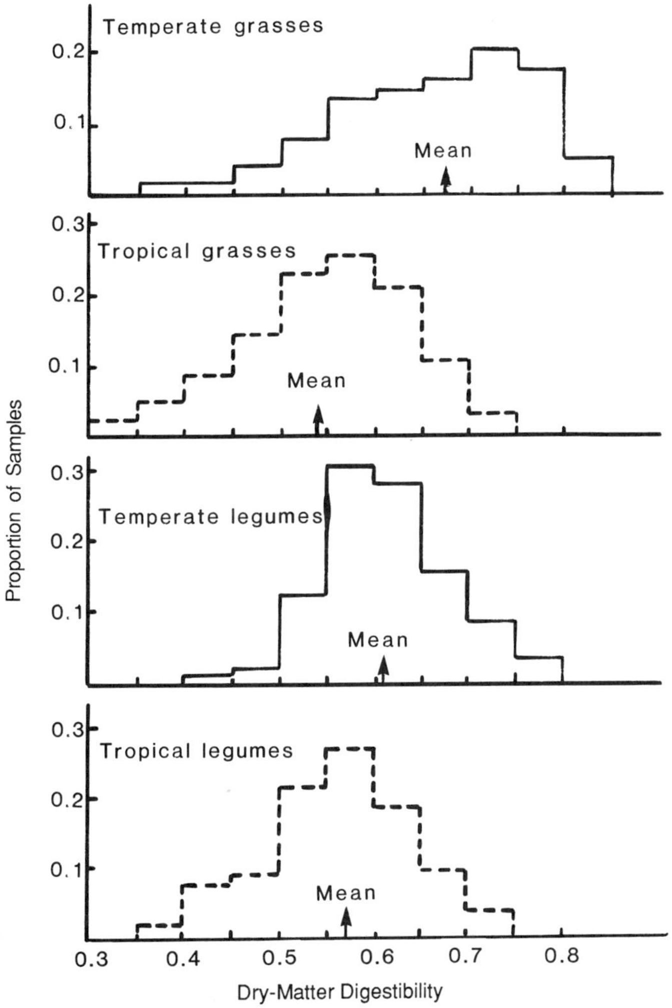

Fig. 4.6. *In vivo* dry-matter digestibility of temperate and tropical grasses and legumes (World literature).

contrast, only a small difference (0.04) occurs between the mean DMD of temperate legumes and tropical legumes (Minson and Wilson, 1980). The lower digestibility of tropical grasses is caused by differences in anatomical structure associated with the different photosynthetic pathways (Laetsch, 1974) and the higher temperature at which tropical grasses are normally grown.

Leaf blades of tropical grasses have more vascular bundles per unit cross-sectional area and hence more sites for lignification (Chonan, 1978), a closely packed cell structure (Carolin *et al.*, 1973), and suberized, thick-walled bundle sheaths that have a high resistance to penetration by rumen microflora (Hanna *et al.*, 1973). All these anatomical factors lead to a low potential digestibility of the leaves of tropical grasses. The stems of tropical grasses also have a higher proportion of vascular bundles than those of temperate forages. It has been shown that when grown under the same conditions anatomical differences between temperate and tropical grasses cause a 0.07 difference in DMD (Wilson *et al.*, 1983).

Tropical grasses are normally grown in climates with high temperatures and high potential transpiration rates. For monthly regrowths of temperate and tropical grasses, DMD was closely related ($r = 0.89$) to mean temperature (x, expressed in °C) at which the forages were grown ($\text{DMD} = 0.915 - 0.0114x$) (Minson and McLeod, 1970). Studies in controlled-temperature rooms show that DMD of both temperate and tropical grasses declines at a mean rate of 0.006 for each °C rise in temperature and this accounts for about half the 0.13 difference in DMD between tropical and temperate grasses grown in the field (Wilson and Minson, 1980). This difference is associated with a lower soluble carbohydrate concentration in tropical grasses (Wilson and Ford, 1971; Forde *et al.*, 1976).

The DMD of temperate and tropical legumes is similar. This is to be expected since both temperate and tropical legumes have the same photosynthetic pathway and leaf anatomy.

2. Temperate Species

The DMD of *Lolium perenne* is higher than that of *Dactylis glomerata* when they are compared at the same stage of growth (Table 4.4). The mean difference is 0.046, with a tendency for larger differences at younger stages of growth. The higher DMD of *L. perenne* is associated with a higher level of soluble carbohydrates and a lower proportion of leaf and crude protein (Table 4.4). The high digestibility of *L. perenne* has also been found in many *in vitro* studies of forage digestibility (Dent and Aldrich, 1963; Green *et al.*, 1971).

Regularly cut *Trifolium repens* has a higher DMD (0.77) than *Lolium* species (0.71) or cultivars of *D. glomerata* (0.68) (Michell, 1973b).

3. Tropical Species

When tropical grasses are cut every month or as mature regrowths there are consistent differences in DMD, with the same ranking order at

TABLE 4.4

Difference in Dry-Matter Digestibility, Composition, and Yield of *Lolium perenne* and *Dactylis glomerata*

	L. perenne	D. glomerata	Reference
DMD coefficients			
Spring growths	0.74	0.69	Minson *et al.* (1960b), Jarrige and Minson (1964)
Monthly regrowths	0.76	0.71	Minson *et al.* (1960b), Jarrige and Minson (1964)
Bimonthly regrowths	0.70	0.66	Minson *et al.* (1960b), Jarrige and Minson (1964)
Hays	0.65	0.62	Castle *et al.* (1962)
Dried regrowths	0.70	0.66	Michell (1973b)
Mean	0.71	0.67	
Mean compositions			
Soluble carbohydrates (g/kg)	128	69	Jarrige and Minson (1964)
Lignin (g/kg)	37	50	Jarrige and Minson (1964)
Crude protein (g/kg)	164	184	Minson *et al.* (1960b)
Leaf (g/kg)	480	650	Minson *et al.* (1960b)
Mean yield (kg/ha)	2540	2410	Minson *et al.* (1960b)

TABLE 4.5

Dry-Matter Digestibility (*in Vivo*) of Five Species of Tropical Grass, Each Cut as 1-Month Regrowths[a]

	Dry-matter digestibility		
	Monthly regrowths	Mature regrowths	Mean
Setarias sphacelata var. *splendida*	0.65	0.58	0.62
Digitaria decumbens	0.63	0.57	0.60
Chloris gayana	0.61	0.54	0.58
Panicum maximum	0.61	0.52	0.57
Pennisetum clandestinum	0.60	0.52	0.56
Mean	0.62	0.55	0.59

[a]From Minson (1972).

different stages of growth. The largest difference in one study is between *Setaria sphacelata* var. *splendida* and *Pennisetum clandestinum* (Table 4.5). Differences also exist between species of *Digitaria* harvested at different stages of growth. One species, *D. setivalva*, is markedly superior to two other lines, a difference that could not be attributed to any major differences in morphology or chemical composition (Table 4.6).

B. Cultivar Differences

1. Temperate Grasses

Large differences in DMD have been found between cultivars of *L. perenne* when first growths in spring are cut on the same day (Minson *et al.*, 1960b; Dent and Aldrich, 1963). The cultivar S.23 had a DMD 0.10 higher than the cultivar S.24 (Fig. 4.7) due to a delay in growth and flow-

TABLE 4.6

Mean Dry-Matter Digestibility and Composition of Regrowths of Three *Digitaria* Species[a]

Species	DMD	Yield (kg/ha)	Leaf (g/kg)	NDF (g/kg)	Lignin (g/kg)
D. setivalva	0.58	6530	270	680	64
D. pentzii	0.53	7070	330	700	67
D. smutsii	0.53	6050	330	710	65

[a]From Minson (1984).

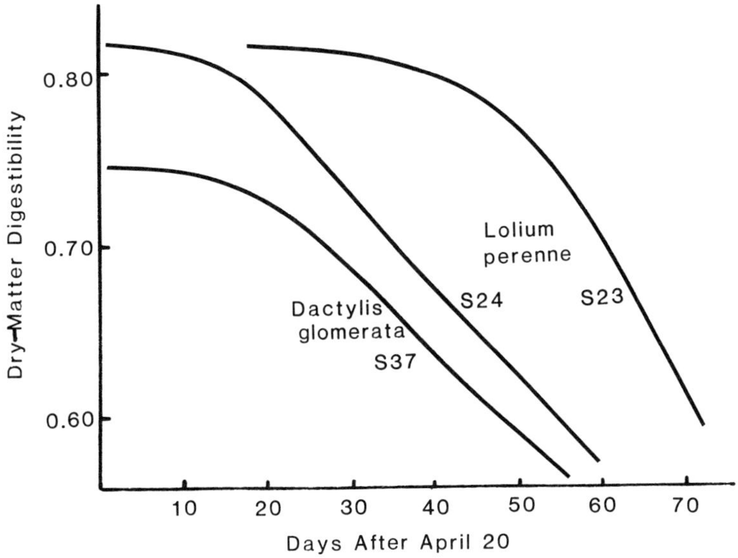

Fig. 4.7. Dry-matter digestibility of first growths of three contrasting grasses. Adapted from Minson *et al.* (1960b) and Jarrige and Minson (1964).

ering of S.23. There is a tendency for early-heading cultivars to have a high DMD if cut at heading because dry-matter yields are low and the forage has grown for less time (Fig. 4.8).

In contrast to primary spring growth, only small differences in DMD occur between cultivars harvested at regular intervals (Dent and Aldrich, 1963). The range in DMD among 26 cultivars of *L. perenne* was only 0.038, with a different ranking order of the cultivars at two sites. There was little difference in DMD between *L. perenne* cultivars S.23 and S.24, which is in agreement with other studies (Minson *et al.*, 1960b; Chestnutt, 1966). Diploid and tetraploid cultivars of *Lolium multiflorum* also had similar DMD when compared at the same yield (Thomson, 1971). Differences between cultivars of other species were usually small; *D. glomerata,* 0.020; *Festuca pratensis,* 0.026; and *Phleum pratense,* 0.026 (Dent and Aldrich, 1963). A difference of 0.036 in DMD was found between *D. glomerata* cultivars Currie and Grasslands Apanui (Michell, 1973b).

2. Tropical Grasses

The DMD of tropical grasses has been compared as monthly and mature regrowths. Differences in DMD between introductions of *Panicum* spp. were usually small with no evidence of any interaction between

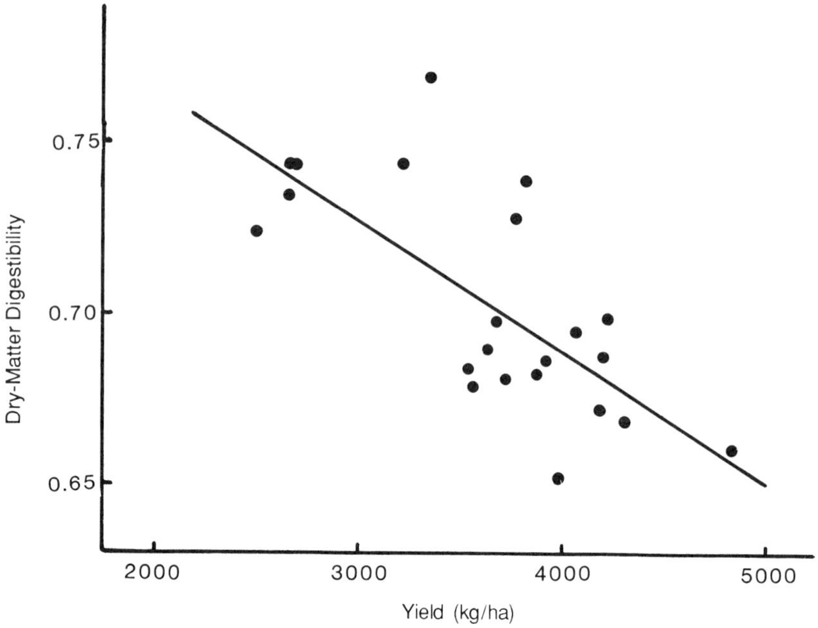

Fig. 4.8. Effect of yield on the dry-matter digestibility of *Lolium perenne* cut 10 days after 50% head emergence. Adapted from Dent and Aldrich (1963).

DMD and the regrowth period (Table 4.7). *Panicum maximum* was more leafy than *P. coloratum* but this had no effect on the DMD. In other studies the *P. maximum* cultivars Coloniao and Serdang had similar DMD (Devendra, 1977).

The first cultivar released on the basis of higher DMD was Coastcross 1, a bred cultivar of *Cynodon dactylon*. This had a mean DMD of 0.60 compared with 0.54 for Coastal Bermuda, the cultivar it was aimed to replace (Burton *et al.*, 1967). When rotationally grazed at monthly intervals by steers, Coastcross 1 had a mean DMD of 0.53 compared with 0.49 for Coastal Bermuda grass and this resulted in a higher growth rate of cattle on the Coastcross 1 (0.72 versus 0.60 kg/day) (Utley *et al.*, 1974).

Another cultivar of *C. dactylon*, Tifton 44, had a mean DMD of 0.66 compared with 0.63 for Coastal Bermuda grass. When grazed by steers, liveweight gain was 0.80 kg/day compared with 0.67 kg/day for animals grazing Coastal Bermuda grass (Utley *et al.*, 1978). In other studies larger differences in DMD were found between five lines of *C. dactylon*. The ranking order for DMD was Tifton 68 (0.65) > S-83 (0.58) = Callie (0.58) > S-16 (0.57) > Coastal (0.55), with no interaction between DMD and length of the regrowth period (Holt and Conrad, 1986).

TABLE 4.7

Differences in Dry-Matter Digestibility of *Panicum* Species and Cultivars under Two Management Systems[a]

	Dry-matter digestibility			
Species	Monthly regrowths	Mature regrowths	Mean	Leaf (g/kg)
P. coloratum				
CPI 13372[b]	0.59[1,c]	0.49[1]	0.54[1]	320
Kabulabula	0.60[1]	0.51[1,2]	0.56[2]	300
Burnett	0.61[1]	0.53[2]	0.57[3]	390
P. maximum				
Hamil	0.60[1]	0.51[1]	0.56[1]	520
Coloniao	0.61[1]	0.53[2]	0.57[2]	580

[a]From Minson (1971a).
[b]CPI, Commonwealth Plant Introduction number.
[c]Values with the same numbered superscripts are not significantly different ($p = 0.05$).

The stem of *Pennisetum glaucum* has a much lower DMD than the leaf (0.40 versus 0.57) and reducing stem height increases the proportion of leaf and DMD (Burton *et al.*, 1968). Forage from drawf plants is higher in DMD than forage from tall plants at all stages of growth, with the possible exception of 4 weeks after anthesis (Hanna *et al.*, 1979).

The most dramatic example of improving DMD by breeding is the incorporation of the brown midrib genes in lines of *Zea mays, Sorghum bicolor,* and *P. glaucum*. In 1964 it was discovered that the stalks of a brown midrib mutant of *Z. mays* contained less lignin (Kuc and Nelson, 1964). Some of these brown midrib mutants had a higher DMD, a difference that also applied to the leaves but not to the grain (Barnes *et al.*, 1971). Genotypes bm_3 and bm_1bm_3 were consistently higher in DMD than normal or bm_1 genotypes, the stem of bm_3 plants being approximately 0.10 higher in DMD. The higher DMD of the bm_3 mutant has been confirmed in many subsequent comparisons (Lechtenberg *et al.*, 1972; Muller *et al.*, 1972; Hartley and Jones, 1978; Stallings *et al.*, 1982; Weller *et al.*, 1984).

Brown midrib mutants also occur in *S. bicolor* and *P. glaucum*. Inclusion of the gene in *S. bicolor* decreased the lignin content of the stem from 103 to 85 g/kg DM and increased DMD from 0.396 to 0.487 (Fritz *et al.*, 1981). The DMD was also increased in leaf blade (0.049) and leaf sheath (0.032) due to a lower concentration of lignin, phenolic acids, and etherated syringyl moieties (Akin *et al.*, 1986). The improvement in DMD was attributed to an improvement in the microbial degradation of marginally digestible tissues (Akin *et al.*, 1986). In *P. glaucum*, a brown midrib gene also increases DMD and reduces the lignin concentration (Cherney *et al.*, 1988).

The DMD of cultivars of *Z. mays* varies with the proportion of grain in the forage. In American studies, the DMD of late varieties is 0.04 to 0.06 lower than early- and medium-maturity cultivars when cut at the same time, a difference that is associated with a smaller proportion of grain in the late-maturing cultivar (Nakui *et al.*, 1980). However, in English studies, there were only small differences in DMD when the proportions of leaf, stem, and grain were changed by using different cultivars and planting densities (Phipps and Weller, 1979).

3. Temperate Legumes

No difference occurs in DMD between *Medicago sativa* cultivars Vernal and DuPuits when grown in four regions and managed in different ways (Matches *et al.*, 1970). Similar DMD values were also reported for the *Medicago sativa* cultivars Anchor, Saranac, Thor, Vernal and Washoe (Wilson *et al.*, 1978). The only evidence of a difference in DMD in temperate legumes occurs in *Trifolium subterraneum*, in which the cultivar Woogenellup has a higher DMD than the cultivar Yarloop (0.58 versus 0.55) (Hume *et al.*, 1968).

4. Tropical Legumes

The only evidence that DMD differs between cultivars was found with *Stylosanthes guianensis* (Gardener *et al.*, 1982). The leaves of the cultivar Cook had a lower digestibility (0.65) than the leaves of the cultivar Oxley (0.68) but there was no difference in DMD of the stems.

C. Plant Parts

1. Temperate Grasses

The DMD of all plant parts is similarly high at an immature stage of growth, but at maturity there are large differences in DMD between the different fractions (Fig. 4.9). Similar differences in DMD between plant parts have been found with *D. glomerata* (Pritchard *et al.*, 1963; Mowat *et al.*, 1965a,b), *L. multiflorum* (Wilman *et al.*, 1976), and *L. perenne* (Terry and Tilley, 1964; Johnston and Waite, 1965; Hides *et al.*, 1983). *Phleum pratense* flowers later than other temperate species and the DMD of the stem fraction is relatively high (Terry and Tilley, 1964), with only small differences between leaf and stem fractions (Pritchard *et al.*, 1963; Mowat *et al.*, 1965a). Results with *Bromus inermis* are conflicting. Only

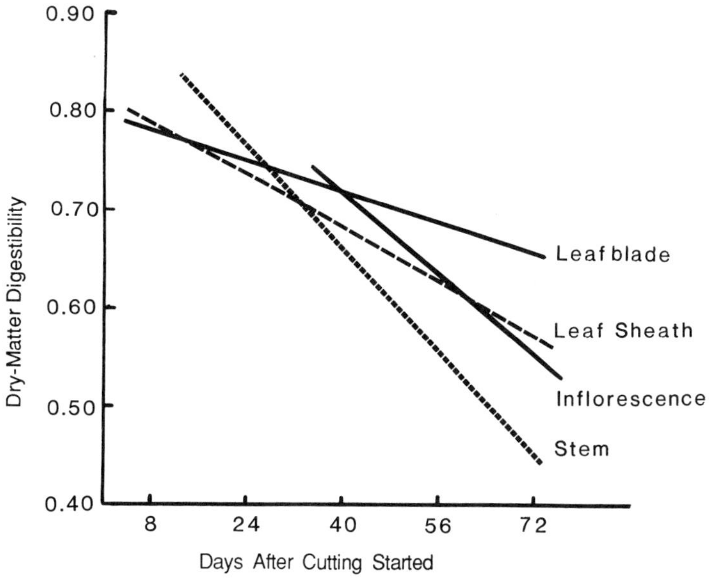

Fig. 4.9. Dry-matter digestibility of different parts of *Dactylis glomerata* at different stages of maturity. Adapted from Terry and Tilley (1964).

small differences were found between leaf (0.68). and stem (0.64) in one study (Mowat *et al.*, 1965a), but in another trial (Pritchard *et al.*, 1963) the DMD pattern was similar to that shown in Fig. 4.9.

2. Tropical Grasses

At an immature stage of growth, the DMD of leaf and stem of *P. glaucum* are similar, but when mature, the stem is much less digestible than the leaf (Monson *et al.*, 1969). Studies with mature *Z. mays* have given conflicting results. In the U.S. and The Netherlands the leaf is usually more digestible than the stalk (Deinum, 1976; Weaver *et al.*, 1978), but in English studies the stem is the most digestible fraction (Phipps and Weller, 1979). Studies with sheep and cattle have shown separated leaf to be less digestible than the stem of many tropical grasses cut at various stages of growth (Laredo and Minson, 1973, 1975b; Poppi *et al.*, 1981a), but the sample of leaf did not include a light fraction lost during mechanical separation.

3. Temperate Legumes

The leaf and stem of *M. sativa* have similar DMD when cut at a young stage of growth, but there are large differences between plant parts when mature (Tilley and Terry, 1969; Mowat *et al.*, 1965a). The DMD of leaf remains almost constant as plants mature, but large decreases occur in DMD of stem.

4. Tropical Legumes

Leaf and stem of *Desmodium* spp. harvested every 6 weeks have similar DMD (Stobbs and Imrie, 1976), but with 8- to 16-week regrowths of *M. atropurpureum* the leaf is more digestible than the stem. The leaf is also more digestible than the stem in *Stylosanthes* spp. and *Lablab purpureus,* with the largest difference occurring with mature forage (McIvor, 1979; Gardener *et al.*, 1982; Hendricksen and Minson, 1985a).

5. Variation within a Sward

Grazing animals tend to select the upper parts of the plant and differences in DMD occur between the top and bottom of a sward. When *B. inermis, D. glomerata,* and *P. pratense* are cut at a young stage of growth there is no difference in DMD between top and bottom of the flowering stem, but when mature the top segment is more digestible (Pritchard *et al.*, 1963). The DMD also decreases from top to bottom of the stem of the legumes *M. sativa, Onobrychis viciifolia,* and *L. purpureus* but, unlike the grasses, the difference occurs at all stages of growth (Terry and Tilley, 1964; Hendricksen and Minson, 1985a).

Leaves at the top of a plant usually have a higher DMD than older leaves. This pattern has been found in *P. glaucum* (Burton *et al.*, 1964) and *L. purpureus* (Hendricksen and Minson, 1985a).

D. Stage of Growth

As grasses grow there is a reduction in the proportion of leaf lamina and an increase in leaf sheath, stem, and inflorescence. In the previous section it was shown that there are only small differences in DMD between these fractions when first produced and that changes in morphological and reproductive state have no immediate effect on DMD (Fig. 4.7). The initial plateau of DMD is followed by a period of rapid decline. In the third phase DMD declines slowly as the more digestible components are lost by death and decay. These three phases of plant growth will now be considered.

1. Digestibility Plateau

Growth of forage in the spring is often very rapid and digestibility remains relatively constant for several weeks. This effect was first found in *D. glomerata* and *L. perenne* in a study covering two seasons (Fig. 4.7).

The digestibility remains at a high level until flower heads start to emerge and yield of dry matter reaches about one-third of the maxium. Subsequent studies have demonstrated digestibility plateaus in many forages: *Avena sativa* (Taji, 1967), *B. inermis* (Mowat *et al.*, 1965a), *D. glomerata* (Murdock *et al.*, 1961; Minson *et al.*, 1964; Mowat *et al.*, 1965c; Sheehan, 1969), *Festuca arundinacea* (Minson *et al.*, 1964), *L. multiflorum* (Minson *et al.*, 1964; Sheehan, 1969), *L. perenne* (Demarquilly and Jarrige, 1964; Wilson and McCarrick, 1967; Sheehan, 1969), *M. sativa* (Mowat *et al.*, 1965c), and *P. pratense* (Mowat *et al.*, 1965c; Langille and Calder, 1968).

The plateau in DMD is unexpected since changes occur in spring growths which are normally associated with a depression in DMD. There is a decrease in the proportion of leaf lamina and crude protein (Minson *et al.*, 1960b) and an increase in the concentration of cellulose, hemicellulose, and lignin (Jarrige and Minson, 1964). The only component that remains constant during this period is the soluble carbohydrate fraction (Jarrige and Minson, 1964; Tilley and Terry, 1969).

2. Digestibility Fall

A study of a wide range of spring growths of green forages, hays, and silages fed at Cornell University showed that DMD decreases 0.0048 each day cutting is delayed after April 30(x) (Reid *et al.*, 1959a).

$$\text{DMD} = 0.85 - 0.0048x$$

The rates of fall in DMD for spring growths of different forages are presented in Table 4.8. The mean rates of fall in DMD of temperate grasses and legumes are 0.0047/day and 0.0033/day, respectively. Growth and flowering in *P. pratense* become progressively later with increasing latitude and this increases the rate of decline in DMD: 0.0045/day in The Netherlands (latitude 51°) and 0.0064/day in Tromso, Norway (latitude 69°) (Deinum *et al.*, 1981). Curves have now been published which relate DMD to the age of spring growth of many different forages (Green *et al.*, 1971; Corrall *et al.*, 1979). These must be adjusted for the effect of latitude (Mellin *et al.*, 1962; Harkess and Alexander, 1969) and seasonal differences.

The rapid fall in DMD is associated with an increase in the proportion of leaf sheath, stem, and flowering head (Minson *et al.*, 1960b; Terry and Tilley, 1964), a reduction in the proportion of crude protein, and a rise in the cellulose, hemicellulose, and lignin (Jarrige and Minson, 1964; Waite *et al.*, 1964), but none of these factors is necessarily causative.

The rate of fall in DMD for regrowths is much smaller than that for spring growths. In the temperate forages *D. glomerata* and *L. perenne*,

TABLE 4.8

Fall in Dry-Matter Digestibility Coefficient of Spring Growths of Temperate Forage

Species	Fall in DMD/day		Reference
	Mean	Range	
Temperate Grasses			
Agrostis gigantea	0.0033	—	Colovos *et al.* (1966b)
Avena sativa	0.0058	0.0037–0.0078	Meyer *et al.* (1957), Taji (1967)
Bromus inermis	0.0040	0.0030–0.0050	Colovos *et al.* (1961), Mowat *et al.* (1965c), Calder and MacLeod (1968), Reid *et al.* (1978b)
Dactylis glomerata	0.0049	0.0033–0.0095	Minson *et al.* (1960b, 1964), Murdock *et al.* (1961), Spahr *et al.* (1961), Mowat *et al.* (1965c), Haenlein *et al.* (1966), Brown *et al.* (1968), Calder and MacLeod (1968), Colburn *et al.* (1968), Sheehan (1969), Reid *et al.* (1978b)
Festuca arundinacea	0.0052	0.0018–0.0104	Reid *et al.* (1967a, 1978b), Pendlum *et al.* (1980)
Lolium multiflorum	0.0041	0.0040–0.0042	Minson *et al.* (1964), Sheehan (1969)
Lolium perenne	0.0052	0.0043–0.0075	Minson *et al.* (1960b), Wilson and McCarrick (1967), Sheehan (1969), Reid *et al.* (1978b), Anderson (1982)
Mixed pasture	0.0031	—	Paquay *et al.* (1970)
Phalaria arundinacea	0.0034	—	Colovos *et al.* (1969)
Phleum pratense	0.0046	0.0029–0.0072	Swift *et al.* (1952), Kivimae (1959, 1965), Lloyd *et al.* (1961), Mellin *et al.* (1962), Mowat *et al.* (1965c), Colovos *et al.* (1966a), Heaney *et al.* (1966a), Wilson and McCarrick (1967), Brown *et al.* (1968), Calder and MacLeod (1968), Langille and Calder (1968)
Poa pratensis	0.0041	—	Reid *et al.* (1964b)
Mean for temperate grasses	0.0047		
Temperate Legumes			
Medicago sativa	0.0037	0.0028–0.0048	Martz *et al.* (1960), Weir *et al.* (1960), Sphar *et al.* (1961), Mowat *et al.*(1965c), Davis *et al.* (1968), Calder and MacLeod (1968)
Trifolium pratense	0.0033	—	Kivimae (1959)
Trifolium repens	0.0014	—	Davis *et al.* (1968)
Mean for temperate legumes	0.0033		

the rate of fall in DMD of regrowths is only a third of the rate found with spring growths. For *Trifolium pratense* the DMD of the second cut declined at 0.0021/day compared to 0.0033/day for primary growths (Kivimae, 1959). The mean rate of fall in DMD of regrowths of tropical forages is 0.0026/day, with no evidence of any difference between species (Table 4.9). This is less than for temperate grasses possibly because the tropical grasses start from an initially lower level.

3. Digestibility of Mature Pasture

The phase of rapid fall in digestibility is followed by a period when DMD remains relatively constant. The DMD of mature forage depends on the proportion of low-digestibility stem and high-digestibility grain. Most forages have small seeds and, when mature, DMD is less than 0.5 (Fig. 4.9). A limited, but very important, group of forages has large seeds which offset the decrease in the DMD of other plant parts (Demarquilly, 1970a). As *Z. mays* approaches maturity, there is a rapid increase in yield of grain, and digestibility remains almost constant (Corrall *et al.*, 1977; Weller *et al.*, 1984) unless there is frost and forage is left standing in the field for a long time (Calder *et al.*, 1977). *Zea mays* has the highest digestibility of all the grain crops at maturity. When compared at the same yield of dry matter the ranking order for digestibility is *Z. mays* > *Hordeum vulgare* > *Triticum aestivum* > *A. sativa* > *Secale cereale* (Corrall *et al.*, 1977).

E. Soil Fertility

1. Soil Nitrogen

Applying fertilizer N increases DM yield, protein, and water content of forage and reduces the proportion of leaf. These changes might be expected to have a major effect on DMD but most studies have shown only small differences (Table 4.10). There is no consistent pattern in the response to fertilizer N, both increases and decreases in DMD occurring with both young and mature forage. The higher crude protein in the N-fertilized forage is offset by a reduction in the level of soluble carbohydrate (Alberda, 1965; Raymond and Spedding, 1965; Wilson and Mannetje, 1978).

2. Soil Sulfur

Low levels of available soil sulfur (S) reduce DMD (Rees *et al.*, 1974). Application of fertilizer S to S-deficient *D. decumbens* increased DMD by 0.05 (Rees *et al.*, 1974) and energy digestibility by 0.10–0.12 (Rees *et*

TABLE 4.9

Fall in Dry-Matter Digestibility of Regrowths of Tropical Forage

Species	Fall in DMD/day Mean	Range	Reference
Brachiaria brizantha	0.0020	—	Coward-Lord *et al.* (1974)
Brachiaria ruziziensis	0.0027	—	Coward-Lord *et al.* (1974)
Cenchrus ciliaris	0.0025	0.0017–0.0030	Minson and Milford (1968a), Playne (1970b), Combellas and Gonzalez (1972a), Coward-Lord *et al.* (1974), Minson and Bray (1985)
Chloris gayana	0.0019	0.0013–0.0027	Milford (1960b), Milford and Minson (1968b), Minson (1972)
Cynodon dactylon	0.0030	0.0021–0.0039	Miller *et al.* (1965a), Combellas *et al.* (1972), Olubajo *et al.* (1974), Gutierrez *et al.* (1980), Holt and Conrad (1986)
Cynodon plectostachyus	0.0021	—	Olubajo *et al.* (1974)
Digitaria spp.	0.0030	0.0006–0.0065	Butterworth (1961), Minson (1972), Coward-Lord *et al.* (1974), Strickland (1974), Chenost (1975), Klock *et al.* (1975), Ventura *et al.* (1975), Sleper and Mott (1976), Gutierrez *et al.* (1980), Minson (1984), Minson and Hacker (1986)
Hemarthria altissima	0.0040	—	Coward-Lord *et al.* (1974)
Hyparrhenia rufa	0.0038	—	Coward-Lord *et al.* (1974)
Panicum spp.	0.0020	0.0013–0.0034	Ishizaki and Stanley (1967), Minson (1972), Coward-Lord *et al.* (1974), Olubajo *et al.* (1974)
Paspalum dilatatum	0.0020	—	Minson (1972)
Pennisetum clandestinum	0.0020	0.0018–0.0022	Said (1971), Minson (1972)
Pennisetum hybrids	0.0023	0.0000–0.0040	Coward-Lord *et al.* (1974), Olubajo *et al.* (1974), Ogwang and Mugerwa (1976)
Setaria sphacelata	0.0018	0.0016–0.0021	Minson and Milford (1968a), Hacker and Minson (1972), Minson (1972)
Sorghum sudanensis	0.0034	—	Reid *et al.* (1964a)
Mean	0.0026 ± 0.003		

al., 1980). This is due to enhanced fiber-digesting capability of the rumen microorganisms and not to any change in the anatomy of the forage (Akin and Hogan, 1983). The extent of the improvement in DMD varies with the extent of the S deficiency in the animal, sheep being more suspectible to S deficiency than cattle (Bird, 1974).

TABLE 4.10

Dry-Matter Digestibility of Forage at Two Levels of Fertilizer Nitrogen

Species	Mean DMD		Reference
	Low N	High N	
Bromus inermis	0.65	0.67	Markley *et al.* (1959), Colovos *et al.* (1961), Calder and MacLeod (1968)
Cenchrus ciliaris	0.48	0.52	Minson and Milford (1968a), Donaldson and Rootman (1977)
Chloris gayana	0.57	0.60	Milford (1960b), Minson (1973)
Cynodon dactylon	0.49	0.53	Webster *et al.* (1965), Fribourg *et al.* (1971), Herrera (1977)
Dactylis glomerata	0.62	0.64	Poulton *et al.* (1957), Markley *et al.* (1959), Minson *et al.* (1960b), Kane and Moore (1961), Reid *et al.* (1967b), Dent and Aldrich (1968)
Digitaria spp.	0.55	0.55	Chapman and Kretschmer (1964), Minson (1967, 1973); Klock *et al.* (1975)
Eragrostis curvula	0.51	0.52	Van Heerden *et al.* (1974), Holt and Dalrymple (1979)
Festuca arundinacea	0.65	0.68	Reid and Jung (1965a), Reid *et al.* (1967a), Kaiser *et al.* (1974)
Lolium multiflorum	0.66	0.65	Binnie *et al.* (1974), Wilman (1975)
Lolium perenne	0.68	0.68	Minson *et al.* (1960b), Deinum *et al.* (1968), Demarquilly (1970b)
Medicago sativa	0.63	0.66	Calder and MacLeod (1968)
Pennisetum clandestinum	0.60	0.61	Minson and Milford (1968a), Minson (1973)
Pennisetum purpureum	0.62	0.66	Ogwang and Mugerwa (1976)
Phalaris aquatica	0.61	0.62	Chalupa *et al.* (1961), Saibro *et al.* (1978)
Phleum pratense	0.64	0.61	Poulton and Woelfel (1963), Cameron (1967), Colovos *et al.* (1966a), Hogan *et al.* (1967), Calder and MacLeod (1968)
Poa pratensis	0.62	0.65	Reid *et al.* (1964b)
Sorghum sudanense	0.69	0.69	Reid *et al.* (1964a)
Temperate pasture	0.71	0.71	Ferguson (1948), Holmes and Lang (1963), Barlow (1965), McCarrick and Wilson (1966), Hight *et al.* (1968), Blaxter *et al.* (1971)
Mean	0.616	0.625	

3. Soil Phosphorus

Application of fertilizer phosphorus (P) has little or no effect on the DMD of grasses (Rees and Minson, 1982; Playne, 1972a). However, doubling the level of superphosphate applied to a grass/legume pasture grazed by cattle increased the DMD of plucked forage samples from 0.42 to 0.45, possibly due to an increase in the proportion of legume in the sward (Thornton and Minson, 1973a).

4. Soil Calcium

Increasing the calcium (Ca) level in a Ca-deficient soil causes small increases in forage DMD (Rees and Minson, 1976). With *D. decumbens*, DMD was increased from 0.46 to 0.48, a difference attributed to a change in structure of the plant.

5. Soil Potassium

Fertilizer potassium (K) had no consistent effect on the DMD of *B. inermis* and *P. pratense* but increased the DMD of *M. sativa* in one study (Calder and MacLeod, 1968).

6. Soil Zinc

Level of zinc (Zn) in the soil appears to have no effect on DMD. Applying 4 kg Zn/ha increased the Zn concentration in *M. sativa* from 18 to 21 mg Zn/kg DM but had no effect on DMD (Reid *et al.*, 1987).

7. Soil Magnesium

Level of magnesium (Mg) in the soil appears to have little effect on DMD. Addition of 390 kg Mg/ha to a soil low in Mg raised the Mg concentration in *P. pratense* from 0.8 to 1.5 g Mg/kg DM and slightly depressed (0.019) DMD (Reid *et al.*, 1984).

F. Climate

1. Seasonal Variation

The DMD of most forages is low in the middle of the summer (Table 4.11). Seasonal differences in DMD could be caused by changes in temperature, water availability, or light.

2. Temperature

High temperatures generally increase the concentration of fiber in forages (Brown, 1939; Deinum, 1966b) and reduce DMD (Table 4.12). This effect occurs with both tropical and temperate grasses, but with legumes high temperature has a smaller effect on DMD. In field studies with *M. atropurpureum* and *P. maximum* the fall in DMD was 0.003/°C and 0.007/°C, respectively (Wilson *et al.*, 1986).

Frost causes leaf death and shatter in tropical plants, leading to a depression in DMD (Milford, 1960b, 1967). Even when dead leaves remain on the plant, DMD declines rapidly (Wilson and Mannetje, 1978).

TABLE 4.11

Forages in Which Dry-Matter Digestibilities Are Lowest in Midsummer

Species	Reference
Andropogon gayanus	Haggar and Ahmed (1970)
Cynodon dactylon	Miller *et al.* (1965a), Webster *et al.* (1965), Jolliff *et al.* (1979), Mislevy and Everett (1981), Holt and Conrad (1986)
Dactylis glomerata	Dent and Aldrich (1968), Knight and Yates (1968)
Digitaria spp.	Mislevy and Everett (1981)
Festuca arundinacea	Kaiser *et al.* (1974), Probasco and Bjugstad (1980)
Forbs	Lewis *et al.* (1975)
Lolium perenne	Minson *et al.* (1960b), Dent and Aldrich (1968), Clark and Brougham (1979)
Medicago sativa	Hidiroglou *et al.* (1966)
Paspalum spp.	Mislevy and Everett (1981)
Phleum pratense	Hidiroglou *et al.* (1966)
Setaria sphacelata	Hacker and Minson (1972)
Trifolium repens	Chestnutt (1966), Michell (1973b)

3. Water

It has been suggested that high rate of transpiration is a factor contributing to the low DMD of forages grown at high temperatures (Minson and McLeod, 1970). This could be due to the development of a larger vascular system to convey the greater quantities of water passing through the plant or to wilting that occurs whenever soil is unable to supply water at sufficient speed to meet the potential evapotranspiration. When *P. coloratum* was grown in a controlled environment, soil water stress depressed OMD of both leaf and stem (Pitman *et al.*, 1981), but conflicting results were obtained in two studies using *M. sativa* (Vough and Martin, 1971). In field studies, water-stressed *M. sativa* and grasses had a similar or higher

TABLE 4.12

Mean Change in Dry-Matter Digestibility of Forages for Each °C Rise in Temperature[a]

	Grass		Legume	
Plant part	Tropical	Temperate	Tropical	Temperate
Leaf	−0.0057	−0.0064	+0.0019	−0.0009
Stem	−0.0086	−0.0076	−0.0027	−0.0022
Tops	−0.0060	−0.0056	−0.0028	−0.0021

[a]From Wilson and Minson (1980).

DMD than well-watered plants, possibly due to a reduction in the proportion of stem and higher DMD of the stressed leaves (Vough and Martin, 1971; Snaydon, 1972; Mislevy and Everett, 1981; Wilson, 1982a, 1983).

4. LIGHT

French (1961) suggested that high light intensity may be responsible for the poor quality of tropical forages, but studies with both temperate and tropical grasses show that DMD is increased by high radiation (Garza *et al.*, 1965; Masuda, 1977; Wilson and Wong, 1982).

G. Processing

Cut forage can be processed in many different ways prior to feeding to ruminants. Some of these processes increase DMD, others depress DMD. Forage has been used as a source of nutrients for the growth of yeasts (Henry *et al.*, 1978) and fungi (Zadrazil, 1984). These microbial products have a higher DMD than the original forage but any gain in DMD is at the expense of large losses of forage energy. The following review will be limited to the effects of freezing, drying, ensiling, and pelleting on DMD.

1. FREEZING

The high capital and operating costs of cold storage limit this method of conservation to nutrition studies. Freezing has negligible effect on DMD of temperate and tropical forage (Table 4.13).

TABLE 4.13

Effect of Freezing on the Digestibility of the Dry Matter or Organic Matter in Forage

Species	Digestibility: Fresh	Digestibility: Frozen	Reference
Digitaria decumbens	0.476	0.474	Minson (1966)
Lolium perenne/Trifolium repens	0.776	0.776	Raymond *et al.* (1953b)
Panicum maximum	0.560	0.568	Minson (1966)
Phleum pratense/Trifolium repens	0.721	0.708	Raymond *et al.* (1953b)
Sorghum almum	0.562	0.551	Minson (1966)
Sorghum sudanense	0.665	0.668	Heaney *et al.* (1966b)
Temperate pasture	0.644	0.646	Heaney *et al.* (1966b)
Mean	0.629	0.627	

2. Drying

Drying is the most common method of forage conservation. The main improvements have focused on ways of increasing drying rate and reducing the loss of the digestible components by rain damage.

In 1930 it was found that the DMD of grass grown in Northern England and dried with heated air was similar to forage grown at Cambridge and fed to sheep without drying (Woodman *et al.*, 1930). Subsequent comparisons, using the same forage, showed that drying at 100°C caused only small reductions in DMD (Table 4.14). Drying at 100°C quickly stops loss of soluble carbohydrates by respiration and this temperature is sufficiently low that undesirable oxidative changes are kept to a minimum. Drying at temperatures above 100°C can have an adverse effect on DMD. When dried at 150°C the OMD and energy digestibility of *T. pratense* were depressed 0.05 and 0.04, respectively (Demarquilly and Jarrige, 1970; Beever and Thomson, 1981).

The length of time cut forage is exposed in the field to posssible rain damage can be reduced by barn drying, tripoding, or similar systems. During barn drying large volumes of air are forced with powerful fans

TABLE 4.14

Effect of Drying at Low Temperature (<100°C) on the Digestibility of the Dry Matter, Organic Matter, or Energy of Forage

Species	Digestibility		Reference
	Fresh	Dried	
Brassica oleracea var. *acephala*	0.794	0.769	Pelletier and Donefer (1973)
Digitaria decumbens	0.476	0.476	Minson (1966)
Lolium perenne	0.674[a]	0.681	Beever *et al.* (1971b)
Medicago sativa	0.661	0.667	Donker *et al.* (1975)
Mixed pastures	0.731	0.707	Ekern *et al.* (1965)
	0.644	0.634	Heaney *et al.* (1966b)
	Depression of 0.005		Demarquilly and Jarrige (1970)
Panicum maximum	0.560	0.566	Minson (1966)
Phalaris arundinacea	0.665	0.631	Donker *et al.* (1975)
Sorghum almum	0.562	0.553	Minson (1966)
Sorghum sudanense	0.665	0.689	Heaney *et al.* (1966b)
Trifolium pratense	0.816	0.811	Kivimae (1959)
Mean	0.645	0.640	

[a]Frozen grass.

through partially dried forage. The air temperature is raised above ambient by the waste heat from the motor driving the fan or by supplementary sources (Shepperson, 1960). The OMD of barn-dried hay is lower than that of the same material frozen, the decrease depending on the weather conditions. In good hay-making weather the OMD of barn-dried hay was only 0.02 lower than the corresponding frozen forage, but when the weather was wet the depression in OMD was 0.06 (Shepperson, 1960). When hay is dried on tripods or racks in the absence of rain the decrease in OMD is similar to that found with barn drying (Shepperson, 1960; McCarrick and Wilson, 1966).

Large reductions in digestibility occur when forages are dried in the field during rainy weather. Losses in OMD of 0.05 and 0.14 occurred during a dry and wet English summer, respectively (Shepperson, 1960), whereas in France average losses were 0.06, 0.08, and 0.13 when the weather during drying was dry, <100 mm rain, and >100 mm rain, respectively (Jarrige *et al.*, 1982). Leaf shatter loss in *M. sativa* can be high, even in dry weather, and OMD may be reduced by 0.12 (Jarrige *et al.*, 1982). Field exposure of *L. purpureus* and *Vigna unquiculata* for 8 days depresses DMD by 0.04 (Milford and Minson, 1968a).

Baling of partially dried forage will reduce field exposure and loss of DMD, but this advantage may be negated by subsequent decomposition in the stack or even complete loss of the hay by spontaneous combustion. In one study, baling *M. sativa* containing 288 g water/kg compared with 142 g water/kg depressed DMD by 0.03 (Mathison *et al.*, 1986).

Loss of DM during storage depends on the initial moisture content of hay, the relative humidity of the air, temperature, and duration of the storage period (Greenhill *et al.*, 1961). When hay contains 85–160 g water/kg, DMD remains constant for as long as 3 yr if the hay is stored under cover in a dry area (Cochrane and Radcliffe, 1977). However, when baled hay is left in the field, DMD may be reduced by 0.09 within 9 months (Newbery and Radcliffe, 1974). Loss of DMD is generally confined to the outer layer of the bale but when the bales are left on clay soil, the inner layers may also be affected (Newbery and Radcliffe, 1975).

3. Ensiling

The digestibility of forages cut and immediately ensiled in airtight silos is generally similar or even marginally higher than that of the original forage (Table 4.15). In French studies with temperate grasses and *M. sativa,* ensiling reduced OMD by 0.011 and 0.014, respectively (Jarrige *et al.*, 1982). Increasing the temperature during ensiling from 27 to 51°C has little effect on DMD (Harris and Raymond, 1963). Evacuating air from ensiled forage has no effect on DMD (Wilson *et al.*, 1969a) because the quantity of entrapped air is insufficient to cause appreciable losses of OM.

TABLE 4.15

Digestibility of the Dry Matter or Energy in Frozen Grass or Direct-Cut Silage

Species	Digestibility		Reference
	Grass	Silage	
Dactylis glomerata	0.744	0.793	Harris and Raymond (1963)
Festuca pratensis	0.699	0.698	Harris and Raymond (1963)
Lolium multiflorum	0.724	0.721	Harris and Raymond (1963)
Lolium perenne	0.674	0.720	Beever *et al.* (1971b)
Phleum pratense	0.750	0.739	Harris and Raymond (1963)
Mean	0.718	0.734	

Additives may be used to improve silage fermentation and their potential effect on digestibility should be considered. Silage produced by adding mineral or organic acids has a similar OMD to that of the original forage (Jarrige *et al.*, 1982; Anderson, 1982) or is similar to silage made without additives (Chamberlain *et al.*, 1982; Rooke *et al.*, 1988). Fermentation is reduced by adding formaldehyde but this treatment leads to a small reduction in OMD and ED (Barry and Fennessy, 1973; Thomson *et al.*, 1981; Kaiser *et al.*, 1982b).

Wilting forage before ensiling reduces seepage losses and improves fermentation but these advantages may be outweighed by loss of digestible material during wilting. The DMD and ED of wilted silage is generally similar to that of the same forage when frozen but large reductions sometimes occur (Harris and Raymond, 1963; Beever *et al.*, 1971b). Wilted silage generally has a higher digestibility than direct-cut silage (Jarrige *et al.*, 1982; England and Gill, 1983).

4. Pelleting

In laboratory studies it has been shown that cellulose digestion of forages and wood can be increased by very fine grinding (Dehority and Johnson, 1961; Fukazawa *et al.*, 1982). However, when fed to ruminants, ground and pelleted forages have a lower digestibility than chopped forage (Table 4.16), with a larger decrease for grasses (0.06) than for legumes (0.02). The reduction in digestibility is associated with the shorter time that ground and pelleted forages are retained in the rumen (Blaxter and Graham, 1956; Alwash and Thomas, 1971; Laredo and Minson, 1975b). The extent of the depression in digestibility is related to the fineness of grinding and the level of feeding. The finer the forage is ground, the greater the depression (Blaxter and Graham, 1956; Rodrique and Allen,

TABLE 4.16

Effect of Grinding or Pelleting on the Digestibility of the Dry Matter, Organic Matter, or Energy of Forages[a]

Species	Digestibility: Chopped	Digestibility: Ground or pelleted	Reference
Grasses			
Bromus inermis	0.608	0.538	Kennedy (1985)
Cenchrus ciliaris	0.471	0.398	Minson and Milford (1968a)
Chloris gayana	0.410	0.362	Laredo and Minson (1975b)
Dactylis glomerata	0.637	0.557	Heaney *et al.* (1963)
Digitaria decumbens	0.539	0.486	Minson (1967), Minson and Milford (1968a), Laredo and Minson 1975b)
Festuca arundinacea	0.719	0.636	Tetlow and Wilkins (1974)
Lolium multiflorum	0.725	0.692	Milne and Campling (1972), Osbourn *et al.* (1981)
Lolium perenne	0.739	0.674	Greenhalgh and Reid (1973), Osbourn *et al.* (1981), Tetlow and Wilkins (1974)
Pennisetum clandestinum	0.620	0.576	Minson and Milford (1968a)
Phalaris arundinacea	0.529	0.458	Kennedy (1985)
Phleum pratense	0.598	0.533	Heaney *et al.* (1963), Lloyd *et al.* (1960), Osbourn *et al.* (1981)
Setaria sphacelata	0.493	0.423	Minson and Milford (1968a), Laredo and Minson (1975b)
Temperate pasture	0.686	0.626	Alwash and Thomas (1971), Blaxter and Graham (1956), Rodrique and Allen (1960), Uden (1988), Westra and Christopherson 1976)
Mean grasses	0.609	0.549	
Legumes			
Medicago sativa	0.618	0.590	Heaney *et al.* (1963), Kenned (1985), Kromann and Meye (1966), Magill (1960), Meye *et al.* (1959a), Osbourn *et a* (1981), Thomson *et al.* (197
Onobrychis viciifolia	0.480	0.512	Wainman and Blaxter (1972)
Trifolium pratense	0.606	0.594	Beever *et al.* (1981), Kennedy (1985), Lloyd *et al.* (1960), Osbourn *et al.* (1981)
Mean legumes	0.609	0.587	

[a]Results have been omitted where the chopped feed was selectively eaten or where the chemical analysis inc cates that chopped and pelleted feeds probably had a different origin.

1960; Wilkins *et al.*, 1972) and the shorter time forage is exposed to the microflora in the rumen (Rodrique and Allen, 1960).

The digestibility of pelleted forages is also reduced by changes in rumen pH. Pelleting and grinding depress pH (Alwash and Thomas, 1971; Dafaala and Kay, 1980) and at low pH rate of forage digestion is reduced (Terry *et al.*, 1969). The lower rumen pH is possibly caused by the shorter time spent eating and ruminating ground forage (Kick *et al.*, 1937), which is believed to affect the quality of saliva entering the rumen. The depression in rumen pH is greatest for forages high in soluble carbohydrate but is also affected by the buffering capacity of the forage (Osbourn *et al.*, 1981). Legumes contain more ash and have a buffering capacity twice that of grasses and this could account for the smaller depression in digestibility when legumes are ground and pelleted (Table 4.16).

Hay wafers and cobs are made by compressing long, coarsely chopped, lacerated or coarse ground hay into bondless packages usually 20–50 mm thick and 50–250 mm in diameter or square. The digestibility of wafers and cobs will depend on the DMD of the original hay and the particle size of the processed forage (Fig. 4.10).

H. Detrimental Factors

Digestibility is reduced by the presence of compounds which suppress the activity of the rumen microflora or protect the cell wall from the microflora.

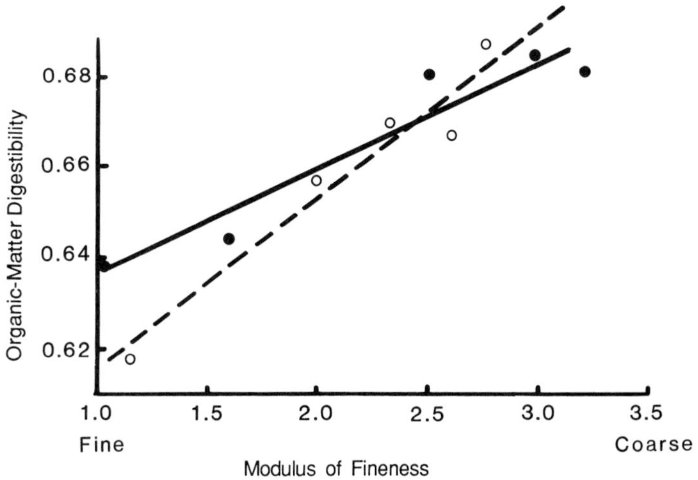

Fig. 4.10. Relation between modulus of fineness and organic-matter digestibility of *Lolium perenne* by sheep (——— •) and cattle (— — — ○). Adapted from Wilkins *et al.* (1972).

1. SILICA

Minerals are present throughout the cell walls and when the organic matter is removed by ashing, the main features of the cell walls and vascular bundles are still clearly visible in electron micrographs (Jones *et al.*, 1963). The deposit in the cell wall was believed to be polymeric silicic acid that might limit microbial degradation of the structural carbohydrates in a similar manner to lignin (Van Soest and Jones, 1968). Using forages containing different levels of silica grown in different parts of the U.S.A, it was shown that silica accounted for 58% of the variance in DMD that could not be attributed to differences in the proportion of cell walls and lignin (Van Soest and Jones, 1968). When the experiment was repeated using forages grown in the same environment, silica had no effect on OMD over and above that accounted for by differences in cell wall and lignin (Minson, 1971b), indicating that silica may not protect cell walls from microbial digestion.

Subsequent work has shown that very little of the silica in plants is present in the cell wall (McManus *et al.*, 1977) and that it is the presence of soluble silica that depresses OMD (Smith and Nelson, 1975). The depressing effect of soluble silica appears to be associated with a direct effect on the activity of the enzymes involved in forage digestion and is not due to any protection of the cell wall (Shimojo and Goto, 1989).

2. TANNIN(S)

The term tannin is used to describe polyphenolic substances that have the capacity to bind proteins and inhibit enzymes (Wong, 1973; McLeod, 1974). Low-tannin lines of *Lespedeza juncea* have a higher DMD than normal or high-tannin lines (Donnelly and Anthony, 1969, 1970, 1973), an effect caused by differences in tannin level in leaves (Table 4.17). *Lotus pedunculatus* contains condensed tannins, the level varying inversely with soil fertility (Barry and Manley, 1984). Samples containing 46 and 106 g/kg DM total condensed tannin had OMD of 0.73 and 0.66, respectively. The effect of tannin on protein digestibility is considered in Chapter 6.

3. VOLATILE OILS

Some browse shrubs contain essential oils which inhibit rumen bacteria. The essential oils of *Artemisia tridentata* and *A. nova* depress rate of cellulose digestion, gas production, and volatile fatty acid concentrations in rumen fluid (Nagy *et al.*, 1964; Nagy and Tengerdy, 1967) because the bacteria are unable to adapt to the inhibitory action of the essential oils (Nagy and Tengerdy, 1968). Not all essential oils inhibit digestion; extracts of essential oils from *Bothriochloa intermedia* have no effect on

TABLE 4.17

Dry-Matter Digestibility of High- and Low-Tannin Lines of *Lespedeza juncea*[a]

	Tannin level			
	Leaves		Stems	
	Low	High	Low	High
Tannin (g/kg)	33	79	19	25
Crude protein (g/kg)	239	209	99	92
DMD	0.66	0.46	0.46	0.43

[a]From Donnelly and Anthony (1973).

DMD or the production of volatile fatty acids and ammonia (McLeod, 1969).

4. Insecticides and Herbicides

Pesticides used to control undesirable insects and plants in pastures may depress digestibility when present at very high concentration (1 g/kg DM). However, at this high level digestibility is not affected by Baygon, Black-Leaf 40, Malathion, and 2,4-D, and none of the materials tested inhibit digestion at 0.1 g/kg DM (Schwartz *et al.*, 1973). The levels required for inhibition are very high and unlikely to be encountered in farm practice.

III. PREDICTION OF FORAGE DIGESTIBILITY

A. Methodology

Farm advisors, plant breeders, and agronomists require rapid methods for predicting DMD and related parameters. The *in vivo* DMD of a forage may be predicted from regressions of the type DMD $= a + bx$, where x is a characteristic of the forage or feces. These regressions are derived from the results of *in vivo* digestibility studies with cut forage and therefore may give biased estimates of digestibility if applied to animal species or conditions of feeding different from those used to derive the regression. Because forage digestibility is related to many different factors, it is not possible for one or even a combination of factors to account for all differences in digestibility. The variation that cannot be accounted for by a regression is expressed as the residual standard deviation (RSD) and is the *minimum* error that must be applied to any digestibility figure predicted from the regression. The RSD is used as the basis for comparing the value of different criteria for predicting DMD since, unlike the correlation coefficient (r), the RSD is not affected by the range in *in vivo* DMD of the forage samples used to derive the regression.

B. Character

1. Leafiness

The DMD of first-growth forages in New York state was found to be related to the proportion of leaf (X) (Reid *et al.*, 1959a).

$$\text{DMD} = 0.408 + 0.4X, \; r = 0.95$$

When different forage species are compared, this type of equation can be misleading. For example, *D. glomerata* is leafier than *L. perenne* but has a much *lower* digestibility (Minson *et al.*, 1960b). Another approach is to predict digestibility from a multiple regression which includes leaf, stem, dead material, and a stage of growth index, but different equations are required for early- and late-maturing species and cultivars (Jones and Walters, 1969).

2. Age of Forage

The DMD of spring growths of temperate forages can be predicted from cutting date (Section II,E,2). The relation is affected by latitude (Fig. 4.11) and must be adjusted for forage species and cultivar (Fig. 4.7) and spring temperature. A recent development in England has been the prediction of D value from meteorological factors that influence forage growth and forage cultivar (Thompson *et al.*, 1989). Different regressions are required for early-, medium-, and late-flowering *Lolium* species.

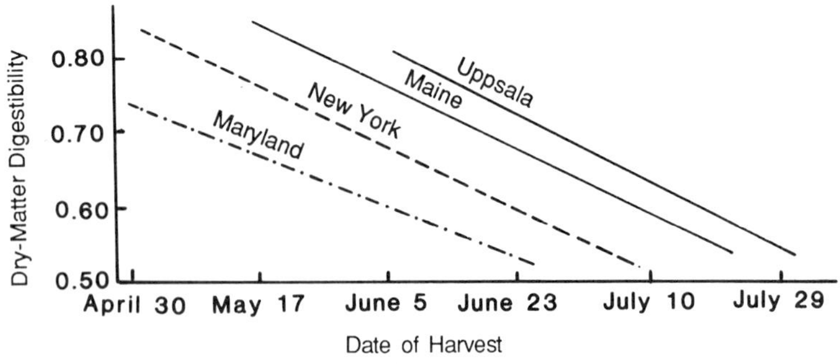

Fig. 4.11. Relation between date of harvest at four sites and digestibility of first-growth forage. Adapted from Kane and Moore (1959), Reid *et al.* (1959a), Mellin *et al.* (1962), and Kivimae (1966).

TABLE 4.18

Regressions for Predicting Dry-Matter Digestibility from the Crude Protein Content of the Forage (g/kg DM)

Species	Regressions	RSD	r	Reference
Medicago sativa	DMD = 0.462 + 0.00080X	±0.039	0.40	Sullivan (1964)
Grasses (temperate)	DMD = 0.635 + 0.00040X	±0.061	0.61	Sullivan (1964)
All forages	DMD = 0.607 + 0.00035X	±0.071	0.24	Sullivan (1964)
Legumes (temperate and tropical)	DMD = 0.357 + 0.00135X	±0.040	0.84	McLeod and Minson (1976)

C. Chemical Analysis

Many different chemical fractions have been used to predict DMD, OMD, ED, and ME (Minson, 1982). The prediction of DMD and OMD will be reviewed with special emphasis on the way estimates of digestibility can be biased by forage species and environmental conditions.

1. Crude Protein

In 1936 an equation was published for estimating the starch equivalent of temperate forage from the crude protein (CP) concentration (Watson

TABLE 4.19

Regressions for Predicting Organic-Matter Digestibility from the Crude Protein Content of the Forage (g/kg DM)

Species	Regression	RSD	r	Reference
Medicago sativa				
May cuts	OMD = 0.566 + 0.00081X	±0.020	—	Minson and Brown (1959)
June cuts	OMD = 0.330 + 0.00169X	±0.037	—	Minson and Brown (1959)
All cuts	OMD = 0.530 + 0.00061X	±0.052	—	Minson and Brown (1959)
Trifolium pratense				
First cut	OMD = 0.436 + 0.00169X	±0.020	0.93	Kivimae (1960)
Second cut	OMD = 0.518 + 0.00101X	±0.028	0.62	Kivimae (1960)
Grasses (temperate)				
April cuts	OMD = 0.818 − 0.00006X	±0.036	—	Minson and Kemp (1961)
May cuts	OMD = 0.675 + 0.00060X	±0.040	—	Minson and Kemp (1961)
June cuts	OMD = 0.545 + 0.00125X	±0.051	—	Minson and Kemp 1961)
July cuts	OMD = 0.546 + 0.00099X	±0.051	—	Minson and Kemp (1961)
August cuts	OMD = 0.543 + 0.00101X	±0.036	—	Minson and Kemp (1961)
September cuts	OMD = 0.567 + 0.00084X	±0.050	—	Minson and Kemp (1961)
October cuts	OMD = 0.498 + 0.00110X	±0.051	—	Minson and Kemp (1961)
November cuts	OMD = 0.410 + 0.00142X	±0.065	—	Minson and Kemp (1961)
All grasses	OMD = 0.597 + 0.00083X	±0.062	—	Minson and Kemp (1961)

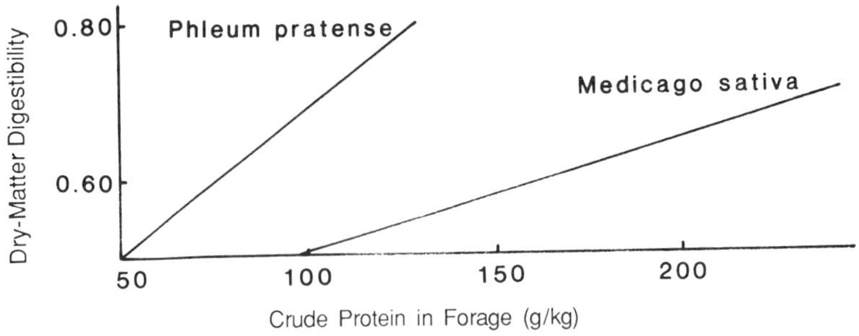

Fig. 4.12. Difference between grass and legume in the relation between dry-matter digestibility and crude protein concentration. Adapted from Phillips and Loughlin (1949).

and Horton, 1936). Regressions were subsequently published relating both DMD and OMD to CP concentration of different forages (Tables 4.18 and 4.19).

Regressions which include both legumes and grasses have large RSDs (Table 4.18). This is caused by the large difference in digestibility of grasses and legumes when they contain the same concentration of CP (Fig. 4.12). A large RSD is also associated with regressions which include grasses cut at different times of the year (Table 4.19) and this is caused by a large seasonal difference in the relationship between digestibility and CP content (Fig.4.13).

Only a small proportion of the RSD of regressions based on a wide range of forages is caused by errors in measuring *in vivo* digestibility and chemical composition (McLeod and Minson, 1976). As a corollary, the RSD cannot be significantly reduced by increasing the number of animals used to determine the *in vivo* digestibility of the forages or by repeated analysis of the samples of forage.

2. Crude Fiber

The first regressions relating digestibility of feedstuffs to crude fiber (CF) were published by Axelsson (1938) and by Jarl (1938). Subsequent work confirmed that DMD and OMD are negatively correlated with the CF concentration in grass, legume, hay, and silage (Minson, 1982). These regressions have RSD values varying from ±0.022 to ±0.062. Changes in CF concentration have a greater effect on the DMD of grasses than on that of legumes (Fig 4.14).

3. Cellulose and Xylose

In 1943 it was found that the OMD of a wide range of feedstuffs was negatively correlated with cellulose content ($r = -0.808$) (Lancaster,

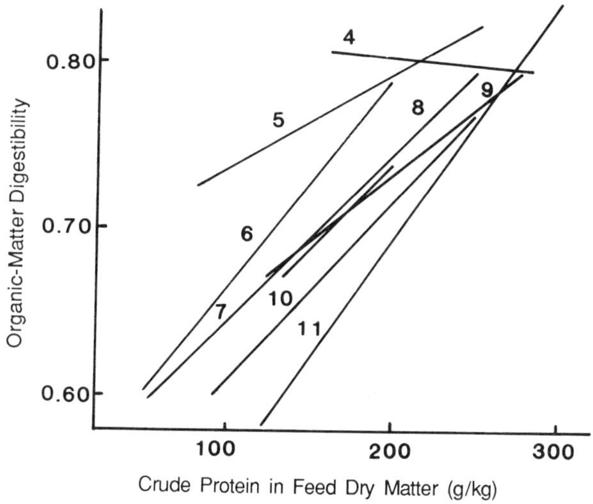

Fig. 4.13. Relation between organic-matter digestibility of grass cut at different times of the year in England and crude protein content. Figures on lines indicate month of harvest. Adapted from Minson and Kemp (1961).

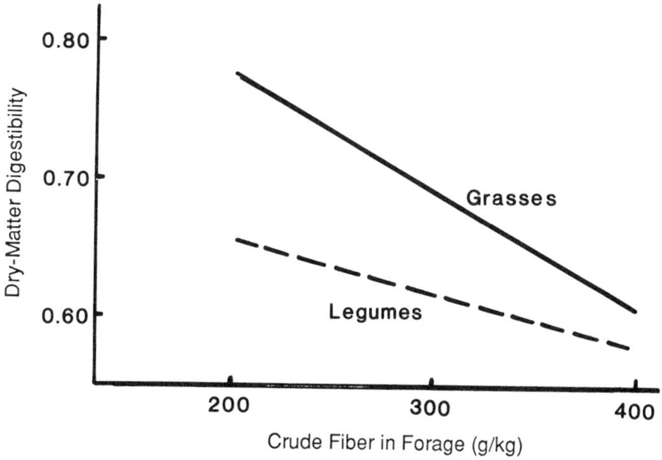

Fig. 4.14. Relation between dry-matter digestibility and crude fiber in grasses and legumes. Adapted from Sullivan (1964).

1943). Subsequent work showed that separate regressions were required for prediction the DMD of grasses and legumes (Sullivan, 1964).

The possibility of predicting DMD of both grasses and legumes from the xylose content of forages was examined by Burdick and Sullivan (1963). The regression had a low RSD but no attempt appears to have been made to develop this method for routine forage evaluation.

4. Neutral Detergent Fiber (NDF)

This fiber fraction contains all the hemicellulose, cellulose, lignin, and some of the ash in forage and is negatively correlated with DMD (Van Soest, 1965b). Regressions relating OMD to NDF in grasses and legumes have RSDs of ±0.051 and ±0.024, respectively (Bosman, 1970). The regressions for grasses and legumes are markedly different, legumes being 0.16 less digestible than grasses with the same concentration of NDF (Fig. 4.15).

5. Acid Detergent Fiber (ADF)

This fraction contains cellulose, lignin, and some of the ash in the forage (Van Soest. 1965b). For a wide range of grasses and legumes DMD and ADF are related ($r = -0.79$) (Van Soest, 1963) and the regression based on this data has a RSD of ±0.09. The main advantage of ADF

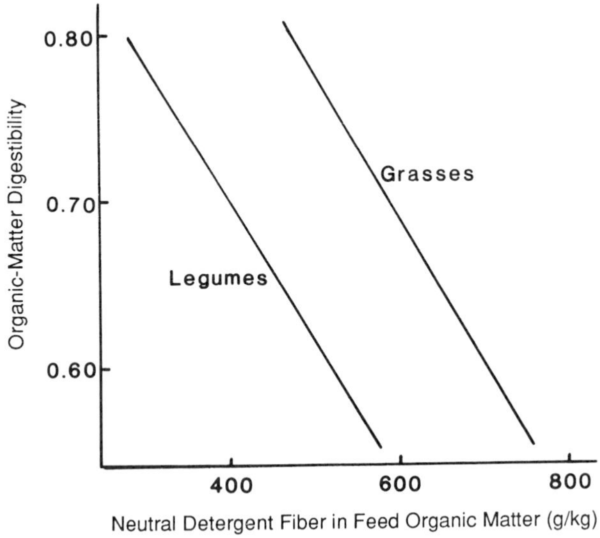

Fig. 4.15. Relation between organic-matter digestibility and neutral detergent fiber in grasses and legumes. Adapted from Bosman (1970).

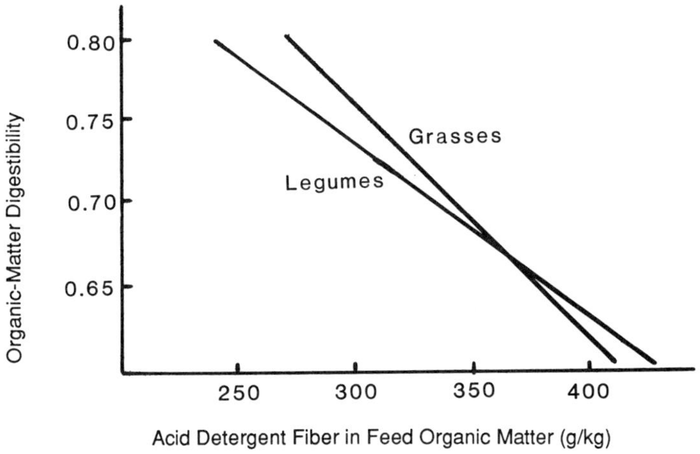

Fig. 4.16. Relation between organic-matter digestibility and acid detergent fiber in grasses and legumes. Adapted from Bosman (1970).

over NDF is the absence of large differences between the regressions for grasses and legumes, and this allows the same equation to be used for predicting the DMD of grasses and legumes or mixtures of species (Fig. 4.16). By increasing the hydrolysis time from 1 to 2 hr the RSD for temperate forages is reduced from ±0.056 to ±0.036 (Clancy and Wilson, 1966). This has been described as modified acid detergent fiber (MADF) and is used for advisory work in the United Kingdom. With tropical farages, the RSD is similar when using ADF and MADF (McLeod and Minson, 1972).

6. LIGNIN

The DMD and OMD of forage are negatively correlated with the lignin content of forages (Lancaster, 1943; Richards *et al.*, 1958; Sullivan, 1962, 1964; Allinson and Osbourn, (1970). Many equations have been published for predicting digestibility from lignin concentration in different forage species or groups of species (Minson, 1982). The most important feature of these regressions is that they account for a higher proportion of the variance in digestibility and the RSD is usually lower than for regressions based on other chemical fractions. This low RSD only applies if grasses and legumes are not included in the same regression (Fig. 4.17).

7. METHOXYL

The lignin molecule contains methoxyl groups and this has been used as a measure of the relative quantity of lignin in forage (Richards and

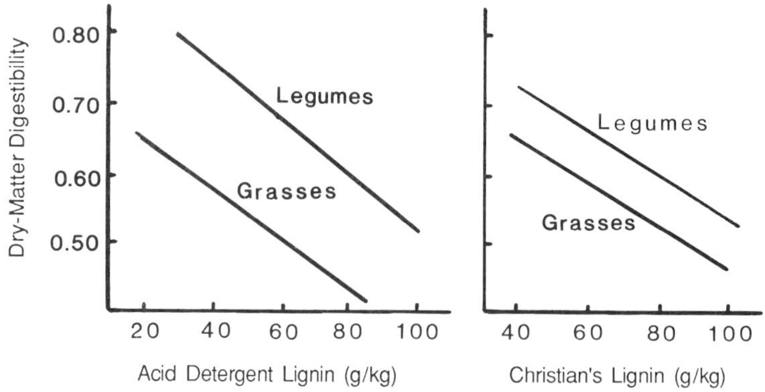

Fig. 4.17. Effect of two methods of determining lignin on the relation between dry-matter digestibility and lignin in grasses and legumes. Adapted from Sullivan (1964) and McLeod and Minson (1976).

Reid, 1952). First reports were very encouraging, with a high correlation (r = 0.99) between methoxyl and DMD, but in subsequent studies regressions had RSD values between ±0.02 and ±0.04 (Anthony and Reid, 1958; Richards *et al.*, 1958; Kivimae, 1960, 1966; Shearer, 1961). There are large differences in OMD between *P. pratense* and *T. pratense* when they contain similar levels of methoxyl (Kivimae, 1960).

8. Multicomponent Equations

The digestibility of forages is generally predicted from just one fraction but, in an attempt to reduce the RSD, equations have been developed for predicting digestibility from two or more components. Van Soest (1965a) showed that DMD could be estimated by summation of separate estimates of the digested cell contents and digested cell wall. When tested with 19 legumes and grasses, of known *in vivo* DMD, the regression had an RSD of ±0.027. In another study using 30 forages, the RSD was ±0.037 and including silica improved the prediction of DMD (Van Soest and Jones, 1968). Gaillard (1966) published a multicomponent equation for predicting the OMD of a wide range of grasses and legumes from the lignin, cellulose, hemicellulose, and anhydrouronic acid concentration in forage. This had an RSD of ±0.032 compared with ±0.043 and ±0.047 for regressions based on lignin or crude fiber, respectively.

D. Bioassay

Three bioassay techniques have been developed for predicting digestibility: *in vitro* digestibility using rumen microorganisms, *in vitro* digestibility using an enzyme preparation, and the nylon bag or *in sacco* technique. All three techniques are usually calibrated with forage samples of known *in vivo* digestibility (DMD or OMD) and results are quoted as estimated *in vivo* digestibility. Occasionally only analytical results are presented with or without correction for differences between runs (Tilley and Terry, 1963; Dhanoa and Deriaz, 1984), values being described as *in vitro* or *in sacco* digestibilities. These unconverted analytical values are useful for ranking forage in plant breeding and selection programs but they can differ from *in vivo* digestibility coefficients by 0.10 or more (Milne, 1977).

1. Rumen Fluid–Pepsin Technique

The common features of these methods are fermentation of a ground forage sample with rumen microorganisms in a buffered medium under controlled conditions of anaerobiosis, temperature, and pH (Shelton and Reid, 1960). When first used, cellulose digestibility was measured and values were converted, using an appropriate regression, into *in vivo* OMD (Pidgen and Bell, 1955), DMD (Baumgardt *et al.*, 1962; Baumgardt and oh, 1964), or DE (Hershberger *et al.*, 1959).

In 1958 it was found that "by using a 48 hour digestion period with whole rumen fluid inoculum and then adding a 24 hour digestion with 2% pepsin–HCl, estimates of dry matter digested were found to be significantly correlated (r = 0.79) with the *in vivo* dry matter digestion coefficient" (Clark, 1958). The inclusion of the acid–pepsin stage removes the crude protein remaining after the microbial fermentation and reduces the RSD from ±0.044 to ±0.020 (Tilley *et al.*, 1960). A full description of the method was published by Tilley and Terry (1963).

The two-stage *in vitro* method is considered to be the most precise technique for estimating digestibility of forage DM and OM (Oh *et al.*, 1966; Braver and Eriksson, 1967; Bosman, 1967, 1970; Newman, 1972; Morgan, 1974; Scales *et al.*, 1974; Tinnimit and Thomas, 1976; Aerts *et al.*, 1977; Thomas *et al.*, 1980; Deinum *et al.*, 1984) and is suitable for samples containing different proportions of grass and legume (Monson and Reid, 1968; McLeod and Minson, 1969b). The method has been used to evaluate grasses, legumes and *Brassica oleracea* var. *acephala* (Dent, 1963), shrubs and trees (Wilkins, 1966; McLeod, 1973), and in grass breeding (Cooper *et al.*, 1962; Burton *et al.*, 1967). It has also been used to estimate the digestibility of grazed forage (Wilson *et al.*, 1971), modified to determine true digestibility (Van Soest *et al.*, 1966) and to measure potential digestibility (Wilkins, 1969; Prins *et al.*, 1981).

The method of preparing forage for *in vitro* analysis can affect digestibility. Samples of forage frozen in liquid nitrogen before grinding have a higher digestibility (0.03) than samples dried at 60–100°C, due to a loss of soluble carbohydrate during oven drying. This loss occurs at all drying temperatures, and drying at 100°C had been recommended for the routine preparation of samples for determination of *in vitro* digestibility (Cochrane and Brown, 1974). After drying, forage should be ground to pass a 1-mm screen but not so fine as to disrupt the cell wall, as occurs with ball milling (Tilley and Terry, 1963). There are problems in preparing silage for *in vitro* analysis because drying causes a loss of volatile acids (Minson and Lancaster, 1963) and depresses *in vitro* digestibility (Trinder and Hall, 1972). This problem can be prevented by using homogenized undried silage (Alexander and McGowan, 1966).

Many regressions have been published relating *in vivo* DMD to *in vitro* DMD. For forages of low digestibility separate regressions are required for different forage species (Troelsen and Bell, 1968; McLeod and Minson, 1969a; Goldman *et al.*, 1987), and methods of conservation (Bosman, 1970). Another problem encountered is the variation in the cellulolytic activity of the rumen liquor (Troelsen and Hanel, 1966), which causes week-to-week differences in *in vitro* digestibility (Dhanoa and Deriaz, 1984). Both problems are avoided by including, in each run, samples of known *in vivo* digestibility as similar as possible to the forage being tested. Cattle are more efficient digesters of forage than sheep (Fig. 4.2) and if the test forage is to be eaten by cattle, then DMD of the standards should have been determined with cattle. Alternatively, the estimated sheep DMD can be converted to cattle DMD by addition of the average difference in DMD between the two species (i.e., 0.024) (Section I,B).

2. *In Vitro* Cellulase Technique

Many fungi produce cellulase and other enzymes that hydrolyze forage carbohydrates. In 1963, the first attempt was made to replace rumen fluid with cellulase in the *in vitro* technique, but the correlation coefficient with DMD was only 0.68 (Donefer *et al.*, 1963). Subsequent work, using a different cellulase, was more successful with a correlation between OMD and cellulase residue of 0.92 and an RSD of ±0.032 (Jarrige *et al.*, 1970). The error was further reduced when the cellulase incubation was preceded by a 24-hr incubation with acid–pepsin (Jones and Hayward, 1975). The two-stage *in vitro* cellulase technique has proved successful with tropical grasses (Adegbola and Paladines, 1977; Goto and Minson, 1977), tropical legumes (McLeod and Minson, 1978), temperate silages (Dowman and Collins, 1977; Aufrere, 1982), temperate hays (Adamson and Terry, 1980; Aufrere, 1982; De Boever *et al.*, 1988), and other feedstuffs (Clark and Beard, 1977; Dowman and Collins, 1982; De Boever *et al.*, 1986, 1988).

Many highly active cellulases are now commercially available. Those derived from *Trichoderma reesei* (formerly *T. viride*) have generally been found to give the highest DM disappearance (Jones, 1976b) and regressions with the lowest RSD (Jones and Hayward, 1975; McQueen and Van Soest, 1975). Onozuka SS (P1500) cellulase was widely recommended but has been replaced by Onozuka FA, an enzyme marketed specifically for forage analysis. This has twice the activity of Onozuka SS and contains cellulase, hemicellulase, and pectinase. Modification of the technique has been described by Clarke *et al.*, (1982) and by McLeod and Minson (1982). The technique is standardized with samples of known *in vivo* digestibility as similar as possible to the forages being analyzed.

3. *In Sacco* Technique

The *in sacco,* silk bag, or nylon bag technique was first used by Quin *et al.* (1938) to study digestion within the rumen and later adapted for predicting forage *in vivo* digestibility (Demarquilly and Chenost, 1969). The technique has predicted OMD of grass hay, silage, and pelleted forages with a lower error than the *in vitro* rumen fluid technique (Aerts *et al.*, 1977). In their study Aerts *et al.* used 150 × 80 mm bags made from nylon cloth with a maximum pore size of 50 μ. Each bag was filled with 3 g of forage that had been ground through a 1-mm screen. Fifty bags were attached to a 0.6-kg ring and suspended for 48 hr in the rumen of a cow fed good hay. After incubation the bags were washed in running water, incubated for 48 hr at 39°C in darkness with an HCl–pepsin solution, washed, dried, and the organic matter determined. Forages of known *in vivo* digestibility, as similar as possible to those being analyzed, were included so that *in vivo* digestibility could be predicted (Fig. 4.18). When no *in vivo* standards are included the results are described as "dry matter disappearance" (Miller *et al.*, 1965a) and this prevents any confusion with *in vivo* digestibility.

The *in sacco* method has been used to study the effect of age of forage regrowth (Lowrey *et al.*, 1968), seasonal changes (Miller *et al.*, 1965a; Lowrey *et al.*, 1968; Monson *et al.*, 1969; Neathery, 1972), species and clonal variation (Burton *et al.*, 1964; Lowrey *et al.*, 1968; Monson *et al.*, 1969; Cote *et al.*, 1983), breakdown of seed in the rumen (Playne *et al.*, 1972), alkali treatment (McAnally, 1942), and the effect of gamma radiation (McManus *et al.*, 1972a). The rate of disappearance of forage *in sacco* varies with the pore size of the cloth bags (Lindberg and Knutsson, 1981), but this should not affect the prediction of *in vivo* DMD provided standard forages of known *in vivo* digestibility are included in each run.

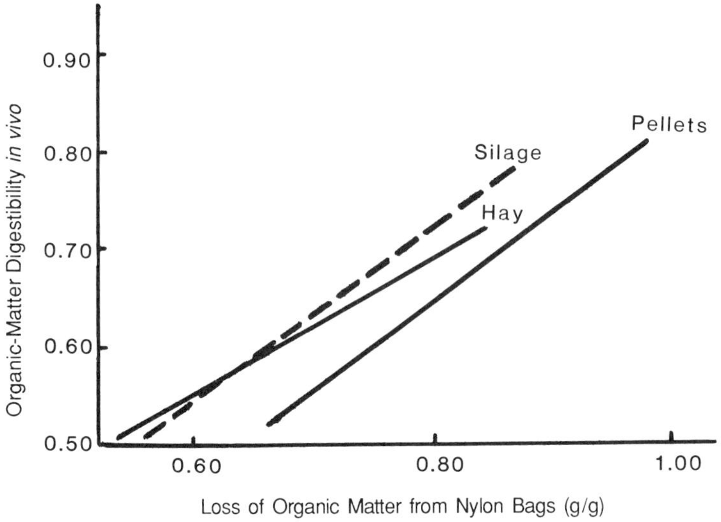

Fig. 4.18. Relation between *in vivo* organic-matter digestibility and organic-matter disappearance *in sacco*. Adapted from Aerts *et al.* 1977).

E. Physical Measurement

The digestibility of forages can be predicted from many different physical parameters including near infrared reflectance, energy required to grind dry samples, density of ground forage, leaf strength, and water content. These will now be considered.

1. Near Infrared Reflectance (NIR)

In 1976 it was shown that the *in vivo* DMD of forage could be predicted from the reflectance from ground samples of near-infrared radiation (Norris *et al.*, 1976). Reflectance was measured at more than 1000 different wavelengths using a scanning NIR spectrometer, and 9 wavelengths which accounted for a high proportion of the variation in DMD were selected. A multiple regression relating DMD to reflectances at all 9 wavelengths had an RSD of ±0.036. Subsequent studies with different forages show that other wavelengths were required for minimum RSD (Table 4.20). When the instrument is calibrated with forage of the type or form of the forage to be analyzed the error in predicting DMD is small (±0.006), but it increases to about ±0.03 when DMD is predicted for forage species that are *not* included in the calibration (Minson *et al.*, 1983). It is now recommended that NIR should only be used to predict the nutritive value of forage of the type that is used to calibrate the instrument

TABLE 4.20

Wavelengths Selected and Error of Predicting Dry-Matter or Organic-Matter Digestibility

Species	Standard deviation of prediction	Wavelengths	Reference
Agropyron cristatum	±0.016[a] *(in vitro)*	1539, 1623, 1679, 1779, 1827, 1947, 2207, 2251, 2311	Gabrielsen *et al.* (1988)
Brassica oleracea var. *acephala*	±0.018 *(in vitro)*	1982, 2170, 2256	Allison (1983)
Bromus inermis	±0.018[a] *(in vitro)*	1219, 1699, 2187, 2247, 2339	Gabrielsen *et al.* (1988)
Cynodon dactylon	— *(in vitro)*	1300, 2170	Barton and Burdick (1979)
Cynodon dactylon	±0.025 *(in vitro)*	1714, 2097, 2187	Burdick *et al.* (1981)
Forages	±0.027 *(in vitro)*	1437, 1641, 1675, 1754, 1786, 1958	Shenk *et al.* (1979)
Temperate forages	±0.036	1512, 1596, 1666, 1758, 1868, 1992, 2100, 2210, 2266	Norris *et al.* (1976)
Temperate forages	±0.056 *(in vitro)*	1680, 1940, 2100, 2180, 2230, 2310	Winch and Major (1981)
Temperate forages	±0.042 (Cattle)	1468, 1668, 2188, 2208, 2328	Redshaw *et al.* (1986)
Temperate forages	±0.034 (Sheep)	1468, 2048, 2228	Redshaw *et al.* (1986)
Temperate forages	— (Sheep)	1338, 2100, 2236, 2282	Lindgren (1983)
Temperate forages	±0.030 *(in vitro)*	1680, 1818, 1982, 2100, 2310, 2336	Bengtsson and Larsson (1984)

[a]Standard error of calibration.

and that it should only be used to predict forage characteristics that can be determined in the laboratory (Marten, 1985). Thus, NIR can be used to predict *in vitro* digestibility but not *in vivo* digestibility.

2. Grinding Energy

As plants mature there is an increase in fiber concentration and there is an associated increase in the quantity of energy required to grind dried samples through a 1-mm screen (Chenost, 1966). Regressions relating OMD and DMD to grinding energy have RSDs of 0.031 and 0.042, respectively (Chenost, 1966; Winter and Collins, 1987). In a comparison between two cultivars of *L. multiflorum,* the one with the higher DMD had the lower grinding energy (Table 4.21). However, when separated leaf and stem of tropical grasses were compared the stem fraction had a higher grinding energy value than the leaf fraction even when there was no difference in DMD (Laredo and Minson, 1973). It appears that grinding

TABLE 4.21

Differences in Grinding Energy and Dry-Matter Digestibility of Two Cultivars of *Lolium multiflorum*[a]

	Cultivar or genotype	
	RVP	Bb 1277
Dry-matter digestibility		
Leaf	0.79	0.80
Stem	0.69	0.72
Grinding energy		
Leaf	459	426
Stem	1251	1065

[a]Adapted from Hides *et al.* (1983).

energy can be a useful indicator of relative DMD provided the forages compared have a similar proportion of leaf.

3. Tensile Strength

Equipment has been developed for measuring the breaking strain of grass leaves (Evans, 1967a). When applied to eight temperate grasses, leaf strength was positively correlated with the proportion of cellulose (Evans, 1967b) but no data has been published on the relation between tensile strength and DMD. Leaf tensile strength is an inherited characteristic in *L. perenne* (Wilson, 1965).

4. Bulk Density

The bulk density of finely ground (<60 mesh) forage has been determined by mixing a 0.5-g sample with kerosene in a graduated tube and allowing it to settle for 2 hr (Arora and Das, 1974). Dry-matter digestibity was closely related to bulk density, with an RSD of ±0.04 (Fig. 4.19).

5. Water Content

In a study of 28 fresh temperate forages DMD was negatively related to the dry-matter concentration of the forage (X, expressed in g/kg) (Reid *et al.*, 1959b).

$$\text{DMD} = 0.874 - 0.00104X, \text{ RSD} = \pm 0.8, \ r = \pm 0.042$$

In subsequent studies with *D. decumbens*, DMD and dry-matter concentration were poorly correlated ($r = 0.309$) (Chenost, 1975). This is to

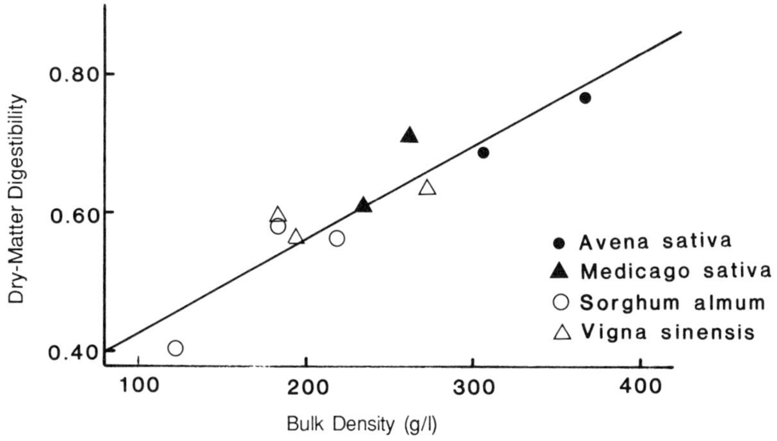

Fig. 4.19. Prediction of dry-matter digestibility from bulk density of finely ground forage. DMD = 0.303 + 0.00131BD; RSD ± 0.043; r = 0.91. Adapted from Arora and Das. (1974).

be expected since the dry-matter concentration of forage varies with the level of external moisture, time of day, and level of nitrogen fertilizer, all factors which have no direct effect on DMD.

F. Fecal Analysis

The digestibility of forage selected by grazing ruminants may be determined from regressions relating forage digestibility to fecal composition. Many different fecal components have been used to predict digestibility including nitrogen, chromogen (pigments), normal-acid fiber, crude fiber, and methoxyl.

1. Fecal Nitrogen

The first equations for predicting DMD from the N concentration in feces were published by Lancaster (1947) and by Gallup and Briggs (1948). When regressions were based on a wide range of temperate forages, the RSD varied between ±0.029 and ±0.057 (Raymond *et al.*, 1954; Minson and Raymond, 1958; Minson and Kemp, 1961). Some of these regressions are compared in Fig. 4.20.

Large differences occur between fecal nitrogen regressions derived from forage cut at different times of the year (Minson and Raymond, 1958; Greenhalgh and Corbett, 1960; Greenhalgh *et al.*, 1960; Minson and Kemp, 1961; Vercoe *et al.*, 1962; Hadjipieris *et al.*, 1965; Reid *et al.*, 1967a,b; Chenost, 1985). This seasonal difference is caused by the higher

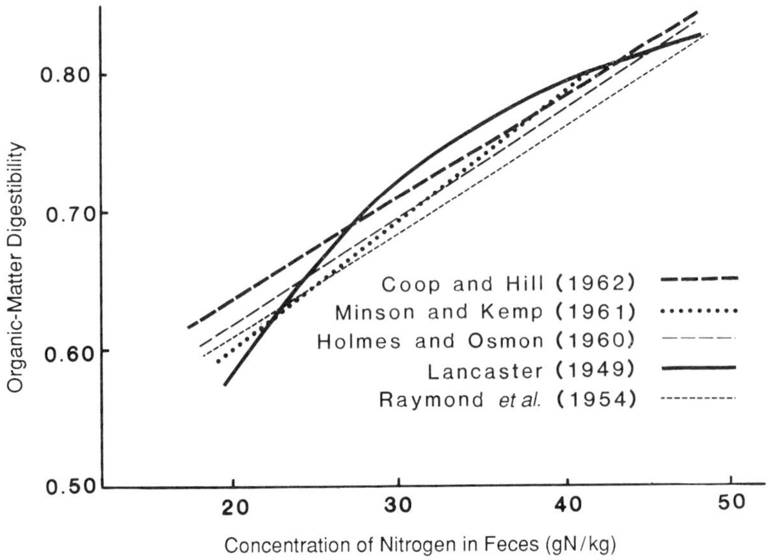

Fig. 4.20. Relation between organic-matter digestibility and fecal nitrogen concentration for temperate forages.

concentration of N in forage grown in the autumn (Minson and Kemp, 1961) and differences between fecal nitrogen regressions for leaf and stem fractions (Lambourne and Reardon, 1962). This bias may be overcome by using regressions applicable to different months of the year (Minson and Kemp, 1961).

The OMD of forages varies with level of feeding and animal species (Section I,B), and any change in efficiency of digestion will alter the N concentration of the feces and hence predicted OMD. Fortunately, the changes in predicted digestibility are similar to the observed *in vivo* difference in digestibility that occurs with changes in level of feeding (Raymond *et al.*, 1959) or between cattle and sheep (Chenost, 1986).

2. Chromogen

Chromogen is the pigment fraction which absorbs light at 406 m (Reid *et al.*, 1950) and in 1952 it was shown that "the direct computation of digestibility from the chromogen concentration of the faeces" was possible (Reid *et al.*, 1952). When tested with temperate forages in the United Kingdom and New Zealand the regression had a larger RSD than regressions based on fecal N (Raymond *et al.*, 1954; Kennedy *et al.*, 1959). Chromogen regressions have a seasonal bias similar to fecal N regressions (Minson and Raymond, 1958; Kennedy *et al.*, 1959) and fecal chromogen is now rarely used to predict digestibility.

3. Fecal Fiber (CF)

The OMD of temperate forages is negatively correlated with the CF concentration in the feces and the regression has an RSD of ±0.034, only slightly higher than the corresponding fecal N regression (Raymond *et al.*, 1954). However, increasing the level of feeding decreases the fiber concentration in the feces and OMD, but the OMD predicted from fecal fiber concentration is increased. This leads to a large bias when predicting OMD of grazed forage (Raymond *et al.*, 1956) and the method is no longer used.

4. Fecal Methoxyl

The methoxyl content of feces is significantly and negatively correlated with DMD ($r = -0.851$) and the regression has an RSD of ±0.036 (Richards *et al.*, 1958). Methoxyl can be determined by different methods but the method has little effect on the RSD of regressions for predicting DMD (Shearer, 1961). Fecal methoxyl concentration does not appear to be used for routine forage evaluation.

G. Internal Indigestible Markers

A major cause of error when predicting the digestibility of forage from the composition of the diet or feces is that full allowance cannot be made for the effect on digestibility of level of feeding, difference between sheep and cattle, or physical form (Section I,C). All these problems would be solved if a marker could be found that is naturally present in the forage, completely recovered in the feces, and simply determined. Digestibility of forage could then be calculated as follows:

$$\text{DMD} = \frac{Y - X}{Y}$$

where X = concentration of marker in forage dry matter and Y = concentration of marker in feces dry matter. The potential markers studied have included silica, acid-insoluble ash, lignin, methoxyl, chromogen, indigestible fiber, and *n*-alkanes.

1. Silica

Most of the silica in forage is usually recovered in feces; the highest values occur with diets containing high levels of silica (Jones and Handreck, 1965). However, incomplete recoveries of silica in the feces have

TABLE 4.22

Comparison of Organic Matter Measured *in Vivo* at Two Levels of Feeding with Digestibility Predicted from Acid-Insoluble Ash (AIA) and Potentially Indigestible Cellulose (PIC)[a]

Species	Level of feeding (g/kg)	*In vivo*	Organic-matter digestibility Predicted from AIA	Predicted from PIC	*In vitro*
Lolium perenne	15	0.67	0.68	0.68	
	25	0.63	0.65	0.65	0.70
Medicago sativa	15	0.58	0.62	0.59	
	25	0.56	0.59	0.58	0.57

[a]Adapted from Penning and Johnson (1983a).

been reported by McManus *et al.*, (1967), possibly caused by an accumulation of large silica particles in the rumen.

2. Acid -Insoluble Ash (AIA)

Shrivastava and Talapatra (1962) used as a marker forage ash that was insoluble in hydrochloric acid. The average recovery was 99.8% and in a subsequent study recovery was 96.7 ± 6.7% (Van Keulen and Young, 1977). Good agreement was obtained between *in vivo* OMD and predicted OMD of *L. perenne* and *M. sativa* at two levels of feeding (Table 4.22). Unsatisfactory results were reported in Irish studies with cattle and sheep, with 127% of the AIA recovered in the feces. Even when allowance was made for this excess excretion, AIA was inferior to *in vitro* and acid detergent fiber as methods for predicting *in vivo* DMD (Wilson and Winter, 1984).

3. Lignin

Lignin is the most indigestible organic component in forage and should be the ideal marker. However, the quantity of lignin in forage varies with the method of analysis (McLeod and Minson, 1974) and recovery of lignin in the feces varies from 0.51 to 1.05 (Streeter, 1969). It has been suggested that lignin as an indigestible marker should be viewed with caution (Fahey and Jung, 1983).

4. Chromogen

The chromogen in 36 temperate pasture samples, conserved in different ways, was claimed to be indigestible and completely recovered in the feces (Reid *et al.*, 1950). However, more chromogen was found in the feces of cattle fed *D. glomerata* than was present in the forage (Kane and Moore, 1961), possibly due to an error in measuring chromogen concentration. It is now known that chromogen is unstable in the presence of light and is converted into pheophytin (Kane and Jacobson, 1954; Lancaster and Bartrum, 1954; Scaut, 1959; Kane and Moore, 1961). Chromogen is no longer used to estimate digestibility.

5. Potentially Indigestible Fiber

By digesting forage for prolonged periods *in sacco* (Penning and Johnson, 1983a), with cellulase (Penning and Johnson, 1983b), or with rumen fluid (Cochran *et al.*, 1986) it is possible to produce a fiber fraction that is indigestible and appears to meet the requirements of an ideal marker. Very encouraging results were obtained when this analysis was applied to *L. perenne* and *M. sativa* (Penning and Johnson, 1983b) (Table 4.22) but not when applied to *F. arundinacea*, when only half the marker was recovered in the feces (Cochran *et al.*, 1986). In another study, recovery was variable; *Cenchrus ciliaris*, 0.96; *C. dactylon*, 0.78; *L. perenne*, 1.30; *S. bicolor*, 0.96 (Lippke *et al.*, 1986). Further work is required to determine the cause of variable recoveries of indigestible fiber.

6. N-Alkanes

The cuticle of forage contains long-chain alkanes of uneven chain length (C_{19}–C_{35}) which are chemically discrete compounds and relatively indigestible. The recovery of alkanes in *L. perenne* increases with rising chain length: C_{27}, 0.71; C_{29}, 0.74; C_{31}, 0.85; C_{33}, 0.89; and C_{35}, 0.93 (Mayes *et al.*, 1986). If these recovery coefficients are shown to be constant for all forages and levels of feeding, then alkanes could be used to estimate forage digestibility. The role of alkanes in estimating forage intake was considered in Chapter 3.

IV. INCREASING DIGESTIBILITY

This section will consider the possibilities for improvement in forage digestibility by management, supplements, plant breeding, chemical treatment, and plant fractionation.

A. Management Strategies

1. Forage Species

It was shown in Section II,B that temperate grasses generally have a higher DMD than tropical grasses, so, where possible, temperate grasses should be sown in preference to tropical species. In many subtropical areas dairy farming relies on temperate pastures grown in the cooler parts of the year (Rees *et al.,* 1972). There are also differences in digestibility between grass species, and where the winters are sufficiently mild *L. perenne* should be sown in preference to *D. glomerata.*

2. Forage Cultivars

Digestibility declines as forage matures (Section II,D) and the most effective way of avoiding a low DMD is to maintain forage in a young vegetative stage of growth by regular cutting or grazing. This is particularly important in spring when DMD falls rapidly due to an increase in the stem fraction (Fig. 4.8). Any technique which prevents stem elongation and flowering will reduce this fall in digestibility. Applying mefluidide to *F. arundinacea* inhibits floral development (Glenn *et al.,* 1980) and slightly increases DMD (0.548 to 0.572) (Robb *et al.,* 1982). Mefluidide also suppresses seed-head formation in *B. inermis, D. glomerata, Phalaris arundinacea,* and *Poa pratensis,* the extent depending on the date of application, digestibility usually being increased if mefluidide is applied before culm elongation (Glenn *et al.,* 1987). Any improvement in DMD is usually accompanied by a decrease in yield of DM (Sheaffer and Marten, 1986).

Forage growth and maturation are stopped when desiccants are applied and if they are used at an early stage of growth the dried forage might be expected to have a high DMD. When paraquat was used to chemically cure *Agropyron* and *Bromus* spp. DMD was slightly higher than in naturally cured grass (Wallace *et al.,* 1966). However, Australian studies showed no improvement in DMD (Romberg *et al.,* 1969; Pullman and Allden, 1971) or liveweight gain (Pullman and Allden, 1971), the potential advantage of halting maturation being outweighed by leaf loss, decomposition, and reduced yield.

3. Plant Density

By planting *Z. mays* at high density, the size of individual stems is reduced but this has no effect on silage DMD (Goering *et al.,* 1969; Wilkinson and Phipps, 1979; Nicholson *et al.,* 1986). In one study high-density planting caused a small fall in digestibility, possibly because it increases the proportion of rind in the stem (Masaoka and Takano, 1980). Stem diameter had no effect on the DMD of *M. sativa* (Mowat *et al.,* 1967).

4. Stimulating Regrowth

Digestibility of pasture falls as it matures (Section II,E) but can be improved by removing the mature growth and allowing new growth to occur. The mature forage can be removed by cutting, grazing, or burning. Burning of an *Andropogon* spp. pasture increased DMD by 0.055 (Smith *et al.*, 1960).

5. Applying Fertilizers

Nitrogenous fertilizer stimulates growth but has no consistent effect on DMD when both fertilized and unfertilized pastures are cut at the same time (Section II,F). However, forage is usually grazed or cut when it reaches a predetermined yield or height, so fertilized forage will be utilized at a younger stage of growth, when DMD will usually be higher. Digestibility will also be improved by applying fertilizer sulfur, provided the forage is deficient in sulfur (Section II,F).

6. Selective Grazing

In Section II,D it was shown that digestibility of forage usually declines from top to bottom of the plant. Allowing animals to selectively graze will increase the DMD of the forage ingested. The extent of the improvement depends on the variation within the sward and the level of selection allowed by the stockman.

As pastures are grazed down there is a decrease in digestibility (Raymond *et al.*, 1956; Blaser *et al.*, 1960). With light grazing, selected forage had an OMD approximately 0.03 higher than when the pasture was heavily grazed (Raymond *et al.*, 1956). The traditional method of exploiting this variation in quality is to lightly graze pastures with animals that have a high nutrient requirement and follow with animals with a lower requirement. The advantage of this leader-and-follower grazing system is illustrated in Table 4.23. In another study, conducted in spring, there was little difference in OMD of forage eaten at the beginning and at the end of grazing, but later in the year there were large differences in DMD as pasture was grazed down (Tayler and Deriaz, 1963).

B. Supplements

The digestibility of forage diets may be improved by feeding energy supplements or by supplying specific nutrients if these are deficient.

1. Energy Supplements

Energy supplements generally have a higher DMD than forage, so feeding energy supplements generally improves DMD of the diet. The extent of the improvement depends on the proportion of energy supplement in the mixed diet, the DMD of the forage and supplement, and the effect of the supplement on the activity of rumen microbes. With late-cut hay of low digestibility, feeding an energy supplement increases

TABLE 4.23

Dry-Matter Digestibility of Forage Selected by Leader (Upper Sward Levels) and Follower (Lower Sward Levels) Cattle Rotationally Grazing Two Forages[a]

	Dry-matter digestibility		
Time	Leaders	Followers	Difference
Dactylis glomerata			
April–May	0.717	0.656	0.061
June–July	0.705	0.658	0.047
August–September	0.606	0.544	0.062
Festuca arundinacea			
April–May	0.700	0.620	0.080
June–July	0.683	0.649	0.034
August–September	0.607	0.557	0.050
Mean	0.670	0.614	0.056

[a]Adapted from Blaser *et al.* (1960).

digestibility, but with early-cut hay the effect is much smaller (Forbes *et al.*, 1967a; Arriaga-Jordan and Holmes, 1986) (Fig. 4.21). Linear increases in digestibility have been found when the following forages were supplemented with grain: *Chloris gayana* (Tagari and Ben-Ghedalia, 1977), *C. dactylon* and *D. glomerata* (Jones *et al.*, 1988), *P. clandestinum* (Campbell *et al.*, 1969), mixed prairie grass (Rittenhouse *et al.*, 1970), dried grass (Orskov and Fraser, 1975), and temperate grass silage (Beever *et al.*, 1984). The energy supplement may be another forage cut at an immature stage of growth. The DMD of mixtures is directly proportional to the quantities of the two forages, unless one forage is deficient in an essential nutrient (Minson and Milford, 1967; Bowman and Asplund, 1988a).

Feeding energy supplements may have an adverse effect on the rumen microbes and may depress digestibility (Schneider and Flatt, 1975). This depression in DMD only occurs when the energy supplement reduces pH in the rumen (Terry *et al.*, 1969) or when protein is deficient in the mixed diet (Fick *et al.*, 1973).

2. Protein and Mineral Supplements

Digestibility of forage may be slightly depressed by a deficiency of an essential nutrient, and supplying the deficient nutrient will increase digestibility. This subject will be considered in later chapters dealing with individual nutrients.

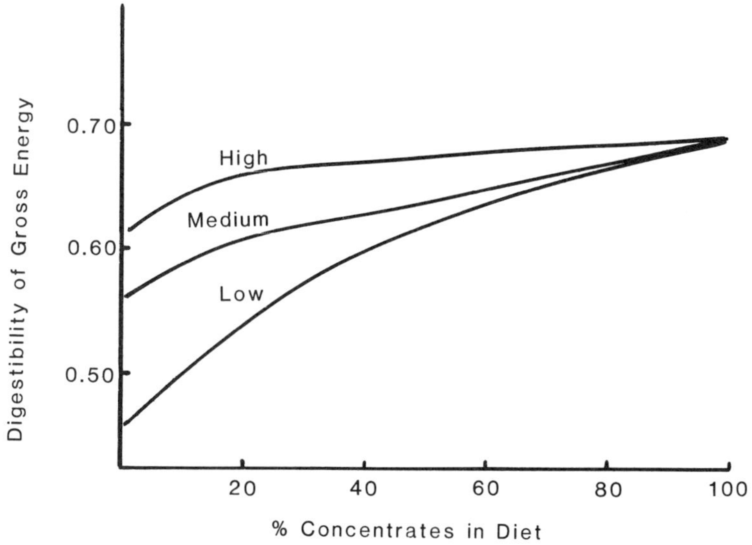

Fig. 4.21. Effect of energy supplements on the energy digestibility of different-quality hays. Adapted from Blaxter and Wilson (1963).

C. Breeding and Selection

Digestibility of forages may be increased by breeding and selection, provided there is a range of DMD and that some of this variation is of genetic origin (Cooper and Breese, 1980; Hacker, 1982). In many species the range in DMD exceeds 0.10 and about half of this variation is genetic in origin (Table 4.24). This variation has been exploited to produce commercial varieties of high digestibility in several forage species (Section II,C).

D. Chemical Treatment

Many chemical treatments which increase the digestibility of cereal straws have been developed (Sundstol and Owen, 1984). Some of these have been applied to forages and will now be reviewed. The effects on digestibility of drying, freezing, ensiling, or pelleting forages were considered in Section II,G.

1. Sodium Hydroxide

Homb (1984) described developments in the sodium hydroxide (NaOH) treatment of cereal straws from the first German studies in 1891.

TABLE 4.24

Variation in Dry-Matter Digestibility and Its Heritability in Forage

Species	DMD range	Heritability	Reference
Agropyron intermedium	0.03	—	Thaden *et al.* (1975)
Agropyron spp.	0.06	0.36–0.76	Coulman and Knowles (1974)
Andropogon gerardii	0.07	0.72	Ross *et al.* (1975)
Bromus inermis	0.12	—	Christie and Mowat (1968)
	0.07	1.06	Ross *et al.* (1970)
Cenchrus ciliaris	0.11	0.25–0.77	Bray and Pritchard (1976)
	0.09	—	Lovelace *et al.* (1972)
Chloris gayana	0.14	0.35–0.48	Quesenberry *et al.* (1978)
Cyamopsis tetragonoloba	0.11	—	Das *et al.* (1975)
Cynodon dactylon	0.24	0.27–0.69	Burton and Monson (1972)
	0.06	—	Andrews and Croft (1979)
	0.11	—	Holt and Conrad (1986)
Dactylis glomerata	0.17	—	Breese (1970)
	0.19	—	Lovelace *et al.* (1972)
	—	0.49–0.91	Stratton *et al.* (1979)
Digitaria decumbens	0.13	—	Klock *et al.* (1975)
	0.06	—	Schank *et al.* (1977)
Digitaria milanjiana	0.04	—	Minson and Hacker (1986)
Hemarthria altissima	0.28	—	Schank *et al.* (1973)
Lespedeza juncea	0.08	—	Donnelly and Anthony (1969)
Medicago sativa	0.18	—	Allinson *et al.* (1969)
Panicum maximum	0.29	—	Quesenberry *et al.* (1978)
Panicum virgatum	0.04	—	Vogel *et al.* (1981)
Phalaris aquatica	0.09	—	Clements *et al.* (1970)
	0.12	0.54–0.77	Burton and Monson (1972)
	0.12	0.60	Oram *et al.* (1974)
Phalaris arundinacea	0.03	—	O'Donovan *et al.* (1967)
	—	0.58–0.77	Hovin *et al.* (1974, 1976)
	0.09	0.02–0.63	Marum *et al.* (1979)
Setaria sphacelata	0.10	—	Hacker (1974b)
Setaria sphacelata	0.04	—	Bray and Hacker (1981)
Setaria sphacelata var. *splendida*	0.07	—	Hacker and Minson (1972)
Sorghum bicolor	0.12	—	Stratton *et al.* (1979)
Zea mays	0.10	—	Barnes *et al.* (1971)
Zea mays	0.03	—	Deinum and Bakker (1981)

In the traditional Beckmann method, straw was soaked in a solution containing 15–20 g NaOH/liter for 12 hr and the remaining alkali washed out and discharged into rivers, where it caused pollution. This problem has now been overcome by the "dry" alkali treatment in which straw is sprayed with a concentrated solution of NaOH. After curing for several days the processed straw, including the residual alkali, can be fed (Wilson and Pigden, 1964; Rexen and Knudsen, 1984). Alkali treatment partly solubilizes the hemicellulose, swells the cellulose (Theander and Aman, 1984), and breaks the bonding in the intercellular layers, allowing access of the rumen microorganisms (Spencer and Akin, 1980; Spencer *et al.*, 1984). The extent of the increase in digestibility depends on the quantity of NaOH used (Jayasuriya and Owen, 1975) and the temperature and time allowed for the reaction (Ololade *et al.*, 1970).

The extent of the increase which has been recorded depends on the method used to estimate digestibility. In many upgrading studies, digestibility is estimated by the *in vitro* technique, but improvements in OMD are much smaller when digestibility is determined *in vivo* (Jayasuriya and Owen, 1975; Van Eenaeme *et al.*, 1981). This is because NaOH increases the rate of passage of forage through the animal and reduces the rate of digestion (Berger *et al.*, 1980).

The DMD *(in vitro)* of *Lotus corniculatus, M. sativa,* and *P. pratense* has been improved by adding 60 g NaOH/kg DM, the largest improvement occurring with mature forage (Pidgen *et al.*, 1966; Thomas, 1978). This effect of forage maturity on the extent of the improvement in digestibility applies to both temperate (Fig. 4.22) and tropical grasses (Spencer

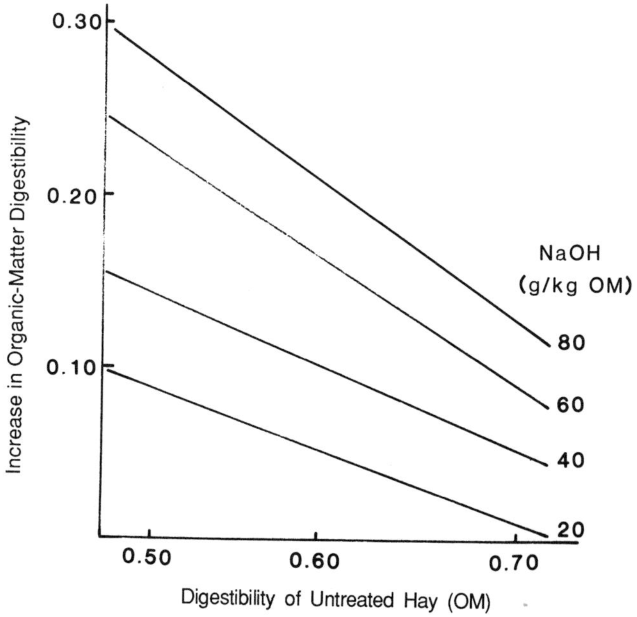

Fig. 4.22. Increase in organic-matter digestibility *(in vitro)* of different-quality temperate grass hays by treatment with sodium hydroxide. Adapted from Mwakatundu and Owen (1974).

and Amos, 1977). Other forages improved by treating with NaOH include *Agrostis* spp./*L. perenne* (Mwakatundu and Owen, 1974), *C. dactylon* (Spencer and Amos, 1977), straw of *L. perenne* (Anderson and Ralston, 1973), stem of *M. sativa,* and stover of *Z. mays* (Ololade *et al.*, 1970). Adding NaOH at a rate of 30 g/kg forage increased the DMD *(in vivo)* of pelleted *C. dactylon* by 0.066 (Utley *et al.*, 1982) but had no effect on the OMD *(in vivo)* of pelleted *M. sativa,* even when NaOH and KOH were added at 60 g/kg (Mathews and McManus, 1976). With mature pasture containing *Themeda* and *Cymbopogon* species, 33 and 67 g/kg NaOH increased DMD *(in vivo)* by 0.10 and 0.13, respectively (Meissner *et al.*, 1973).

2. Ammonia

More than 50 yr ago ammonia was used to upgrade straws in Germany (Sundstol and Coxworth, 1984) and ammonia has now replaced NaOH for treating roughages in Norway (Homb, 1984). Ammonia is applied in aqueous or gaseous form, the latter giving a more even distribution. The increase in OMD *(in vitro)* of *Z. mays* stover depends on the moisture content of the roughage, temperature, and time the reaction is allowed to continue. Optimum conditions are 30–40 g NH_3/kg DM, 600 g H_2O/kg DM, and 90°C for 6–12 hr (Oji *et al.*, 1979).

When anhydrous ammonia was applied to hay at a rate of 30 g/kg DM, without additional heat, the following increases in DMD *(in vivo)* were found: *D. glomerata,* 0.071 (Moore *et al.*, 1983); *F. arundinacea,* 0.185 (Buettner *et al.*, 1982); grass/legume silage, 0.134 (Moore *et al.*, 1986); and mixed temperate hay, 0.142 (Wylie and Steen, 1988). The DMD of *Z. mays* stover was increased 0.066 when treated with anhydrous ammonia at 20 g/kg DM, 0.021 of the improvement being attributed to an increase in CP content of the forage (Saenger *et al.*, 1982). In studies with mature *Panicum virgatum* and *Sorghastrum nutans,* ammonia (30 g/kg DM) and heat (85°C for 23 hr) increased DMD *(in vivo)* by 0.102 and 0.076, respectively (Gates *et al.*, 1987).

Anhydrous ammonia is inconvenient to handle and urea can be used as a source of ammonia, the conversion relying on the hydrolysis of the urea by microbial and plant ureases present in the forage (Hadjipanayiotou, 1982, 1984; Williams *et al.*, 1984). The DMD of *C. dactylon* was increased 0.177 by applying 56 g urea/kg DM at the time of cutting compared with 0.200 for an equivalent quantity of alkali applied as anhydrous ammonia after baling (Craig *et al.*, 1988).

Improvements in DMD with ammonia or urea appear to be caused both by changes in the structure of the forage and by an increase in the nitrogen content. There are, however, reports that some animals consuming treated hay develop neurological symptoms, a subject discussed in Chapter 6.

3. Oxidizing Agents

Sodium chlorite is used as a delignifying agent in the preparation of wood pulp for paper making. In the laboratory, treatment with sodium chlorite reduces lignin concentration and increases digestibility *(in vitro)* of cell walls of both *D. glomerata* and *D. decumbens* (Goering *et al.*, 1973; Ford, 1978). There appears to be almost no limit to the improvement in digestibility that can be achieved with oxidizing agents. If all the lignin is removed, DMD *(in vitro)* of leaf and stem of *D. decumbens* would rise to 0.97 and 0.83, respectively (Ford, 1978). Delignification of *D. glomerata* and *M. sativa* also increases the rate of cellulose digestion (Darcy and Belyea, 1980; Belyea *et al.*, 1983).

E. Physical Treatment

In 1962 it was shown that the DMD *(in vitro)* of straw is increased to 0.95 following exposure to gamma radiation emitted by cobalt-60 (Pritchard *et al.*, 1962). Subsequent work with *L. corniculatus, M. sativa,* and *P. pratense* showed that with low levels of radiation the improvement in DMD *(in vitro)* depended on the maturity of the forage. When the forage is immature, DMD is slightly depressed by the treatment but with mature forage there are large increases in DMD (Fig. 4.23). In all these studies

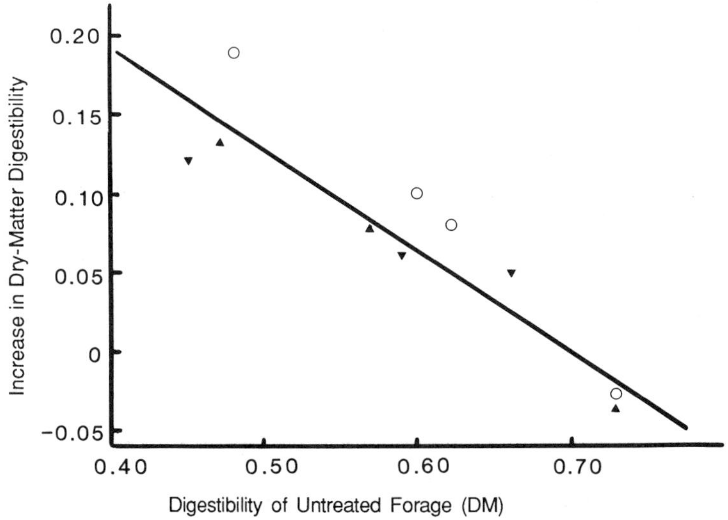

Fig. 4.23. Increase in dry-matter digestibility of different-quality forages by gamma irradiation. (▼) *Phleum pratense;* (O) *Medicago sativa;* (▲) *Lotus corniculatus;* $y = 0.45 - 0.63x$. Adapted from Pigden *et al.* (1966).

TABLE 4.25

Effect of Fractionation of *Medicago sativa* on the Dry-Matter Digestibility of Residual Press Cake[a]

Processing	Dry-matter digestibility	
	Experiment 1	Experiment 2
Unprocessed	0.635	0.645
Once roller pressed	—	0.596
Twice roller pressed	0.552	0.598
Thrice roller pressed	—	0.598
Four times roller pressed	—	0.559

[a]From Raymond and Harris (1957).

DMD was measured *in vitro*, but when irradiated *T. aestivum* straw was fed to sheep, DMD was reduced due to the shorter time the irradiated forage was retained in the rumen (McManus *et al.*, 1972b).

F. Forage Fractionation

Forages contain high levels of fiber and cannot be efficiently used by pigs and poultry. However the protein and soluble carbohydrates can be mechanically separated from the fiber in a form which is suitable for these animals (Tilley and Raymond, 1957; Wilkins, 1976). The level of extraction varies from the production of a low-protein juice as an adjunct to green crop drying to an exhaustive extraction with leaf protein concentrate as the main product (Jones, 1976a).

All processes result in two fractions, one with a low fiber content suitable for nonruminants and the other a higher-fiber "press cake" with a DMD that is lower than that of the original forage (Raymond and Harris, 1957; Greenhalgh and Reid, 1975). In one study the press cake had a higher DMD than the original forage (Walker *et al.*, 1982), but usually there is a depression in DMD which is related to the rigor of the extraction process (Table 4.25).

V. CONCLUSION

The availability of energy in forage is usually measured as dry-matter digestibility (DMD), which is closely related to other energy parameters: organic-matter digestibility, digestible organic matter, total digestible nutrients, digestible energy, and metabolizable energy.

The DMD of forage varies with the proportion of cell contents and cell-wall constituents. The cell contents are very digestible, while cell-wall digestion depends on the degree of lignification, the activity of the rumen microbes, and the time the forage is retained in the rumen. Forages are digested less efficiently by sheep than by cattle and are digested less at high levels of feeding, when ground and pelleted, or when deficient in protein.

In the routine *in vivo* evaluation of forages DMD is usually measured with sheep fed artificially dried chopped forage, offered just below maximum intake to prevent selection. The *in vivo* DMD of forages varies with species, cultivar, stage of growth, soil fertility, climate, processing, and the presence of factors detrimental to microbial activity. Differences in DMD are associated with variations in the histological, chemical, and physical composition of the forage.

Many laboratory methods have been developed for predicting DMD and related parameters from forage analysis for leaf proportion, crude protein, fiber, lignin, near-infrared reflectance, grinding energy, leaf strength, bulk density, and water concentration. Forage DMD can also be estimated by *in vitro* techniques using rumen fluid or cellulase, by the *in sacco* method, and from nitrogen, chromogen, or fiber concentration in the feces. Also used to estimate DMD are naturally occurring indigestible tracers including silica, acid-insoluble ash, lignin, chromogen, potentially indigestible fiber, and *n*-alkanes.

The DMD of forage can be increased by plant breeding and selection and treating with sodium hydroxide, ammonia, oxidizing agents, and gamma radiation. Grazed forage can be improved by sowing higher DMD species and cultivars, by adopting management strategies that keep the pastures at an immature stage of growth, and by allowing animals to select a high-DMD diet by avoiding overstocking.

5

Energy Utilization by Ruminants

I. ENERGY AVAILABLE IN FORAGE

Ruminant production from forage usually depends on voluntary intake (Chapters 2 and 3), concentration of digestible energy (Chapter 4), or a combination of both. However, there are situations where production cannot be accounted for by these factors. In these cases differences in production are caused by variations in the efficiency of utilizing the energy released from the forage during digestion.

Forage digestibility is expressed in terms of digested energy or related functions (Chapter 4). Some of this energy is lost as methane and urine and it is now accepted that the energy value of feeds and the energy requirements of ruminants should be expressed not as digestible energy, but as metabolizable energy (ME), which takes into account the energy lost in the feces, urine, and as methane (Fig. 5.1). Equations for predicting ME from DE and DM were presented in Chapter 4. In this chapter factors affecting the utilization of ME will be reviewed and methods of improving efficiency of utilization considered.

II. EFFICIENCY OF UTILIZATION OF METABOLIZABLE ENERGY

A. Nutrients Absorbed

Nutrients absorbed from the digestive tract include volatile fatty acids (VFA), amino acids, fatty acids, glucose, minerals, and vitamins. These are used in the synthesis of the many different compounds found in meat, milk, and wool, and to replace nutrients used for maintaining life processes including reproduction. If one of these nutrients is deficient and cannot be synthesized by the animal, then production will be limited by the deficient nutrient and the ME in the forage will be inefficiently utilized by the animal.

Acetic, propionic, and butyric acid are the main end products of anaer-

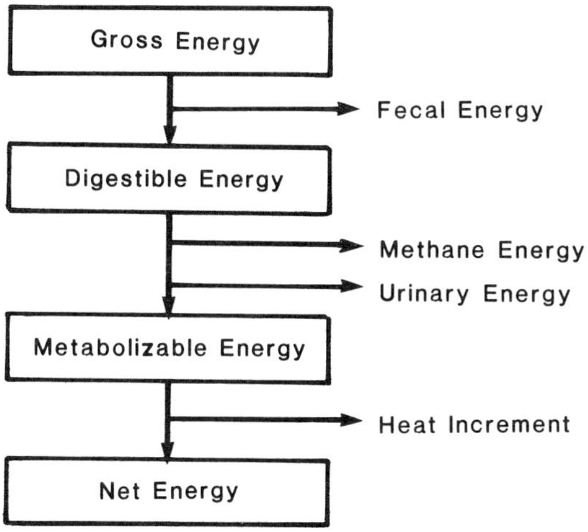

Fig. 5.1. Relationship between gross energy, digestible energy, metabolizable energy, and net energy.

obic fermentation of forage in the rumen and the form in which most of the energy in forage is absorbed by ruminants (Table 5.1). In studies with different proportions of the three acids, there is a fall in efficiency of utilization of ME for fattening (k_f) as the proportion of propionic acid decreases (Blaxter, 1962; Hovell *et al.*, 1976; Hovell and Greenhalgh, 1978). Acetic and butyric acids can only be used for fattening with high efficiency if there is an adequate supply of propionate or glucose (Annison and Armstrong, 1970). The amino acids absorbed from the small intestine can be an important source of glucose and when large quantities of casein are fed via an abomasal catheter different mixtures of VFA are utilized with the same efficiency (Orskov *et al.*, 1979). It now appears that the efficiency of utilization of the ME in forage will only be related to VFA proportions when insufficient protein or glucose is absorbed from the forage (MacRae and Lobley, 1982).

The efficiency of utilization of ME is usually quoted as a coefficient *k*, with a suffix m, l, or f depending on whether *k* refers to the use of the feed for maintenance, lactation, or fattening/growth (ARC, 1980). For most feeds $k_m > k_l > k_f$, with values depending on the digestibility of the diet. Studies published by ARC (1980) showed that as the proportion of the gross energy that is metabolizable rose from 0.4 to 0.7, k_m increased from 0.64 to 0.75 and k_l increased from 0.56 to 0.66 (ARC, 1980). With forages of similar digestibility there is no difference in k_m (Armstrong,

TABLE 5.1

Molar Proportions of Volatile Fatty Acids Produced from Different Forages in the Rumen

Species	Molar porportions			Reference
	Acetic	Propionic	Butyric	
Chloris gayana	0.75	0.16	0.09	Holmes *et al.* (1966)
Dactylis glomerata	0.65	0.23	0.12	Bath and Rook (1965), Milford and Minson (1965b, 1966), Tilley *et al.* (1960)
Heteropogon contortus	0.79	0.15	0.06	Playne and Kennedy (1976)
Lolium perenne	0.61	0.25	0.14	Bath and Rook (1965), Milford and Minson (1965b, 1966), Rook (1964), Tilley *et al.* (1960)
Lolium multiflorum	0.60	0.25	0.15	Bath and Rook (1965)
Medicago sativa	0.64	0.22	0.14	Bath and Rook (1965)
Setaria sphacelata	0.73	0.19	0.08	Holmes *et al.* (1966)
Stylosanthes humilis	0.75	0.17	0.08	Playne and Kennedy (1976)
Trifolium pratense	0.65	0.21	0.14	Bath and Rook (1965), Rook (1964)
Trifolium repens	0.65	0.22	0.15	Tilley *et al.* (1960)
Veldt	0.71	0.19	0.10	Topps *et al.* (1965)

1964; Corbett *et al.*, 1966; Tudor and Minson, 1982). No k values appear to have been published for wool production. Most studies of ME utilization from different forages have concentrated on the determination of k_f.

B. Forage Species

1. Legume versus Grass

Lambs fed *Trifolium repens* grew faster than when fed *Lolium perenne* (McLean *et al.*, 1962; Joyce and Newth, 1967; Ulyatt, 1969). This difference was associated with a higher voluntary intake of the legume, but it is possible that the difference was also associated with a higher k_f of the ME absorbed from the legume. When fed the same quantities of forage ME, lambs receiving the *T. repens* grew 28% faster and produced 38% more wool than when given *L. perenne* (Joyce and Newth, 1967; Rattray and Joyce, 1969). The ME was used for growth with an efficiency of 0.33 and 0.51 for *L. perenne* and *T. repens*, respectively (Rattray and Joyce, 1974), a difference associated with higher crude protein in the legume (Rattray and Joyce, 1974) and higher proportion of propionic acid in the rumen (Ulyatt, 1969). The proportion of organic matter digested in the rumen was similar for the legume and the grass (Ulyatt, 1969).

Studies with mature sheep and cattle have shown little or no difference in k_f between legumes and grasses. When *L. perenne* and *Trifolium resupinatum* were fed to mature sheep both forages had a k_f value of 0.41 (Margan *et al.*, 1989). With cattle fed *L. perenne* and *T. repens* the mean values for k_f were 0.42 and 0.46, respectively (Cammell *et al.*, 1986).

2. Differences between Genera

Lambs grazing *L. perenne* retained more energy and had better-quality carcasses than when grazing *Dactylis glomerata* (Milford and Minson, 1966). This difference was caused by a lower intake (Chapter 2), lower dry-matter digestibility of the *D. glomerata* (Chapter 4), and lower efficiency of utilization of the digested nutrient. Lambs fed equal quantities of digestible energy of the two species grew faster and retained 21% more energy on the *L. perenne* diet (Milford and Minson, 1965b). However, in respiration calorimetry studies there was no difference in efficiency of utilization; k_f was 0.51 and 0.54 for *L. perenne* and *D. glomerata,* respectively (Armstrong, 1964), possibly due to the use of mature sheep in the determination of k_f.

Differences in k_f have also been found between species of tropical grasses. When *Digitaria decumbens* and *Setaria sphacelata* were fed to young cattle in slaughter studies k_f values were 0.28 and 0.17, respectively (Tudor and Minson, 1982). However when separated leaf from these two grasses was fed to mature sheep in respiration calorimeters, the k_f values for both forages were the same (Margan *et al.*, 1989).

C. Stage of Growth

As forages mature there is a decrease in k_f (Breirem, 1944; Blaxter, 1962; Armstrong, 1964). This depression in k_f is associated with a decline in crude protein (Table 5.2) and an increase in the proportion of indigestible residues that require energy for their mastication and propulsion through the alimentary tract. The relation between k_f and metabolizable energy for the forages listed in Table 5.2 is shown in Fig. 5.2. Of the total variation in k_f, 37% can be accounted for by difference in ME, 13% by variations in crude protein, and none by differences in the proportions of volatile fatty acids. This lack of effect of VFA proportions is associated with the small range in propionic acid proportion in these studies (0.17 to 0.24). The maximum k_f recorded for a forage was 0.541. This was slightly lower than the k_f of 0.588 found when feeding grain of *Zea mays* (Blaxter

TABLE 5.2

The Efficiency of Utilization for Growth (k_f) of Metabolizable Energy of Forages

Species	Metabolizability of energy	k_f	Protein (g/kg DM)	Propionic acid in VFA	Reference
Dactylis glomerata	0.65	0.541	226	0.186	Armstrong (1964)
	0.60	0.539	175	0.203	
	0.56	0.461	140	0.194	
Digitaria decumbens	0.46	0.410	100	—	Margan *et al.* (1989)
	0.44	0.277	119	—	Tudor and Minson (1982)
Lolium perenne	0.66	0.525	185	0.227	Armstrong (1964)
	0.61	0.535	152	0.219	
	0.62	0.467	138	0.222	
	0.52	0.336	96	0.208	
	0.71	0.518	177	0.238	
	0.62	0.502	155	0.229	
Lolium perenne					
First harvest	0.62	0.450	121	—	Blaxter *et al.* (1971)
Third harvest	0.57	0.340	184	—	Blaxter *et al.* (1966)
Lolium perenne	0.64	0.420	173	—	Cammell *et al.* (1986)
	0.47	0.410	150	—	Margan *et al.* (1989)
	0.60	0.329	205	—	Rattray and Joyce (1974)
Medicago sativa	0.47	0.284	169	—	Thomson and Cammell (1979)
Whole	0.48	0.435	170	—	Margan *et al.* (1985)
Stem	0.40	0.305	100	—	
Mixed temperates					
Summer	0.52	0.430	133	0.241	Corbett *et al.* (1966)
Autumn	0.49	0.320	161	0.171	Corbett *et al.* (1966)
Mixed temperates					
First harvest	0.66	0.540	208	—	MacRae *et al.* (1985
Third harvest	0.55	0.430	232	—	MacRae *et al.* (1985
Phleum pratense	0.59	0.476	113	0.207	Armstrong (1964)
	0.55	0.433	89	0.178	
	0.46	0.431	70	0.180	
Setaria sphacelata	0.46	0.410	100	—	Margan *et al.* (1989)
	0.44	0.169	144	—	Tudor and Minson (1982)
Temperate grass	0.63	0.262	138	—	Blaxter and Graham (1956)
	0.62	0.242	84	—	Wainman and Blaxter (1972)
	0.55	0.400	101	—	Wainman *et al.* (1972)
Trifolium repens	0.63	0.510	259	—	Rattray and Joyce (1969)
	0.61	0.460	274	—	Cammell *et al.* (198
Trifolium resupinatum	0.54	0.410	190	—	Margan *et al.* (1989
Trifolium subterraneum	0.62	0.490	268	—	Graham (1969)
Mean	0.56	0.42	157	0.207	

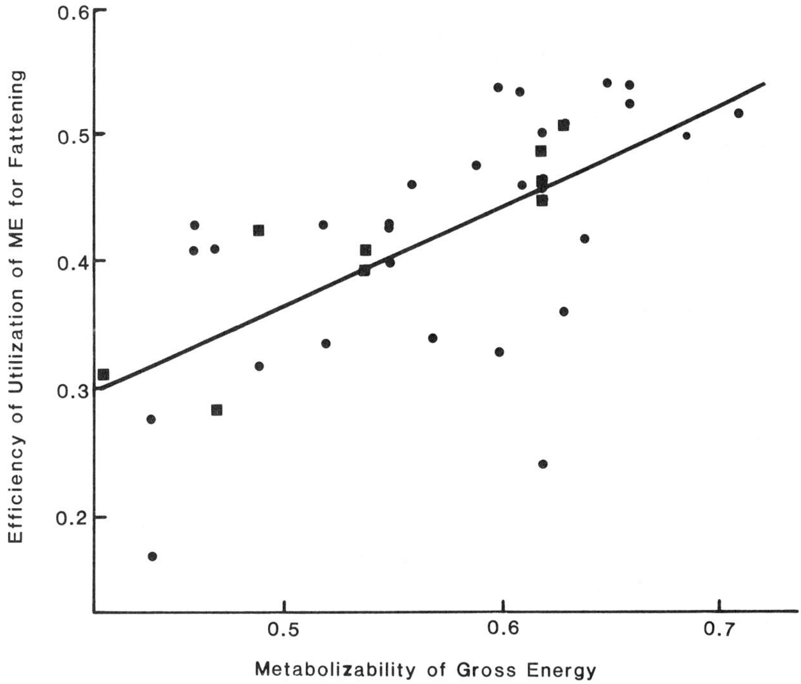

Fig. 5.2. Effect of metabolizable energy concentration in forages on the efficiency of utilization of metabolizable energy for fattening. (●) Grass; (■) legume; $y = 0.766$ ME $- 0.015$; $r = 0.61$. Adapted from data in Table 5.3.

and Wainman, 1964). When compared at the same level of ME there was no difference in k_f between legumes and grasses or between temperate and tropical grasses.

D. Season of the Year

In temperate regions graziers have often observed reduced performance of ruminants grazing autumn forage, particularly in lambs after weaning (Clarke, 1959). This lower production is partly due to the lower intake of metabolizable energy, but the k_f is also low for autumn-grown forage (Table 5.2). This depression in k_f is associated with lower soluble carbohydrate concentration and higher crude protein in autumn grasses (Corbett *et al.*, 1966; Blaxter *et al.*, 1971; Beever *et al.*, 1978; MacRae *et al.*, 1985). Autumn forage produces a low proportion of propionic acid in the rumen due to low soluble carbohydrate levels (Tilley *et al.*, 1960), so the low k_f may be a direct result of the shortage of propionic acid or glucose. In a study by MacRae *et al.* (1985), casein was infused into the abomasum to avoid deamination in the rumen. This increased the k_f of autumn grass from 0.45 to 0.57, the same value as for spring forage. Although autumn grass contains higher levels of crude protein than spring grass, it is extensively deaminated in the rumen and less protein is available for absorption as amino acids from the intestines (see Chapter 6). MacRae *et al.* (1985) have also shown that protein in autumn grass is less well absorbed from the small intestine (0.27) than protein in spring grass (0.75). No studies have been conducted in tropical regions to investigate whether there are similar differences in k_f between forage grown in the wet and dry seasons.

E. Effect of Drying

Studies by Ekern *et al.* (1965) in which temperate grass was dried at 102°C showed that drying increased the proportion of forage energy lost in the feces (0.024) and urine (0.005) but reduced methane production (0.002). Metabolizable energy as a proportion of forage energy was therefore depressed by 0.027. However, this depression was completely offset by the higher k_f of the dried grass (0.551 versus 0.504), so the dried grass had a slightly higher net energy value than fresh forage.

F. Pelleting

Grinding and pelleting reduce the time forages are retained in the rumen and cause a reduction in digestible and metabolizable energy (Chapter 4) (Table 5.3). This loss of energy is more than offset by an improvement in k_f (Table 5.3) and the net energy, growth rates, and wool production from pelleted forage are often higher than those from chopped material (Thomson and Cammell, 1979). The magnitude of the improvement is small with immature forage where the initial k_f is high, but large increases in k_f are found when mature forage is pelleted (Osbourn *et al.*, 1976; ARC, 1980). The improvement in k_f following pelleting is associated with (1) changes in the distribution of species of bacteria in the rumen (Thorley *et al.*, 1968), (2) reduced production of VFA within the rumen (Black and Tribe, 1973), (3) an increase in postruminal digestion of energy (Black and Tribe, 1973), (4) reduced deamination of dietary protein and production of microbial protein (Coelho da Silva *et al.*, 1972b; Osbourn *et al.*, 1976; Faichney and Teleki, 1988) (Chapter 6), and (5) a reduction in the energy cost of eating and ruminating (Osuji *et al.*, 1975).

TABLE 5.3

Effect of Pelleting on Efficiency of Utilization of Metabolizable Energy of Forages

Species	Metabolizability of gross energy	k_f	Protein (g/kg)	Propionic acid in VFA	Reference
Medicago sativa					
Chopped	0.47	0.284	169	0.25	Thomson and Cammell (1979)
Pelleted	0.42	0.533	188	0.24	
Chopped	10.79[a]	0.324	—	—	Smith *et al.* (1976)
Pelleted	10.52[a]	0.458	—	—	
Chopped	0.46	0.519	—	—	Forbes *et al.* (1925)
Ground	0.46	0.542	—	—	
Temperate forage					
Chopped	0.62	0.242	84	—	Wainman and Blaxter (1972)
Pelleted	0.53	0.541	112	—	
Temperate grass					
Chopped	0.63	0.362	138	—	Blaxter and Graham (1956)
Pelleted	0.59	0.435	140	—	
Chopped	0.55	0.400	101	—	Wainman *et al.* (1972)
Pelleted	0.52	0.519	110	—	
Mean					
Chopped	0.55	0.355	123		
Pelleted	0.50	0.505	138		

[a]Metabolizable energy (mJ/kg DM).

III. IMPROVING THE EFFICIENCY OF ENERGY UTILIZATION

In the previous sections it was shown that any process which increases the quantity of propionic acid or amino acid absorbed may raise the efficiency of the utilization of ME. Absorption of these glucose precursors can be increased by changing the pattern of rumen fermentation in three ways: feeding an ionophore, feeding grain, or increasing the level of soluble carbohydrates in the forage by plant breeding. Efficiency can also be improved by feeding supplementary protein, a method considered in Chapter 6.

A. Use of Ionophores

The ionophore antibiotics monensin and lasalocid increase the proportions of propionic acid in the rumen and reduce the loss of methane (Potter *et al.*, 1974; Rowe, 1985; Demeyer *et al.*, 1986); they also increase the level of plasma glucose (Potter *et al.*, 1976). Dietary protein degradation in the rumen is also reduced and there is an increase in the protein flowing to the small intestine (Poos *et al.*, 1979; Rowe, 1983). This reduced protein degradation and ammonia concentration in the rumen is an advantage with many forage diets, but if the crude protein concentration in forage is low then feeding an ionophore may fail to improve production (Jacques *et al.*, 1987).

Steers fed ground *Medicago sativa* containing 30 ppm monensin grew 25% faster (1.15 versus 0.92 kg/day) than animals eating the same quantity of legume containing no monensin (Dinius *et al.*, 1978). This increase was associated with a rise in the proportion of propionic acid in the rumen VFA from 0.188 to 0.224 and a decrease in the rumen ammonia concentration from 68 to 34 mg/liter. With cattle grazing temperate and tropical forages, feeding monensin each day increased rumen propionic acid and growth rate (Tables 5.4 and 5.5). A daily dose of 200 mg monensin appears to achieve the optimum response and this level was used in a study of cattle grazing 22 temperate pastures and 11 tropical forages in different parts of the U.S. (Potter *et al.*, 1986). The unsupplemented animals gained, on average, 0.58 and 0.57 kg/day on the temperate and tropical pastures, respectively, and monensin increased gain by 0.09 and 0.10 kg/day, respectively.

Feeding monensin in quantities greater than 200 mg/day depressed growth (Table 5.5) due to a lower forage intake. In experimental work ionophores are fed each day to grazing ruminants together with a grain supplement, but for commercial use a safe slow-release device that can be placed in the rumen is required. An ionophore has now been developed

TABLE 5.4

Effect of Feeding Monensin on the Molar Proportions of Volatile Fatty Acids in the Rumen of Cattle Grazing Temperate Forage[a]

Monensin (mg/day)	Molar proportion		
	Acetic	Propionic	Butyric
0	0.70	0.19	0.10
50	0.69	0.21	0.09
100	0.68	0.22	0.09
200	0.67	0.24	0.08
300	0.65	0.25	0.08
400	0.64	0.27	0.08

[a]Adapted from Potter *et al.* (1976).

that has little adverse effect on food intake when fed in excess (Rowe, 1983).

B. Grain Supplementation

Feeding grain of *Z. mays* increases the k_f of hay-based diets in direct proportion to the level of grain in the diet, with little difference in response between sheep and cattle (Fig. 5.3). This improvement in efficiency is atttributed to a fall in the proportion of acetic acid in the rumen VFA, reduced heat of fermentation, and a smaller quantity of energy used to chew the supplemented hay (Blaxter and Wainman, 1964). It is

TABLE 5.5

Effect of Level of Monensin on the Growthof Cattle Grazing Temperate and Tropical Forage (kg/day)

Species	Monensin (mg/day)						Reference
	0	50	100	200	300	400	
Cynodon dactylon	0.56	0.73	0.78	0.71	—	—	Oliver (1975)
Temperate pasture	0.54	0.55	0.60	0.63	0.60	0.58	Potter *et al.* (1976)
	0.79	—	—	0.89	—	—	Wilkinson *et al.* (1980)
	0.66	—	—	0.72	—	—	Wagner *et al.* (1984)
Mean	0.64			0.74			

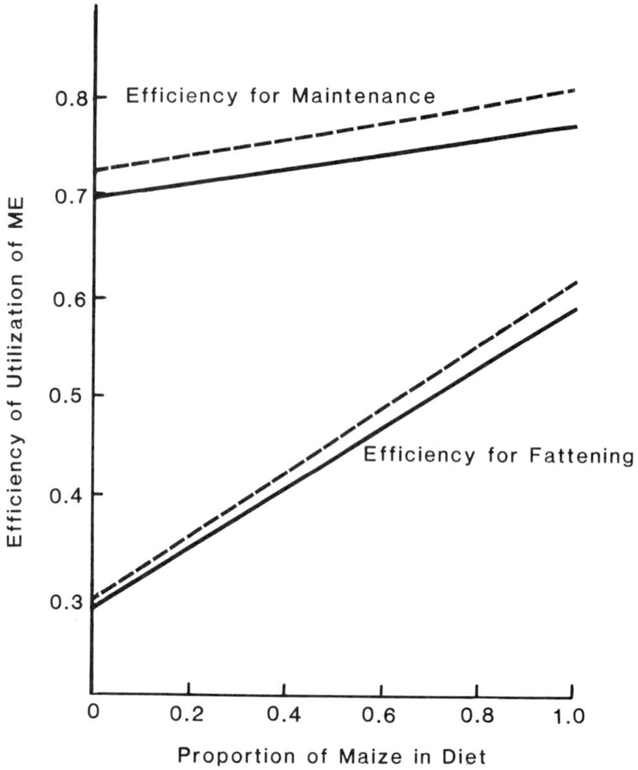

Fig. 5.3. The effect of *Zea mays* grain in a hay diet on the efficiency with which metabolizable energy is used for maintenance and for fattening in ruminants (— — —) cattle; (———) sheep. Adapted from Blaxter and Wainman (1964).

probable that the quantity of amino acid absorbed from the large intestine was also improved but this was not measured.

C. Plant Breeding

Raising the level of soluble carbohydrate in forages should increase the proportion of propionic acid in the rumen, reduce methane loss, and increase the quantity of protein leaving the rumen. Large varietal differences in soluble carbohydrates occur in *D. glomerata, L. perenne* (Cooper, 1962), and *L. multiflorum* (Bugge, 1978), so selection and breeding for higher soluble carbohydrates should be possible. The concentration of soluble carbohydrates in forage is generally negatively correlated with the protein concentration (Hacker, 1982), but there is sufficient genotypic independence to allow breeders to simultaneously select for both high soluble carbohydrates and high protein (Vose and Breese, 1964).

IV. CONCLUSION

Differences in ruminant production from forages can usually be accounted for by variation in voluntary intake, energy digestibility, or both factors combined. In some comparisons these two parameters fail to account for all the differences in production and it is necessary to consider efficiency of utilization of the digested or metabolizable energy for fattening or growth (k_f). For a range of forage species, cultivars cut at different stages of growth and at various times of the year had a mean k_f of 0.42, varying from 0.17 to 0.54. Differences in ME concentration accounted for 37% of this variation, and crude protein concentration in the forage accounted for 13%. The proportion of propionic acid in the rumen volatile fatty acids (0.17 to 0.24) had no effect on k_f.

Pelleting forage increased mean k_f from 0.36 to 0.50. The largest improvement occurred with forages with a low k_f. Feeding ionophores increases growth rate of cattle grazing temperate and tropical forage by 0.09 and 0.10 kg/day, respectively. Feeding grain supplements increases k_f. Raising the soluble carbohydrate concentration in forages by breeding and selection may improve k_f.

6

Protein

I. PROTEIN IN RUMINANT NUTRITION

A. Function of Protein

Protein is necessary for the production of milk, muscle, wool, and hair and to replace the protein unavoidably lost during the maintenance of body weight. Protein contains 22 different amino acids and ruminants may be unable to synthesize some of these amino acids at a sufficient rate to meet their optimum requirements. These are described as indispensable or essential amino acids and must be absorbed from the digesta in the small intestine (McDonald *et al.*, 1988). The proportions of essential amino acids present in milk, muscle, and wool proteins are shown in Table 6.1, together with the composition of the protein available for absorption from the small intestine of sheep fed forage. There are large differences between the three tissues in their relative requirements for the various indispensable amino acids; the synthesis of muscle protein requires almost twice the quantity of arginine as does milk protein. The synthesis of wool protein requires very large quantities of the sulfur amino acids methionine, cysteine, and cystine.

In monogastric animals the quantity of indispensable amino acids available for absorption from the small intestines can be readily determined by analyzing the feed. This simple approach cannot be applied to ruminants due to the presence of rumen microbes which modify the quantity and proportions of amino acids available for absorption (Fig. 6.1).

Three major changes occur to CP in the rumen:

1. Degradation of CP to ammonia. If the ammonia produced is present in high concentrations it is absorbed, leading to a net loss of CP.
2. Microbial protein is synthesized from nonprotein nitrogen and sulfur present in the rumen. This can lead to a net gain of CP, with more CP entering the duodenum than is eaten.
3. Microbes synthesize protein which has a different amino acid profile from that of the degraded dietary protein (Weston and Hogan, 1971; Beever *et al.*, 1981).

TABLE 6.1

Molar Proportion of Amino Acids in Milk, Muscle, and Wool Protein Compared with That in Digesta Protein Leaving the Rumen of Sheep Fed *Phalaris aquatica*[a]

Amino acid	Milk	Muscle	Wool	Digesta
Indispensable[b]				
Arginine	0.042	0.077	0.100	0.047
Histidine	0.026	0.033	0.010	0.018
Isoleucine	0.075	0.060	0.045	0.049
Leucine	0.011	0.080	0.080	0.079
Lysine	0.087	0.010	0.030	0.056
Methionine, cysteine	0.042	0.044	0.160	0.027
Phenylalanine	0.055	0.050	0.030	0.052
Threonine	0.047	0.050	0.077	0.042
Tryptophan	0.015	0.014	0.012	0.015
Valine	0.070	0.055	0.057	0.059
Dispensible	0.569	0.563	0.601	0.444

[a]Adapted from Hogan (1974).
[b]These amino acids are indispensable for the pig and chicken. It is not known how many may be indispensable for the ruminant.

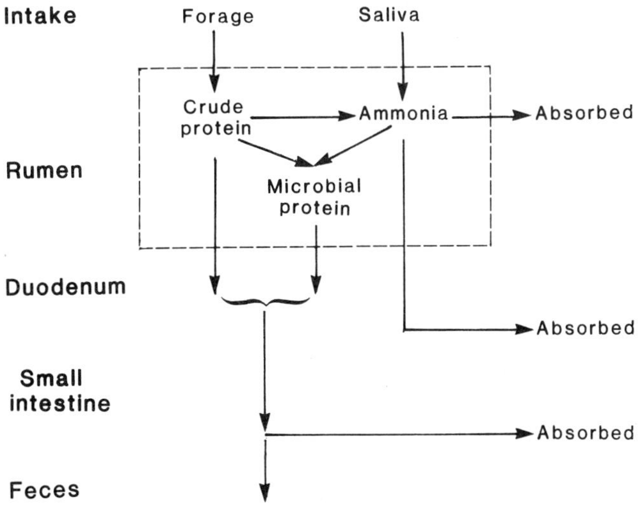

Fig. 6.1. Simplified flow diagram of crude protein in ruminants.

The value of forage as a *potential* source of amino acids can be determined by analyzing for total nitrogen (N) and sulfur (S). Forages contain N in many forms other than amino acids (Ferguson and Terry, 1954; Mangan, 1982; McDonald, 1982) and these may be converted to amino acids by the rumen microbes. The term crude protein (CP) is used to describe all forms of N present in the plant. The amino acids in plants and animals usually contain, on average, 160 g N/kg DM, and hence the CP content is calculated as 6.25 N.

Sulfur (S) amino acids are an essential but small component of all protein. The ratio of S to N in animal products varies between 0.055 and 0.068 (ARC, 1980), so the requirements for N and S are closely correlated. The two elements will be considered together in this chapter.

B. Effect on Production

Grazing and pen-feeding studies have shown that ruminant production from forages may be limited by a deficiency of protein and that production can be improved by providing supplementary protein. These studies will be briefly summarized according to the form of production.

1. Milk

Milk production of cows fed forage is improved by increasing the quantity of amino acids absorbed from the small intestine (Table 6.2). The mean increase in milk production is 1.1 kg/day, equivalent to a 10% improvement. The response is similar for fresh grass and silage, for different concentrations of CP in the diet, and is independent of the level of milk production. The rise in milk production is accompanied by a small but consistent rise in milk protein concentration and a 15% increase in the total quantity of milk protein secreted (Table 6.2). In the case of silage it has been demonstrated that the amount of protein entering the small intestine limits milk production and that methionine is a major amino acid limiting the synthesis of milk and its constituents (Rogers *et al.*, 1979). Increasing the supply of energy by infusing glucose into the abomasum of cows fed silage has no effect on milk production or composition (Rogers *et al.*, 1979).

This deficiency of indispensable amino acids in forage is not limited to lactating cows. Ewes fed a mixed pasture or *Lolium perenne*, containing 219 and 162 g CP/kg DM, respectively, produced more milk when casein was infused into the abomasum or a fish-meal supplement was fed (Barry, 1980; Penning and Treacher, 1981). Perhaps the most spectacular protein response reported is a 68% increase in milk production by ewes from

TABLE 6.2

Effect of a Casein Supplement Protected from Rumen Degradation with Formaldehyde or Infused into the Abomasum on Milk Production and Composition

Species	Forage CP (g/kg)	Milk production (kg/day)		Milk protein (g/kg)		Reference
		Control	Casein	Control	Casein	
Chloris gayana	197	12.3	14.7[a]	33.0	35.4	Stobbs *et al.* (1977)
(grazed)	145	9.6	10.1[a]	37.0	38.0	Flores *et al.* (1979)
Lolium perenne	—	14.9	16.6[a]	32.5	33.3	Wilson (1970)
(grazed)	—	13.4	13.1[a]	40.0	41.5	Wilson (1970)
Lolium perenne (grazed)	275	15.1	15.9[a]	36.0	38.0	Minson (1981)
Temperate pasture (fresh indoors)	175	16.1	18.1	31.3	32.3	Rogers *et al.* (1980)
	—	10.9	11.2	29.2	30.0	Rogers and McLeay (1977), Rogers *et al.* (1979)
Temperate silage	—	7.5	8.6		+1.5	Rogers and McLeay (1977)
		9.9	11.5	26.1	27.9	Rogers and McLeay (1977), Rogers *et al.* (1979)
Temperate silage	—	8.8	10.2	28.8	30.2	Rogers *et al.* (1979)
	204	6.7	7.9	27.7	29.0	Rogers and McLeay (1977), Rogers *et al.* (1979)
		3.6	4.2	33.4	34.6	Rogers *et al.* (1979)
Mean		10.7	11.8	32.3	33.7	

[a]Casein treated with formaldehyde.

supplementation with urea of a mature *Iseilema* spp. hay containing 56 g CP/kg DM (Stephenson *et al.*, 1981). This increase in milk production was probably due to an improvement in total quantity of CP entering the small intestine.

2. GROWTH

When fed a *Heteropogon contortus* hay containing 25 g CP/kg DM, pregnant cows lost 0.8 kg/day. By feeding a mixture of urea and sulfur, VI of the hay was increased by 48% and loss in weight was reduced to 0.3 kg/day (Lindsay *et al.*, 1982). In a study with beef cows fed immature *Bromus inermis*, a CP concentration of 134 g/kg DM was insufficient to achieve maximum growth; feeding a low-degradability protein supplement (0.11 kg/day) increased the rate of gain from 0.91 to 1.06 kg/day (Anderson *et al.*, 1988).

Growth rate and carcass composition of lambs eating a temperate pasture containing 182 g CP/kg DM was dramatically improved by abomasal infusion of casein (Barry, 1981). Liveweight was increased from 79 to 99 g/day and empty-body gain from 62 to 82 g/day. Carcass and whole-body protein content were increased 10 g/kg by casein infusion and fat content was reduced approximately 25 g/kg. Of the total energy deposition, 27% and 41% was in the form of protein in control and protein-infused lambs, respectively. This shows that forage with a relatively high CP content is unable to meet the requirements of young lambs (up to 25 kg liveweight) and that the protein : fat ratio in these animals can be manipulated by changes in protein absorption.

3. WOOL

Wool production by sheep is related to the quantity of protein entering the small intestine (Black *et al.*, 1973). Feeding protein protected from microbial degradation in the rumen by treatment with formaldehyde increases wool growth (Ferguson *et al.*, 1967; Reis, 1969). Wool production was increased 28 and 61% with sheep grazing a *Phalaris aquatica*/*Trifolium repens* pasture when they were supplemented with unprotected and protected casein, respectively (Langlands, 1971).

C. Digestion of Protein

The simplest measure of the availability of protein to animals is the proportion of CP in the feed not appearing in the feces; that is the digestible CP (DCP) content of the diet. Digestible CP is only a relative guide to the quantity of CP available to the ruminant since no allowance is made for changes occurring in the rumen.

In 1954 Dijkstra showed that, for fresh forage, there was a simple linear relation between DCP (g/kg DM) and CP (g/kg DM), with only a small difference between spring and autumn grass:

$$\text{Spring grass DCP} = 0.954\text{CP} - 35.5$$
$$\text{Autumn grass DCP} = 0.954\text{CP} - 41.0$$

Subsequent studies with a wide range of forages grown in different countries have confirmed this general relation and shown that DCP can be predicted from the CP concentration in forage with an error of approximately ± 10 g/kg DM (Minson, 1982). This relation between DCP and CP is not affected by differences in fiber content of the forage (Elliott and Fokkema, 1960). However, leukoanthocyanins in *Onobrychis viciifolia* depressed DCP by 25 g/kg DM (Thomson *et al.*, 1971), and tannin in *Lotus pedunculatus* reduced DCP by 23–38 g/kg DM compared with that predicted from Dijkstra's regression (Barry and Manley, 1984). Tannins are also present in the leaves of many browse trees and shrubs and the DCP in these plants is much lower than in grasses with a similar level of CP (Robbins *et al.*, 1987).

Artificial drying also reduces the DCP of forage (Dijkstra, 1954), the extent of the depression depending on the temperature. Using a conveyor drier with an inlet temperature of 145°C, the DCP of *Trifolium pratense* was depressed 22 g/kg DM below that of the fresh forage (Beever and Thomson, 1981). Natural heating of silage will also depress DCP. With *Medicago sativa,* silage made without heating or allowed to heat to 35 or 60°C, DCP was 115, 77, and 73 g/kg DM, respectively (Weiss *et al.*, 1986b). Formaldehyde is used as a silage additive and when applied to *L. perenne* during silage making reduced DCP by 17 g/kg DM (Thomson *et al.*, 1981).

The parameter DCP fails to take into account changes in the quantity and quality of protein that occur in the rumen, and nutrient requirements of ruminants are now expressed in terms of CP, with or without a measure of the rumen degradability of the CP (NRC, 1978, 1984; ARC, 1980, 1984).

D. Conversion within the Rumen and Absorption

Protein metabolism in the rumen has been studied extensively and the earlier work has been reviewed by Hungate (1966), Smith (1969), Allison (1970), and Hobson (1988). More recent work has quantified the degradation of CP within the rumen, the extent of microbial CP synthesis, and the associated changes in the quantity of amino acids absorbed from the small intestine. These aspects will be considered in the next three sections.

1. Degradation of Crude Protein

The protein in forage can be degraded to amino acids by the mixed microbial population within the rumen (Allison, 1970). This hydrolysis is not important provided all the amino acids are used for the synthesis of microbial protein and any excess is passed out from the rumen and absorbed from the small intestine. However these amino acids together with soluble forms of organic and inorganic N in forage and saliva can be converted to ammonia by rumen microbes. This ammonia may be used to produce microbial protein within the rumen but there is a limit to the quantity of ammonia that can be used in this way. Excess ammonia is absorbed from the rumen and intestine and converted into urea in the liver. Some urea will be recycled to the rumen via saliva and the remainder will be excreted in the urine.

Pathways involved in the conversion of CP in the rumen are shown as a simplified diagram (Fig. 6.1). More detailed information, interaction with other components of the diet, and the numerical values for the different pathways have been published elsewhere (Nolan and Leng, 1972; Black *et al.*, 1982; Murphy *et al.*, 1986). Methods used to measure the flow of undegraded forage CP and microbial CP from the rumen have been described (Hogan and Weston, 1970; Hogan, 1981).

The extent of degradation of CP has now been measured *in vivo* for a wide range of forages (Table 6.3). Most of the CP in fresh forage and silage is degraded in the rumen, with only 25% (on average) CP passing unchanged into the small intestine. The extent of CP degradation varies but there is insufficient information available to decide whether this variation is caused by true differences between forage species and stage of growth or to experimental error and the use of different experimental techniques.

The mean rumen degradation of CP in hays and artificially dried forages is about 0.65 (Table 6.3). This is lower than for fresh forage, possibly due to the formation during drying of complexes between CP and soluble carbohydrates. Resistant CP complexes can also be formed with tannin. In fresh *Lotus corniculatus* the presence of tannin reduces the CP degradation to less than 0.50 (Waghorn *et al.*, 1987). The leukoanthocyanins in *O. viciifolia* also reduce the degradation of CP by rumen microbes (Thomson *et al.*, 1971).

Cell walls of forage are ruptured during eating, exposing the CP to microbial degradation. Over half the CP in immature *T. pratense* is released when chewed by cattle (Reid *et al.*, 1962), but the nature of the CP released varies with forage species (Mangan *et al.*, 1976). Most of the soluble N released from *L. perenne*, *M. sativa*, and *T. pratense* is protein,

TABLE 6.3

Effect of Increasing Levels of Crude Protein in Various Forages on the Quantity of Forage Crude Protein Undegraded in the Rumen, Converted into Microbial Crude Protein, and Gained or Lost as Ammonia[a]

Forage	Dietary CP (g/kg DM)	Nonammonia CP Leaving rumen (g/kg DM)			Lost as ammonia (g CP/kg DM)	Efficiency of microbial protein synthesis (g CP/kg ARDOM)	Reference
		Ungraded	Microbial	Total			
Fresh forage							
Blue grama rangeland	64[b]	45	34	79	−15	129	Funk *et al.* (1987)
Calluna vulgaris (frozen)	66	—	—	110	−44	—	MacRae *et al.* (1979)
Blue grama rangeland	73[b]	54	44	98	−25	141	Funk *et al.* (1987)
Calluna vulgaris (frozen)	80	50	71	121	−41	213	Mayes and Lamb (1982)
Agrostis/Festuca spp. (frozen)	89	—	—	140	−51	—	MacRae *et al.* (1979)
Blue grama rangeland	100[b]	55	45	100	0	98	Funk *et al.* (1987)
Blue grama rangeland	104[b]	53	43	96	8	129	Funk *et al.* (1987)
Temperate pasture	136[b]	24	117	141	−5	308	Corbett *et al.* (1982)
Temperate pasture	159[b]	49	74	123	36	171	Corbett *et al.* (1982)
Medicago sativa	192[b]	4	162	166	26	306	Corbett *et al.* (1982)
Bromus catharticus	202[b]	—	—	195	7	—	Cruickshank *et al.* (1985)
Bromus catharticus	202	—	—	195	7	—	Cruickshank *et al.* (1985)
Medicago sativa	212[b]	17	115	132	80	210	Corbett *et al.* (1982)
Phalaris aquatica	212[b]	21	112	133	79	241	Corbett *et al.* (1982)
Phalaris aquatica	226[b]	16	123	139	87	244	Corbett *et al.* (1982)
Lolium multiflorum × *L. perenne*	238	65	115	180	58	277	MacRae and Ulyatt (1974), Ulyatt *et al.* (1975)
Lolium perenne	242[b]	—	—	188	54	—	Cruickshank *et al.* (1985)
Trifolium repens	244	68	103	171	73	205	MacRae and Ulyatt (1974), Ulyatt *et al.* (1975)
Trifolium repens	254	44	119	163	91	274	Beever *et al.* 1987
Lolium perenne	262	75	77	152	110	148	MacRae and Ulyatt (1974), Ulyatt *et al.* (1975)

TABLE 6.3

(continued)

Forage	Dietary CP (g/kg DM)	Nonamonia CP Leaving rumen (g/kg DM) Ungraded	Microbial	Total	Lost as ammonia (g CP/kg DM)	Efficiency of microbial protein synthesis (g CP/kg ARDOM)	Reference
Trifolium repens	285[b]	—	—	202	83	—	Cruickshank *et al.* (1985)
Medicago sativa	332[b]	—	—	191	142	—	Cruickshank *et al.* (1985)
Mean fresh forage	180	(43)	(90)	146	34	206	
Silage							
Lolium perenne/ Trifolium pratense	135	36	97	133	2	158	Overend and Armstrong (1982)
Lolium perenne	138	29	85	114	24	152	Rooke *et al.* (1985)
Lolium perenne/ Trifolium repens	138	73	63	136	2	158	Rooke *et al.* (1985)
Lolium perenne	148	16	89	105	43	174	Rooke *et al.* (1982)
Lolium perenne	165	78	61	139	26	119	Thomson *et al.* (1981)
Lolium perenne	200	34	79	111	89	135	Siddons *et al.* (1979)
Mean for silages	154	44	79	123	31	150	
Dried forage							
Heteropogon contortus	46	—	—	76[b]	−30	—	Hunter and Siebert (1980)
Digitaria decumbens	56	—	—	97[b]	−41	—	Hunter and Siebert (1986b)
Lolium perenne × *L. multiflorum*	64	—	—	128	−64	—	Weston and Hogan (1968a)
Caucasian bluestem	69	22	40	62	7	—	Bowman and Asplund (1988b)
Triticum aestivum (straw)	73[b]	38	41	79	−6	151	Walker *et al.* (1975)
Lolium perenne × *L. multiflorum*	78	—	—	131	−53	—	Weston and Hogan (1968b)
Avena sativa	79	—	—	126	−47	—	Hogan and Weston (1969)
Triticum aestivum (hay)	99[b]	33	57	90	9	129	Walker *et al.* (1975)
Bromus inermis	114	45	71	116	−2	197	Kennedy *et al.* (1982)

Lolium perenne × *L. multiflorum*	114	—	—	144	−30	—	Weston and Hogan (1968b)
Avena sativa	115	—	—	157	−42	—	Hogan and Weston (1969)
Bromus inermis	125	35	97	132	−7	224	Kennedy and Milligan (1978)
Temperate grass (spring)	131	—	—	172	−41	—	MacRae *et al.* (1985)
Temperate grass (autumn)	138	—	—	151	−13	—	MacRae *et al.* (1985)
Lolium perenne × *L. multiflorum*	142	—	—	168	−26	—	Weston and Hogan (1968b)
Trifolium subterraneum	148[b]	—	—	147	1	—	Weston and Hogan (1971)
Medicago sativa	152	31	79	110	42	214	Kennedy *et al.* (1982)
Trifolium alexandrinum	158[b]	—	—	158	0	—	Weston and Hogan (1971)
Temperate grass	166	—	—	172	−6	—	MacRae *et al.* (1972)
Medicago sativa	169	47	89	136	33	185	Mathers and Miller (1981)
Avena sativa	192	—	—	229	−37	—	Hogan and Weston (1969)
Medicago sativa	212[b]	85	91	176	36	194	Walker *et al.* (1975)
Avena sativa	217	—	—	185	32	—	Hogan and Weston (1969)
Trifolium subterraneum	223[b]	89	69	158	65	125	Walker *et al.* (1975)
Trifolium subterraneum	226[b]	—	—	193	33	—	Weston and Hogan (1971)
Medicago sativa	230	94	40	134	96	—	Nolan and Leng (1972)
Trifolium alexandrinum	250[b]	—	—	211	39	—	Weston and Hogan (1971)
Trifolium subterraneum	257[b]	—	—	255	3	—	Weston and Hogan (1971)
Avena sativa	261	—	—	268	−7	—	Hogan and Weston (1969)
Lolium perenne × *L. multiflorum*	270	—	—	225	45	—	Weston and Hogan (1968b)
Trifolium alexandrinum	269[b]	—	—	192	77	—	Weston and Hogan (1971)
Avena sativa	299	—	—	262	37	—	Hogan and Weston (1969)
Avena sativa	322	—	—	264	58	—	Hogan and Weston (1969)
Mean for dried forage	170	(52)	(67)	165	5	177	
Mean for all forages	170	(46)	(81)	152	18	187	

[a]Data adapted.
[b]Organic matter assumed to be 900 g/kg dry matter.

but with *O. viciifolia* no protein is released due to the presence of condensed tannins. Chewing also increases the rate of degradation of forage protein in the rumen (Bailey, 1962).

Once the masticated forage is in the rumen, the extent of CP degradation will depend on the activity of the rumen microbes and the time the forage is in the rumen. Degradation of CP is low in forages that are low in S (Millard *et al.*, 1987) and CP (Rooke *et al.*, 1985; Bowman and Asplund, 1988b). Addition of a protein supplement to formaldehyde-treated silage increases the quantity of CP degraded in the rumen (Rooke *et al.*, 1985). Supplementary urea increases the degradation of the CP in mature hay (Bowman and Asplund, 1988b).

Both bacteria and protozoa appear to be responsible for CP degradation. Soluble proteins are primarily degraded by bacteria but protozoa may contribute to the degradation of insoluble particulate proteins (Hino and Russell, 1987). Removing protozoa from the rumen reduces the degradation of CP by as much as 65%, possibly because the forage is retained for a shorter time in the rumen (Ushida *et al.*, 1986).

The degradation of CP is reduced by any factor that reduces the time forage is retained in the rumen and exposed to microbial deamination. A 50% increase in the intake of *M. sativa* doubled the quantity of undegraded forage CP leaving the rumen (Ulyatt *et al.*, 1984; Merchen *et al.*, 1986). Cold exposure has reduced CP degradation (Kennedy *et al.*, 1982). With *B. inermis* and *M. sativa* the degradation of CP was reduced from 0.70 to 0.62 and 0.55 to 0.45 when sheep were kept at normal and low temperatures, respectively. In another study, 78% of the CP in *B. inermis* pellets was degraded in the rumen of sheep kept at 22–25°C and this was reduced to 54% when the sheep were kept at 2–5°C (Kennedy and Milligan, 1978). Grinding and pelleting maximizes exposure of CP to rumen microbes, but this effect is more than offset by other changes in the rumen, with the net effect that CP degradation is reduced by pelleting (Table 6.4).

2. Synthesis of Microbial Crude Protein

Crude protein is synthesized by rumen microbes from the amino acids and ammonia released into the rumen by degradation of the CP in forage and saliva. The quantity of microbial CP produced varies with the quantity of N released and the quantity of energy available for microbial CP synthesis (ARC, 1984).

Data have been published on the quantity of microbial CP synthesized by sheep fed many types of forage. This data has been used to calculate the quantity of microbial CP produced from each kilogram of forage DM (Table 6.3). Average production of microbial protein is 81 gm CP/kg for-

TABLE 6.4

Effect of Pelleting on the Flow of Undergraded Dietary Protein and Microbial Protein from the Rumen and on the Efficiency of Microbial Protein Synthesis[a]

	Lolium multiflorum		*Phleum pratense*	
	Chopped	Pellets	Chopped	Pellets
Total amino acid (g/day) consumed	122	116	137	126
At proximal duodenum				
Undegraded feed	36	62	44	67
Microbial	88	81	87	75
Total	124	143	131	142
Efficiency of microbial protein synthesis (g/kg ARDOM)[b]	172	196	203	180
Degradation of dietary crude protein	0.70	0.46	0.68	0.47

[a]Adapted from Beever *et al.* (1981).
[b]ARDOM, apparent rumen digested organic matter.

age DM eaten, with values ranging from 34 to 162 g/kg DM. The highest production of microbial CP is usually associated with immature, fresh, high-digestibility forage, whereas production of microbial CP is low with dried, mature forages. This is to be expected because the synthesis by microbial cells is energy dependent and the efficiency of conversion of dietary to microbial CP depends on the rate of release of energy compared with that of amino acids and ammonia during forage degradation in the rumen (Hogan, 1982).

Efficiencies of microbial CP synthesis are usually expressed as grams of CP produced from each kilogram of organic matter apparently digested in the rumen (ARDOM). For the forages examined, mean efficiency was 187 g CP/kg ARDOM, but values ranged from 98 to 308 g/kg ARDOM (Table 6.3). With fresh forages the mean efficiency was 206 g/kg ARDOM compared with 177 and 152 g/kg ARDOM for dried forage and silage, respectively (Table 6.3). The higher efficiency of microbial CP synthesis from fresh forage has been attributed to the larger quantity of volatile fatty acid produced from each unit of OM apparently digested in the rumen (Walker *et al.*, 1975).

Low efficiencies of microbial CP synthesis are generally found with forages containing less than 100 g CP/kg DM, possibly because there are insufficient amino acids and ammonia to match the energy available to the rumen microbes. For optimum efficiency of microbial CP production the diet should contain at least 170 g rumen-degradable CP for each kilogram of rumen-degradable organic matter (McMeniman and Armstrong, 1977). The efficiency of microbial CP synthesis and ruminant production on low-CP diets has not always been improved by supplementing with urea (Beever *et al.*, 1971a; Bowman and Asplund, 1988b). This lack of response may be due to a deficiency of other nutrients essential for microbial growth including S (Thomas *et al.*, 1952b), branch-chain fatty acids (Allison *et al.*, 1962; Hume, 1970; Russell and Sniffen, 1984), or essential amino acids such as methionine (Loosli and Harris, 1945).

The efficiency of microbial CP production also depends on the proportion of protozoa in the rumen microbial population. Defaunation, that is, killing the protozoa, increased the efficiency of microbial CP production in sheep fed *M. sativa* from 186 to 360 g/kg ARDOM (Ushida *et al.*, 1986). Increasing the rate of passage of forage through the rumen by subjecting the sheep to cold stress increased the efficiency of microbial CP production by 32% (Kennedy and Milligan, 1978). However, increasing the level of feeding reduced efficiency of microbial synthesis by 11 and 24%, in sheep fed *B. inermis* (Kennedy and Milligan, 1978), and *M. sativa* (Ulyatt *et al.*, 1984) respectively. This reduction in microbial synthesis was possibly due to the smaller quantity of amino acids and ammonia available in the rumen due to the lower degradation of the forage CP.

3. Digestion in the Small Intestine

The quantity of CP (nonammonia) leaving the rumen is the sum of the undegraded forage CP, microbial CP, and endogenous nonammonia CP. When fresh forages and silages contain more than about 130 g CP/kg DM, less nonammonia CP leaves the rumen than enters as CP in the forage (Fig. 6.2). The excess forage CP is absorbed as ammonia (Table 6.3) and excreted in the urine. Dried forages are usually more resistant to microbial degradation within the rumen and a smaller proportion of the forage CP is lost as ammonia than with fresh forage. This is very noticeable when the forage contains high levels of CP (Fig. 6.2). Conversely with fresh forages containing less than 130 g CP/kg DM there is a net gain in CP within the rumen (Fig. 6.2).

Of the nonammonia CP entering the duodenum about 80% consists of true protein and much of the remaining CP is present as nucleic acids (Smith, 1975). The amino acid composition of this true protein varies between forages (Hogan and Weston, 1981), and during digestion in the small intestine the S amino acids cysteine and methionine are absorbed less efficiently than the other amino acids (Armstrong and Hutton, 1975). The net absorption coefficient for the true protein in the small intestine is

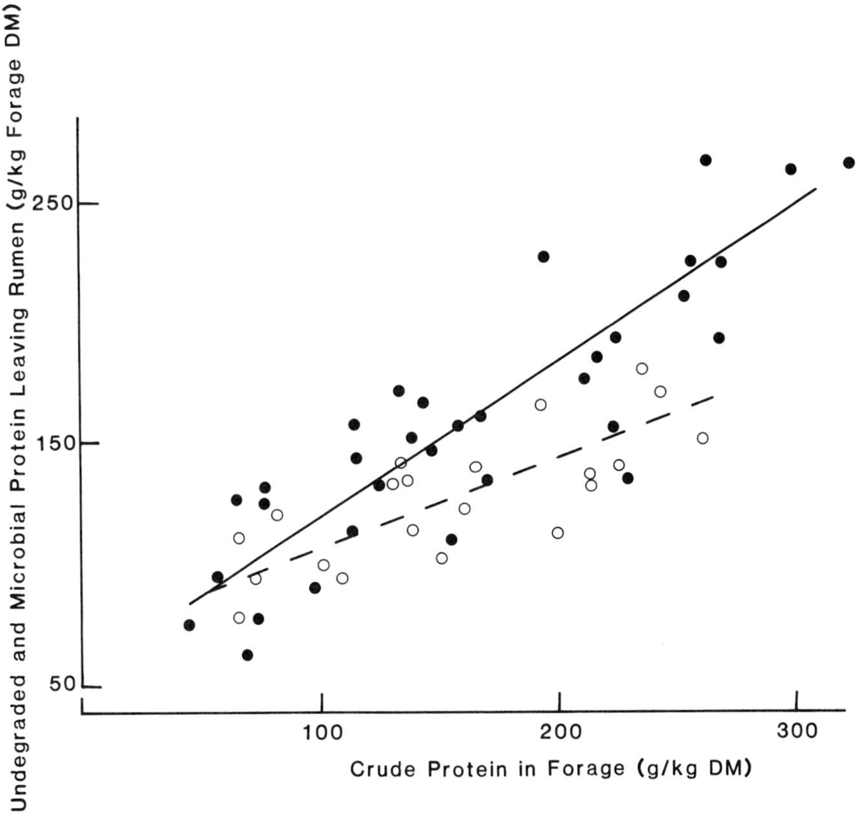

Fig. 6.2. Relation between the undegraded dietary and microbial crude protein leaving the rumen and the crude concentration in forage. (●——●) Dried forage; (○– – –○) fresh forage and silage. Adapted from Table 6.3.

about 0.7, so each kilogram of nonammonia CP passing into the small intestine yields about 560 g of amino acids (Lindsay *et al.*, 1980). The undigested CP passes into the large intestine, where it is either deaminated or excreted in the feces.

For many of the forages shown in Table 6.3 there is data on the quantity of nonammonia CP digested in the small intestine. The quantity digested is positively correlated to the CP in the forage with no difference between fresh and dried forages (Fig. 6.3). It appears that although many changes to CP occur in the rumen, the measurement of CP in the forage can still provide a useful indication, in many situations, of the relative quantity of amino acids that may be absorbed by ruminants.

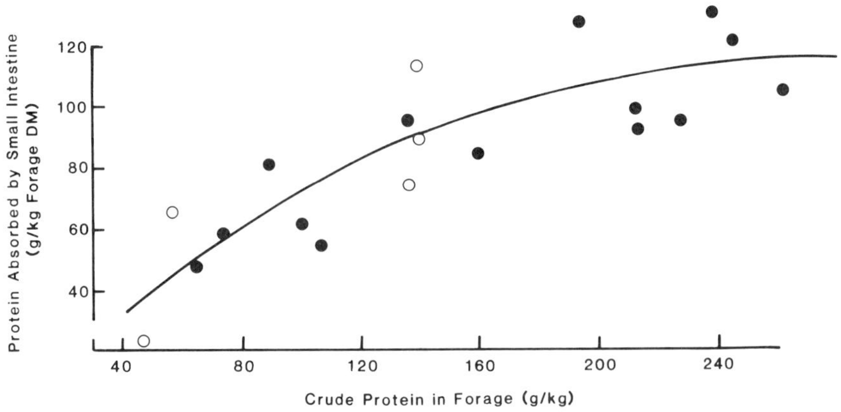

Fig. 6.3. Relation between nonammonia crude protein absorbed in the small intestine and crude protein concentration in forage. (○) Dried; (●) fresh. Derived from data in Table 6.3.

E. Requirements

In the previous section it was shown that the quantity of amino acids absorbed by ruminants is not directly related to the quantity of CP in the forage but is modified during passage through the rumen. The main features of these changes are a loss of CP from diets high in CP and an increase when the forage is low in CP (Table 6.3). The extent of the change depends on the degradability of the feed and whether the amino acids and ammonia released into the rumen from the forage and saliva

TABLE 6.5

Dietary Crude Protein Concentrations (g/kg DM) and Optimum Degradability Recommended for Three Breeds of Dairy Cattle

Breed	Liveweight (kg)	Milk production (kg/day) 0	10	20	30	Reference
Jersey	400	101(1.00)[a]	130(0.77)	146(0.69)	—	ARC (1980)
		—	130	140	150	NRC (1978)
Ayrshire	500	101(1.00)	120(0.84)	139(0.72)	147(0.68)	ARC (1980)
		—	130	130	140	NRC (1978)
Friesian	600	101(1.00)	112(0.89)	133(0.76)	142(0.71)	ARC (1980)
		62(1.00)	94(0.82)	118(0.73)	120(0.76)	ARC (1984)
			130	130	130	NRC (1978)

[a]Optimum degradability in parentheses.

TABLE 6.6

Recommended Concentrations of Dietary Crude Protein (g/kg DM) and Optimum Crude Protein Degradability for Beef Cattle

Liveweight (kg)	Weight gain (kg/day)	Crude protein requirements		
		ARC (1980)	ARC (1984)	NRC (1984)
300	0	100 (1.0)[a]	62 (1.0)	
	0.25	100 (1.0)	—	96
	0.75	100 (1.0)	—	132
	1.25	100 (1.0)	92 (1.0)	183
500	0	100 (1.0)	—	
	0.25	100 (1.0)	—	85
	0.75	100 (1.0)	—	108
	1.25	100 (1.0)	—	130

[a]Optimum degradability in parentheses.

exceed the energy available to the microbes. Recognition of these principles led to the development of new systems for protein rationing of housed ruminants (Burroughs *et al.*, 1975; Roy *et al.*, 1977; Vérité *et al.*, 1979; ARC, 1980, 1984; NRC, 1984).

Recommended dietary levels of protein are quoted either as CP/kg DM with or without an optimum CP degradability coefficient (Tables 6.5–6.7).

TABLE 6.7

Dietary Crude Protein Concentration and Optimum Degradability Required by Fattening Lambs Eating Diets with a High Metabolizable Energy Content[a]

Liveweight (kg)	Liveweight gain (kg/day)	Crude protein required (g/kg)	Optimum degradability coefficient
20	0	100	1.00
	0.1	138	0.73
	0.2	154	0.65
30	0	100	1.00
	0.1	111	0.90
	0.2	121	0.82
	0.3	123	0.82
40	0	100	1.00
	0.1	100	1.00
	0.2	103	0.97
	0.3	102	0.98

[a]Adapted from ARC (1980).

Including an optimum degradability coefficient removes the need for a safety margin and reduces the recommended requirement for CP (Table 6.5).

II. CRUDE PROTEIN IN FORAGE

A. Mean Concentration

The CP concentration has been determined for many forages grown in many different ways. The results of these analyses are scattered throughout the scientific literature and in recent years this information has been summarized by the International Network of Feed Information Centers for different countries or regions of the world. The CP content of these forages varied from <30 to >270 g/kg DM with a mean of 142 g/kg DM (Fig. 6.4).

B. Species and Cultivar Differences

1. Grass versus Legume

A major cause of the large variation in CP concentration of forages is the difference between legumes and grasses. The mean concentration of CP in legumes was 170 g/kg DM compared with 115 g/kg DM for grasses, a superiority that applied to both temperate and tropical forages (Fig. 6.5).

2. Temperate versus Tropical Forages

Temperate grasses generally contain more CP than tropical grasses, with mean concentrations of 129 and 100 g/kg DM, respectively (Fig. 6.5).

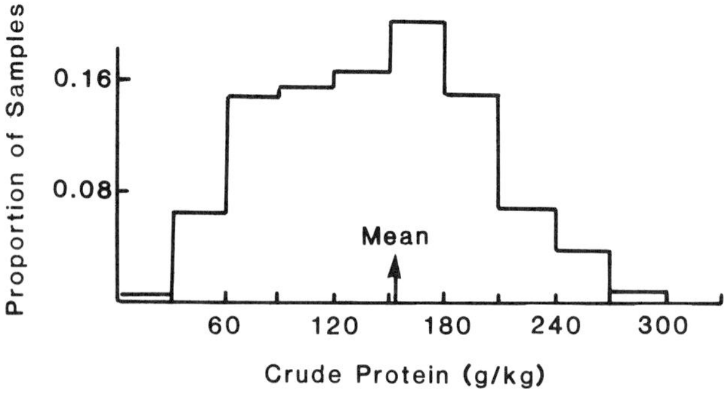

Fig. 6.4. Crude protein concentration in a selection of all forages (world data).

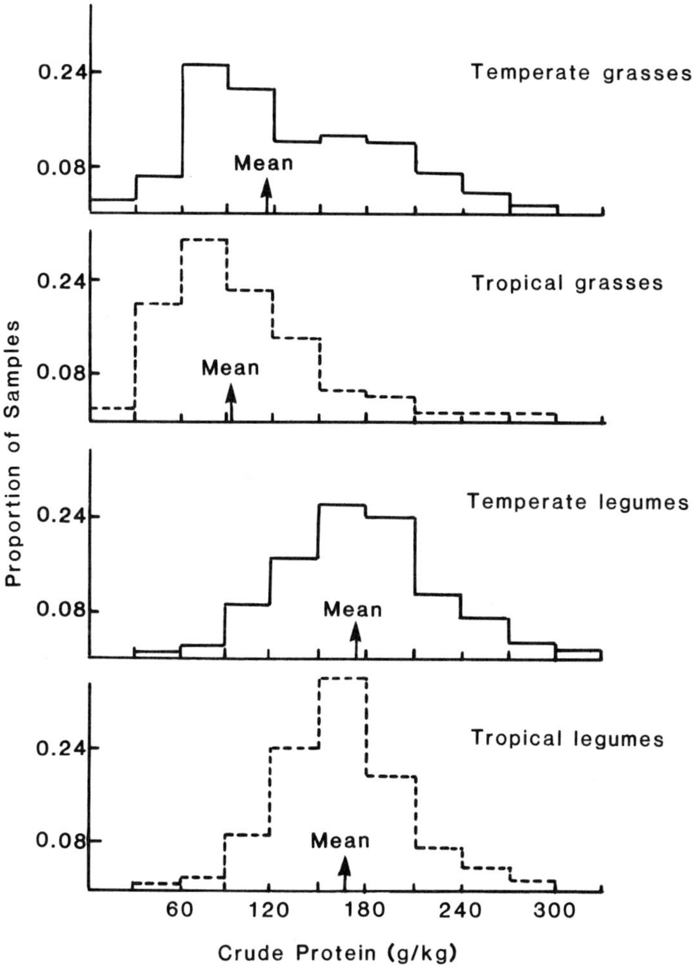

Fig. 6.5. Crude protein concentration in temperate and tropical grasses and legumes (world data).

Of particular concern is the high proportion of tropical grass samples (21%) that contained less than 60 g CP/kg DM, the minimum quantity required to meet the requirements of rumen bacteria. Only about half the tropical grass samples would meet the CP level recommended for the maintenance of cattle (Table 6.6). This low level of CP in tropical grasses is associated with the C_4 pathway of photosynthesis, a high proportion of stem, and large vascular bundles in the leaves (Wilson and Minson, 1980). Temperate and tropical legumes have the same leaf anatomy, C_3 pathway of photosynthesis, and similar levels of CP; mean concentrations of CP were 175 and 166 g/kg DM, respectively (Fig. 6.5).

3. Grass Species

Differences in CP concentration between grass species, independent of flowering date, have been found between S.37 *Dactylis glomerata* and S.24 *L. perenne*. These two grasses have a similar flowering date and yield of DM in the spring, but *D. glomerata* contained more CP (5 to 20 g/kg DM) than *L. perenne* (Fig. 6.6). However, any advantage of the high CP content of *D. glomerata* may be offset by a lower production of microbial protein in the rumen due to the low levels of digestible energy and water-soluble carbohydrates (Jarrige and Minson, 1964; Dent and Aldrich, 1968; Tilley and Terry, 1969). With monthly and bimonthly regrowths, the differences in favor of *D. glomerata* were 23 and 28 g CP/kg DM, respectively (Minson *et al.*, 1960b). A similar difference between *D. glomerata* and *L. perenne* was also reported by Dent and Aldrich (1968).

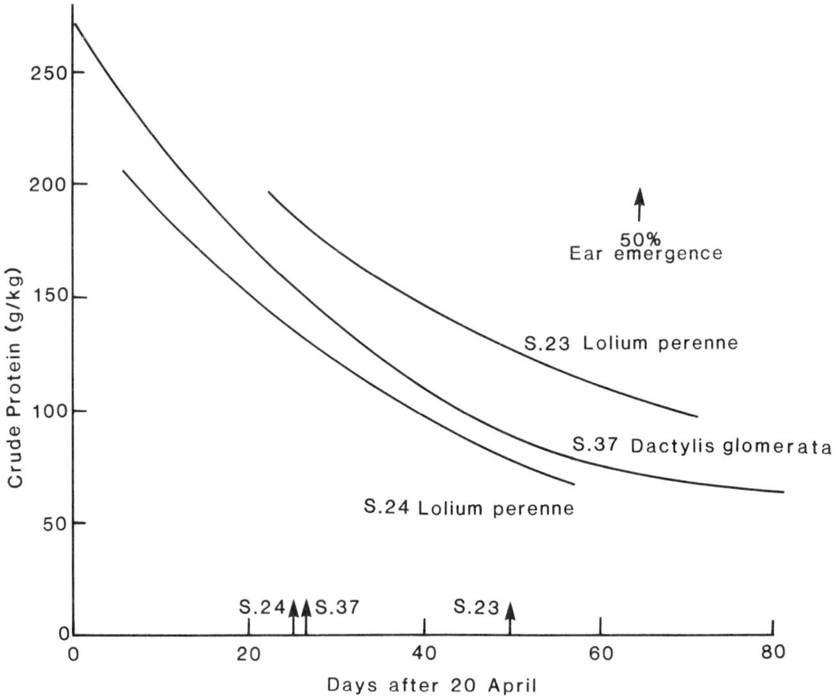

Fig. 6.6. Crude protein concentration in primary growths of three forages as affected by harvest date. (Adapted from Minson *et al.*, 1960b.)

Comparisons between forages are usually made on regularly cut or grazed swards in an attempt to avoid the confounding effect of stage of growth on CP concentration. When compared in this way, differences in CP concentration between temperate species are usually small (Table 6.8).

4. CULTIVARS

Cultivars of grass species differ in CP when cut on the same day (Fig. 6.6), but this can often be attributed to differences in stage of maturity. The ranking order for CP concentration depends on whether cultivars are compared at the same cutting date, stage of development, or dry-matter digestibility (Table 6.9). There may be large differences between cultivars in CP concentration when compared on the same date, but these differences disappear or are even reversed when the cultivars are compared at ear emergence or at the same dry-matter digestibility.

When compared at the same stage of maturity, differences in CP concentration were found between cultivars of *D. glomerata* (16 g CP/kg DM) and *B. inermis* (10 g/CP kg DM) but not between cultivars of *M. sativa* or *Phleum pratense* (Fulkerson *et al.*, 1967).

The difficulty in comparing cultivars which have different growth rates in the spring has been avoided by comparing their CP concentration in monthly regrowths taken throughout the season (Dent and Aldrich, 1963). The maximum difference in CP between cultivars of four temperate grass species was only 22 g/kg (Table 6.10), indicating that where forage is regularly cut there is little scope for selecting cultivars that have a higher CP content.

C. Plant Parts

Leaf blades in forage have approximately double the CP concentration of the leaf sheath and stem fractions. This applies to all forages (Table 6.11). In some studies with *M. sativa* the difference in CP between leaf and stem fractions tends to increase as the plant matures, due to a faster rate of fall in CP concentration in the stem (Terry and Tilley, 1964; Thom and Smith, 1980). However, in another study plant maturity had no effect on the difference in CP between leaf and stem of *B. inermis, D. glomerata, M. sativa,* and *Phleum pratense* (Mowat *et al.*, 1965a).

D. Stage of Growth

Age has a large influence on the CP concentration of forage. Increasing the period of growth of *L. multiflorum* from 2 to 10 weeks depressed CP concentration from 188 to 69 g/kg DM (Fagan, 1928). In *Pennisetum*

TABLE 6.8

Crude Protein Concentration (g/kg DM) of Regularly Cut or Grazed Temperate Grasses

Species	Sullivan *et al.* (1956)	Dent and Aldrich (1963)	Minson *et al.* (1964)	Davies and Morgan (1982)
Agrostis gigantea	159	—	—	—
Arrhenatherum elatius	178	—	—	—
Bromus inermis	158	—	—	—
Dactylis glomerata	183	214	199	211
Festuca arundinacea	177	—	194	183
Festuca pratensis	—	206	175	—
Lolium perenne	—	198	—	184
Phalaris arundinacea	186	—	—	—
Phleum pratense	156	204	169	195
Poa pratensis	180	—	—	—

TABLE 6.9

Crude Protein Concentration (g/kg DM) in Cultivars of *Dactylis glomerata* and *Lolium perenne* Compared on the Basis of Three Different Criteria[a]

Species and cultivar	Criteria used for comparison		
	Same day (in May)	Ear emergence	0.7 DMD[b]
Dactylis glomerata			
S.345	128	202	198
Scotia	131	144	142
S.37	144	144	144
S.143	195	150	185
Mean	150	160	167
Lolium perenne			
S.24	123	135	110
Taptoe	129	115	108
S.321	151	120	98
S.23	174	130	110
Mean	144	125	106

[a]Adapted from Jones and Walters (1969).
[b]DMD, dry-matter digestibility coefficient.

TABLE 6.10

Crude Protein Concentration of Different Cultivars of Four Grasses Cut at Monthly Intervals[a]

Species	Numbers of cultivars examined	Crude protein (g/kg DM)		
		Lowest cultivar	Highest cultivar	Difference
Dactylis glomerata	14	208	224	16
Festuca pratensis	11	200	211	11
Lolium perenne	26	188	210	22
Phleum pratense	19	194	213	19

[a]Adapted from Dent and Aldrich (1963).

TABLE 6.11

Crude Protein Concentration of Leaf and Stem of Forage

Species	Crude protein (g/kg DM)		Reference
	Leaf	Stem	
Bothriochloa spp.	95	47	Dabo *et al.* (1988)
Bromus inermis	124	55	Mowat *et al.* (1965a)
Cynodon dactylon	165	92	Burton *et al.* (1968)
Dactylis glomerata	148	59	Gueguen and Fauconneau (1960), Johnston and Waite (1965), Mowat *et al.* (1965a)
Desmodium intortum	233	107	Rotar (1965)
Desmodium uncinatum	212	110	Rotar (1965)
Festuca pratensis	140	64	Gueguen and Fauconneau (1960)
Lablab purpureus	240	127	Hendricksen and Minson (1985a,b)
Lespedeza juncea	233	103	Donnelly and Anthony (1973)
Lolium multiflorum	104	59	Fagan (1928)
Lolium perenne	126	54	Johnston and Waite (1965)
Medicago sativa	290	128	Holter and Reid (1959), Terry and Tilley (1964), Mowat *et al.* (1965a), Heinrichs *et al.* (1969), Jarrige (1980), Thom and Smith (1980)
Melilotus officinalis	230	83	Sotola (1946)
Phleum pratense	108	55	Fagan (1928), Waite and Sastry (1949), Mowat *et al.* (1965a)
Stylosanthes spp.	176	89	Little *et al.* (1984)
Trifolium pratense	216	82	Fagan (1928)
Trifolium subterraneum	199	101	Hume *et al.* (1968)
Zea mays	152	64	Johnson *et al.* (1966), Jarrige (1980)
Mean	177	82	

purpureum cut every 4 or 12 weeks the CP concentrations were 98 and 61 g/kg DM, respectively (Paterson, 1933). This fall in CP is caused by two factors. As forage matures there is an increase in the proportion of leaf sheath and flowering stem, which have a lower CP than the leaf fraction. The other cause is a fall in the CP of all fractions as they mature. These two effects are illustrated in Fig 6.7. These changes with maturity

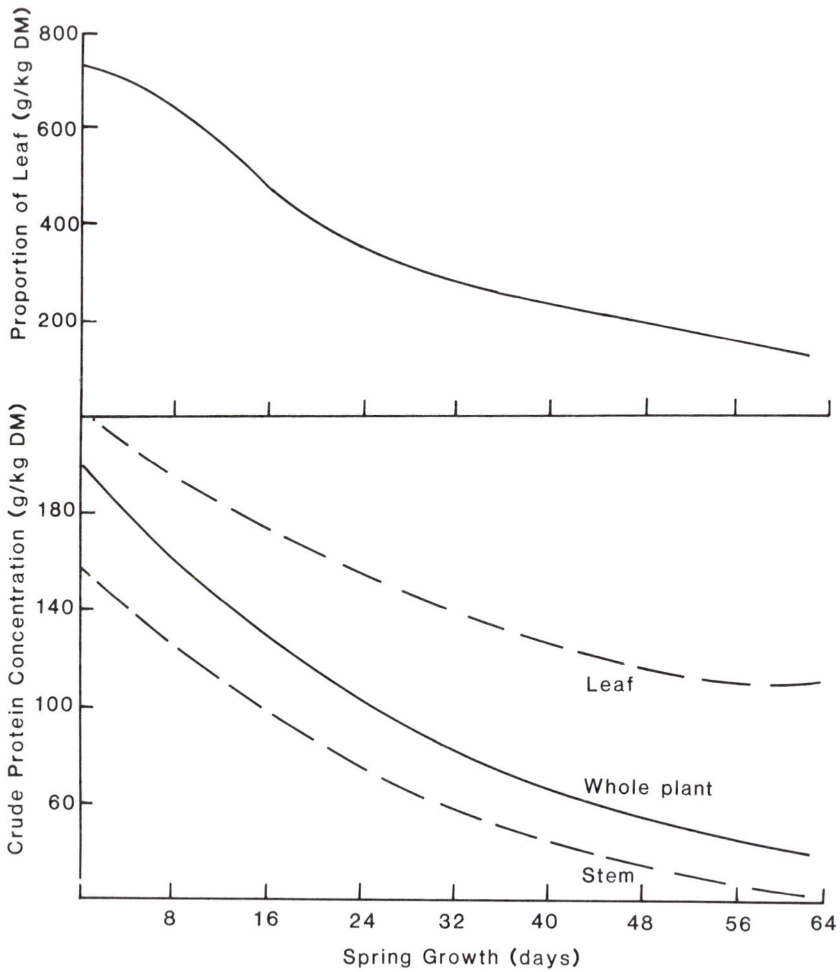

Fig. 6.7. Effect of forage maturity on the proportion of leaf blade and crude protein concentrations in leaf blade, stem (including sheath), and whole plant of *Phleum pratense*. (Adapted from Waite and Sastry, 1949.)

have also been observed in *M. sativa* (Terry and Tilley, 1964; Thom and Smith, 1980; Buxton *et al.*, 1985) and *T. pratense* (Buxton *et al.*, 1985).

The rate of fall in CP has been calculated between prebudding and flowering in legumes and between preshooting and full flowering in grasses (Kivimae, 1959). These rates of change in CP concentration of a range of forages are shown in Table 6.12. The mean rate of fall in CP with increasing maturity was 2.2 g/kg DM/day, with no significant difference between the rate of fall of temperate and tropical forages. The rate of fall in CP varies with the initial CP concentration. When the CP content of *L. multiflorum* was raised to different levels by applying fertilizer N, the subsequent decline in CP was almost twice as fast for the high-CP forage (Fig. 6.8).

E. Soil Fertility

The concentration and composition of CP in forage are affected by the level of available N in the soil solution. In 1928 Fagan showed that the CP concentration of *L. multiflorum* could be increased by fertilizer N, the largest response occurring in the leaf.

The composition of CP is affected by the level of available soil N. The proportion of nitrate N rises rapidly following the application of fertilizer N but then falls at a rate depending on the quantity of fertilizer applied (Fig. 6.8). The effect of fertilizer on CP concentration in forage will be considered in Section IV,B.

The sulfur-containing amino acids, methionine and cysteine, are required for protein synthesis. These amino acids are produced from sulphur (S) absorbed from the soil solution through the root system (Smith and Siregar, 1983). When soil is deficient in S there is a marked decline in true protein concentration in the forage and the leaves are pale, a sign normally associated with a deficiency of soil N. There is also an increase in the level of nitrate in the forage (Adams and Sheard, 1966).

F. Climate

1. Seasonal Variation

Before considering the various climatic factors that may affect the CP concentration of forage, seasonal changes in CP of frequently cut swards will be examined.

The CP concentration in regularly cut and fertilized temperate grasses is lowest in midsummer (170 g/kg DM) and highest in autumn (230 g/kg

TABLE 6.12

Mean Rate of Fall in Crude Protein Concentration as Forages Mature

Species	Fall in protein (g/kg DM/day)	Reference
Arachis glabrata	0.7	Romero *et al.* (1987)
Bothriochloa spp.	1.8	Dabo *et al.* (1988)
Brachiaria mutica	1.5	Appelman and Dirven (1962)
Bromus inermis	2.9	Fulkerson *et al.* (1967)
Cenchrus ciliaris	3.1	Combellas and Gonzalez (1972a)
Cynodon dactylon	1.7	Burton *et al.* (1963), Miller *et al.* (1965a), Gomide *et al.* (1969a), Combellas *et al.* (1972)
Dactylis glomerata	2.7	Minson *et al.* (1960b, 1964), Austenson (1963), Haenlein *et al.* (1966), Brown *et al.* (1968), Sheehan (1969)
Digitaria decumbens	1.6	Vicente-Chandler *et al.* (1961), Appleman and Dirven (1962), Gomide *et al.* (1969a)
Festuca arundinacea	1.9	Austenson (1963), Minson *et al.* (1964)
Festuca pratensis	2.1	Minson *et al.* (1964)
Lolium multiflorum	2.5	Fagan (1928), Minson *et al.* (1964), Sheehan (1969)
Lolium perenne	2.2	Minson *et al.* (1960b), Austenson (1963, Sheehan (1969)
Medicago sativa	3.1	Kivimae (1959), Terry and Tilley (1964), Fulkerson *et al.* (1967), Thom and Smith (1980), Buxton *et al.* (1985), Romero *et al.* (1987)
Melinis minutiflora	2.1	Gomide *et al.* (1969a)
Panicum maximum	2.9	Gomide *et al.* (1969a)
Panicum maximum var. *trichoglume*	3.1	Combellas and Gonzalez (1972b)
Pennisetum clandestinum	2.8	Gomide *et al.* (1969a), Said (1971)
Pennisetum purpureum	3.0	Paterson (1933), Butterworth (1965), Gomide *et al.* (1969a)
Phalaris arundinacea	3.4	Colovos *et al.* (1969)
Phleum pratense	2.7	Waite and Sastry (1949), Homb (1953), Mellin *et al.* (1962), Austenson (1963), Minson *et al.* (1964), Fulkerson *et al.* (1967), Brown *et al.* (1986), Langille and Calder (1968)
Stylosanthes humilis	0.9	Fisher (1969)
Trifolium pratense	2.2	Kivimae (1959), Buxton *et al.* (1985)
Trifolium subterraneum	1.1	McLaren and Doyle (1986)
Tripsacum laxum	2.4	Appelman and Dirven (1962)
Mean	2.2	

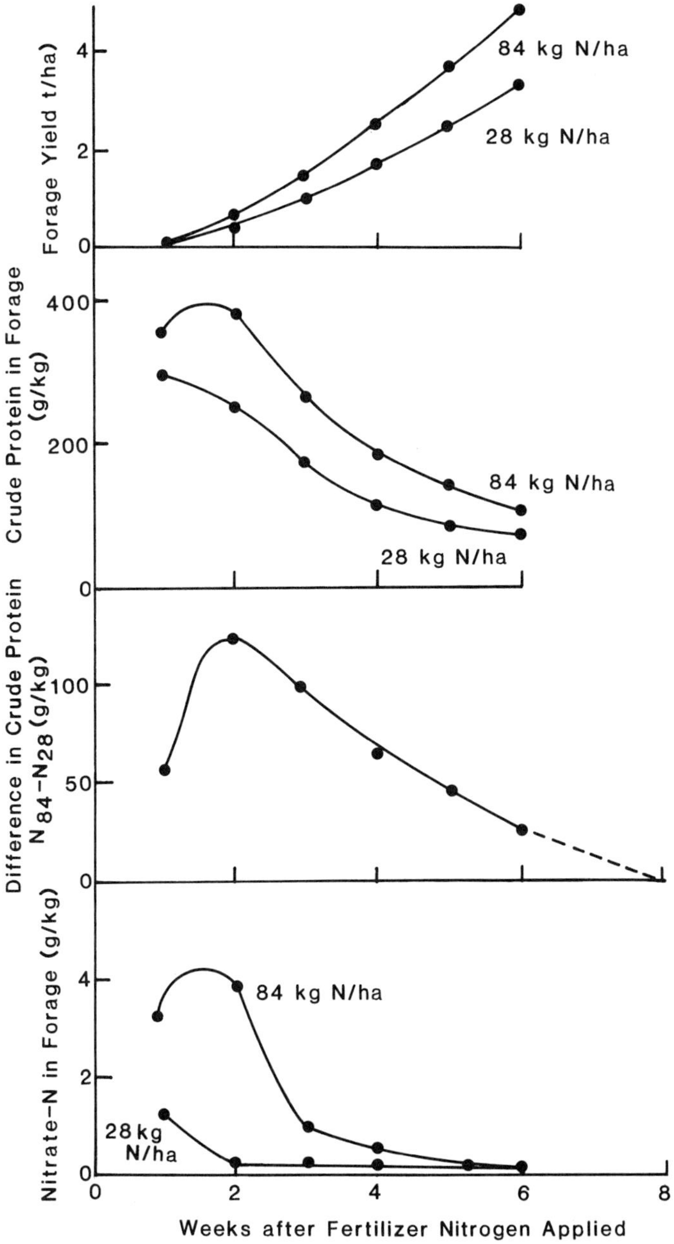

Fig. 6.8. Effect of level of fertilizer nitrogen and time of cutting on the yield, crude protein, and nitrate concentration in *Lolium multiflorum*. (Adapted from Wilman, 1965.)

DM), a rise associated with an increase in the proportion of leaf in the forage (Minson *et al.*, 1960b, 1964). Similar seasonal changes in CP concentration were found in other studies with temperate forages (Woodman *et al.*, 1927; Chestnutt, 1966; Dent and Aldrich, 1968; Metson and Saunders, 1978b; Thompson and Warren, 1979) and the tropical grasses *Cynodon dactylon* (Herrera, 1979) and *Setaria sphacelata* (Bray and Hacker, 1981). The magnitude of the seasonal variation of CP in *C. dactylon* is influenced by the quantity and regularity of fertilizer N applied; at high levels of fertilizer N, the difference between summer and winter is much greater (Herrera, 1979).

2. Temperature

The low CP concentration in summer, mentioned in the previous section, does not appear to have been caused by the high temperature in summer. In controlled-environment studies *L. perenne* was grown at different temperatures; when grown at 10 and 25°C the mean CP concentration of leaves was 196 and 252 g/kg, respectively (Alberda, 1965). A similar increase in CP due to growth at high temperatures was found with *B. inermis* (Smith, 1970a), *L. corniculatus* (Smith, 1970b), *M. sativa* (Smith, 1969, 1970b,c; Walgenbach *et al.*, 1981), *Melilotus officinalis* (Smith, 1970b), *P. pratense* (Smith, 1970a), *Trifolium hybridum* and *T. pratense* (Smith, 1970b). In contrast to these results, growth at higher temperatures resulted in a fall in the CP of *L. perenne* in one study (Deinum, 1966a).

3. Light Intensity

Seasonal variation in CP concentration is probably caused in part by differences in light intensity. The CP concentration in forage is decreased by high light intensity (Bathurst and Mitchell, 1958; Burton *et al.*, 1959; Alberda, 1965; Deinum, 1966a); this reduction is associated with an increase in forage yield and dilution of the available CP. When the forage is supplied with adequate N, high light intensity slightly increases the CP concentration of *L. perenne* (Alberda, 1965).

4. Water Stress

Soil moisture stress has an inconsistent effect on the CP concentration of forages (Deinum, 1966a; Walgenbach *et al.*, 1981). If the period of stress is long then both growth and N uptake are reduced, causing a decrease in CP concentration in the forage, but in less severe situations only growth is reduced and the CP concentration in the forage is increased.

G. Conservation

1. Drying

Rapid drying has little or no effect on the CP concentration of forage provided there is no leaf loss. The CP in fresh and dried (70°C) *D. glomerata* was 265 and 262 g/kg DM, respectively (Ohyama, 1960). Drying in the sun for 3 days had no adverse effect on the CP concentration (263 g/kg DM). Although drying has no effect on the total CP in forage it changes the proportion of CP that is water soluble. Heat drying generally reduces CP solubility, whereas drying in the sun increases CP solubility (Ohyama, 1960).

Field drying has little or no effect on the CP concentration of hay made in both good and poor drying conditions provided there is no leaf loss (Wilman and Owen, 1982). However, where machinery is used to condition forage during the final stages of hay making, leaf loss occurs and CP concentration of the hay is reduced. The greatest leaf loss occurred in a dry season when CP concentration was reduced from 119 to 94 g/kg DM compared with a decrease from 94 to 86 g/kg DM when hay was made in a wet season (Shepperson, 1960). Loss of leaf is reduced when drying is completed by barn drying or stacking the half-dried forage on tripods in the field. Both methods reduce the loss of CP (Table 6.13). The time *M. sativa* hay is exposed to leaf loss in the field can be reduced by applying potassium carbonate when the crop is cut (Tullberg and Minson, 1978). This treatment increased the CP concentration of the hay from 183 to 192 g/kg (Nocek *et al.*, 1988).

TABLE 6.13

Crude Protein Concentration in Temperate Grass Hay Made in Various Ways, Compared with the Original Forage[a]

Treatment	Crude protein (g/kg DM) weather conditions: Dry	Wet	Mean
Frozen at cutting	119	94	106
Barn dried	113	93	103
Tripod dried	108	90	99
Swath made	94	86	90

[a]Adapted from Shepperson (1960).

When hay of *L. perenne* × *L. multiflorum*, *M. sativa*, and *T. repens* was stored in the laboratory for 9 months at −18, 7, 21, and 36°C and various moisture contents up to 205 g/kg DM there were no significant changes in the CP concentration (Greenhill *et al.*, 1961). Under farm conditions the situation is very different. In South Australia, rectangular bales, round bales, and fodder rolls are left in the field for up to 9 months (Newbery and Radcliffe, 1974). During this time CP concentration decreased from 226 to 169 g/kg DM, the largest decreases occurring in the rectangular bales.

2. ENSILING

The CP concentration of silage is almost the same as that of the fresh forage ensiled (Russell *et al.*, 1978; Flores *et al.*, 1986). However during ensiling there is an increase in the proportion of ammonia, nonprotein N, and free amino acids (Table 6.14). These changes could cause a reduction in the quantity of CP passing from the rumen into the small intestine (Table 6.3). By wilting *L. perenne* before ensiling, degradation of CP to ammonia was reduced and the proportion of the CP present as true protein was increased from 0.30 to 0.45 (McDonald, 1982). Other aspects of the effect of method of conservation on the nitrogeneous components of forages have been reviewed by McDonald and Whittenbury (1973).

III. DIAGNOSIS OF PROTEIN DEFICIENCY

A. Clinical Signs

Protein is probably only second to energy as the nutrient most commonly limiting animal production from forage, but there appears to be no

TABLE 6.14

Effect of Ensiling on the Total Crude Protein and Its Composition in *Medicago sativa*[a]

Component	Fresh	Ensiled
Total crude protein (g/kg DM)	225	219
Crude protein fraction		
Ammonia (g/kg DM)	6	24
Nonprotein (g/kg DM)	61	124
Free amino acids (g/kg DM)	17	41

[a]Adapted from Flores *et al.* (1986).

specific method of visually diagnosing this deficiency in animals. The main effect of a CP deficiency is to reduce forage intake (Chapter 2) and efficiency of utilization of metabolizable energy (Chapter 5). The visual signs of a protein deficiency are no different from those caused by any other nutritional factor which limits intake. The main signs are slow growth, low milk production, and poor reproduction.

B. Forage Analysis

Samples of forage for analysis may be cut, plucked by hand, or collected from animals fitted with an esophageal fistula. The use of an esophageally fistulated animal overcomes the problem of selective grazing, but the samples are contaminated with salivary N (Hoehne *et al.*, 1967). The quantity of salivary N is usually small and for many purposes can be ignored (Galt and Theurer, 1976).

1. Crude Protein Determination

The CP concentration in forage may be determined by the Kjeldahl procedure but for rapid estimation of protein a range of chemical and physical methods is now available (Lakin, 1978), including near-infrared reflectance (NIR) spectroscopy (Norris *et al.*, 1976; Shenk *et al.*, 1979). This method is rapid and provides an estimate of CP with a low error if the instrument is calibrated with samples similar to those being analyzed. However, when used to predict the CP concentration in forage species or plant fractions that were not used to calibrate the instrument, the predicted values for CP may be seriously biased (Minson *et al.*, 1983).

2. Crude Protein Degradability

Earlier in this chapter it was shown that the CP in forage is partly degraded in the rumen but this potential loss is partially or completely offset by the synthesis of microbial protein. In feeds with a high CP concentration, a high degradability may lead to a loss of ammonia from the rumen and poor utilization of the protein (Chalmers and Synge, 1954). Conversely, a low degradability of the CP can cause a deficiency of ammonia required for the synthesis of microbial protein (ARC, 1980).

Many different methods have been used to determine CP degradability of small samples of forage (Table 6.15). These methods have been reviewed (Jarrige, 1980; Osbourn and Siddons, 1980; Cottrill, 1982) so only a summary of the present position is warranted. The degradability of CP has been determined by measuring CP solubility in boiling water, mineral salt solutions, or autoclaved rumen fluid; by rate and extent of hydrolysis by protease or proteolytic rumen bacteria; and by measurement of the

TABLE 6.15

Methods Used to Determine the Degradability of Crude Protein in Small Samples of Feed

Method of analysis	Reference
Protein solubility in	
autoclaved rumen fluid	Wohlt *et al.* (1973), Crawford *et al.* (1978), Waldo and Goering (1979)
McDougal's artificial saliva	Crooker *et al.* (1978), Chamberlain and Thomas (1979)
Burrough's solution (10%)	Wohlt *et al.* (1973), Crawford *et al.* (1978)
NaCl 1.00 *M*	Annison *et al.* (1954),
0.15 *M*	Crooker *et al.* (1978)
NaOH 0.02 *N*	Craig and Broderick (1981)
Boiling water	Waldo and Goering (1979)
Protein hydrolysis by	
Protease	Chamberlain and Thomas (1979), Laycock *et al.* (1985)
Proteolytic rumen bacteria	Laycock *et al.* (1985)
Net ammonia production by rumen fluid (39°C)	Crooker *et al.* (1978), Chamberlain and Thomas (1979), Raab *et al.* (1983)
In sacco	
Nylon	Bailey (1962), Aii and Stobbs (1980), Lindberg and Varvikko (1982), Dove and McCormack (1986)
Terylene	Playne *et al.* (1972)
Polyester	Crawford *et al.* (1978), Mehrez and Orskov (1977), Nocek (1985), Dhanoa (1988)

ammonia produced when forage is incubated *in vitro* with rumen fluid. The degradation of CP has also been determined by measuring the rate and extent of protein lost from forage samples enclosed in synthetic cloth bags hung in the rumen.

Protein degradability by the different laboratory and *in sacco* methods has been compared. High correlations were found between the solubility of CP in a mineral buffer solution and the disappearance of CP from dacron bags suspended in the rumen (Crawford *et al.*, 1978). Comparisons of this type assume that the CP lost in the *in sacco* technique is constant for each feed, but it is now known that it is affected by the diet fed to the rumen-fistulated animals (Siddons and Paradine, 1981). At present there is no information on which of the methods shown in Table 6.15 provides the best estimate of the proportion of forage CP that flows into the duodenum.

C. Fecal Analysis

Feces are the residue of forage eaten, metabolic secretions, and microbial organic matter. The possibility of predicting the CP concentration of forage (g/kg DM) from the CP concentration in feces (g/kg OM) was first examined by Raymond (1948). The relation between the two factors for sheep was:

$$\text{Forage CP} = 8.8 + 0.80 \text{ fecal CP}$$

A similar equation was derived for sheep in Australia (Fels *et al.*, 1959):

$$\text{Forage CP} = 6.6 + 0.93 \text{ fecal CP}$$

Using plucked forage samples and feces produced by grazing cattle and published data Moir (1960a) derived a logarithmic equation:

$$\text{Forage CP} = 8.4\,(\text{fecal CP} - 55)^{0.60}$$

When the feces of grazing cattle contained less than 80 g CP/kg the diet was deficient in CP and production was improved by feeding a urea supplement (Winks *et al.*, 1979). However, when animals have access to forage high in phenols, fecal CP levels are inflated and the method cannot be used to predict the CP concentration of the diet (Robbins *et al.*, 1987). This same reservation applies to forage legumes that contain tannins or other phenolic compounds that depress CP digestibility.

D. Tissue Analysis

1. Rumen Ammonia

Ammonia is produced in the rumen by degradation of the CP entering the rumen in forage and saliva. A high ammonia concentration in the rumen is indicative of high levels of degradable protein in the diet. Conversely, a low level of rumen ammonia is found when the quantity of degradable CP is low, either because the CP has a low degradability coefficient or because the total CP concentration is low. When stock ate mature tropical forage, rumen ammonia concentration was positively correlated ($r = 0.58$) with the CP concentration in the diet (Playne and Kennedy, 1976).

Samples of rumen fluid for ammonia analysis are usually collected from animals fitted with rumen fistulae; they are then centrifuged, acidified, and stored at 20°C until analysed (Wiedmeier *et al.*, 1986). Ammonia concentration tends to be lowest in the midventral sac of the rumen and samples collected through an esophageal tube underestimate mean ammonia concentration in the rumen by about 13% (Wiedmeier *et al.*, 1986).

A number of studies, reviewed by Hoover (1986), have demonstrated that maximum microbial growth rate and organic-matter digestibility are achieved when the concentration of rumen ammonia is 10 to 60 mg N/liter, but in other studies 70 to 76 mg N/liter was required for maximum microbial activity. When *Digitaria decumbens* and *H. contortus* containing 50 and 25 g CP/kg, respectively, were supplemented with various levels of urea and sulfur, maximum microbial fermentation occurred when rumen ammonia concentrations reached 60–80 and 45 mg N/liter, respectively (Hunter and Siebert, 1985a; Boniface *et al.*, 1986). However, the optimum rumen ammonia concentration for maximum intake exceeded 120 mg N/liter (Boniface *et al.*, 1986), indicating that the critical levels of rumen ammonia are different for maximum microbial growth and maximum voluntary intake.

2. Blood Urea

Blood samples are usually easier to collect than samples of rumen fluid and it would be very convenient if blood urea could be used to predict ammonia level in the rumen. The concentrations of blood urea and rumen ammonia are related but the correlation is low, possibly because the concentration of blood urea is also affected by the level of energy in the diet (Haaland *et al.*, 1977).

IV. PREVENTING PROTEIN DEFICIENCY

A. Supplementary Protein

When the CP concentration in forage is low the deficiency can be alleviated by feeding additional protein, nonprotein nitrogen, or a combination of both forms of supplement. Nonprotein nitrogen (NPN) is the cheapest form of supplementary CP but is only effective under a limited range of conditions (McDonald, 1968a; Kempton, 1982; Hunter and Vercoe, 1984).

1. Nonprotein Nitrogen

Microbes present in the rumen have the ability to convert NPN into microbial CP which the animal can absorb from the small intestine (Section I,D). The NPN will only be converted into microbial CP if other nutrients required by the microbes are available in sufficient quantities. These include sulfur (Loosli and Harris, 1945; Hume and Bird, 1970; Coombe *et al.*, 1971), branch-chain acids (Allison *et al.*, 1962), specific amino acids and peptides (Clark and Petersen, 1988), and energy in the quantity (Chalupa, 1972) and form that can be used by the rumen microbes in the synthesis of CP (Johnson, 1976).

Low-CP forage sometimes contains sufficient quantities of all the nutrients except ammonia, required by rumen microbes, and only supplementary urea is required to improve production. For example, the loss in weight of weaner cattle grazing mature tropical grass was reduced from 0.14 to 0.06 kg/day when supplemented with 25 g urea/day (Alexander *et al.*, 1970). Including urea in the drinking water (1.2 g/liter) increased the milk production of ewes and the growth rate of their lambs when offered mature pastures containing *Astrebla* spp. and *Iseilema* spp. (Table 6.16).

Maximum efficiency of microbial protein production will, in theory, only be achieved if ammonia is released from urea at the same slow rate as the energy from the forage (Johnson, 1976). This concept has led to the development of many different forms of slow-release urea (Forero *et*

TABLE 6.16

Effect on Production of Urea Added to the Drinking Water Offered to Ewes Grazing Mature Tropical Grasses[a]

Measurement	Treatment: Control	Urea
Milk yield (ml/day)		
7 weeks	372	444
12 weeks	318	414
Lamb growth (g/day)		
5–9 weeks	89	113
9–12 weeks	33	104
Lamb liveweight (kg)		
5 weeks	6.2	6.2
12 weeks	9.4	11.6

[a]Adapted from Stephenson *et al.* (1981).

TABLE 6.17

Effect of Supplementary Molasses and Urea on the Performance of Yearling Cattle Grazing Mature Tropical Pasture[a]

Treatment	Liveweight change[b] (kg/day)
No supplements	−0.15
Molasses (0.23 kg/day)	−0.12
Molasses (0.23 kg/day) + urea (56 g/day)	+0.02

[a]Adapted from Winks *et al.* (1979).
[b]Mean for 3 yr.

al., 1980; Makkar *et al.*, 1988), but there is no convincing evidence that these are better than normal urea when supplementing grazing ruminants.

Urea is often fed with molasses, which not only acts as an attractive carrier for the urea, but supplies both energy and S. When fed alone molasses has little effect on the loss of weight of animals on feeds low in CP, but when given in combination with urea cattle are able to maintain weight throughout the winter (Table 6.17). Urea–molasses is generally considered to be a "survival" supplement for ruminants grazing mature, low-CP forages. For example, pregnant cows lost 0.82 kg/day when eating mature *H. contortus* containing 25 g CP/kg, but when fed urea and S this loss was reduced to 0.31 kg/day (Table 6.18). This improvement was partly due to a 48% increase in the quantity of forage eaten. Practical

TABLE 6.18

Effect of Supplementary Nonprotein Nitrogen and Sulfur, *Medicago sativa*, and Protected Cottonseed Meal on the Liveweight Change of Pregnant Cows Eating Mature *Heteropogon contortus*[a]

Supplement	Intake (kg/day)			Liveweight change (kg/day)
	Supplement	Forage[b]	Total	
No supplement	0	4.24	4.24	−0.82
Urea + sulfur	0.06	6.27	6.33	−0.31
Medicago sativa	1.79	6.68	8.47	0.40
Urea + sulfur + bypass cottonseed meal	1.01	8.12	9.13	0.75

[a]Adapted from Lindsay *et al.* (1982).
[b]25 g CP/kg DM.

aspects of feeding urea–molasses have been reviewed by Alexander (1972).

2. BYPASS PROTEIN

Supplementing with NPN usually only reduces the rate of loss in weight, converting a submaintenance diet to, at best, a maintenance diet. Further improvement in production can only be obtained by feeding a natural protein which is only slowly degraded by the rumen microbes and bypasses the rumen (Chalupa, 1975). The importance of bypass protein in preventing protein deficiency is illustrated by the results shown in Table 6.18. In this study feeding a small quantity of cottonseed protein, which had been protected against excess degradation with formaldehyde, increased total food intake 44% above that of the treatment with the urea and S supplement. The bypass protein changed a loss of 0.31 kg/day to a gain of 0.75 kg/day. *Medicago sativa* hay was less effective in stimulating intake of the mature forage and increasing growth, because the CP was unprotected from degradation in the rumen. Other studies have shown a similar advantage of natural proteins over urea when supplementing low-CP forages (Nelson and Waller, 1962; Williams *et al.*, 1969; Oltjen *et al.*, 1974).

The benefit of bypass protein is not limited to mature low-CP forages but also occurs with animals grazing immature high-CP pastures. This is illustrated by the higher milk production of grazing cows fed protected casein (Table 6.2). However, it would be wrong to assume that a high proportion of bypass protein is always beneficial. When formaldehyde-protected soybean meal was substituted for unprotected soybean meal in a supplement fed to cows receiving *L. perenne* silage (163 g CP/kg DM) there was no effect on milk production or composition (Castle and Watson, 1984a). Gill and England (1984) concluded that with silage of poor fermentation quality and low CP (103 g CP/kg DM), factors other than degradability have a greater influence on the response to protein supplementation.

The optimum degradability of protein supplements can be calculated if information is available on the quantity of dry matter and CP in the forage ingested, the degradability of the forage CP, and the form and level of animal production (Satter and Roffler, 1977; ARC, 1984). With grazed forage these factors are constantly changing, so the optimum degradability of supplementary CP will fluctuate.

B. Fertilizer Nitrogen

As indicated in Section III,E the CP concentration in forage can be increased by applying fertilizer N. With pastures containing both grasses and legumes, fertilizer N depresses the proportion of legume. The net result is that fertilizer N has little effect on the CP concentration of mixed pasture until the rate exceeds about 94 kg/ha/cut (Ferguson and Terry, 1956).

With pure grass pastures the magnitude of the increase in CP concentration depends on the time interval between applying the N to the stubble and the next harvest (Fig. 6.8). Maximum increase in CP concentration occurred 2 weeks after applying the fertilizer and then decreased because the fertilized grass grew faster. By the tenth week there was little or no difference in CP concentration in the grass fertilized at different levels (Fig. 6.8). In some studies with tropical grasses the interval between applying fertilizer N and harvesting exceeded 2–3 months and the fertilizer had little or no effect on CP concentration (Oakes and Skov, 1962; Henzell, 1963; Oakes, 1966). The magnitude of the increase in CP concentration depends on the extent of growth after the fertilizer is applied. In the dry season, fertilizer N increased the CP concentration of *C. dactylon*

TABLE 6.19

Effect on Crude Protein Concentration of Applying Fertilizer Nitrogen to Mature Forage Several Weeks Prior to Harvesting[a]

Species	Fertilizer nitrogen (kg/ha)	Crude protein (g/kg)		Reference
		Control	Increase	
Cynodon dactylon	47	75	25	Barnes (1960)
Cynodon dactylon	50	51	38	Kretschmer (1964)
	100	51	84	
	150	51	96	
	200	51	100	
Digitaria decumbens	50	35	46	Kretschmer (1964)
	100	35	65	
	150	35	84	
	200	35	101	
Digitaria decumbens	58	43	36	Minson (1967)
Lolium perenne	58	100	20	Ferguson (1948)
Panicum maximum	53	90	23	Owen (1964)
Temperate forage	26	101	10	Lewis (1941)
	53	101	17	
	28	68	7	Moon (1954)
	51	68	16	
	80	67	19	
Mean	86	62	46	

[a]Data adapted.

by 46 g/kg DM, compared to a rise of only 16 g/kg DM in the wet season (Herrera, 1979).

When fertilizer N is applied to mature forage a few weeks before harvesting, the CP concentration is raised to a higher level than if the same quantity of fertilizer is applied when the forage is immature and rapidly growing (Table 6.19). On average the CP concentration is increased by 46 g/kg DM when the fertilizer application is 86 kg N/ha. A similar quantity of fertilizer applied at the start of growth would have increased the CP concentration by about 10 g/kg DM (Imperial Chemical Industries, 1966). It is clear that rainfall plays a major role in the magnitude of the response to late application of fertilizer N (Moon, 1954; Owen, 1964). If the soil is dry and no rain falls after the fertilizer is applied, the absorption of the additional N is low and the increase is small. Conversely, if rainfall is high after application. absorption of N is high and the forage would initially have a high CP content, but with increased growth dilution of the extra CP would occur. The extra CP produced by late applications of fertilizer N contains a smaller proportion of true protein and a higher proportion of nitrate N than the CP in the unfertilized forage (Lewis, 1941; Ferguson and Terry, 1956).

C. Use of Legumes

Legumes have a higher CP concentration than grasses (Fig. 6.5) so including a legume in the pasture would reduce the probability of a CP deficiency. Legumes also have the ability to fix atmospheric nitrogen, increase soil N, and improve the concentration of CP in grasses growing in association with the legumes (Table 6.20). An increase in the proportion

TABLE 6.20

Effect of a Companion Legume on the Crude Protein Concentration of Grass (g/kg DM)

Species	Treatment: Grass alone	Grass plus legume	Reference
Temperate grasses with *Lotus corniculatus*	136	140	Abdalla *et al.* (1988)
Tropical grasses and legumes	42	65	Stobbs (1966)
Tropical grasses with *Lotononis bainesii*	93	101	Jones *et al.* (1967a)
Tropical grasses with *Macroptilium atropurpureum*	93	116	Jones *et al.* (1967a)
Mean	91	106	

of legume in the diet will raise the concentration of CP and increase the quantity of CP leaving the rumen. The *total* quantity of CP leaving the rumen will be further increased by the higher voluntary intake of diets containing legume (Chapter 2).

The CP in *Leucaena leucocephala* is only slowly degraded in the rumen (Aii and Stobbs, 1980) and when fed as a supplement to cows grazing *Chloris gayana* increases milk production to the same level as achieved by feeding protected casein (Flores *et al.*, 1979). *Lotus corniculatus* contains tannins which also reduce CP degradation (Ulyatt *et al.*, 1976; Section IV,E,2) and increase the quantity of CP absorbed from the intestine.

D. Chemical and Physical Treatment

Many methods have been used to improve the quantity of amino acids absorbed from forage diets. These may be divided into two main groups: methods that increase the total CP of the forage by the addition of urea or ammonia and those that increase the quantity of amino acids that pass from the rumen to the small intestine.

1. Urea and Ammonia

The CP concentration of hay can be increased by applying granular urea at the time of cutting or during baling (Van de Riet *et al.*, 1988). When applied to *L. perenne* at the time of cutting, granular urea increased the CP from 136 to 156 g/kg DM, with 23% of the urea being retained by the forage. Applying the urea during baling proved less successful. Only 16 and 5% of the urea was recovered when the hay was baled at a high and low level of moisture, respectively.

Both aqueous and anhydrous ammonia have been used to increase the digestibility and CP concentration of straws and other low-quality forages (Sundstol *et al.*, 1978). The retention of anhydrous ammonia varied with the moisture content of the forage and the quantity of ammonia applied (Moore *et al.*, 1985). In laboratory studies with *D. glomerata* containing 500 g DM/kg, all the added ammonia (36 g/kg DM) was retained, but when the forage was very dry (900 g DM/kg) only 52% of the ammonia was retained. Sixty-three days after applying the ammonia, between 37 and 50% of the nitrogen was still present as ammonia.

Large-scale studies of ammoniation gave similar results to those reported from laboratory studies, with poor retention of ammonia by dry hay (Knapp *et al.*, 1975). Hay made from *T. pratense/P. pratense* retained only 20% of the applied ammonia. In another study using hay made from *Festuca arundinacea*, CP concentration was increased from 79 to 169 kg DM by applying anhydrous ammonia at 30 g/kg DM (Buettner *et al.*,

1982). When the hays were fed to sheep, N retention was similar for all treatments.

Treating high-quality grasses such as cereal grain hay, sorghum hay, and immature grass hay with ammonia can produce compound(s) which cause hyperexcitability characterized by sudden galloping in circles (Weiss *et al.*, 1986a; Perdok and Leng, 1987). This disorder only occurs in cereal straws when the temperature of the ammoniated straw exceeds 70°C during treatment (Perdok and Leng, 1987). With hay it is more likely in a dry season when hay contains a high level of readily available carbohydrates.

Anhydrous ammonia has also been used to increase the CP concentration of silage made from *D. glomerata*/*T. repens* pasture (Moore *et al.*, 1986). When applied at a rate of 30 g/kg DM, most of the ammonia was retained, raising the CP concentration from 100 to 228 g/kg, but this treatment had no effect on the growth rate of steers.

2. Formaldehyde Treatment

Treatment of forages with the correct quantity of formaldehyde can, by cross-linking and modifying amino acids and other functional groups, decrease protein solubility and microbial degradation in the rumen without impairing the subsequent absorption of amino acids from the small intestine (Ferguson *et al.*, 1967). When formaldehyde was applied to a mixed dried forage (70% *Trifolium alexandrinum* and 30% *L. perenne*) at 10 g/kg DM, degradation of CP in the rumen was reduced, the quantity of nonammonia CP digested in the intestine was increased by 60%, and wool production was improved by 15% (Table 6.21). In another study, treating *M. sativa* hay (171 g CP/kg DM) with formaldehyde increased wool production by 6 and 11% when the hay was fed at restricted and ad libitum levels of intake, respectively, and improved the efficiency of utilization of digestible energy for growth (Barry, 1976). The CP in *M. sativa* was rendered indigestible when formaldehyde was applied at 20 g/kg DM, resulting in a larger loss of CP in the feces and lower retention of N by sheep (Yu, 1978).

Favorable responses have been reported when grass hays were treated with limited quantities of formaldehyde. When formaldehyde was added to *C. dactylon* hay at 15 g/kg DM the flow of CP from the rumen was increased by 50% and absorption of amino acids was improved (Amos *et al.*, 1976). Fresh forage has also been treated with formaldehyde (Beever *et al.*, 1987). Spraying formaldehyde onto cut *L. perenne* and *T. repens* several hours before feeding increased by 25% the quantity of CP entering the small intestine. This increase was caused by reduced degradation of dietary CP in the rumen and an increase in the efficiency of microbial CP synthesis. It is probable this response is partly due to changes in the rumen microbial population.

Addition of formaldehyde to legume-dominant forage before conserving reduced fermentation temperature, production of NPN, and volatile fatty acids (Barry and Fennessy, 1972; Thomson *et al.*, 1981). It also reduced the proportion of CP that was soluble in hot water (Waldo, 1975) and mineral buffer solubility and rate of loss from porous bags suspended in the rumen (Siddons *et al.*, 1982). When treated silage was fed to sheep the concentrations of ammonia and iso- and *n*-valeric acid in the rumen were reduced, indicating that CP degradation during the combined ensiling plus rumen fermentation processes was reduced by formaldehyde (Barry and Fennessy, 1973).

Adding formaldehyde to *L. perenne* during ensiling increased the quantity of CP leaving the rumen from 78 to 115 g/kg forage DM and the absorption of nonammonia CP from the intestine from 119 to 148 g/kg forage (Thomas *et al.*, 1981). With silage made from *D. glomerata*, addition of formaldehyde increased the growth rate of heifers from 0.37 to 0.57 g/day (Waldo, 1975).

3. Dehydration

Artificial drying reduces the deamination of CP in the rumen and increases the quantity of nonammonia CP entering the small intestine (Beever *et a.*, 1976; Beever and Thomson, 1981; Amos *et al.*, 1984). However, this advantage may be offset by a lower absorption of amino acids from the small intestine. With *L. perenne* dried at 103°C absorption of CP from the small intestine was increased 32% (Beever *et al.*, 1974), but

TABLE 6.21

Effect of Feeding Forage Treated with Formaldehyde on Crude Protein Absorption and Wool Production in Sheep[a]

Measurement (g/day)	Untreated	Treated	Change (%)
Organic matter intake	610	606	−1
Crude protein intake	179	180	1
Nonammonia CP			
Leaving rumen	126	187	48
Digested in intestine	90	144	60
Excreted in feces	36	43	19
Wool production	10.6	12.2	15

[a]Adapted from Hemsley *et al.* (1970).

when *T. pratense* was dried at 145°C absorption of amino acids was reduced by 20% (Beever and Thomson, 1981).

4. Pelleting

Pelleting reduces CP degradation in the rumen and increases the flow of nonammonia CP from the rumen (Coelho da Silva *et al.*, 1972b) (Table 6.4). The additional CP in the rumen increases the quantity of amino acids flowing to, and absorbed from, the small intestine. It has been suggested that the increased absorption of protein may contribute to some of the improvement seen in the net energy value of pelleted diets (Beever *et al.*, 1981).

E. Plant Breedings and Selection

The quantity of amino acids absorbed from the small intestines might be improved by forage breeding and selection. Four different approaches are possible. The least obvious but probably the most promising methods are to increase the voluntary intake and digestibility of forage, approaches already considered in Chapters 2 and 4. The other two methods are selection for high CP and lower degradation of CP in the rumen.

1. Selection for Crude Protein

Varietal and genotypic differences have been found in the grasses *C. dactylon, D. glomerata, L. perenne, Phalaris arundinacea, P. aquatica,* and the legumes *Medicago falcata, M. media,* and *M sativa* (Cooper, 1973). Although there is scope for selection, the differences are small (Section II, B, 4) and the CP concentration in grasses is negatively correlated with yield of dry matter (Cooper, 1973) and concentration of soluble carbohydrate (Hacker, 1982). For example, the CP concentration in *Pennisetum glaucum* was improved by increasing the proportion of leaf and reducing the length of the stem internodes. This gain in CP concentration was offset by a lower carrying capacity and the total liveweight gain per hectare was not improved when the high- and low-CP lines were grazed by cattle (Burton *et al.*, 1969). The CP concentration in the leaf, sheath, stem, and the head of *P. aquatica* was improved by selection but the yield of dry matter and CP was depressed by 20 and 13%, respectively (Clements, 1973).

2. Selection for Tannin

The presence of tannin in legumes decreases CP degradation in the rumen (McLeod, 1974; Barry *et al.*, 1986; Barry and Blaney, 1987). Low levels of tannin will increase the quantity of amino acids absorbed from the intestines, but high levels of tannin will reduce voluntary intake and depress production. The optimum tannin concentration in forage has not been determined but will probably vary with the level of CP in the forage, the energy available for the synthesis of microbial CP, and the extent that tannin depresses voluntary intake of the forage. It will be necessary to know the level of soil S because tannin concentration in *L. corniculatus* is depressed by the application of fertilizer S (Lowther *et al.*, 1987).

In *L. corniculatus* there are large differences in tannin content between cultivars (Lowther *et al.*, 1987). The semi-erect cultivars have lower concentrations of condensed tannins (1.3 to 8.4 g/kg DM) than the erect cultivars (11.6 to 39.0 g/kg). Large differences in tannin levels have also been found in *Lespedeza juncea* (Donnelly and Anthony, 1969), so it should be possible to select a cultivar of these legumes with the optimum level of tannin for maximum animal production. There are no reports of tannins being present in grasses but it may be possible to transfer the gene or genes responsible for tannin formation into grasses and reduce the rate of CP degradation in the rumen.

F. Antibiotic Additives

Two groups of compounds have been used to increase the quantity of amino acids absorbed from the small intestine. These are ionophores including monensin, lasalocid, salinomycin, and ICI 139603, and the glycopeptide antibiotic, avoparcin (Rowe, 1985). The use of ionophores for improving efficiency of energy utilization has already been discussed in Chapter 5. Feeding ionophores reduces the protozoal population in the rumen (Poos *et al.*, 1979; Habib and Leng, 1986), the level of rumen ammonia (Dinius *et al.*, 1976), and the degradation of dietary CP (Poos *et al.*, 1979; Russell and Martin, 1984; Rowe, 1985). This advantage is partly offset by a reduction, of about 20%, in the synthesis of microbial protein (Rowe, 1985). When forage contains large quantities of potentially degradable CP, feeding an ionophore increases the quantity of amino acids absorbed from the intestine. In grazing studies, summarized by Rowe (1985), including an ionophore in a grain supplement increased the mean liveweight gain of cattle from 0.72 to 0.83 kg/day. The largest improvements occurred when the control group had the lowest growth rate. These improvements were probably due to both an improved absorption of amino acid and a rise in the efficiency of utilization of the metabolizable energy in the forage (Chapter 5). Avoparcin caused a decrease in the degradability of dietary CP in the rumen (Jouany and Thivend, 1986) and increased the efficiency of absorption of amino acids from the small intestine (Rowe, 1985).

TABLE 6.22

Digestion of Crude Protein in *Medicago sativa* Hay by Faunated and Defaunated Sheep[a]

Crude protein (g/day)	Faunated	Defaunated
Intake	142	138
Entering duodenum	151	210
Microbial[b]	83	105
Undegraded	67	94
Digested in intestine	101	156
Fecal output	50	54

[a]Adapted from Ushida *et al.* (1986).
[b]Mean of three different methods.

The presence of protoza in the rumen appears to impart no nutritional advantage to ruminants and there is an increasing body of evidence that rumen protoza reduce the protein leaving the rumen (Bird and Leng, 1985). In a comparison of untreated and defaunated sheep fed *M. sativa* hay the absence of protozoa reduced CP degradation and increased the quantity of microbial protein leaving the rumen (Table 6.22). The higher absorption of amino acids in defaunated sheep fed marginal diets improved wool growth by 37% (Bird and Leng, 1985). Defaunation of lambs grazing *Avena sativa* resulted in an increased in body weight gain (23%) and wool growth (19%) (Bird and Leng, 1984), but there was only a small response to defaunating grazing ewes, except in a drought year (Bird and Leng, 1985). In these experiments the grazing animals were maintained free of protozoa for the duration of the trial. Although defaunation did not always improve production, it never depressed production. The methods at present used to defaunate ruminants cannot be applied in the grazing industry and there is a need for an acceptable antiprotozoal chemical.

V. CONCLUSION

Milk, meat, and wool are synthesized by ruminants from amino acids and other compounds absorbed from the small intestine. These amino acids come from forage protein that escapes microbial degradation in the rumen and from rumen microbes that flow from the rumen into the small intestine. Microbial protein is synthesized from the protein and nonprotein nitrogen and sulfur present in the forage. For practical purposes all forms of protein or potential protein are considered as a single fraction, crude protein (CP), defined as 6.25 × nitrogen concentration.

A high proportion of the CP in fresh forage and silage is degraded by the rumen microbes (mean 0.73). The resulting amino acids and ammonia together with salivary CP are used in the synthesis of microbial CP within the limits set by the energy released from the forage. With fresh forages containing less than about 130 g CP/kg DM these changes in the rumen increase the quantity of amino acids absorbed from the intestine, but a net loss of amino acids occurs with forages high in CP. The extent of these changes in the rumen is dependent on many plant and animal factors. Published data on the CP requirements of ruminants show that all forms of production can be limited by a deficiency of CP in forage and that with young lambs and lactating animals the high degradability of the CP in fresh forage and silage can also be important.

The concentration of CP in forage is affected by many plant, environmental, and management factors. Legumes contain more CP than grasses (170 versus 115 g/kg) and temperate grasses more than tropical grasses (129 versus 100 g/kg). Leaves have a higher CP than stems, and CP concentration declines as forage matures. Temperature and light intensity have little effect on the CP concentration in adequately fertilized forage, but the effect of water deficiency on CP depends on the duration of the stress. Drying, particularly at high temperatures, will reduce CP degradation by rumen microbes. Ensiling has no effect on CP concentration but increases the proportion of nonprotein nitrogen.

A deficiency of CP in ruminants leads to low production with no specific symptoms. Deficiencies of CP may be identified by analyzing a sample of the forage eaten or the feces produced and comparing it with published CP requirements. Lambs or lactating animals have a high requirement for amino acids and a measure of potential degradation could be helpful. Many methods have been published, but these have not been related to the quantity of undegraded forage passing into the small intestine and there is conflicting evidence on the concentration of rumen ammonia required for maximum microbial activity.

Protein deficiency can be prevented in many ways. Supplements of nonprotein N can only convert a submaintenance forage diet into one that will maintain ruminants. The CP of conserved forage can be increased by adding urea or ammonia, but their value to the animal is usually limited unless an energy supplement is also provided.

Higher levels of production can only be achieved with supplements containing bypass protein of low rumen degradability. The CP of forage can be increased by applying fertilizer nitrogen a few weeks before it is grazed or cut or by including a legume.

The quantity of CP absorbed from the small intestine can sometimes be improved by reducing the degradation of forage CP in the rumen. This can be achieved by the use of forages containing condensed tannin, by drying, or by treatment with formaldehyde. The degradation of forage protein can also be reduced by defaunating or feeding antibiotics.

7

Calcium

I. CALCIUM IN RUMINANT NUTRITION

A. Function of Calcium

Cattle contain approximately 12 g calcium (Ca) per kilogram liveweight (ARC, 1965), while for sheep the value is slightly higher (15 g/kg) (Grace, 1983b). Most of the Ca is present in bone (98.5–99.2%) (Grace, 1983b), which acts as a reserve which can be drawn on to maintain a relatively uniform level of Ca in the blood supplying the tissue of the animal. Only 0.33% of the Ca is present in blood, with 0.30% in the plasma and 0.03% in the red blood cells (Grace, 1983b). Calcium is required for normal blood clotting, rhythmic heart action, neuromuscular excitability, enzyme activation, and permeability of membranes.

Low levels of Ca in the blood of lactating animals can lead to ''parturient paresis'' or milk fever, a problem first described in the eighteenth century (Underwood, 1981). The frequency of the disease has risen as milk yields have increased. In the United Kingdom about 9% of cattle are affected annually (Allen and Davies, 1981). The disease is most likely to occur where the prepartum diet is high in Ca (Boda and Cole, 1954) and is caused by a failure of the hormone system to respond with sufficient speed to the increased Ca demands of lactation by bone resorption and increased absorption of Ca from the digestive tract (Braithwaite, 1976). Management systems to reduce this disease will be considered in Section IV,C.

Deficiencies of dietary Ca in sheep fed high-grain diets weaken bones, deform teeth, and slow growth (Franklin and Johnstone, 1948; Franklin, 1950). Forage diets are rarely deficient in Ca (McDonald, 1968b), with the possible exception of forage grown on leached soils in the humid tropical or subtropical regions (see Section II,G). High levels of oxalate, which reduces the absorption of Ca (see Section I,D), can result in Ca deficiency to ruminants despite high levels of Ca in the forage.

Where diets are low in phosphorus (P), increasing the Ca level exacerbates the P deficiency, reducing growth rate (Theiler *et al.*, 1927, 1937) and conception rate (Hignett and Hignett, 1951; O'Moore, 1952). However, where diets contain adequate levels of P, a high level of Ca in the diet increases both conception (Hignett and Hignett, 1951) and growth rate (Theiler *et al.*, 1937; Vipperman *et al.*, 1969). When the Ca content of the diet of cattle exceeded 7.5 g/kg, growth rates were depressed (Dowe *et al.*, 1957) due to an induced zinc deficiency (Fontenot *et al.*, 1964). Phosphorus is normally absent from the urine of sheep but when the diet contains excess P (>6 g/kg DM) large quantities of P are excreted in the urine. This causes a high incidence of urinary calculi, a problem that may be reduced by increasing the Ca content of the diet (Emerick and Embry, 1963).

These interactions between Ca and P are obviously of practical importance and attempts have been made to take them into account by using the ratio of Ca:P in the diet (Otto, 1938; Wise *et al.*, 1963; Young *et al.*, 1966a) as an indicator of diet sufficiency. As already indicated, the effect of increasing Ca in the diet varies with the P concentration and using the Ca:P ratio fails to help our understanding of these interactions. It would appear more desirable to consider the concentrations of both Ca and P and to ensure that these meet the recommended allowances than to consider the Ca:P ratio. This view was first propounded by Theiler *et al.* (1927) and more recently, in The Netherlands, by the Committee on Mineral Nutrition (1973). They said that "some non-Dutch literature attaches significance to the ratio of Ca:P in the ration. Recent research suggests that the ratio has no special significances. As long as the cow receives enough Ca and P and has an adequate supply of Vitamin D, a wide range of ratios has no effect on production, growth and health." In this chapter and in the following one on phosphorus the subject will be reviewed without further considering Ca:P ratios.

The absorption and metabolism of Ca is under strong hormonal control. The hormones involved and their interaction are beyond the scope of this volume and readers are referred to reviews of this topic (Borle, 1974; Brathwaite, 1976). Vitamin D is also involved in the regulation of Ca and P metabolism but does not appear to limit Ca metabolism in the grazing animal. For information on Vitamin D, readers are referred to reviews by DeLuca (1979) and by Horst (1986).

B. Voluntary Intake and Digestibility

The voluntary intake (VI) of diets low in Ca can be increased by raising the level of Ca in the diet, provided Ca is the primary nutrient limiting intake. With 8-week-old lambs fed a pelleted diet containing 0.7 g Ca/kg additional Ca increased voluntary intake by 34% (Field *et al.*, 1975). With older animals fed similar low-Ca diets only 12 and 6% increases have been reported following addition of Ca (Benzie *et al.*, 1960; Sevilla and Ternouth, 1980), possibly due to the higher body reserves of Ca and a lower requirement.

Calcium increases the animal requirements for P, and when diets are low in P then feeding additional Ca will exacerbate the deficiency. Voluntary intake of 8-week-old lambs was depressed 35% by a Ca supplement when the diet contained 1.1 g P/kg DM (Field *et al.*, 1975). For a 6-month-old lambs the depression in voluntary intake was 7% (Sevilla and Ternouth, 1980).

When sheep receiving *Digitaria decumbens* were fed a Ca supplement there was no effect on VI or OM digestibility (Rees and Minson, 1976). With cattle fed a 71% pasture/hay diet, raising the proportion of Ca from 4.9 to 10.9 g/kg DM had no effect on the digestibility of the dry matter, organic matter, and crude protein; it increased the apparent absorption of P, K, Na, and Mg but depressed the apparent absorption of higher fatty acids and Ca (Hartmans, 1971).

C. Effect on Production

The most dramatic effects of a Ca deficiency are found in sheep fed high-grain diets containing 0.5 to 1.1 g Ca/kg DM. On these rations dental development is defective, with exaggerated irregularities of the incisor and molar teeth. This problem may be completely eliminated by feeding 1% of ground limestone (Franklin, 1950). On a similar low-Ca diet the growth rate of weaner lambs was doubled when a limestone supplement was fed, an improvement associated with a rise in serum Ca. Growth rate of 5-month-old wethers was increased fivefold when a straw/concentrate/starch diet containing 0.8 g Ca/kg DM was supplemented with Ca (Benzie *et al.*, 1960). The Ca supplement improved the ash percentage of the skeleton and increased the length of the long bones.

Silage made from *Zea mays* can be low in Ca and feeding Ca supplements to cows in both the dry and the lactation periods has increased milk production (Kim and Evans, 1975). However, most grazed forages contain sufficient Ca and feeding a Ca supplement to ewes grazing hill pastures in Scotland had no long-term effect on either liveweight or lambing percentage, but there was a suggestion that premature loss of teeth was reduced (Gunn, 1969).

In studies with grazing ewes and lambs, applying calcium carbonate at 1.25 tons/ha increased liveweight by 5 kg/ewe, fleece weight by 0.5 to 0.6 kg/ewe and, in the third year, lamb liveweight at weaning by 16% (O'Connor *et al.*, 1981). This response appeared to be due to increased pasture production rather than to any improvement in pasture quality. Similar results have been reported in other studies with fertilizer Ca (Edmeades *et al.*, 1987).

D. Efficiency of Absorption

The ruminant has only a limited capacity to excrete calcium in the urine and the quantity of fecal endogenous Ca is relatively constant (Braithwaite, 1976). The animal therefore attempts to absorb from the digestive tract only the quantity of Ca it actually requires. The "availability" of Ca in feeds depend on the needs of the animal and is rarely limited by a characteristic of the forage except where oxalate levels are high. Lambs have a high requirement for Ca and absorbed 70% of Ca in a synthetic diet with a low level of Ca but only 40% of the Ca when the Ca level was high (Hodge, 1973). Average absorption of Ca in these feeds dropped from 66% in the first 35 days of the study to 46% in the second half, when the lambs were older and had a lower requirement for Ca. In another study, young wethers absorbed 39% of the Ca in a mixed diet compared with only 14% absorbed by mature sheep (Braithwaite and Riazuddin, 1971). Infusing calcium chloride into the jugular vein, thus reducing the need for absorbed Ca, provided confirmation that these differences are cause by differenes in the animals' requirement for Ca and were not due to changes in efficiency of absorption with age. Calcium absorption of infused wethers was depressed from 30 to 8% (Braithwaite, 1978).

In most studies the quantity of Ca in the feed exceeds the Ca required by the animal. Calcium absorbed in excess of requirements is excreted in the feces and urine. When these losses are taken into account the results are expressed as true availabilities as opposed to apparent availabilities. This correction increases the value for Ca availability but fails to overcome the problem that the values obtained rarely represent the potential availability of Ca in the forage. With this reservation the information on apparent and true availabilities of Ca in forage diets in presented in Table 7.1.

Only a few values have been published for the true availability of Ca in forage and these are on average 24% higher than those for apparent availability. Of particular interest is the effect of the requirements of animals on Ca availability. When a mixed temperature pasture was fed to dry cows only 2% of the Ca was apparently available (Lomba *et al.*, 1970), but when a similar forage was fed to lactating cows the apparent availability increased to 23% (Hutton *et al.*, 1967) (Table 7.1).

This principle not only applies to apparent availability but also to true

TABLE 7.1

Apparent and True Availabilities of Calcium in Forages

Forage	Availability coefficient		Reference
	Apparent	True[a]	
Dactylis glomerata	0.10	0.45	Hansard *et al.* (1957)
D. glomerata (early)	0.38	—	Rook and Balch (1962)
D. glomerata/Lolium perenne (early)	0.27	—	
D. glomerata/Lolium perenne (late)	0.14	—	
D. glomerata			
Low nitrogen	0.10	—	Stillings *et al.* (1964)
High nitrogen	0.08	—	
Festuca pratensis	—	0.34[b]	Gueguen and Demarquilly (1965)
Lespedeza juncea	0.12	0.43	Hansard *et al.* (1957)
Lolium perenne	—	0.31[b]	Gueguen and Demarquilly (1965)
L. perenne	0.26	—	Joyce and Rattray (1970b)
Medicago sativa	0.08	0.46	Hansard *et al.* (1957)
M. sativa	—	0.27[b]	Gueguen and Demarquilly (1965)
M. sativa (silage)	0.08	—	Ivan *et al.* (1983)
Mixed temperate pasture	0.23	—	Hutton *et al.* 1967)
Mixed temperate pasture	0.02	—	Lomba *et al.* (1970)
Phleum pratense		0.36[b]	Gueguen and Demarquilly (1965)
Trifolium repens	0.27	—	Joyce and Rattray (1970b)
Zea mays (silage)	0.33	—	Ivan *et al.* (1983)
Mean	0.13	0.37	

[a]Potential availabilities of Ca are probably higher than the true availabilities listed because ruminants restrict absorption of Ca according to their needs.
[b]Assumed that endogenous loss of Ca is 40 mg/kg liveweight.

availability. With young cattle both the apparent and the true availabilities of Ca in forages are higher than when measured with mature cattle (Table 7.2).

Some of the Ca in forage may be present in the form of calcium phytate or crystals of calcium oxalate (McKenzie and Schultz, 1983). Calcium phytate is not readily digested by monogastric animals but microorganisms in ruminants are able to hydrolyze phytate and the true availability of Ca present in calcium phytate is similar to that of monocalcium phosphate (Tillman and Brethour, 1958). Oxalates are present in high concentrations in some tropical grasses (e.g., *Cenchrus ciliaris, Setaria sphacelata*) and have been suggested as a cause of Ca deficiency in Latin America (Kiatoko *et al.*, 1978) and transit tetany in Australia. Mineral balance studies with cattle fed *C. ciliaris* and *S. sphacelata* containing calcium oxalate crystals showed that the apparent availability of Ca was

TABLE 7.2

Availability Coefficient of Ca in Forages When Measured with Young and Mature Cattle[a]

Forage	Apparent availability		True availability	
	Young steers	Mature steers	Young steers	Mature steers
Dactylis glomerata	0.26	−0.07	0.51	0.39
Lespedeza juncea	0.17	0.08	0.50	0.36
Medicago sativa	0.18	−0.01	0.41	0.31
Mean	0.20	0	0.47	0.35

[a]From Hansard *et al.* (1957).

about half that of calcium carbonate and 20% lower than in *Aristida* spp./*Bothriochloa* spp. mixture and *Triticum aestivum* containing very little oxalate (Blaney *et al.*, 1982). *Medicago sativa* contains 20 to 33% of the Ca in the form of oxalate that is unavailable to ruminants (Ward *et al.*, 1979). The availability of Ca is improved when the rumen microorganisms are adapted to metabolize calcium oxalate (Barry and Blaney, 1987), but the adaption period has not been determined.

Where pastures are closely grazed, sheep and cattle can ingest 75 and 600 kg/yr of soil, respectively (Healy, 1972). This soil apparently has a beneficial effect on Ca metabolism, halving the quantity of Ca excreted in the feces and increasing the quantity of Ca retained by sheep (Grace and Healy, 1974).

E. Requirements for Calcium

In theory, the quantity of dietary Ca required each day can be estimated from the absolute requirements of the animal for the particular level of production and the efficiency of absorption of Ca from the digestive tract. Both parameters are extremely variable. As indicated in Section I,A, ruminants have large reserves of Ca in the bone that can be used to reduce the effect of any Ca deficiency in the diet. The figures for available Ca do not represent the potential availability of the Ca in forages and it is probable that the published Ca requirements of ruminants are overestimated.

Requirements for Ca are usually expressed in terms of grams per day, but to be of any value in the interpretation of analytical data from forages the requirements are required in qualitative terms, i.e., grams Ca per kilogram DM. Qualitative requirements can be derived from published quantitative requirements (ARC, 1980) if assumptions are made on the quantity of forage dry matter required to achieve a particular level of

TABLE 7.3

Calcium Requirements of Beef Cattle Fed ad Libitum Expressed as Dietary Concentrations (g/kg DM)[a]

Liveweight (kg)	Liveweight gain (kg/day)		
	0	0.5	1.0
200	1.2	3.1	4.4
400	1.6	2.6	3.3
600	1.7	2.5	3.0

[a]Adapted from ARC (1980).

TABLE 7.4

Calcium Requirements of Lactating Cows Fed ad Libitum Expressed as Dietary Concentrations (g/kg DM)[a]

Liveweight (kg)	Breed	Milk yield (kg/day)			
		0	10	20	30
400	Jersey	1.6	2.9	4.3	—
500	Ayrshire	1.7	2.5	3.4	4.2
600	Friesian	1.7	2.6	3.4	4.2

[a]Adapted from ARC (1980).

production (Tables 7.3, 7.4, 7.5, and 7.6). For most forms of production, sufficient Ca will be supplied by forage containing about 4.0 g/kg DM, but for rapidly growing lambs higher levels are needed.

II. CALCIUM IN FORAGE

A. Mean Concentration

The Ca concentrations of 1263 forage samples from a wide range of environments have been published in the scientific literature. These had a mean Ca content of 9 g/kg DM and varied from 1 to 40 g/kg DM (Fig. 7.1). Of these samples 31% contained less than 3 g Ca/kg DM. A similar proportion of 1123 samples of forage collected in Latin America (McDowell *et al.*, 1977) and 103 Caribbean grass samples (Devendra, 1977) contained less than 3.0 g Ca/kg DM. A study of tropical forages grown in

TABLE 7.5

Calcium Requirements of Castrated Lambs Fed ad Libitum Expressed as Dietary Concentrations (g/kg DM)[a]

Liveweight (kg)	Liveweight gain (kg/day)			
	0	0.1	0.2	0.3
20	0.9	3.3	6.6	—
30	1.1	2.5	3.7	5.3
40	1.1	2.2	3.2	4.3

[a]Adapted from ARC (1980).

TABLE 7.6

Calcium Requirements of Lactating Ewes Fed ad Libitum Expressed as Dietary Concentrations (g/kg DM)[a]

Liveweight (kg)	Milk yield			
	0	1.0	2.0	3.0
40	1.2	2.5	3.6	—
75	1.4	2.2	3.0	3.7

[a]Adapted from ARC (1980).

Kenya gave a mean value of 3.9 g Ca/kg DM with approximately 35% of the samples below 3.0 g Ca/kg DM (Howard *et al.*, 1962). Other data have also been published on the Ca content of forages grown in various parts of South Africa (Henrici, 1934; Du Toit *et al.*, 1934, 1935, 1940; Van Wyk *et al.*, 1951).

Temperate forages generally contain more Ca than those grown in the tropics (Fig. 7.2). Samples of silage and hay sent for analysis in Scotland had mean levels of 6.2 ± 1.7 and 5.3 ± 1.3 g Ca/kg DM, respectively (Hemingway *et al.*, 1968). Only 12% of the silage samples contained <4.0 g Ca/kg DM compared with 44% of the hay samples. Hay grown in Ireland had a mean Ca concentration of 8 g/kg DM (Wilson *et al.*, 1968), which is similar to that reported by the Pennsylvania State Forage Test Service for over 9500 forage samples grown in 5 yr (Adams, 1975). Mean level of Ca was 8.4 g/kg DM with mean values of 4.0 and 11.5 g/kg DM for grass and legume hays, respectively.

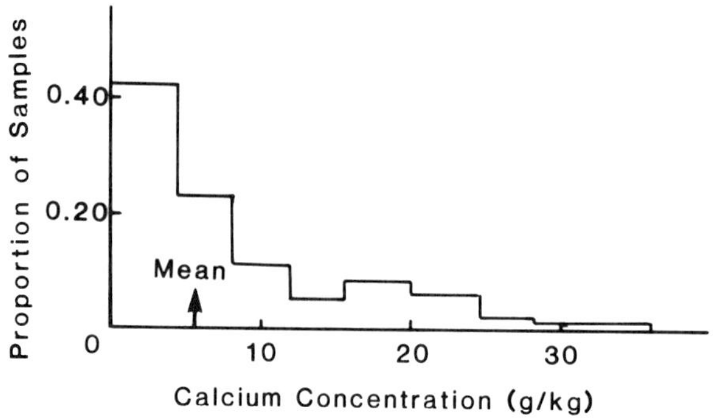

Fig. 7.1 Calcium concentration in a range of forages (data from the world literature).

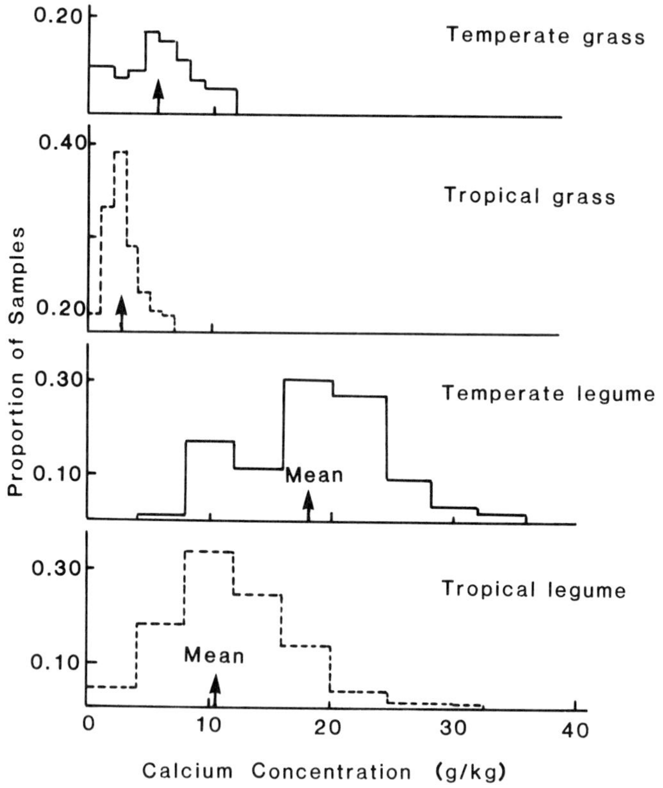

Fig. 7.2 Calcium concentration in temperate and tropical grasses and legumes (data from the world literature).

B. Species Differences

There are many reports that legumes contain higher concentrations of Ca than grasses in both temperate and tropical climates and these differences are illustrated in Fig. 7.2. The Ca concentrations in some of the more common forages are presented in Table 7.7. Temperate and tropical legumes had mean Ca concentrations of 14.2 and 10.1 g/kg DM, respectively compared with 3.7 and 3.8 g/kg DM for the corresponding grasses (Table 7.7).

TABLE 7.7

Mean Concentrations of Calcium in Temperate and Tropical Grasses and Legumes

Forage	Calcium concentration (g/kg DM) Mean	Range	Reference
Temperate grasses			
Avena sativa	2.8	1.8–3.7	Sotola (1937), Van Wyk *et al.* (1951)
Dactylis glomerata	4.4	1.1–8.1	McNaught (1959), Coppenet and Calvez (1962), Whitehead (1966), Patil and Jones (1970), Ando *et al.* (1972), Rosero *et al.* (1980), Forbes and Gelman (1981), Currier *et al.* (1983)
Festuca arundinacea	3.0	2.7–3.3	Coppenet and Calvez (1962), Patil and Jones (1970)
Festuca pratensis	5.2	3.3–7.1	Coppenet and Calvez (1962), Gueguen and Demarquilly (1965)
Hordeum vulgare	3.4	3.3–3.5	Sotola (1937), Van Wyk *et al.* (1951)
Lolium perenne	5.9	2.5–12.5	Fagan (1928), McNaught (1959), Butler *et al.* (1962), Coppenet and Calvez (1962), Gueguen and Demarquilly (1965), Whitehead (1966), Wilson and McCarrick (1967), Joyce and Rattray (1970b), Patil and Jones (1970), Cooper (1973), Reay and Marsh (1976), Forbes and Gelman (1981), Whitehead *et al.* (1983)
Phleum pratense	3.8	2.8–6.0	Coppenet and Calvez (1962), Gueguen and Demarquilly (1965), Wilson and McCarrick (1967), Patil and Jones (1970)
Triticum aestivum	2.5	2.0–3.0	Sotola (1937), Van Wyk *et al.* (1951)
Zea mays	2.3	2.2–2.4	Van Wyk *et al.* (1951), Coppock *et al.* (1972)
Mean	3.7	1.8–9.2	

(*continues*)

TABLE 7.7

(*continued*)

Forage	Calcium concentration (g/kg DM)		Reference
	Mean	Range	
Tropical grasses			
Brachiaria decumbens	3.3	2.0–3.9	Long *et al.* (1970), Perdomo *et al.* (1977), Poland and Schnabel (1980)
Chloris gayana	3.4	2.7–3.8	Harvey (1952), Ando *et al.* (1972), Long *et al.* (1970)
Cynodon dactylon	5.0	3.4–7.4	Jones (1963a), Gomide *et al.* (1969b), Perdomo *et al.* (1977)
Digitaria decumbens	3.9	1.4–6.3	Appelman and Dirven (1962), Gomide *et al.* (1969b), Devendra (1977), Perdomo *et al.* (1977), Poland and Schnabel (1980)
Digitaria pentzii	4.2	2.5–5.8	Devendra (1977), Minson (1984)
Digitaria setivalva	4.2	3.3–5.2	Devendra (1977), Minson (1984)
Digitaria smutsii	4.7	3.6–5.8	Devendra (1977), Minson (1984)
Heteropogon contortus	4.0	2.9–5.2	Jones (1963a), Playne (1969a)
Panicum coloratum	3.4	3.3–3.4	Innes (1947), Devendra (1977)
Panicum maximum	3.7	3.4–4.2	Gomide *et al.* (1969b), Long *et al.* (1970), Perdomo *et al.* (1977)
Paspalum dilatatum	5.3	2.7–7.3	McNaught (1959), Ando *et al.* (1972)
Paspalum notatum	2.6	2.0–3.2	Ando *et al.* (1972), Rojas *et al.* (1987)
Pennisetum clandestinum	4.0	3.9–4.2	Gomide *et al.* (1969b)
Pennisetum purpureum	4.3	2.3–8.2	Innes (1947), Appelman and Dirven (1962), Gomide *et al.* (1969b), Long *et al.* (1970), Devendra (1977)
Setaria sphacelata	3.4	1.2–5.5	Long *et al.* (1970), Bray and Hacker (1981)
Tripsacum laxum	2.2	0.9–3.6	Innes (1947), Appelman and Dirven (1962)
Mean	3.8	0.9–8.2	

TABLE 7.7

(continued)

Forage	Calcium concentration (g/kg DM) Mean	Range	Reference
Temperate legumes			
Medicago sativa	13.6	9.5–16.7	Sotola (1937), Van Wyk *et al.* (1951), Andrew and Norris (1961), Gueguen and Demarquilly (1965), Davies *et al.* (1968), Coppock *et al.* (1972)
Trifolium pratense	14.8	11.3–20.6	Fagan (1928), De Groot (1963), Davies *et al.* (1968), Reay and Marsh (1976)
Trifolium repens	14.3	11.5–17.3	Andrew and Norris (1961), De Groot (1963), Davies *et al.* (1968), Joyce and Rattray (1970b), Forbes and Gelman (1981), Whitehead *et al.* (1983)
Mean	14.2	9.5–20.6	
Tropical legumes			
Centrosema pubescens	10.2	7.8–12.5	Andrew and Norris (1961), Andrew and Robins (1969b)
Desmodium intortum	9.0	8.9–9.0	Dougall and Bogdan (1966), Andrew and Robins (1969b)
Desmodium uncinatum	10.0	8.4–11.6	Andrew and Norris (1961), Andrew and Robins (1969b), Dougall and Bogdan (1966)
Neonotonia wightii	9.8	8.6–10.9	Van Wyk *et al.* (1951), Andrew and Robins (1969b)
Lablab purpureus	13.2	3.1–19.8	Hendricksen and Minson (1985b)
Macroptilium lathyroides	9.8	8.1–11.6	Andrew and Norris (1961), Andrew and Robins (1969b)
Stylosanthes humilis	8.6	8.2–8.9	Andrew and Robins (1969b), Playne (1969a)
Mean	10.1	3.1–19.8	

C. Cultivar Difference

Clones or cultivars of the same forage species frequently differ in Ca concentration. In *Lolium perenne* large differences have been found between the cultivars Hora (5.4 g Ca/kg DM), Perma (4.8 g Ca/kg DM), Premo (3.8 g Ca/kg DM), and Barvestra (3.0 g Ca/kg DM) (Forbes and Gelman, 1981). In another study of eight cultivars of *L. perenne* the highest concentration was found in Sceempter (5.0 g Ca/kg DM) and the lowest in North Irish (4.1 g Ca/kg DM), with similar ranking in two seasons (Whitehead, 1966).

Large differences have been reported between the cultivars of *Dactylis glomerata* (Whitehead, 1966). Cultivar S.143 contained 4.6 g Ca/kg DM compared with a mean of 3.3 Ca/kg DM in seven other cultivars. Large differences have also been reported between progenies of *S. sphacelata* (Bray and Hacker, 1981). In other grasses there is at present no evidence of consistent difference in the Ca levels between cultivars of *Phleum pratense* (Whitehead, 1966; Jones and Walters, 1969), *Digitaria* spp. (Devendra, 1977; Rojas *et al.*, 1987), or *Paspalum notatum* (Rojas *et al.*, 1987).

Small differences have been reported in the Ca concentrations of cultivars of *M. sativa* (Matches *et al.*, 1970). When compared in four American States and 2 yr the cultivar Vernal had a higher concentration of Ca than Du Puits (16.8 and 15.8 g Ca/kg DM). In *Trifolium repens* large consistent differences between clones have been found when grown at five different sites (Robinson, 1942). The high-Ca line contained 25.7 g Ca/kg DM compared with 16.8 g Ca/kg DM for the low line. Varietal differences have also been found in *Trifolium pratense* with higher levels in early- (12.6 g Ca/kg DM) than in late- (9.4 g Ca/kg DM) flowering varieties (Davies *et al.*, 1968). None of these differences in the legumes are of practical importance because the levels of Ca are several times higher than those required by ruminants.

D. Plant Parts

The leaf fraction of forage has a higher Ca concentration than the stem fraction, with mean levels of 8.4 and 4.1 g Ca/kg DM, respectively (Table 7.8). The difference in Ca concentration between leaf and stem fractions was not constant for the same species. The stem fraction of *P. pratense* contained similar levels of Ca in different studies, but the concentration in the leaf fraction varied between sites (Table 7.8). This difference may have been caused by the stage of growth of the forage used in the study. At an immature stage of growth the leaf fraction contained twice the Ca

TABLE 7.8

Calcium Concentration in Leaf and Stem Fraction of Forages

Forage	Calcium concentration (g/kg DM)		Reference
	Leaf	Stem	
Temperate grasses			
Bromus inermis	4.0	2.0	Pritchard *et al.* (1964)
Dactylis glomerata	5.7	2.7	Fleming (1963)
Festuca pratensis	8.7	3.7	Fleming (1963)
Lolium multiflorum	12.3	2.9	Fagan (1928)
Lolium perenne	8.7	3.0	Fleming (1963)
Lolium perenne	5.0	2.6	Laredo and Minson (1975a)
Phleum pratense	8.8	2.1	Fleming (1963)
Phleum pratense	4.2	2.0	Pritchard *et al.* (1964)
Mean temperate grasses	7.2	2.6	
Tropical grasses			
Andropogon gayanus	2.9	2.0	Haggar (1970)
Chloris gayana	8.3	3.4	Laredo (1974)
Chloris gayana	4.4	2.9	Poppi *et al.* (1981a)
Digitaria decumbens	6.9	3.7	Laredo (1974)
Digitaria decumbens	3.7	1.2	Poppi *et al.* (1981a)
Panicum maximum	6.8	2.1	Laredo (1974)
Pennesetum clandestinum	5.6	2.5	Laredo (1974)
Setaria sphacelata	6.1	2.2	Laredo (1974)
Mean tropical grasses	5.6	2.5	
Temperate legumes			
Trifolium pratense	21.0	11.0	Fleming (1963)
Tropical legumes			
Lablab purpureus	16.3	11.8	Hendricksen *et al.* (1981)
Lablab purpureus	14.8	8.9	Hendricksen and Minson (1980)
Lablab purpureus	14.8	10.2	Hendricksen and Minson (1985a)
Mean tropical legumes	15.3	10.3	
Overall mean	8.4	4.1	

concentration of the stem fraction in both *Bromus inermis* and *P. pratense* (Pritchard *et al.*, 1964). With increasing maturity there was a rise in the Ca concentration in the leaf and a decrease in that in the stem so that after flowering the Ca concentration in the leaf fraction was over 10 times that in the stem fraction. Within a pasture of *Lablab purpureus* the highest Ca concentrations were found in the lowest layers of the plant, with a large difference between leaf and stem fractions. Not only did the tops contain lower levels of Ca, but there was very little difference in Ca concentration between leaf and stem (Hendricksen and Minson, 1980, 1985a).

E. Stage of Growth

As forage matures there is an increase in the proportion of stem, which generally contains less Ca than the leaf fraction (Section II,D). Forage for silage is generally cut at a younger stage of growth than forage cut for hay and usually contains more Ca (6.2 versus 5.3 g/kg DM) (Hemingway *et al.*, 1968). In other studies there has been no change in Ca concentration with increasing maturity (Sen and Mabey, 1965; Wilson and McCarrick, 1967; Davies *et al.*, 1968; Whitehead and Jones, 1969).

F. Soil Fertility

The Ca concentration of forages also depends on the quantity of exchangeable Ca in the soil and the levels of other elements, particularly nitrogen and potassium. The addition of Ca usually has only a relatively small effect on forage Ca, but where soils are depleted of Ca, especially in the tropics, very large increases can occur. Nine legumes grown on a very poor low humic gley soil contained only 3.8 g Ca/kg DM but increased to 20.0 g Ca/kg DM following the application of calcium carbonate (Andrew and Norris, 1961). Addition of calcium carbonate to the same soil raised the Ca concentration of *D. decumbens* from 2.2 to 3.8 g/kg DM (Rees and Minson, 1976). In Puerto Rico, calcium carbonate at 18 tons/ha increased the Ca in three tropical grasses from 3.6 to 5.0 g/kg DM (Abruna *et al.*, 1964). Increases of this size are rather exceptional and usually only small increases occur following liming (Beeson, 1946). For example, in New Zealand liming increased the Ca concentration of *L. perenne* and *T. repens* from initial values of approximately 5 and 11 g/kg DM to 6 and 13 g/kg DM, respectively (Edmeades *et al.*, 1983). Flooding of soils causes large increases in the level of Ca in forage. *Dactylis glomerata* on nonflooded and flooded soil contained 2.2 and 4.0 g Ca/kg DM, respectively, a difference that was associated with a 30% depression in growth on the flooded soil (Currier *et al.*, 1983).

Nitrogen fertilizer decreased Ca in first cuts of temperate pastures but raised it in the third cut (Rahman *et al.*, 1960). With frequently cut or grazed grass swards, fertilizer nitrogen increases the Ca concentration of the forage (Stewart and Holmes, 1953; Whitehead, 1966; Mudd, 1970), but with forage cut less frequently fertilizer nitrogen has no consistent effect (Weinmann, 1950; Burton *et al.*, 1959; Vincente-Chandler, 1959a,b; Stillings *et al.*, 1964; Reid and Jung, 1965a; Whitehead, 1966; Miaki, 1970; Behaeghe and Carlier, 1974). Applying fertilizer nitrogen to grass/legume swards reduces the Ca concentration of mixed forage (Walker *et al.*, 1953; Reith *et al.*, 1964; Hartmans, 1975; Rodger, 1982). This decrease is caused by a depression in the legume content of the mixed pasture. When mixtures were separated into grass and legume it was found that fertilizer nitrogen had no consistent effect on the Ca concentration in either component (Walker *et al.*, 1953; Rahman *et al.*, 1960; Hemingway, 1961a,b; Whitehead *et al.*, 1983).

When potassium fertilizer was applied to pasture low in nitrogen, Ca concentration was increased (Stewart and Holmes, 1953). In pastures fertilized with nitrogen, potassium fertilizer depressed Ca concentration (Stewart and Holmes, 1953; McNaught, 1959; Reith *et al.*, 1964; Smith, 1971; Balasko, 1977; Whitehead *et al.*, 1978). Urine is high in both nitrogen and potassium and, as might be expected, urine patches in grazed pastures have a lower Ca concentration than surrounding forage (Joblin and Keogh, 1979). Phosphorus fertilizer generally decreases Ca concentration of grasses although the effect is small (Playne, 1972a; Balasko, 1977; Rees and Minson, 1982).

G. Climate

Forage Ca is generally low during periods of active growth and high when lower temperature or lack of moisture slow growth. With temperate grasses Ca is low in spring and high in autumn and winter (Ferguson, 1932; Thomas and Trinder, 1947; Waite *et al.*, 1952; Hemingway, 1961a; Coppenet and Calvez, 1962; Whitehead, 1966; Fleming and Murphy, 1968; Reid *et al.*, 1970; Fleming, 1973; Hartmans, 1975; Thompson and Warren, 1979; Whitehead *et al.*, 1983). In contrast, one study in France failed to show seasonal variation (Gueguen and Demarquilly, 1965) and, in New Zealand, higher levels of Ca were found during periods of active summer growth of *L. perenne*, *T. pratense*, and *T. repens* (McNaught *et al.*, 1968; Reay and Marsh, 1976; Edmeades *et al.*, 1983).

During periods of active growth in tropical grasses, Ca concentrations are generally lower than when growth is slow (Theiler *et al.*, 1924; Du Toit *et al.*, 1934; Daniel and Harper, 1935; Taylor, 1941; Weinmann, 1950; Jones, 1963a). Experimental evidence on the effect of temperature on Ca concentration of forage is confusing. There was no consistent difference in Ca concentrations of five temperate legumes grown at 15/10°C and 32/27°C (Smith, 1970b), but in a subsequent study highest Ca concentrations occurred in plants grown at the lower temperature (Smith, 1971). Conversely, higher Ca concentrations occurred in *L. perenne* when grown at 20°C than when grown at 10°C (Dijkshoorn and Hart, 1957). These variable responses are not confined to different species but also occur within genotypes of the same species. When two lines of *Festuca arundinacea* were grown at 25/20°C and 17/12°C the higher temperature increased the Ca concentration in one line but depressed it in the other (Sleper *et al.*, 1980). Shading lowered temperature and increased the Ca concentration in *Cynodon dactylon* when soil nitrogen was low but had little effect when high levels of fertilizer nitrogen were applied (Burton *et al.*, 1959).

III. DIAGNOSIS OF CALCIUM DEFICIENCY

A. Forage Species

Chronic Ca deficiency in ruminants fed grass is very rare and never occurs on legume-based pastures. Most grasses appear to contain adequate Ca, but the availability of the Ca may be reduced by oxalic acid, which is high in some young, actively growing, well-fertilized, tropical grasses (Jones and Ford, 1972a,b).

B. Forage Analysis

There are large differences in Ca concentration between leaf and stem fractions of forage (Section II,D) and the Ca level in the forage eaten will depend on the degree of selection allowed and the behavior of the animal. Samples for analysis may be plucked, simulating the grazing animal, or collected using an animal fitted with an esophageal fistula. Saliva contains very little Ca and samples collected via the esophageal fistula have the same Ca concentration as the forage selectively grazed (Langlands, 1966; Little, 1975). The results of the forage analysis can be compared with the animals' requirements (Tables 7.3–7.6) to assess whether the diet contains sufficient Ca and hence whether the animal needs to draw on Ca reserves in the bone. Rate of gain in weight or milk secretion will only be reduced by a severe and protracted deficiency of Ca (Field, 1983).

C. Clinical Signs

Forages are rarely deficient in Ca, but diagnostic criteria have been developed for high-grain diets that are low in Ca. The most characteristic sign is the irregularity of the teeth of lambs reared on Ca-deficient diets (Franklin, 1950). Not only are the teeth irregular, but they are soft and wear is often distorted. In extreme cases bone strength is reduced and there is an increase in number of fractures.

Acute Ca deficiency occurring at the time of parturition is characterized by the classical signs of Downer's disease or milk fever described by Allen and Davies (1981): decline in appetite, dullness, lethargy, hocks straight, paddling from one hind foot to the other, possible grinding of teeth. An affected animal sometimes becomes very excitable and difficult to restrain and sweats profusely. Finally the animal becomes recumbent, first lying on its sternum, with characteristic curvature of the neck, and makes unsuccessful attempts to rise. Eventually it lies on its side and becomes comatose, with dilated pupils and dry muzzle. The rumen becomes tympanitic and its contents may be regurgitated. Unless treated, usually with calcium borogluconate, the animal dies from respiratory failure as a result of rumen tympany or from the inhalation of rumen contents.

D. Fecal Analysis

The principle pathway for the excretion of Ca is via the feces, and analysis of feces can provide a simple guide to the approximate Ca concentration of the forage eaten. With animals fed forages containing 1.5 to 15.4 g Ca/kg, the forage Ca concentration (Y) was related to the Ca concentration in the fecal organic matter (X) by the equation $Y = 0.189 + 0.4049X$ (Moir, 1960b).

Some of the Ca in the diet of lactating cows in secreted in the milk and 62–86% of the dietary Ca was excreted in the feces (Rook and Balch, 1962; Hutton *et al.*, 1967). With lactating cows there is a different relation between Ca concentration in the forage (Y) and Ca concentration in fecal organic matter (X): $Y = 0.92 + 0.247X$ (Moir, 1960b).

E. Blood Analysis

Serum Ca levels of cattle in the United Kingdom have a mean value of 99 mg/liter with a standard deviation of ±4.4 mg/liter (Topps and Thompson, 1984), a value slightly higher than previously reported for cows in 13 herds (95 ± 4 mg Ca/liter) (Rowlands and Pocock, 1976). In lactating cows suffering from parturient paresis serum Ca falls below 60 mg/liter (Boda, 1956), but as previously mentioned this disease is due to a failure of the hormone system to adjust to the high Ca demand of lactation and not to a deficiency of Ca in the diet. Serum Ca may also be depressed by deficiency of vitamin D. When animals were kept indoors and away from direct sunlight for 28 weeks, serum Ca declined to 62 mg/liter although the diet contained sufficient Ca and P (5 g Ca and 4 g P/kg DM) (Franklin and Reid, 1948).

There appear to be no reports of low serum Ca associated with forage. The only reports of low serum Ca were associated with a milk-substitute diet low in Ca. Serum Ca levels in lambs were depressed from normal values of 100 and 120 mg/liter to 79 mg/liter within 5 weeks of feeding the Ca-deficient diet and then fluctuated between 70 and 90 mg/liter (Hodge *et al.*, 1973). The Ca-deficient lambs grew 37% more slowly than the controls. Serum levels of only 5.4 mg Ca/liter have been found in lambs fed a Ca-deficient grain diet for 300 days, and they grew 53% more slowly than lambs that were supplemented with limestone (Franklin and Johnstone, 1948).

The level of serum Ca is affected by several factors in addition to the level of Ca in the diet, but the changes are generally small and probably have no effect on production. Exposure to acute cold depressed Ca level by 13 mg/liter (Sykes *et al.*, 1969), while both within-day and seasonal differences have been reported (Shirley *et al.*, 1968; Hewitt, 1974; Thompson *et al.*, 1978). The level of Ca in the blood is negatively correlated with the P concentration ($r = -0.79$) (Wiener and Field, 1969). Increasing the P content of the diet raises the P level in the blood and depresses the level of Ca (Benzie *et al.*, 1959; Chicco *et al.*, 1973a; Reed *et al.*, 1974a,b; Pless *et al.*, 1975; Belonje, 1978), particularly when the level of Ca in the diet is low (Table 7.9). Serum Ca is

TABLE 7.9

Effect of Dietary Calcium and Phosphorus on the Levels of Plasma Calcium and Phosphorus in 6-Month-Old Lambs[a]

Calcium level in diet[b]	Low		High	
Phosphorus level in diet[c]	Low	High	Low	High
Plasma Ca (mg/liter)	90	75	89	86
Plasma P (mg/liter)	42	76	34	76

[a]From Sevilla and Ternouth (1980).
[b]Low and high Ca = 1.35 and 8.2 g/kg DM, respectively.
[c]Low and high P = 0.72 and 4.5 g/kg DM, respectively.

not affected by excitement of the animal (Gartner *et al.*, 1965) and only to a small extent by the larvae of *Ostertagia circumcincta* (Coop *et al.*, 1977).

F. Bone Analysis

A low-Ca diet leads to reduced bone development and a reduction in bone strength. Lambs reared on Ca-deficient synthetic diets had femurs with low specific gravity, low ash and Ca in the dry fat-free bone, and low Ca in the bone ash (Hodge *et al.*, 1973). However, with grazing cattle the Ca level in dry fat-free rib bone was not related to the Ca or P concentration in the pasture (Cohen, 1973b).

IV. PREVENTING CALCIUM DEFICIENCY

A. Supplementary Calcium

It is rarely necessary to provide forage-fed ruminants with an additional source of Ca, but if a supplement is required calcium carbonate is usually offered. With Ca-deficient grain diets the addition of 1% limestone to the diet restores serum Ca to normal levels (Franklin and Johnstone, 1948). Various forms of limestone have been fed to eliminate the Ca deficiency in grain-fed animals, including material ground to pass 20–40 mesh, 40–120 mesh, and unground limestone (Franklin and Johnstone, 1948). Many forms of P supplement contain Ca, including bone meal, rock phosphate, and calcium phosphate. The true availability of the Ca in some of these sources has been determined with both young and mature steers (Table 7.10). The lower apparent availability of the supplements when fed to the mature animals is caused by their lower Ca requirement (Braithwaite, 1976).

B. Fertilizers

The Ca concentration of forages can be increased by applying fertilizer Ca (Section II,F). The most important sources of fertilizer Ca are (1) calcium carbonate ($CaCO_3$) marketed as chalk, ground chalk, screened chalk, or ground limestone, (2) calcium hydroxide ($Ca(OH)_2$) sold as hydrated lime or slaked lime, and (3) calcium oxide (CaO) marketed as burnt or quick lime (Cooke, 1972). Other sources are ground magnesium limestone, lump chalk, marl, and calcareous sea sands. The rate at which the Ca in these fertilizers becomes available to the forage depends on particle size (Cooke, 1972).

TABLE 7.10

True Availability Coefficient of Calcium in Different Calcium Supplements[a]

Supplement	True availability Young steers (180 kg)	True availability Mature steers (490 kg)	Mean
Bone meal	0.68	0.55	0.62
Calcium carbonate (C.P.)	0.51	0.40	0.46
Ground limestone (com.)	0.45	0.37	0.41
Calcium chloride (C.P.)	0.60	0.53	0.56
Dibasic calcium phosphate (C.P.)	0.64	0.50	0.57
Monobasic calcium phosphate (C.P.)	0.61	0.56	0.58
Dicalcium phosphate (A)	0.58	0.49	0.54
Dicalcium phosphate (B)	0.56	0.38	0.47
Dicalcium phosphate (C)	0.60	0.56	0.58
Dicalcium phosphate (D)	0.60	0.51	0.56
Dicalcium phosphate (E)	0.58	0.55	0.56
Dicalcium phosphate (Defluorinated com.)	0.55	0.40	0.48
Mean	0.58	0.48	0.53

[a]From Hansard *et al.* (1957).

C. Parturient Paresis

This disease is caused by failure of the hormone system to adjust to the Ca required for milk production and is associated with feeding high-Ca diets prior to parturition (Boda and Cole, 1954). The aim of any management system must be to reduce the intake of Ca during the last 2 weeks of pregnancy so the animals rely less on Ca absorption from the diet and can more readily increase Ca resorption from the bone at parturition (Goings *et al.*, 1974; Braithwaite, 1976). Legumes contain 3–4 times as much Ca as grass (Table 7.7), so they should be avoided in the 2 weeks before parturition. When cows were fed either *M. sativa* or a low-Ca diet prior to calving, parturient paresis was 33 and 0%, respectively (Boda and Cole, 1954). It is interesting to note that when cows have been supplemented with magnesium the incidence of parturient paresis is reduced to 5% compared with 15% for unsupplemented animals (Young and Rys, 1977).

D. Plant Breeding

Ca concentration varies in grasses and high heritabilities show that breeding and selection for high levels of Ca are possible (Table 7.11).

TABLE 7.11

Variation and Heritability of Calcium in Grasses

Species	Variation in Ca (g/kg DM)	Heritability: Broad sense	Heritability: Narrow sense	Reference
Dactylis glomerata	1.1–6.5	0.76–0.83	—	Currier *et al.* (1983)
Lolium perenne	9.3–12.5	0.34	—	Butler *et al.* (1962)
Lolium perenne	3.4–5.6	—	0.78	Cooper (1973)
Pennisetum glaucum	—	0.62–0.84	—	Gupta and Sehgal (1971)
Phalaris arundinacea	3.3–5.3	0.64	0.56	Hovin *et al.* (1978)
Setaria sphacelata	1.5–3.6	—	0.28	Bray and Hacker (1981)

Similar variability has been found in legumes (Robinson, 1942; Davies *et al.*, 1968; Hill and Jung, 1975), but the Ca level was high relative to ruminant requirements. If it were shown conclusively that the excess Ca in legumes had a detrimental effect on production (other than parturient paresis), then the level of Ca could possibly be reduced by plant breeding and selection.

V. CONCLUSION

Calcium is essential for many body functions. It is a major constituent of bone, from which it can be withdrawn when the diet is low in Ca. Feeding Ca supplements and raising the Ca level of the forage by applying fertilizer Ca has no effect on ruminant production. Parturient paresis, or hypocalcemia, is caused by feeding legumes high in Ca prior to calving and can be prevented by feeding grasses which contain less Ca. Leaf contains twice as much Ca as stem fraction. Stage of growth, fertilizer, and climate have no consistent effect on Ca levels in forage. Calcium concentration in the diet may be determined from plant samples of forage and from the Ca concentration in the feces. Calcium deficiency in ruminants is determined by analyzing the blood serum for Ca.

8

Phosphorus

I. PHOSPHORUS IN RUMINANT NUTRITION

A. Function of Phosphorus

Phosphorus (P) has long been recognized as an important essential mineral element concerned with bone development, growth, reproduction, and energy transfer. Clinical signs of P deficiency in cattle are unthriftiness, fragile bones, and botulism as a result of bone chewing (Underwood, 1981). Sheep are less susceptible to low levels of dietary P but can be adversely affected (McMenimen, 1976; Ozanne *et al.*, 1976; Ternouth and Sevilla, 1984; Read *et al.*, 1968a).

Cattle contain approximately 6.3 g P/kg liveweight (ARC, 1965). Most of the P is in the skeleton (75–80%) (ARC, 1965) with less than 0.4% in the blood serum. It has been claimed that P has more known functions than any other mineral element in the animal body. In addition to combining with calcium (Ca) to form bones and teeth, P is found in every cell of the body and is essential in many metabolic processes including the buffering of body fluids. It is required by the rumen microbes for fermentation of forage (Komisarczuk *et al.*, 1984) and synthesis of microbial protein (Breves *et al.*, 1985).

Ruminants attempt to maintain P at a relatively constant concentration in the blood plasma (40–70 mg/liter) (Whitten, 1971). This is achieved by absorption of P from the digestive tract and release of P from the bone to balance the loss of P from plasma in the course of metabolic processes. Excess P is passed via the saliva into the digestive tract and excreted in the feces. Increased excretion via the urine only occurs when diets are relatively low in calcium (Vipperman *et al.*, 1969) or when animals are fasted and energy intake is restricted (Siebert and Cameron, 1978).

B. Appetite and Dry-Matter Digestion

Forages that have received fertilizer P are generally preferred by sheep (Reid and Jung, 1965b; Ozanne and Howes, 1971) and cattle (Staten,

1949). In the classic South African studies by Theiler *et al.*, (1924) P deficiency depressed voluntary intake (VI) of cattle by 29% when they were offered an *Eragrostis tef* hay containing 0.7 g P/kg DM. This effect on VI has been confirmed in studies with sheep (Field *et al.*, 1975; Sevilla and Ternouth, 1980; Ternouth and Sevilla, 1984; Budhi and Ternouth, 1988) and cattle (Kleiber *et al.*, 1936; Little, 1968; Butcher *et al.*, 1979; Bass *et al.*, 1981; Gartner *et al.*, 1982; Call *et al.*, 1986). Most of these diets contained less than 1.0 g P/kg DM, but where the Ca level was high, feeding supplementaty P increased the VI of diets containing higher levels of P [1.3 g P/kg DM, (Field *et al.*, 1975); 1.7 g P/kg DM, (Wise *et al.*, 1963)]. This depression in VI is associated with low plasma inorganic P (Section III,H).

When sheep were fed *Stylosanthes humilis* (0.8 g P/kg DM, 8.2 g Ca/kg DM) P supplement increased the VI of sheep by 15%, but with *Heteropogon contortus* (0.8 g P/kg DM and 2.9 Ca/kg DM) a P supplement failed to increase VI (Playne, 1966a). This difference appeared to be associated with the higher Ca in the legume and hence increased demand for P. This conclusion is supported by the observation that the intake of a diet low in P is depressed by feeding a Ca supplement (Theiler *et al.*, 1937; Otto, 1938), but when diets contain more than 1.7 g P/kg DM, VI is not affected by high levels of Ca and the ratio of Ca:P has no nutritional significance (Theiler *et al.*, 1927; Haag *et al.*, 1932) (Chapter 7, Section I,A). Legumes contain high levels of protein but there is no evidence that this increases the demand for P and depresses the VI of diets low in P (McLachlan and Ternouth, 1985).

The digestibility of the dry matter and energy in mixed diets appears to be unaffected by low P concentrations [0.7 g P/kg DM (Kleiber *et al.*, 1936); 0.9 g P/kg DM (Gartner *et al.*, 1982); 1.2 g P/kg DM (Witt and Owens, 1983)]. When cattle were fed a diet containing 1.2 g P/kg DM, the P concentration in the rumen was 208 mg P/liter (Witt and Owens, 1983), well in excess of the 20–80 mg P/liter required for maximum cellulose digestion *in vitro* (Hall *et al.*, 1961; Chicco *et al.*, 1965). Increasing the level of P by applying fertilizer P had no effect on the digestibility of the dry matter in *Festuca arundinacea* (Reid and Jung, 1965a) and *Digitaria decumbens* (Rees and Minson, 1982).

C. Effect on Production

1. Death

In extreme cases P deficiency can indirectly lead to botulism, a disease caused by cattle eating carcass debris infected with *Clostridium botuli-*

num (Underwood, 1981). This depraved appetite is a direct result of a P deficiency. In one field study 12% of unsupplemented cows grazing low-P forage died of botulism compared with only 1% when a P supplement was fed (Theiler *et al.*, 1928). It has also been suggested that the depraved appetite leads to greater losses of stock through eating poisonous plants (Theiler *et al.*, 1928).

High levels of P, especially in the presence of relatively low levels of Ca, can cause health problems in sheep. When the concentration of P is equal to or exceeds that of Ca, there is a high incidence of urinary calculi with mortality of 50% when the concentration of P is twice that of Ca (Emerick and Embry, 1963).

2. Fertility

First-calf heifers deficient in phosphorus in South Africa were found to have a lower calving percentage than those receiving a P supplement (51 versus 80%) (Theiler *et al.*, 1928). In recent work at the same site, P deficiency halved the conception rate of cows and reduced average weaning weight by 22% (Read *et al.*, 1986b). In Texas, calving percentages were raised from 57 to 91% by feeding a bone-meal supplement, to 96% with a disodium phosphate supplement, and to 98% by fertilizing the pasture with triple superphosphate. The adverse effect of P deficiency on calving percentage appeared to be mainly due to the smaller body size of the deficient cows, 377 kg compared with 471 kg for cows receiving P (Black *et al.*, 1949). A low liveweight is generally considered to be the cause of reduced fertility of P-deficient cows (Ward, 1968; Holmes, 1981), but a direct effect of P is also involved. Estrus in P-deficient cows was restored within 10 to 35 days of starting to dose each day with disodium phosphate (O'Moore, 1952). A survey of dairy herds in England indicated that breeding efficiency might be impaired when P intake was low and Ca intake was high (Hignett and Hignett, 1951). However, when this hypothesis was tested in a controlled experiment there was no evidence of a relationship between the Ca:P ratio of the diet and fertility in diary heifers (Littlejohn and Lewis, 1960). In Ireland it was found that when pastures are deficient in P then excess Ca can aggravate the low fertility found with the low P but that high Ca has no detrimental effect when P in the forage is normal (O'Moore, 1960a). It appears that a high Ca intake increases the requirement for P (O'Moore, 1952), possibly by increasing the P required for bone growth (Braithwaite, 1984).

The conception of cows grazing pastures of *S. humilis*/*H. contortus* increased from 66 to 84% when superphosphate was applied (Edye *et al.*, 1971). This improvement was accompanied by an increase in the legume and nitrogen content of the diet and a higher liveweight of the cows on the fertilized pasture compared with the controls (392 versus 346 kg) (Williams *et al.,* 1971). It is impossible to decide whether the higher conception rate was caused by the improved liveweight, the supply of P to the cows, or other changes in nutrition associated with the fertilized pasture. In another study, superphosphate application increased estrus of Hereford heifers from 4.3 to 20.7% (Cohen, 1980). A concentration of 1 g P/kg DM is sufficient for beef cattle, and increasing the P concentration of *D. decumbens* from 1 to 3 g/kg DM by applying fertilizer P had no effect on weaning percentage or cow weights (Hodges *et al.,* 1964, 1968).

3. Milk Production

Milk production imposes a severe strain on the P reserves of cows. Feeding a P supplement to animals grazing P-deficient forage has increased milk production by 40%, the rise occurring within a month of starting to feed the P supplement (Theiler *et al.,* 1924). Irish studies showed that milk production could be doubled within 14 days of starting to feed additional P and was maintained at the high level while additional P was fed (Sheehy *et al.,* 1948). In Australia, milk production of cows grazing tropical grass/legume pastures has been increased from 3600 to 4150 kg/cow by feeding a P supplement (Davison *et al.,* 1986). Production of protein and solids-not-fat was also improved.

Feeding P supplements to cows grazing forage low in P improved the growth rate of calves by 10% (Black *et al.,* 1949; Ward, 1968) and 61% (Holmes, 1981). These increases in production were probably due to higher forage intake and milk production by the supplemented cows. In Australia, supplementing cows with P increased milk production by 15% during the wet season but had little effect when cows grazed mature forage (Allan *et al.,* 1972). Supplements are easier to feed in dry weather and, although feeding P at this time has little direct effect on milk production, the carry-over effect increased subsequent milk production by 12% in the following wet season when no P supplements were fed (Allan *et al.,* 1972).

4. Growth Rate

Bone meal fed at the rate of 84 g/day increased the annual growth rate of steers by 77 to 100% (Theiler *et al.,* 1924). The average growth rate of the supplemented steers was 0.35 kg/day compared with 0.18 kg/day for the unsupplemented animals (Table 8.1). The effect of the P supplement was not consistent and varied with time of the year. In the wet season, average growth rates were 0.58 kg/day and 0.39 kg/day for the supplemented and the controls, respectively, while in the dry season the P supplement had little effect, possibly because both groups were generally losing or only maintaining weight due to nutritional factors other than P supply. This seasonal difference in response to supplemental P has been confirmed in studies with cattle grazing both native grass pasture (Van Schalkwyk and Lombard, 1969; Winks *et al.*, 1977) and the legume *S. humilis* (Winks *et al.*, 1977). In the dry season both the control and the P-supplemented animals lost 0.1 kg/day, whereas in the wet season the supplement increased mean growth rate by 17% to 0.82 kg/day (Winks *et al.*, 1977) (Table 8.1). This supplemental response was obtained with unfertilized pastures containing about 1 g P/kg DM, but when the P content of the *S. humilis* pastures was doubled by the application of superphosphate then the P supplement had no effect (Winks *et al.*, 1977). In a similar study, feeding a P supplement (5–7 g P/day) increased by 60–80% the annual growth rate of heifers grazing unfertilized *Stylosanthes* spp. pastures but had no effect with pastures that had been fertilized with P (Coates, 1987). Feeding P supplements in the dry season had little effect on the growth rate of weaner cattle (Winks *et al.*, 1976) (Table 8.1). Phosphorus-supplemented cows grazing low-P pastures (0.7 g P/kg DM) maintained a higher body weight than unsupplemented cows, with a mean difference over four lactations of 27% (Black *et al.*, 1949) (Table 8.1).

Positive responses have also been recorded when sheep have been supplemented with P. Lambs fed a mixed diet containing 0.7 g P/kg DM maintained weight but when the P level of the diet was raised to 3.6 g/kg DM they grew at 0.06 kg/day (Ternouth and Sevilla, 1984). In another study sheep fed *Trifolium subterraneum*/*Vulpia* spp. containing 0.7 g P/kg DM lost 18 g/day but gained 124 g/day when a P supplement was fed (Ozanne *et al.*, 1976). A positive response in wethers has also been obtained with diets of *Acacia aneura* containing 0.7 g P/kg DM. The P-supplemented group had a growth rate of 22 g/day compared with a rate of -1 g/day of the controls and grew 19% more clean wool (McMeniman, 1976). Responses to a P supplement have usually only been obtained where the diet is both low in P and high in Ca.

There are many studies which show no response to P supplements (Cohen, 1972, 1975a; Winks *et al.*, 1976, 1979; Holroyd *et al.*, 1983). There are three possible reasons for these negative results: The diet may contain sufficient P to meet the requirements of the animal; the Ca level in the diet may be low, thus reducing the animals' requirement for P; or the animals may have sufficient P reserves in the bone and thus be able to compensate for the low P level in the diet. The importance of P reserves is illustrated by the improved growth and milk production of cattle many months after the feeding of P supplements has stopped (Van Schalkwyk and Lombard, 1969).

The P content of forage diets may be increased by the application of

TABLE 8.1

Increase in Growth Rate Following Phosphorus Supplementation

Initial liveweight (kg)	Forage	Growth rate (kg/day) Control	Supplemented	Improvement (%)	Reference
Beef cattle			Annual		
136	Native forage	0.22	0.39	77	Theiler *et al.* (1924)
247		0.19	0.38	100	
318		0.14	0.28	100	
313		0.15	0.33	120	Van Schalkwyk and Lombard (1969)
202		0.14	0.41	193	Bisschop (1964)
			Dry season		
140		−0.04	−0.08	—	Winks *et al.* (1976)
257		−0.15	−0.19	—	Winks *et al.* (1977)
257	*Stylosanthes humilis*	−0.05	−0.02	—	
			Wet season		
272	Native pasture	0.20	0.46	130	Turner *et al.* (1935)
Various		0.49	0.59	21	Knox and Watkins (1942)
254		0.25	0.32	28	Norman (1960)
272		0.68	0.80	18	Winks *et al.* (1977)
273	*Stylosanthes humilis*	0.73	0.85	16	
Growing heifers					
275	Native pasture	0.30	0.45	50	Black *et al.* (1943)
Lactating cattle			Weight when calves weaned		
320	Native pasture	364	463	27	Black *et al.* (1949)
Sheep			Pen Studies		
24	Native pasture	0.000	0.062	—	Tenouth and Sevilla (1984)
–		−0.018	0.124	—	Ozanne *et al.* (1976)
22		0.001	0.022	—	McMeniman (1976)

fertilizer P. The weaning weight of calves has been improved by 11% when their dams grazed *S. humilis*/*H. contortus* pasture fertilized with superphosphate (Edye *et al.*, 1972). Applying superphosphate to grass/legume mixtures has also led to increases in both growth rate per animal (Thorton and Minson, 1973a; Winks *et al.*, 1974; Curll, 1977b; Winter *et al.*, 1977b; Eng *et al.*, 1978; Shaw, 1978) and liveweight gain per hectare (Younge and Plucknett, 1965; Evans and Bryan, 1973; Winks *et al.*, 1974; Curll, 1977a,b; Winter *et al.*, 1977b; Eng *et al.*, 1978; Shaw, 1978; Holroyd *et al.*, 1983). In many of these studies superphosphate increased the concentration of P in the forage and it is very tempting to ascribe the higher production to the high P in the forage. In most of these studies superphosphate increased the proportion of legume in the sward and the improved animal production is probably due to the higher legume and protein content of the diet (Evans, 1970).

D. Efficiency of Absorption

The quantity of dietary P absorbed from the upper small intestine is directly related to the quantity of P in the diet and is not related to the need for P (Braithwaite, 1976). Excess P absorbed from the digestive tract is usually excreted in the feces (Barrow and Lambourne, 1962), with only small amounts of P appearing in the urine (see Section III,G). The excretion of excess P in the feces makes it difficult to assess the true availability of dietary P in balance trials, even with the help of isotopic ^{32}P, because endogenous fecal P is so variable. In sheep fed a wide range of feeds, including forages, endogenous fecal P varied from 7 to 64 mg/kg liveweight, with a mean of 34 mg/kg. Corresponding values for cattle were 2 to 40 mg/kg liveweight with a mean of 18 mg/kg (Playne, 1976). Even when fecal loss of P is corrected for the quantity of endogenous P, the true availability of P varies exponentially with the concentration of P in the feed (Fig. 8.1).

The apparent and true availabilities of P in forages will be considered together with the factors that might cause these differences. Most published reports cite apparent absorption of P but it would be unwise to assume that similar conclusions would be drawn if data on true absorption had been available.

1. Forage Species

There is surprisingly little information on the true availability of P in dried forages and even less on fresh forage (Suttle, 1987b). The limited data show that the true availability of P in grass and legumes varies considerably (Table 8.2), but it is impossible to decide whether these differ-

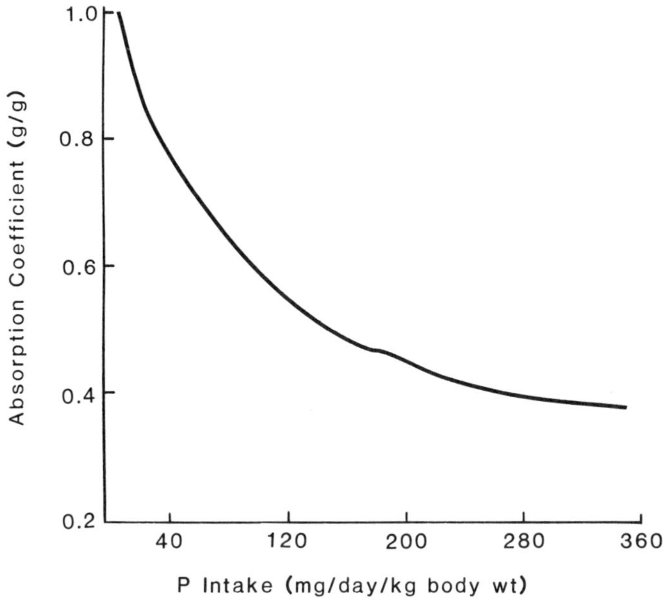

Fig. 8.1 Relationship between phosphorus intake and the efficiency of absorption of dietary phosphorus by sheep. Adapted from Brathwaite (1984).

ences are due to true differences between forages or to experimental bias (Suttle, 1987b). Studies with forages suspended in nylon bags in the rumen have confirmed the high true availability of P. In seeds and pods of *S. humilis* 0.99 and 0.65, respectively, of the P was solubilized (Playne et al., 1972), while in another study 0.78, 0.56, and 0.66 of the P in hays of *Medicago sativa, S. humilis,* and *Chloris barbata,* respectively, disappeared (Playne *et al.,* 1978). However, with *H. contortus* containing 1.2 g P/kg DM, only 0.11 of the P disappeared in 48 hr.

Apparent availability of P in different forages is shown in Table 8.2. These values are lower than the true availabilities due to the excretion in the feces of excess absorbed P. Restricting the level of feeding reduces the retention of P, increases the excretion of P, and hence lowers apparent P absorption (Joyce and Rattray, 1970b). Apparent absorption of P from legumes is higher than that from grasses (0.17 versus 0.04), a difference possibly caused by the higher retention of Ca by sheep fed the legume (Reid *et al.,* 1978a) and higher requirement for P.

2. Protein Content

It has been suggested that P absorption is positively related to the protein content of the forage (Teleni, 1976). However this conclusion is

TABLE 8.2

Apparent and True Availability Coefficients of Phosphorus in Forage

Species	No. samples	Mean P (g/kg DM)	Apparent availability coefficient	True availability coefficient	Reference
Brachiaria decumbens	3	3.3	+0.26	—	Perdomo *et al.* (1977)
Cynodon dactylon	3	2.3	−0.15	—	Butterworth (1966)
Cynodon plectostachyus	1	0.5	+5.25	—	
	3	4.1	+0.37	—	Perdomo *et al.* (1977)
Digitaria decumbens	8	3.6	+0.05	—	Butterworth (1966)
Festuca arundinacea	3	3.1	−0.11	—	Pendlum *et al.* (1980)
Festuca pratensis	22	3.3	—	+0.52[a]	Gueguen and Demarquilly (1965)
Lolium multiflorum	2	4.1	+0.13	—	Lamand *et al.* (1977)
Lolium perenne	17	2.9	—	+0.69[a]	Gueguen and Demarquilly (1965)
	1	4.6	+0.12	—	Joyce and Rattray (1970b)
	1	4.4	+0.24	+0.64	Grace (1981)
Lolium perenne (mixture)	1	4.5	+0.08	—	Grace and Healy (1974)
Lolium perenne/Trifolium repens	17	3.0	+0.34	—	Hutton *et al.* (1967)
Medicago sativa	1	2.6	+0.22	+0.91	Lofgreen and Kleiber (1953)
	1	2.5	+0.31	+0.94	Lofgreen and Kleiber (1954)
	1	2.8	+0.09	+0.56	Grace (1981)
	21	3.4	—	+0.64[a]	Gueguen and Demarquilly (1965)
Panicum maximum	3	3.3	+0.35	—	Perdomo *et al.* (1977)
Phleum pratense	11	2.9	—	+0.64[a]	Gueguen and Demarquilly (1965)
Pennisetum purpureum	4	4.7	+0.20	—	Butterworth (1966)
Sorghum bicolor	1	2.2	−0.17	—	
Trifolium repens	1	3.7	+0.11	—	Joyce and Rattray (1970b)
Pasture (mixed)	14	—	+0.13	—	Lomba *et al.* (1970)
Grass hays	28	2.6	+0.04	—	Reid *et al.* (1978a)
Legume hays	10	2.6	+0.17	—	
Hay	1	2.2	—	+0.74	Field *et al.* (1982)

[a] Assumed that loss of metabolic P = 40 mg/kg liveweight/day. Isotopic techniques were used to measure true availability in other studies.

mainly based on the P retention of calves and sheep when fed high-protein diets. In controlled studies, Tuen *et al.* (1984) found that the absorption of P and excretion of endogenous fecal P were not altered by increasing the protein level of the diet.

3. Calcium Content

Ruminants can normally tolerate a wide range of Ca in the diet without it affecting the true absorption of P, provided the diets are not deficient in P (Lueker and Lofgreen, 1961; Young *et al.*, 1966a). With diets containing 1.2 g P/kg DM, increasing the level of Ca from 1.4 to 4.2 g/kg DM had no effect on the apparent absorption of P (Chicco *et al.*, 1973a). However, apparent absorption of P is related to Ca retention, and when this was increased by infusing calcium chloride into the jugular vein, apparent absorption of P was increased from 0.57 to 0.68 (Braithwaite, 1984).

4. Magnesium Content

The magnesium content of the diet generally appears to have little effect on P absorption (Cohen, 1975b) but the effect depends on the level of P. Increasing the magnesium content of the diet from 5 g/kg DM to the very high level of 55 g/kg DM had no effect on the apparent absorption by lambs of P in diets containing 1.2 g P/kg DM but slightly depressed absorption at the high P level of 3.6 g P/kg (Chicco *et al.*, 1973a). A similar small depression in absorption was found with diets containing 4.5 g P/kg DM (Dutton and Fontenot, 1967). Applying Mg fertilizer had no significant effect on the apparent absorption of P (Reid *et al.*, 1978a).

5. Phytate

Some P in forages occurs in the form of phytates. The proportion of the P present as phytate in 11 forages varied from 30% in *Panicum maximum* to 85% in *Chloris gayana*, with a mean of 52% (Lakke-Gowda *et al.*, 1955). Values as low as 2% have been reported for dried *M. sativa* (Pons *et al.*, 1953). As forages mature a greater proportion of the total P is present as phytate (Lakke-Gowda *et al.*, 1955).

Phytate P is readily available to the rumen microbes (Raun *et al.*, 1956). In the rumen it is completely hydrolyzed and little phytate is present in the intestines (Reid *et al.*, 1947) or feces (Ota, 1960). Inorganic and phytate P appear to be equally available when P levels in the forage are high (Tillman and Brethour, 1958; Dutton and Fontenot, 1967), but it has been suggested that phytate may limit P availability in senescent forage when P level is low (Cohen, 1975b). No evidence has been published supporting this hypothesis.

6. Oxalate

Oxalates are found in many grasses, and although they depress the absorption of Ca they have no effect on the availability of P (Blaney *et al.*, 1982).

7. Aluminium

Addition of large quantities of aluminium to the diet (2 g/kg DM) depressed true availability of P from 30% to zero (Valdivia-Rodriguez, 1979), but no data appears to have been published on the effect of levels of aluminum normally found in forages.

8. Lipids

Forages contain lipids which might affect the absorption of P, but inclusion of 7.5% lipids in a mixed diet had no effect on the apparent or true availability of P (Tillman and Brethour, 1958).

E. Requirement for Phosphorus

The requirements for P, in terms of forage composition, can be calculated from the gross quantities of P required in the diet for different forms and levels of production (ARC, 1980) and estimates of the ad libitum forage intake required to achieve the requred level of production. The P concentrations that are needed in forages to meet the dietary requirements of four classes of ruminant production are shown in Tables 8.3–8.6.

When calculating the requirements for P, ARC (1980) assumed a P absorption coefficient of 0.78 for cattle up to 1 yr, 0.58 for older cattle, 0.73 for lambs up to 1 yr, and 0.60 for mature sheep. Most of these values

TABLE 8.3

Phosphorus Requirements of Beef Cattle Fed ad Libitum, Expressed as Dietary Concentrations (g/kg DM)[a]

Liveweight (kg)	Liveweight gain (kg/day)		
	0	0.5	1.0
200	0.8	1.8	2.3
400	1.4	2.0	2.5
600	1.6	2.0	2.2

[a] Derived from ARC (1980).

TABLE 8.4

Phosphorus Requirements of Lactating Cows Fed ad Libitum, Expressed as Dietary Concentrations (g/kg DM)[a]

Liveweight (kg)	Breed	Milk yield (kg/day)			
		0	10	20	30
400	Jersey	1.4	2.8	4.3	—
500	Ayrshire	1.5	2.4	3.1	3.9
600	Friesian	1.6	2.3	3.1	3.8

[a]Derived from ARC (1980).

were obtained with diets containing adequate levels of P, and it is possible that with diets low in P the efficiency of absorption would be higher (Fig. 8.1) and hence the requirements of high-producing animals on low-P diets could be up to 40% less than are shown in the tables. Dietary P is the main source of P for production but this may, in the short term, be supplemented by P resorption from the bone. Using data of Benzie *et al.* (1959)

TABLE 8.5

Phosphorus Requirements of Castrated Lambs Fed ad Libitum, Expressed as Dietary Concentrations (g/kg DM)[a]

Liveweight (kg)	Liveweight gain (kg/day)			
	0	0.1	0.2	0.3
20	0.8	2.0	3.3	—
30	0.8	1.5	2.1	3.0
40	0.9	1.4	1.8	2.4

[a]Derived from ARC (1980).

TABLE 8.6

Phosphorus Requirements of Lactating Ewes Fed ad Libitum, Expressed as Dietary Concentrations (g/kg DM)[a]

Liveweight (kg)	Milk yield (kg/day)			
	0	1.0	2.0	3.0
40	1.2	2.3	3.4	—
75	1.4	2.1	2.8	3.6

[a]Derived from ARC (1980).

it can be calculated that P released from the bones of pregnant and lactating ewes over a period of 180 days is equal to an additional 0.5 g P/kg DM of feed. The quantity of P available for resorption will depend on previous nutritional history, and differences in the quantity of resorbable P may account for variation in published requirements for dietary P (Webb *et al.*, 1975; Call *et al.*, 1978; Butcher *et al.*, 1979; Little, 1980).

It is concluded that the requirement for P can differ for animals of a similar size and that this could be due to differences in the P available for resorption from the skeleton. Since the quantity of potentially resorbable P is unknown it is often difficult to decide whether grazing animals will respond to supplemental P.

II. PHOSPHORUS IN FORAGE

A. Mean Concentration

The frequency distribution of P concentration in 1823 forage samples from the world literature is shown in Fig. 8.2. The mean concentration was 2.9 g P/kg DM. For 1129 samples of forage collected in Latin America 73% contained less than 3.0 g P/kg DM (McDowell *et al.*, 1977). For the South African veld sampled throughout the year mean P concentration was only 1.0 g/kg DM (Theiler *et al.*, 1924; Van Schalkwyk *et al.*, 1968). A much higher mean concentration of P (4.3 g/kg DM) was found in a study of 40 pastures in Uganda (Long *et al.*, 1970), but a deliberate attempt had been made to restrict sampling to leaf material at the preflowering stage of growth. Where permanently grazed pastures in Kenya were "plucked," mean P concentrations were 1.45 g/kg DM, with 93% of the samples containing less than 2.5 g/kg DM (Howard *et al.*, 1962).

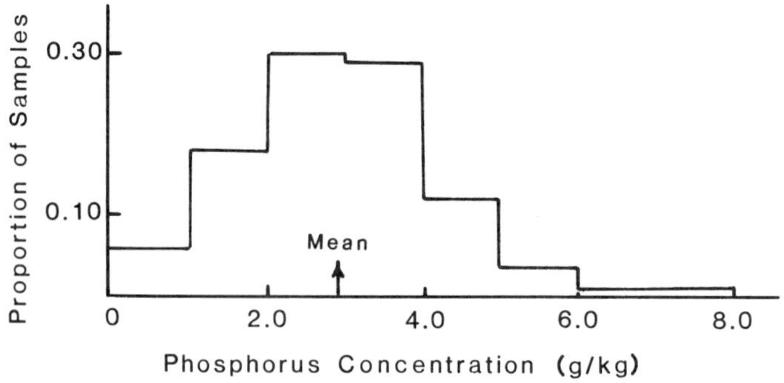

Fig. 8.2 Phosphorus concentration in a range of forages (data from the world literature).

In the temperate zone higher levels of P have usually been reported. Forage samples analyzed by the Pennsylvania State Forage Testing Service over a 5-yr period contained 2.6 g P/kg DM, with a higher concentration in legumes (3.0 g P/kg DM) than in grasses (2.2 g P/kg DM) (Adams, 1975). Samples of old meadow hay produced in County Meath, Ireland, had a mean P content of 2.6 g/kg DM (Wilson *et al.*, 1968) compared with 2.0 g/kg for hay samples grown in Scotland (Hemingway *et al.*, 1968). Silage is usually made from forage at an earlier stage of growth and contained 2.5 g P/kg DM (Hemingway *et al.*, 1968). Grazed pastures in Scotland contained 3.6 g P/kg DM (Thompson and Warren, 1979), a value similar to 1-month regrowths of *Lolium perenne* grown at 14 sites in England and Wales (Whitehead *et al.*, 1978). Forage from pastures in Ireland, on which cows showed symptoms of aphosphorosis, contained only 1.2 g P/kg DM compared with 2.1 g P/kg DM for samples from farms with no aphosphorosis (Sheehy *et al.*, 1948).

B. Species Differences

The forage samples shown in Fig. 8.2 were divided into temperate and tropical grasses and legumes and the frequency distributions plotted (Fig. 8.3). The temperate forages contained more P than the tropical forages, 3.5 versus 2.3 g P/kg DM, respectively, while the legumes had a higher mean concentration of P than the grasses; 3.2 versus 2.7 g P/kg, respectively. Of practical importance is the frequency of forages falling below the level required for a particular level of animal production (Section I,C). For animals requiring forage with 2.5 g P/kg DM, that is, more highly productive lactating cows and ewes, 86% of the temperate legume samples and 54% of the temperate grass samples contained sufficient P. In the tropical forages only 49% of the legume samples and 35% of the grass samples were above this level.

Within each species P concentration varies with cultivar, plant part, stage of growth, soil fertility, and climate.

C. Intraspecific Differences

In spring, the P content of S.23 *L. perenne* is slightly higher than that of S.24 *L. perenne* (Whitehead, 1966), a difference associated with a lower yield and higher proportion of leaf in the later-flowering S.23 (Minson *et al.*, 1960b). However, when regularly cut at monthly or bimonthly intervals, the yield and proportion of leaf are similar for the two varieties and there is no difference in P content (Whitehead, 1966). Phosphorus content was similar for the *M. sativa* cultivars Vernal and DuPuits grown at four sites in the U.S.A. (Matches *et al.*, 1970).

Large differences in P content have been reported among three varieties of *Phleum pratense* grown in Wales (1.7 to 3.3 g/kg) (Patil and Jones, 1970) but, when compared in England, there was no difference among the same three varieties (Whitehead, 1966). This lack of consistency between varieties grown at different sites has also been found with *L. perenne* and *Trifolium repens* (Forbes and Gelman, 1981).

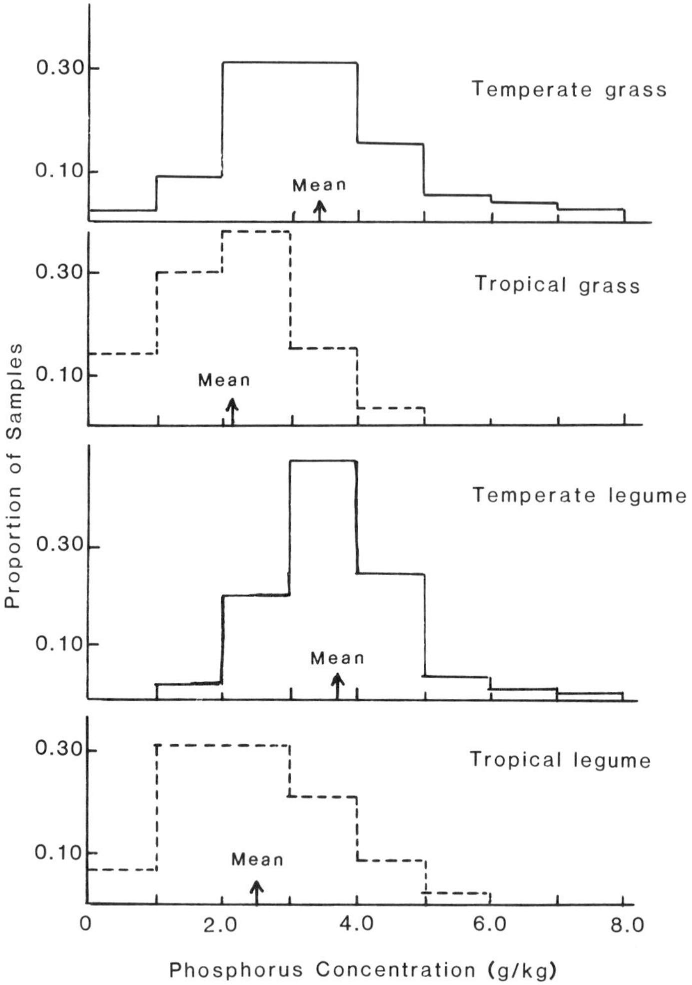

Fig. 8.3 Phosphorus concentration in temperate and tropical grasses and legumes (data from the world literature).

D. Plant Parts

There is no consistent difference in P concentration between leaf and stem fractions of forage plants (Table 8.7). This is partly due to the difference in the stage of growth at the time of the comparison. At immature stages of growth there was no difference between the leaf and stem fractions of four *Stylosanthes* spp., but in mature plants the stem fraction had a lower level of P than the leaf fraction (McIvor, 1979).

Many comparisons have been made between the P concentrations of the inflorescence or seed and other parts of the plant. In most studies the inflorescence of forages had a much higher P content than the remainder of the plant (Table 8.7).

E. Stage of Growth

Phosphorus concentration declines as plants increase in size and advance toward maturity (Fig. 8.4). The rate of fall in P of whole forage varies with species with no consistent difference in rate of fall in P concentration between temperate and tropical forages or between grasses and legumes (Table 8.8). The rate of fall in P concentration varies with the length of the measurement period. A high rate of fall was associated with a short period of active growth, but a slower decline occurred when all stages of maturity were included (Kivimae, 1959). There are also differences between years in the rate of fall of P concentration; the mean rates for four grass species were 0.017 and 0.037 g/kg/DM/day when grown in different years (Whitehead, 1966). Where the forage had been separated into leaf and stem fractions, the rate of decline in P was higher for the stem fraction (Table 8.8).

Mature forage has a low P concentration and when left to stand in the field the rate of decline in P concentration is much lower than that for actively growing forages (Theiler *et al.*, 1924; Norman, 1963; Blue and Tergas, 1969; Haggar, 1970; Playne, 1972b).

F. Soil Fertility

Forage plants absorb P from the soil mainly as orthophosphate (Cooke, 1972) and P concentration in the plant varies with available soil P (Fig. 8.5). The available P depends on the P content of the parent rock and the extent of leaching of soluble P from the soil (Reid and Horvath, 1980); it is increased by the application of P fertilizers. Phosphorus fertilizers, particularly when applied to low-phosphorus soils, cause large increases in the P concentration of forage (Table 8.9).

TABLE 8.7

Phosphorus Concentration of Leaf, Stem, and Inflorescence

Forage	Phosphorus concentration (g P/kg DM)			Reference
	Leaf	Stem	Inflorescence	
Temperate grasses				
Dactylis glomerata	3.2	3.0	4.3	Fleming (1963)
Festuca pratensis	3.0	2.7	4.9	Fleming (1963)
Lolium perenne	3.2	2.7	4.2	Fleming (1963)
	4.1	6.6	—	Laredo and Minson (1975a)
Phleum pratense	1.3	0.8	2.4	Fleming (1963)
Tropical grasses				
Andropogon gayanus	0.7	0.5	—	Haggar (1970)
Bothriochloa insculpta	0.9	0.5	1.5	Du Toit *et al.* (1934)
Chloris gayana	2.6	2.6	—	Laredo (1974)
	2.5	2.9	—	Poppi *et al.* (1981a)
Chrysopogon fallax	0.3	0.2	—	Norman (1963)
Digitaria decumbens	2.6	2.4	—	Laredo (1974)
	2.1	2.4	—	Poppi *et al.* (1981a)
Eragrostis superba	0.7	0.4	1.0	Du Toit *et al.* (1934)
Hyparrhenia hirta	0.7	0.2	0.7	
Panicum maximum	2.6	2.6	—	Laredo (1974)
Pennisetum clandestinum	2.6	3.1	—	
Setaria sphacelata var. *splendida*	2.6	2.7	—	
Sorghum plumosum	0.3	0.1	—	Norman (1963)
Themeda triandra	0.3	0.1	—	
Temperate legumes				
Trifolium pratense	2.9	1.5	4.1	Fleming (1963)
Tropical legumes				
Lablab purpureus	1.8	2.4	—	Hendricksen *et al.* (1981)
Stylosanthes fruticosa	2.8	2.0	3.2	Gardener *et al.* (1982)
Stylosanthes guianensis	3.2	3.2	3.4	
Stylosanthes hamata	2.6	2.0	3.1	
Stylosanthes humilis	0.8	0.5	2.2	Fisher (1969)
	3.5	2.8	4.0	Gardener *et al.* (1982)
Stylosanthes scabra	3.2	2.8	3.9	
Stylosanthes subsericea	3.0	2.8	4.2	
Stylosanthes viscosa	3.0	2.1	3.5	
Other forages				
Brassica spp.	2.9	3.0	—	Hemingway (1960)
Mean	2.2	2.1	3.2	

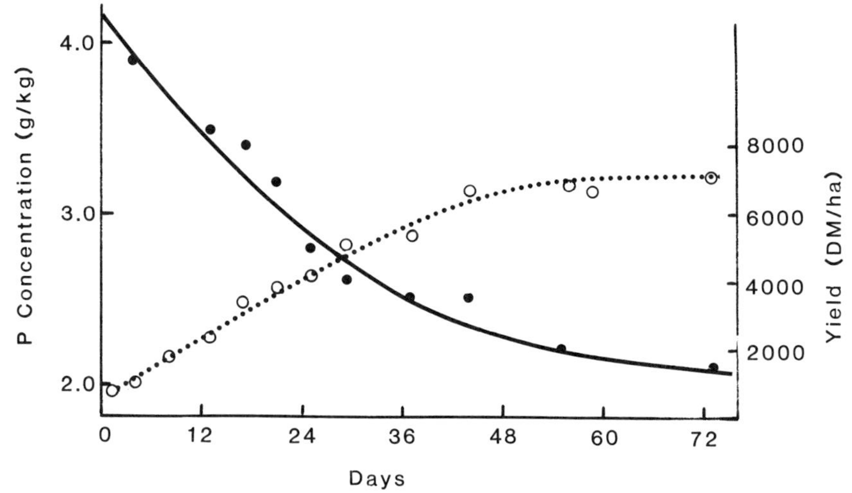

Fig. 8.4 Changes in phosphorus concentration (●———●; ○···○, yield—DM/ha) of temperate grasses with increasing maturity. Adapted from the data of Minson *et al.* (1960b) and Whitehead (1966).

The application of nitrogenous fertilizers stimulates growth and depresses P concentration in a range of forages: *Agrostis tenuis* (Whitehead, 1966), *Brachiaria mutica* (Vincente-Chandler *et al.,* 1959b), *Cynodon dactylon* (Weinmann, 1950), *Dactylis glomerata* (Whitehead, 1966; Rosero *et al.,* 1980), *Digitaria decumbens* (Plucknett and Fox, 1965), *Festuca arundinacea* (Reid *et al.,* 1970; Rosero *et al.,* 1980), *Festuca pratensis* (Whitehead, 1966), *Hyparrhenia rufa* (Tergas and Blue, 1971), *Lolium perenne* (Whitehead, 1966), *Panicum maximum* (Vincente-Chandler *et al.,* 1959a), and *Phleum pratense* (Whitehead, 1966). The extent of the depression in P concentration depends on the quantity of nitrogen fertilizer used and the availability of soil P. The largest depression (3.8 to 1.9 g P/kg DM) followed the application of 900 kg nitrogen/ha to the tropical grass *P. maximum* (Vincente-Chandler *et al.,* 1959a). Even with 84 kg nitrogen/ha, mean P concentration was depressed from 4.0 to 3.4 g/kg DM (Rosero *et al.,* 1980).

The adverse effect of nitrogen fertilizer on P concentration only occurs where forage is cut and removed from the field. When *L. perenne* pastures were intensively grazed, high levels of nitrogen fertilizer (112 kg/ha/month) had no effect on P concentration (Whitehead, 1966).

Urine contains high levels of nitrogen and so might be expected to depress P concentration of forage in urine patches. Forage grown in urine patches had 24% lower levels of P than forage grown in the interexcreta zone (Joblin and Keogh, 1979).

TABLE 8.8

Examples of the Effect of Increasing Maturity and Yield on Phosphorus Concentration during Periods of Rapid Growth

Forage	Decline constant (g P/kg DM/day)	Reference
Temperate grasses		
Bromus inermis	0.039	Reid *et al.* (1970)
Dactylis glomerata	0.030	Whitehead (1966)
Festuca arundinacea	0.045	Reid *et al.* (1970)
Festuca pratensis	0.027	Gueguen and Demarquilly (1965)
Hordeum vulgare	0.019	Dougall (1963)
Lolium perenne	0.030	Whitehead (1966)
Phleum pratense	0.042	Reid *et al.* (1970)
Triticum aestivum	0.047	Dougall (1963)
Tropical grasses		
Brachiaria mutica	0.023	Appelman and Dirven (1962)
Cynodon dactylon	0.050	Dougall (1963)
Digitaria decumbens	0.020	Appelman and Dirven (1962)
Hyparrhenia dissoluta	0.026	Dougall (1963)
Ischaemum aristatum	0.031	Appelman and Dirven (1962)
Ischaemum timorense	0.020	Appelman and Dirven (1962)
Melinis minutiflora	0.020	Gomide *et al.* (1969b)
Panicum maximum	0.018	Gomide *et al.* (1969b)
Pennisetum purpureum	0.032	Gomide *et al.* (1969b)
Pennisetum clandestinum	0.017	Said (1971)
Sorghum sudanense	0.009	Reid *et al.* (1970)
Tripsacum laxum	0.040	Appelman and Dirven (1962)
Urochloa mosambicensis	0.028	Playne (1972b)
Temperate legumes		
Medicago sativa	0.056	Reid *et al.* (1970)
Onobrychis viciifolia	0.028	Whitehead and Jones (1969)
Trifolium hybridum	0.002	Kivimae (1959)
Trifolium pratense	0.018	Whitehead and Jones (1969)
Trifolium repens	0.022	Davies *et al.* (1968)
Tropical legumes		
Stylosanthes guianensis		
Leaf	0.021	McIvor (1979)
Stem	0.032	
Stylosanthes scabra		
Leaf	0.028	
Stem	0.043	
Stylosanthes viscosa		
Leaf	0.035	
Stem	0.060	
Mean	0.030	

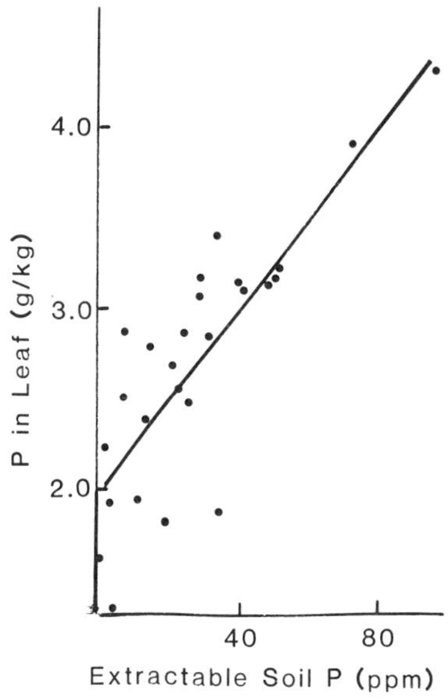

Fig. 8.5 Relation between phosphorus concentration in the leaf of *Desmodium intortum* and extractable phosphorus in soil. Adapted from Plucknett and Fox (1965).

G. Climate

Seasonal differences have been reported in the P concentration of forages (Theiler *et al.*, 1924; Ferguson, 1932; Thomas and Trinder, 1947; Weinmann, 1948, 1950; Jones, 1963a; Norman, 1963; Bisschop, 1964; Playne, 1972b; Gonzalez and Everitt, 1982). These changes are usually caused by periods of rapid growth associated with either an increase in temperature or the arrival of rains. This new growth has a relatively high P concentration, even in areas of P deficiency (Theiler *et al.*, 1924). As forage matures P concentration decreases as described in Section II,E.

Swards of single grass species cut at regular intervals in England, France, and Ireland all showed a similar seasonal variation, with a minimum P concentration in summer (Fig. 8.6). With mixed swards grown at seven sites in New Zealand, no difference was found in P concentration between grasses and clovers; both had a minimum P concentration (2.5 g/kg DM) in summer and maximum value (5.5 g/kg DM) in winter (Metson and Saunders, 1978a).

Very little is known about the seasonal changes in pure swards of legumes. *Medicago sativa* hay made in Oregon had the same P content, irrespective of whether it was first, second, third, or fourth cut (Hanna-

TABLE 8.9

Phosphorus Concentration in Forage Following the Application of Phosphorus Fertilizer

	Fertilizer (kg P/ha/yr)	Phosphorus in forage (g/kg)		Reference
		Control	Fertilized	
Temperate grasses				
Festuca arundinacea	220	3.6	3.8	Reid and Jung (1965a)
Mixed pasture	59	3.3	3.7	Stewart and Holmes (1953)
	49	3.5	3.9	Reith *et al.* (1964)
	38	3.3	3.5	Rodger (1982)
Tropical grasses				
Cynodon dactylon	18	2.3	3.0	Weinmann (1950)
Digitaria decumbens	86	1.6	2.6	Rees and Minson (1982)
Imperata cylindrica	20	1.2	1.9	Stobbs (1969)
Native pasture	20	0.2	0.5	Norman (1962)
	9	1.2	2.1	Jones (1968)
	20	1.0	1.5	Stobbs (1969)
Temperate legumes				
Trifolium subterraneum	51	0.8	1.8	Ozanne and Howes (1971)
Tropical legumes				
Stylosanthes humilis	20	1.6	2.0	Stobbs (1969)
	39	0.6	1.2	Fisher (1970)
	10	0.8	1.8	Winks *et al.* (1977)
Mean	47	1.7	2.4	

way *et al.*, 1981), but in French studies slightly lower levels of P were found in forage cut in July (Gueguen and Demarquilly, 1965). Marked seasonal variation was found in the leaf blades of *Trifolium pratenses*, with lowest levels just after midsummer (Reay and Marsh, 1976).

The lower P concentration in summer is probably due to higher light intensity and temperature. Phosphorus concentration in *C. dactylon* was increased by shading (Burton *et al.*, 1959), and increasing average temperature by 7.5°C decreased the P concentration of *P. pratense* and *T. pratense* by 51 and 20%, respectively (Gross and Jung, 1981). A deficiency of soil moisture could also be important; a severe water shortage reduced P concentration in *L. perenne* from 2.6 to 1.8 g/kg (Whitehead, 1966). Attempts in New Zealand to relate seasonal changes to available soil P were inconclusive (Saunders and Metson, 1971).

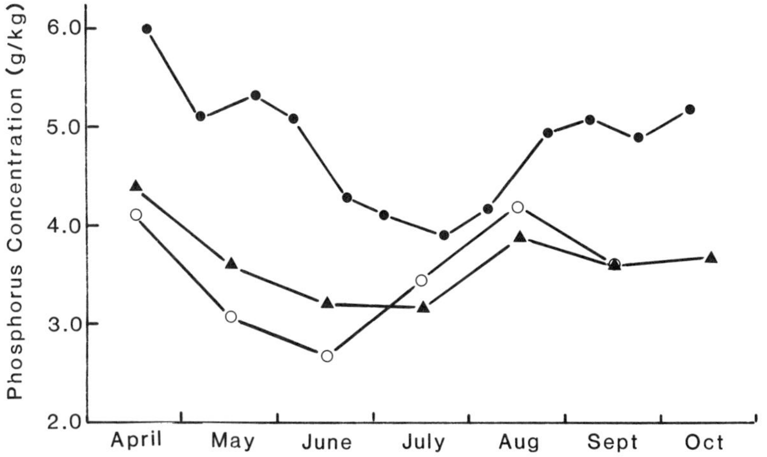

Fig. 8.6 Effect of time of year on the P concentration in grasses. (● ———) 3–4 leaf stage; (▲ ———) monthly regrowths; (○ ———) frequent cutting. Adapted from Coppenet and Clavez (1962), Whitehead (1966), and Fleming and Murphy (1968).

III. DIAGNOSIS OF PHOSPHORUS DEFICIENCY

A. Forage Type

Immature and rapidly growing forage usually contains sufficient P for all forms of production with the possible exception of very high levels of milk production. A major exception among the sown forage species is the *Stylosanthes* spp., which can have low levels of P yet be capable of vigorous growth on soils low in available P.

B. Forage Analysis

Analysis of forages can provide an accurate estimate of their P content and this can be compared with the P required for different forms of production. In farm studies forage P concentrations were well related to field observations of aphosphoresis (Sheehy *et al.*, 1948).

Care must be taken when interpreting forage P analyses. When animals receive forage with less P than the published requirement they will not always respond to the feeding of additional P. The published requirements fail to take into account P released from the bone during periods of high demand. Animals may also fail to respond to supplementary P where some other nutrient is limiting production, and it is helpful to have information on the level of other nutrients that may limit production from forage (Chapter 1).

Where forages are grazed there is always a problem of how best to obtain a sample of the forage selectively grazed. Samples may be collected using esophageally fistulated animals but saliva contamination will inflate the forage P concentrations by 35–273% (Langlands, 1966; Little, 1972). The extent of saliva contamination can be estimated by labeling the grazing animal with ^{32}P and measuring the specific activity of the P in the saliva and bolus (Little *et al.*, 1977).

C. Clinical Signs

A characteristic of P deficiency is "pica," or a depraved appetite. This usually takes the form of osteophagia, or bone chewing, first reported in South Africa (Theiler *et al.*, 1924) and now associated with P deficiency in many other countries (McDowell *et al.*, 1983). Phosphorus-deficient animals also eat stones (McDowell *et al.*, 1983), bark and sticks (Potter and Sparrow, 1973), wood and material (Kleiber *et al.*, 1936), cinders, sacks, and even "the weekly washing waving in the wind" (Green, 1925). Osteophagia is associated with low levels of blood inorganic phosphate (Craven, 1964) and bone chewing by P-deficient animals can lead to death from botulism if the bones are contaminated with *Clostridium botulinum* (Theiler *et al.*, 1924; Underwood, 1981). Bone chewing stopped within 4 weeks of osteophagic cattle starting to graze immature forage containing 2.6 P/kg DM (Theiler *et al.*, 1924). Feeding supplements containing P or supplying water containing superphosphate prevents osteophagia, but feeding calcium carbonate, magnesium sulfate, sodium sulfate, iron sulfate, and sulfur has no effect, indicating that the abnormal behavior is caused by a deficiency of P per se (Theiler *et al.*, 1924).

A deficiency of P will also lead to peg leg, a characteristic form of lameness (Turner *et al.*, 1935; O'Moore, 1952; Barnes and Jephcott, 1955). The main features of peg leg are a creaking of the joints during movement and stiffness, often followed by lameness. Cattle are inclined to walk on their heels and as a result the hooves grow long and may be turned up. When standing the hind legs are usually drawn under the body so that the back is arched. Animals may become so stiff that they remain recumbent all day, making feeble attempts to graze in this position; if forced to rise they exhibit pain (O'Moore, 1952). This behavior, caused by reduced mineralization of the bones, is very common in lactating cows grazing P-deficient pasture (65%) but much less prevalent in steers (3%) (Barnes and Jephcott, 1955).

D. Saliva Analysis

The concentration of P in the saliva of sheep is related to P concentration in the diet (Scott and Buchan, 1988). Saliva of cattle fed a P-deficient diet, with and without supplementary P, contained 180 and 100 mg P/liter, respectively (Clark, 1953). The corresponding values for supplemented and deficient diets in Australian studies were 397 and 215 mg/liter, respectively (Gartner *et al.*, 1982), indicating that there is no critical value of P in saliva below which an animal is P deficient. Many factors other than dietary level of P control the P concentration of saliva (Tomas *et al.*, 1967; Wadsworth and Cohen, 1976; Scott and Beastall, 1978; Cohen, 1980) and more work is required before saliva analysis can be used for diagnosing P deficiency.

E. Rumen Fluid Analysis

The P concentration in the rumen depends on the quantities of P entering with the food and saliva. Saliva contributes about 77% of the P in the rumen (Cohen, 1980) and there is a positive correlation between the P concentration in centrifuged rumen fluid and the inorganic P concentration in the serum ($r = 0.75$) (Tomas *et al.*, 1967).

Phosphorus is present in the rumen as soluble and insoluble P and the observed concentration depends on the method used to prepare the samples for analysis. Rumen fluid collected 6 hr after feeding and strained through silk contained almost twice as much P as samples that had been centrifuged for 20 min (Tomas *et al.*, 1967). Some of the insoluble P is probably available to the animal and a more realistic estimate of P status might be obtained by determining the P concentration in a representative sample of the entire rumen content. Using rumen fluid strained through several layers of cheesecloth, P concentrations of 466, 730, and 924 mg/liter were found in sheep fed diets containing 0.5, 2.7, and 6.3 g P/kg DM and levels of inorganic blood P of 22, 68, and 76 mg/liter (Bclonje, 1978). Levels of rumen P lower than 200 mg/liter have been found in cattle with clinical symptoms of hypophosphorosis (Clark, 1953).

F. Fecal Analysis

In ruminants, the principal pathway for the excretion of P is in the feces (Young *et al.*, 1966b; Scott, 1970; Schneider *et al.*, 1982), and P concentration provides a crude but useful method of identifying low-P diets. For dairy cattle grazing tropical pasture species the P concentration in the forage (Y, expressed in g P/kg DM) is related to the P concentration in the feces (X, expressed in g P/kg OM) as follows: $Y = 0.364X + 0.57$ (Moir, 1960b). With beef cattle fed temperate forage the regression is $Y = 0.370X - 0.22$, $r = 0.91$ (Holechek *et al.*, 1985). A similar relation appears to apply to sheep eating poor-quality forages (Belonje, 1978). As a practical guide, the P concentration in tropical forage is about half that in the feces produced.

Some care needs to be exercised when collecting fecal samples for P analysis. It takes 3 or 4 days for fecal P to adjust to the diet eaten (Belonje and Van den Berg, 1980) and P concentration is lower in morning samples than in afternoon samples of feces (Cohen, 1974).

The level of P in the feces below which a response will be obtained to feeding a P supplement has not been accurately determined. Improved production in response to P supplementation has been reported when feces contained between 2.0 and 4.0 g P/kg DM (Winks *et al.*, 1977; Read *et al.*, 1986d). Above this level, feeding a P supplement did not improve growth rate of cattle (Winks *et al.*, 1977). With senescent forage, fecal P levels were only 1 to 2 g P/kg DM but there was no response to P supplements, probably because factors other than P were limiting production (Winks *et al.*, 1977).

G. Urine Analysis

Most P is excreted in the feces, with only small amounts appearing in the urine (Ammerman *et al.*, 1957; Barrow and Lambourne, 1962; Joyce and Rattray, 1970b; Manston and Vagg, 1970; Scott, 1970; Schneider *et al.*, 1982). However, urine P levels are increased by high levels of parathyroid hormone, a deficiency of dietary energy (Siebert and Cameron, 1978), fine grinding of forage which reduces salivary flow rate (Scott and Buchan, 1987), and feeding grain (Scott, 1970). High levels are sometimes found in the urine of breeds of cattle that have a high metabolic rate or are utilizing body reserves during periods of low energy intake (Siebert and Saunders, 1976). The demand for energy by the developing fetus will also cause a rise in urine P of pregnant cows (Manston and Vagg, 1970) and ewes (Sykes and Dingwall, 1976).

The porportion of dietary P excreted in the urine is generally low and can be increased by many factors other than P content of the diet. As a consequence, analysis of urine for P cannot provide a guide to the P supply or status of ruminants.

H. Blood Analysis

Some fluctuations in blood inorganic P are considered normal (Barnes and Jephcott, 1955), but ruminants attempt to maintain a labile pool of

40–70 mg/liter inorganic P in the plasma (Whitten, 1971). Within this range blood inorganic P may vary in response to different levels of P in the diet (Shirley *et al.*, 1968; Thornton and Minson, 1973a; Cohen, 1974; Ritchie and Fishwick, 1977), fasting or low intake of energy (Siebert and Cameron, 1978), and season of the year (Reed *et al.*, 1974c; Thompson *et al.*, 1978). These fluctuations have little or no effect on the liveweight gain of sheep (McMeniman and Little, 1974; Belonje, 1978) and beef cattle (Murray *et al.*, 1936) or the milk production of cows (Black *et al.*, 1949; Bass *et al.*, 1981). However, when the diet is low in P and the resorption of P from bone is insufficient to compensate for the dietary deficiency, blood inorganic-P concentration can fall to a level where it has adverse effects on VI of sheep (Milton and Ternouth, 1985), lactating cows (Bass *et al.*, 1981), and growing cattle (Belonje, 1978; Gartner *et al.*, 1982).

The level of blood inorganic P is increased by feeding P supplement (Malan *et al,*. 1928; Harris *et al.*, 1956; Bisschop, 1964; Reed *et al.*, 1974a,b; Engels, 1981). Afrikaaner steers with a plasma level of 35 mg P/liter gained only 56 kg/yr compared with 121 kg/yr when the plasma level was raised to 75 mg/liter with a P supplement (Van Schalkwyk and Lombard, 1969). In another study the intake and production of steers was increased by a P supplement once the blood inorganic-P level had fallen to 30–36 mg P/liter (Gartner *et al.*, 1982). With lactating beef cows the intake of *Avena sativa* straw (Y, expressed in kg/day) was related to the P level in the blood (X expressed in mg P/liter) up to a concentration of about 30 mg/liter; $Y = 1.55 + 0.16X$ (Bass *et al.*, 1981).

Blood for the determination of P may be collected from the jugular vein, mammary gland, or femoral artery, but for diagnostic purposes the tail vein is more convenient. Whole blood is treated with buffered trichloracetic acid as a deproteinizing agent prior to analysis (Little *et al.*, 1971; Teleni *et al.*, 1976). Phosphorus levels in blood are decreased by a high intake of dietary Ca if the diet is low in P (Otto, 1938; Wise *et al.*, 1963; Chicco *et al.*, 1973a; Sevilla and Ternourth, 1980) but are increased by cold exposure (Sykes *et al.*, 1969), fasting (Lomba *et al.*, 1970; Siebert and Cameron, 1978), dehydration (Rollinson and Bredon, 1960), or excitement (Gartner *et al.*, 1965, 1969, 1970; Teleni *et al.*, 1976). Fasting and excitement are associated with slaughter and blood collected at abattoirs is likely to have inflated levels of blood P.

The P concentration in blood can respond quickly to changes in diet (Fishwick and Hemingway, 1973), but correcting a low blood P by feeding a P supplement will only improve intake and production if P is the primary limiting nutrient. A popular misconception is that when blood P is low there should also be clinical signs of aphosphorosis. Clinical signs of P deficiency will only occur when animals have been fed a P-deficient diet for many months, so there will be many situations where blood analysis indicates a deficiency and yet there are no clinical signs. Conversely animals can display visual signs of P deficiency yet have normal levels of blood P if they have only been on a diet adequate in P for a short time.

I. Bone Analysis

The skeleton acts as the major reservoir of P. Phosphorus can be deposited in the skeleton when the supply is in excess of requirements and mobolized in times of need. During lactation P can be drawn from these reserves, reducing the weight of bone ash by up to 40% (Benzie *et al.*, 1959) and the concentration of P in rib bone by 17–42% (Little and McMeniman, 1973; Little *et al.*, 1978b). The quantity of bone P depends on the level of dietary P in the previous months and the demands of the animal. Cattle fed for long periods on a high-P diet contained 13% more P than animals of similar weight fed low-P diets (Little, 1983). The level of bone P can only change slowly in response to alterations in P supply. Bone analysis is unable to identify P-deficient diets that have been fed for only a short time. Conversely bone P can be low in animals fed a diet high in P for a short time.

When assessing the need for supplementary P it is desirable to have information on the level of P in the skeleton, the extent to which this P can be mobilized for use in production, and the composition of the diet eaten. Studies with P-deficient diets have shown that the largest changes in ash and P concentration occur in the ribs (Benzie *et al.*, 1959; Little, 1983). Periodic sampling of ribs of live animals apparently causes the animals little inconvenience (Theiler *et al.*, 1937), and routine procedures have been developed for taking biopsy core samples with a 15-mm trephine operated by hand (Little, 1972; Little and Ratcliff, 1979) or electric drill (McDowell *et al.*, 1983). The P concentration of these rib samples has been expressed in five different ways. These will be described and their diagnostic value considered.

1. Weight of Fresh Bone

The P concentration can be measured in fresh bone. In cattle, P concentration of fresh bone decreased from 86 to 74 g/kg after feeding a P-deficient diet for 6 weeks (Little, 1972). Higher values were found in 7-month-old heifers (95 g P/kg), but after 2 yr feeding on diets containing 1.4 and 3.6 g P/kg DM the levels had fallen to 84 and 87 g P/kg, respectively (Call *et al.*, 1978).

These limited data show that P concentration in fresh bone is not constant but varies with both dietary P and age of animal. Insufficient information is available to be able to recommend a critical level of P which indicates that P-deficient diets have been fed for prolonged periods.

2. Volume of Fresh Bone

The specific gravity of bone can vary, and when P concentration is expressed as milligrams P per cubic centimeter instead of milligrams P per gram of fresh bone then differences in P are related to P levels in the diet (Little, 1972). Concentrations exceeding 130–140 mg P/cc are considered normal for sheep (Little and McMeniman, 1973), but with lactating ewes this critical value was never exceeded even when a P supplement was fed (McMeniman and Little, 1974). On P-deficient diets, values of 78 and 80 mg P/cc were found with lactating ewes (McMeniman and Little, 1974; Read *et al.*, 1986a) and 73 mg P/cc was found in 14-month-old lambs (McMeniman and Little, 1974).

The critical value for P in the twelfth rib in cattle is "around 120 mg/cc" with "levels over 150 mg/cc indicating adequacy" (Little and Shaw, 1979). This value may be too low because a level of 169 mg/cc has been found in heifers fed a P-deficient diet for over a year (Hoey *et al.*, 1982). Doubt about the critical value is confirmed by the high P (187 mg/cc) found in the twelfth rib of lactating cows fed a P-deficient diet (Davison *et al.*, 1986). In other studies, levels between 133 and 172 mg P/cc have been found in lactating cows with normal levels of blood P (Read *et al.*, 1986c,d).

3. Fat-Free Dry Bone

Phosphorus concentration in dry bone varies with the quantity of fat present, a confounding factor that may be removed by expressing P content on a fat-free basis. Phosphorus contents of <115 g/kg fat-free dry bone are considered to indicate P deficiency in ruminants (McDowell *et al.*, 1983). This value appears to be based on the observation by Kleiber *et al.* (1936) that in heifers fed a P-deficient supplement, the fat-free bones contained 115 g P/kg DM. This value may be too low since steers fed a P-deficient diet for a year had ribs containing 116 g P/kg fat-free dry weight (Hoey *et al.*, 1982). A much higher critical value of 137 g P/kg DM was tentatively suggested by Cohen (1973a), who later proposed a range of values that varied with the age of the animal: 143, 135, and 127 g P/kg DM for animals in the range 15–27 months (Cohen, 1973b). None of these animals responded to P supplements, so the significance of these critical values is unknown.

4. Bone Ash

The oldest method of expressing the concentration of P is as a percentage of the bone ash. Rib ash of P-deficient steers contained 172 g P/kg DM compared with 180 g in supplemented animals (Kleiber *et al.*, 1936). Higher values were reported for sheep with 182 g P/kg DM for a P-deficient diet and 188 g P/kg DM when the diet contained excess P (Belonje, 1978). McDowell *et al.* (1983) recommended that 176 g P/kg DM should be accepted as the critical level for rib ash.

5. Bone Thickness

A recent development in the assessment of P reserves requires no P analysis and is based on the thickness of compact bone in the biopsy sample (Little, 1983). A compact bone thickness exceeding 3 mm in cattle was considered normal, while <2 mm indicated previous P deficiency. This method needs further study because compact bone thickness will vary with animal size and age.

J. Hair Analysis

Hair samples could provide a useful method of determining the previous level of P nutrition if the composition changes with different levels of dietary P. In one year a significant correlation was found between P concentration in *T. pratense* and P concentration in hair, but in the following year there was no relationship (Anke, 1967). Subsequent work showed that P concentration in hair was poorly related to seasonal differences in P content of pastures (Cohen, 1973a,b).

IV. PREVENTING PHOSPHORUS DEFICIENCY

A. Supplementary Phosphorus

Where forage is deficient in P, feeding supplementary P is often the quickest and sometimes the most economical way of improving animal production. The composition of P supplements and their value relative to dicalcium phosphate has been reviewed by Hemingway and Fishwick (1976). The choice of one or another of these products will depend not only on availability and local price per unit of available P but also on other criteria: dustiness, bulk density, solubility, particle size, granulation, and taste.

1. Bone Meal

Bone meal was the source of supplementary P used in the original South African studies of P deficiency (Theiler *et al.*, 1924) and was greatly favored by graziers. Bone ash contains about 180 g P/kg DM, but the inclusion of varying amounts of organic matter in the bone meal used as supplements reduces the P content: 94 g/kg DM (Reinach *et al.*, 1952),

126 g/kg DM (Ammerman *et al.*, 1957), 136 g/kg DM (Norman, 1960), and 47–75 g/kg DM (Durand, 1974).

Bone meal may be offered ad libitum in self feeders or given by hand dosing. Self feeders vary from split 200-liter drums to a wide range of equipment designed to protect the bone meal from rain and wind (Bisschop, 1964). Hand dosing of individual animals is the most positive method of supplying P and when the animals "are trained one stockman and four assistants can dose from 200–250 cattle per hour with ease" (Bisschop, 1964). The work load can be reduced by dosing every third day and this is as effective as daily dosing in alleviating P deficiency (Reinach *et al.*, 1952).

2. Calcium Phosphate

Precipitated calcium phosphate was examined as an alternative to bone meal in 1924 and was reported to "behave very much like bone-meal, but is more expensive, tasteless, and more troublesome to administer" (Theiler *et al.*, 1924). The P in calcium phosphate has a high availability and is used as the standard against which the availability of other forms of supplemental P are compared. In one comparison of dicalcium phosphate and two samples of bone meal, steers retained 84, 85, and 84% of additional P, respectively (Otto, 1938). Lower retention figures were obtained in American studies with 32 and 30% of P retained from dicalcium phosphate and bone meal, respectively (Ammerman *et al.*, 1957). Where P retention is apparently low in all supplements tested, it is doubtful whether the animals are sufficiently P deficient to be able to detect differences in P availability of the various supplements.

Feed-grade dicalcium phosphate is now produced by an electrothermal process and contains 174–192 g P/kg (Ammerman *et al.*, 1957; Fishwick, 1976).

3. Rock Phosphate

When compared with bone meal "ground mineral phosphates are cheaper, but so much less effective that they cannot compete with bone-meal" (Theiler *et al.*, 1924). There are large differences between different sources of rock phosphate (Hemingway and Fishwick, 1976). Curcao Island phosphate is a raw rock phosphate, naturally low in flourine, which is mined and ground without further processing. It has a similar P availability to dicalcium phosphate in cattle but not in sheep (Ammerman *et al.*, 1957). Mexican rock phosphate also contains P in a highly available form (Webb *et al.*, 1975).

Christmas Island rock phosphate has been used as a P supplement for lactating cows, despite a fluorine content of 15 g/kg (Snook, 1962). When fed over 4 yr at a rate of 50 g/day, butterfat production was doubled although the fluorine content of the ash of the metacarpus increased to 3000 mg/kg compared with 500 mg/kg for the control. When ground to a 100-mesh screen and fed to sheep, the P was only retained with half the efficiency of P in dicalcium phosphate (Fishwick, 1976). Calcining the rock phosphate by heating to 550°C decreased the fluorine content to 2.8 g/kg but failed to improve P retention.

A thermohydrochemical treatment has been described for reducing the fluorine content of rock phosphate to 1.8 g/kg (Hemingway and Fishwick, 1975b). The product, called Triphos, "is manufactured by the chemical reaction of phosphoric acid and soda ash on rock phosphate followed by defluorination by heat treatment to 1500°C and subsequent crushing." Phosphorus retention and blood phosphate levels were similar for Triphos and dicalcium phosphate.

4. Soft Phosphate

During the process of mining rock phosphate in North Central Florida soft phosphate is produced as a byproduct. It contains 88 g P/kg as finely divided phosphate particles together with large amounts of clay, 21 g/kg iron, 66 g/kg aluminium, and 14 g/kg fluorine (Ammerman *et al.*, 1957). When fed to cattle the available P is similar to other P supplements but with sheep less P is retained (Ammerman *et al.*, 1957). It was suggested that low retention may have been due to the iron and aluminium compounds in the soft phosphate combining with the P to form insoluble iron and aluminium phosphates.

5. Tricalcium Phosphate

This manufactured supplement contains 160–185 g P/kg and only 0.1–2.0 g/kg fluorine (Durand, 1974; Fishwick, 1976). The P has a high availability (Fishwick, 1976) and is a safe and convenient source of P for licks where a high Ca level is required (341–365 g Ca/kg).

6. Sodium Phosphate

Both monosodium phosphate (253 g P/kg DM) (Durand, 1974) and disodium phosphate (87 g P/kg DM; Black *et al.*, 1949) have been used as P supplements. Their main advantage over bone-meal and rock phosphorus is their high solubility in water, which allows them to be fed through the drinking water to cattle (Bekker, 1932; Turner *et al.*, 1935; Black *et al.*, 1943, 1949; Barnes and Jephcott, 1955; Durand, 1974) and sheep (McMeniman and Little, 1974). The intestinal availability of monosodium phosphate exceeds 65% even in the presence of excess Ca (Gueguen *et al.*, 1976). The P in disodium phosphate is completely available (Otto,

1938), while 72–94% of the P in monosodium phosphate is retained by sheep (Cohen, 1972). Sodium phosphate has proved a very successful supplement where pastures contain high levels of Ca (Black *et al.*, 1949; Coates, 1987), but care needs to be exercised where the forage is low in Ca since supplemental P can reduce feed intake.

7. Monoammonium Phosphate

Monoammonium phosphate contains 220–235 g P/kg and 101 g N/kg, is very soluble in water, and 91% of the P can be retained by cattle (Otto, 1938). It also has a high level of available P for sheep (Gueguen *et al.*, 1976) but slightly depresses fiber digestion (Fisher, 1978). When compared with dicalcium phosphate the P in monoammonium phosphate is retained to the same extent (Fishwick and Hemingway, 1973). Added to the diet at 12 g/kg it is well accepted by stock but higher levels reduce palatability and nitrogen utilization (Durand *et al.*, 1976).

Ammonium polyphosphate solution containing 147 g P/kg DM is also available and is equal to dicalcium phosphate as a P supplement (Hemingway and Fishwick, 1975a). Its main advantage over other sources of P is that when added to the drinking water it does not reduce potability.

8. Phosphoric Acid

Phosphoric acid prevents osteophagia in P-deficient cattle (Theiler *et al.*, 1924). It may be used as a source of P either by adding to the drinking water (Playne, 1974; Winks *et al.*, 1977; Holm and Payne, 1980) or by mixing with molasses and urea and feeding in troughs (Winks *et al.*, 1976). Added to drinking water at the rate of 5 g P/liter it improved the growth rate of steers in tropical Australia grazing unfertilized native or *S. humilis* pastures in the wet season. However there have been adverse reports on growth and stock health when pastures contain low levels of legume (Winks and Laing, 1972; McMeniman, 1973; Playne, 1974; Holm and Payne, 1980). These adverse effects have been ascribed to depressed food intake caused by a reduction in water intake (Milton and Ternouth, 1978).

9. Superphosphate

Single superphosphate contains 74 g/kg DM of soluble P in the form of monocalcium phosphate ($Ca(H_2PO_4)_2$) and triple superphosphate contains 190 g/kg DM of soluble P (Durand, 1974). Superphosphate has been used as an inexpensive source of P for adding to drinking water (Reinach and Louw, 1952; Snook, 1955; Durand, 1974). Monocalcium phosphate is separated by mixing the single superphosphate with two to four times its weight of water, leaving it overnight, and syphoning off the supernatant. Practical aspects of the production and delivery of P in drinking water have been reviewed (Reinach and Louw, 1952; Bisschop, 1964; Snook, 1955; Durand, 1974).

10. Other Phosphate Sources

There are many other sources of P that can be used as supplements. Phosphoric acid may be neutralized with calcined magnesite or dolomite to produce a P supplement containing 185 g P/kg. The availability of this P is similar to that in dicalcium phosphate (Fishwick and Hemingway, 1973). The availiability of P in vitreous sodium metaphosphate ($NaPO_3$), sodium pyrophosphate ($Na_2H_2P_2O_2$), vitreous calcium metaphosphate ($Ca(PO_3)_2$), and gamma calcium pyrophosphate ($Ca_2P_2O_7$) has been studied (Chicco *et al.*, 1965), and for a complete discussion of this topic the reader is referred to the review by Hemingway and Fishwick (1976).

B. Fertilizer Phosphorus

Phosphorus deficiency in animals is associated with a deficiency of available soil P (Theiler *et al.*, 1924). Applying P fertilizers will increase P content of forage (Table 8.9) and supply the animal with sufficient P if the plant is at an immature stage of growth.

Applying superphosphate to native pasture in South Africa eliminated osteophagia in cattle compared with an incidence of 75% osteophagia in animals grazing unfertilized pasture (Theiler *et al.*, 1924). In studies in Texas, triple superphosphate applied at a rate of 47 kg P/ha increased the P content of the forage, blood P of cows, calving rate from 57 to 98%, and carrying capacity by 50% (Black *et al.*, 1949). The P fertilizer not only overcame the P deficiency in the animals but also improved pasture production so that the financial return per unit area from fertilized pastures was higher than from an equivalent area on which animals were fed a P supplement. Applying superphosphate to a pasture containing the legume *S. humilis* which was low in P increased the growth of steers from 0.73 to 0.91 kg/day. Feeding a P supplement to cattle grazing the P-fertilized pasture failed to increase growth rate, showing that fertilizing at 11 kg P/ha/yr completely overcame the P deficiency in these animals (Winks *et al.*, 1977). Similar results have been found with cattle grazing a *Cenchrus ciliaris*/legume pasture (Kerridge and McLean, 1988).

C. Breeding for Higher Phosphorus Concentration

Genetic variation in P concentration has been reported for many forage species and is sufficient to allow plant breeders to produce varieties with

TABLE 8.10

Mean Phosphorus Concentration and Genotypic Ranges in Concentration

Forage	Mean P (g/kg DM)	Genotypic range (g/kg DM)	Reference
Dactylis glomerata	3.7	2.7–4.0	Butler and Jones (1973)
Hordeum vulgare[a]	3.0	2.1–2.5	Kleese *et al.* (1968)
		1.3–4.3	Rasmusson *et al.* (1971)
Lolium perenne	3.2	2.2–2.7	Jones and Walters (1969)
	—	3.2–4.7	Cooper (1973)
Medicago sativa	2.6	2.7–3.5	Hill and Jung (1975)
Phalaris arundinacea	—	3.7–4.6	Hovin *et al.* (1978)
Phleum pratense	1.9	2.4–2.6	Jones and Walters (1969)
Trifolium repens	4.0	2.4–4.0	Robinson (1942)
	2.8	2.7–3.0	Davies *et al.* (1968)
Triticum vulgare[a]	1.6	2.4–3.1	Kleese *et al.* (1968)
		1.0–3.0	Rasmusson *et al.* (1971)
Zea mays[a]	2.3	2.1–6.4	Gorsline *et al.* (1964)

[a]Leaves.

higher P concentration (Table 8.10). However, no cultivars have been released on the basis of a higher P concentration.

V. CONCLUSION

Phosphorus is an essential nutrient that is often deficient in forage grown on soils derived from parent rock low in P. Phosphorus deficiency reduces forage intake, estrus, conception rate, milk and wool production, growth rate, and survival of ruminants. Deficiencies of P are most prevalent where forage is both low in P and high in Ca. This is probably because the high level of Ca decreases resorption of P from bone or P absorption from the intestinal tract.

The mean P concentration of forages is 2.9 g/kg DM but it is influenced by many factors. Temperate forages generally contain more P than do tropical species (3.5 versus 2.3 g/kg DM) and legumes more than grasses (3.2 versus 2.7 g/kg DM). Differences in P concentration between forage species and cultivars are usually small with no difference between leaf and stem fractions of the same forage. Increased yield and associated maturation depress P concentration by 0.030 g/kg DM/day. High soil P increases P concentration but application of fertilizer N reduces P concentration. Phosphorus concentration is low in forage grown in summer, at high temperatures, and when stressed by drought.

The characteristic clinical signs of P deficiency is the chewing of bones, stones, fence posts, sacks, etc. and, in extreme cases, peg leg, with creaking of the joints, stiffness, and lameness. Potential P deficiency has been diagnosed from P analysis of the forage, feces, rumen fluid, blood, or bone. Phosphorus deficiency in grazing ruminants has been prevented by feeding supplementary P, fertilizer P, or using a combination of both methods.

9

Magnesium

I. MAGNESIUM IN RUMINANT NUTRITION

A. Function of Magnesium

Low absorption of magnesium (Mg) from the diet can cause hypomagnesemia in cattle (Blaxter *et al.*, 1954) and sheep (McAleese and Forbes, 1959), and reduce milk production (Tables 9.1 and 9.2). In extreme cases, low blood Mg can cause hypomagnesemic tetany, an often fatal disease first described by Sjollema (1930) and also known as grass tetany and grass staggers (Grunes *et al.*, 1970). Clinical signs of the disease may occur when serum Mg falls below about 10 mg/liter (Sjollema, 1930; Kemp, 1960) and animals are stressed (Martens and Rayssiguier, 1980). Lactating animals are most prone to the disease, cattle being more susceptible than sheep and goats (Grunes *et al.*, 1970; Mayland and Grunes, 1979). The early work on hypomagnesemia and grass tetany has been reviewed by Allcroft and Green (1938) and by Littledike and Cox (1979). Hypomagnesemic tetany has not been reported in cattle grazing tropical forages (Minson and Norton, 1982).

Cattle contain approximately 0.38 g Mg/kg liveweight (ARC, 1965). About 70% of this Mg is in the bone (ARC, 1965), with only 1% in the extracellular fluid (Martens and Rayssiguier, 1980). Magnesium is required by the muscles of the heart and nervous system and is an activator of about 300 enzymes, including most of those utilizing ADP or catalyzing the transfer of phosphate (Martens and Rayssiguier, 1980).

Ruminants attempt to maintain a constant concentration of Mg in blood plasma (18 to 30 mg/liter) (Grace, 1983a). This can only be achieved if Mg is absorbed continuously from the digestive tract since the mobilizable body reserves of Mg are small (Littledike and Cox, 1979). The low level of reserves was demonstrated by Allcroft (1953) when she showed that serum Mg level fell within 2 days of removing a Mg supplement. This fall in blood Mg occurs even when large quantities of Mg have been fed prior to turning onto lush pastures in the spring (Allcroft, 1953; Ritter *et al.*, 1984). In ruminants, absorption of Mg is mainly localized in the rumen

TABLE 9.1

Effect of Feeding a Magnesium Supplement to Sheep on the Digestibility of *Lolium perenne*/ *Trifolium repens* Hays[a]

	Poor-quality hay			Good-quality hay		
Chemical composition						
Crude protein (g/kg)		122			156	
ADF[b] (g/kg)		444			284	
Mg (g/kg)		1.7			1.9	
Treatment digestibility	No Mg	+Mg[c]	Increase	No Mg	+Mg[c]	Increase
Dry matter	0.63	0.65	0.02	0.68	0.68	0
Crude protein	0.54	0.57	0.03	0.67	0.68	0.01
ADF	0.56	0.61	0.05	0.63	0.65	0.02

[a]Derived from Wilson (1980).
[b]ADF, acid detergent fiber.
[c]2 g Mg/day.

(Martens and Rayssiguier, 1980) and excess Mg is excreted in the urine (ARC, 1980).

B. Appetite and Digestion

Magnesium is required in many energy-transfer reactions within the body and voluntary food intake of sheep and cattle is depressed when the diet is deficient in Mg (McAleese *et al.*, 1961; Martin *et al.*, 1964; Ammerman *et al.*, 1971, 1972; Chicco *et al.*, 1973b). Intake of sheep was decreased by 67% within 3 to 4 days of feeding a deficient diet (Martin *et al.*, 1964; Ammerman *et al.*, 1971) and recovered in a similar period when Mg was fed (Martin *et al.*, 1964; Ammerman *et al.*, 1971, 1972). Part of

TABLE 9.2

Effect of a Magnesium Oxide Drench for 11 Weeks on Milk Yield and Composition of Mature Cows[a]

Treatment	Control	Mg drench (25 g/day)	Increase (kg)	Increase (%)
Milk (liters/day)	2364	2560	196	8
Fat (g/kg)	53	54	1	1
Fat (kg)	125	137	12	10
Days in milk	245	256	11	4

[a]From Young and Rys (1977).

the loss of appetite was caused by a slower digestion within the rumen (Martin *et al.*, 1964). Both *in vivo* and *in vitro* studies showed a lower cellulolytic activity of rumen bacteria in cattle fed Mg-deficient diets (Martin *et al.*, 1964; Ammerman *et al.*, 1971), with cellulose digestibility reduced by 24–30% (Martin *et al.*, 1964).

Magnesium supplements have been shown *in vivo* to increase the digestibility of cellulose (Hubbert *et al.*, 1958), the DMD of poor-quality forage (Bales *et al.*, 1978), and the DMD of young regrowth of *Lolium perenne/Trifolium repens* and *L. perenne* hay (Wilson, 1980). The digestibility by sheep of the DM, acid detergent fiber, and CP in poor-quality *L. perenne/T. repens* was increased by an Mg supplement, but with high-quality hay only small responses were observed (Table 9.1). In contrast, in a study in which the Mg concentration of *Phleum pratense* hay was increased from 0.8 to 1.5 g/kg by applying Mg fertilizer, the DM digestibility coefficient was slightly depressed (Reid *et al.*, 1984).

C. Effect on Production

A deficiency of Mg can depress production in two ways: by an acute deficiency leading to hypomagnesemic tetany or by a chronic subclinical deficiency. In New Zealand, milk production of mature cows was depressed by a subclinical deficiency of Mg (Young and Rys, 1977; Young *et al.*, 1981; Wilson, 1982b). In the first 2 months of lactation dosing with 25 g Mg in the form of an aqueous suspension of Mg oxide increased milk production of cows on temperate pastures by 15–20%, but later in lactation the difference was small, probably because the control herd was no longer Mg deficient (Young and Rys, 1977). For the complete lactation, Mg supplementation increased milk and fat production by 8 and 10%, respectively (Table 9.2). Magnesium may also be supplied as a rumen pellet releasing 2 g/day of Mg and this increased fat production of mature cows by 15% (Table 9.3). Lactating heifers appear to be able to maintain adequate levels of blood Mg and drenching these animals with Mg failed to increase milk production (Young and Rys, 1977; Young *et al.*, 1981). This lack of response might be due to their lower requirement for Mg, but there may be other reasons, including a higher efficiency of Mg absorption by heifers or a greater release of Mg from the bone.

The level of Mg in the plasma may be increased by a subcutaneous injection or rectal infusion of Mg, but neither of these treatments increases milk production (Wilson, 1980). This indicates that the increase in milk production associated with the higher rumen Mg is not due to a simple change in level of plasma Mg but is probably caused by an increase in the digestibility of the forage (Table 9.1) and an increase in forage intake.

TABLE 9.3

Effect of Slow-Release Magnesium Capsules and Magnesium Drench on Yield and Composition of Milk from Mature Cows[a]

Treatment	Control	Mg drench (10 g/day)	Mg capsule (2 g/day)	Advantage of capsule over control (%)
Milk (liters/day)	18.7	18.5	19.0	2[b]
Fat (g/kg)	43.4	44.8	48.0	11[c]
Fat (kg/day)	0.80	0.83	0.92	15[c]
Plasma Mg (mg/liter)	14.6	17.3	17.6	21[c]

[a]From Wilson (1982b).
[b]Not significant; $p > 0.05$.
[c]$p < 0.05$.

D. Efficiency of Absorption

Poor absorption or low availability to the animal of Mg in forage is the most important factor causing Mg deficiency in ruminants (Wilson, 1981). The Mg in forage has a high potential availability to ruminants (Todd, 1961a) and is rapidly released when the forage is digested in the rumen (Playne *et al.*, 1978; Grings and Males, 1987). However, the transfer of the released Mg into the bloodstream of the animal is often low due to the presence in the rumen of other elements and compounds that reduce the solubility and absorption of Mg.

Absorption of Mg by ruminants has been studied by both comparative and direct procedures. The comparative methods assume that, with diets low or deficient in Mg, the level of Mg in the blood is positively correlated with the level of absorption of Mg from the digestive tract. The more direct procedure is to determine the apparent absorption of Mg in conventional indoor digestibility studies by measuring the quantity of Mg in the forage eaten and in the feces excreted. True absorption can be calculated by subtracting the endogenous Mg excreted in the feces. These two approaches have been used to study the different factors controlling the availability and absorption of Mg in forage.

1. Forage Potassium

The incidence of hypomagnesemic tetany is associated with high levels of potassium (K) and a low ratio of sodium (Na) to K in forage (Sjollema, 1932). Feeding high levels of K generally depresses plasma Mg (Suttle and Field, 1969) and absorption of Mg (Kemp *et al.*, 1961; Newton *et al.*, 1972; Fontenot *et al.*, 1973; Tomas and Potter, 1976; Greene *et al.*, 1983;

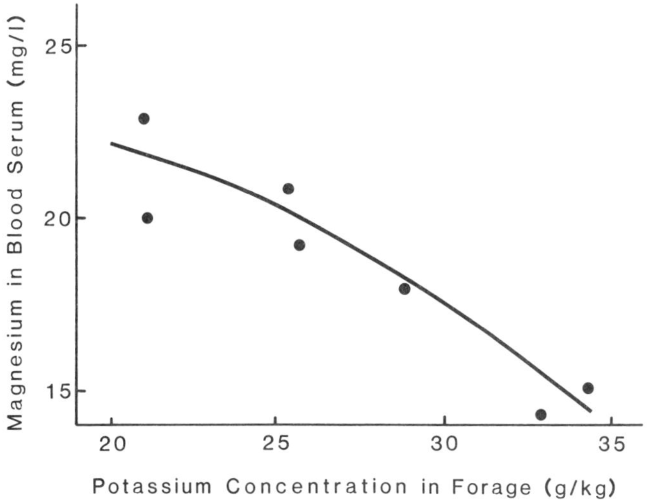

Fig. 9.1 Effect of potassium concentration in temperate forage on blood magnesium. From Kemp (1960.)

Poe *et al.*, 1985), especially when the diets are already low in Mg (Suttle and Field, 1969). The site of this depression of Mg absorption is apparently the rumen (Tomas and Potter, 1976; Martens and Rayssiguier, 1980; Wylie *et al.*, 1985). The fermentation in the rumen can be modified by feeding monensin (Chapter 6) and this reduces the adverse effect of high K on Mg absorption (Greene *et al.*, 1986).

When the K and N concentration in *Lolium multiflorum* was increased by applying fertilizer, serum Mg was reduced from 15 to 7 mg/liter, with clinical signs of hypomagnesemic tetany in two-thirds of the cows (Smyth *et al.*, 1958). The fertilized grass contained slightly more Mg, so the depressed absorption was not due to a low Mg concentration in the diet. In another study, using a mixed pasture, a high concentration of K in the forage depressed serum Mg from 22 to 16 mg/liter (Fig. 9.1).

2. Forage Sodium

The absorption of Mg from the reticulorumen is reduced by a low ratio of Na to K in the rumen (Horn and Smith, 1978; Martens and Rayssiguier, 1980; Martens *et al.*, 1987). This ratio depends not only on the Na:K ratio in the forage but also on the composition of the saliva entering the rumen (Martens and Rayssiguier, 1980). When Na-deficient diets are fed for long periods the level of K in the saliva will increase (see Chapter 10), more K will enter the rumen, and Mg absorption will be depressed, leading to a higher incidence of hypomagnesemic tetany (Martens *et al.*, 1987). In a survey of dairy pastures hypomagnesemic tetany was high on pastures low in Na (Butler, 1963). Feeding an Na supplement has reduced the incidence of hypomagnesemic tetany on farms (Paterson and Crichton, 1960).

3. Forage Protein

The absorption of Mg appears to be depressed by high levels of crude protein (CP) in the diet. In young temperate pasture the apparent absorption coefficient of Mg was only 0.10, but this increased to 0.20 as the forage matured (Kemp *et al.*, 1961). This change in availability was associated with a decrease in CP from 260 to 140 g/kg dry matter. When the CP level of *L. perenne* was increased from 172 to 234 g/kg by applying nitrogen fertilizer, serum Mg was depressed from an initial value of 15 mg/liter to a very low level of 7 mg/liter (Smyth *et al.*, 1958). In another study, raising the CP level of *Dactylis glomerata* from 150 to 260 g/kg reduced the apparent absorption coefficient of Mg from 0.18 to 0.13 (Stillings *et al.*, 1964).

The effect of different protein sources on the availability of Mg was demonstrated by Rook and Balch (1958), who supplemented hay with decorticated ground nut meal and found an apparent absorption of Mg of 0.08 compared with 0.38 when commercial dairy cubes were used. This depression in Mg availability may be due to differences in the level of deamination of the protein in the rumen and the resulting higher pH and concentration of ammonia. Under these conditions, insoluble magnesium ammonium phosphate would be formed (Smith and McAllan, 1967). Raising the ammonia concentration in the rumen by the addition of ammonium salts depressed Mg availability from an initial value of 0.41 to 0.24 (Head and Rook, 1957), but in other work no depression in Mg absorption could be detected (Moore *et al.*, 1972), possibly because the pH was too low for the formation of insoluble magnesium ammonium phosphate.

4. Readily Available Carbohydrates

In 1932 Sjollema, working in The Netherlands, showed that hypomagnesemic tetany could be prevented by feeding sugar-beet pulp or potato residues. The beneficial effect of feeding readily available carbohydrate has been confirmed in subsequent studies (Cunningham, 1934a; Wilson *et al.*, 1969b; Horn and Smith, 1978; Giduck and Fontenot, 1984, 1987).

It appears that the beneficial effect of readily available carbohydrate on Mg absorption is mainly due to its depressing effect on ammonia level in the rumen (Giduck and Fontenot, 1987). This conclusion is supported by earlier studies which showed that raw maize starch drastically reduced the concentration of ammonia in the rumen and increased the level of serum Mg (Head and Rook, 1957). Infusion of volatile fatty acids or hydrochloric acid did not affect Mg absorption, indicating the effect of readily available carbohydrates on Mg absorption is not due to changes in rumen pH or fermentation pattern (Giduck and Fontenot, 1984).

Hypomagnesemic tetany is usually found in spring but will also occur in summer and autumn if the weather is cloudy (Kemp and Hart, 1957; Grunes, 1967). This is probably due to a reduction in the concentration of readily available carbohydrates in the forage. When forage is grown at a low light intensity there is a decrease in the concentration of soluble carbohydrates (Hight *et al.*, 1968; Smith, 1970b).

5. Forage Fat

The level of protein and higher fatty acids in forage is closely correlated (Kemp *et al.*, 1966; Molloy *et al.*, 1973) and it has been suggested that the effect of protein on Mg absorption may be due to high levels of higher fatty acids. When dairy cows were fed supplementary fat (514 g/day), the absorption coefficient for Mg was only 0.08 compared with 0.30 for the control diet (Kemp *et al.*, 1966). Subsequent New Zealand studies found that peanut oil (220 ml/cow/day) depressed the plasma Mg of cows offered *L. perenne* due to the excretion of Mg soaps in the feces (Table 9.4). This detrimental effect of fat on Mg absorption only applies when Mg availability is already marginal; increasing the fat content had no detrimental effect on Mg absorption where the Mg availability in the control diet was high (Grace and Body, 1979).

6. Organic Acids

The concentration of Mg in the blood serum of calves was slightly reduced (2.11 to 1.86 mg/100 ml) by including 1% citric acid in the diet (Burt and Thomas, 1961). Symptoms resembling hypomagnesemic tetany were

TABLE 9.4

Effect of Increasing the Level of Dietary Fat on the Level of Plasma Magnesium and Fecal Soaps[a]

	Control	Peanut oil
Mg in *Lolium perenne* (g/kg)	1.4	1.4
Plasma Mg (mg/liter)	15	12
Fecal composition		
Free fatty acids (mEq/kg DM)	46	70
Fatty-acid soaps (mEq/kg DM)	40	73

[a]From Wilson *et al.* (1969b).

produced in splenectomized cattle following the rapid administration into the rumen of 157 g/100 kg body weight of trans-aconitate or citric acid plus potassium chloride (Bohman *et al.*, 1969). However, in the absence of potassium chloride, the organic acids failed to produce tetany and feeding trans-aconitic or citric acid did not significantly alter the absorption or level of serum Mg (Kennedy, 1968; House and Van Campen, 1971). If organic acids were important in complexing Mg this would only occur shortly after animals started grazing a forage high in organic acid and before rumen microflora populations were established that could effectively degrade the organic acids (Grunes *et al.*, 1985). It is now considered that organic acids in the forage are not the most common nor an essential etiological factor in hypomagnesemic tetany.

7. Stage of Growth

Magnesium deficiency is generally associated with rapidly grown immature forage. When young growths of *L. perenne* × *L. multiflorum* and permanent pasture were fed to cattle, the absorption coefficient for Mg in the first week of growth was only 0.02–0.03. In the second week this increased to 0.21–0.23, a rise that could not be attributed to any change in the quantity of Mg consumed (Rook and Campling, 1962). These levels of absorption are similar to those reported for a range of temperate forages (Table 9.5). This shows that low absorption of Mg can occur when there is a rapid change to lush spring forage (Annison *et al.*, 1959), possibly due to high ammonia and pH of the contents of the reticulorumen and

TABLE 9.5

Apparent Absorption Coefficient of Magnesium in Different Forage Species[a]

Species	Number of samples	Apparent absorption	
		Mean	Sample range
Festuca arundinacea	1	0.04	—
Medicago sativa	2	0.11	0.05–0.18
L. perenne × *L. multiflorum*	3	0.15	0.02–0.23
Phleum pratense	1	0.17	—
Lolium perenne	1	0.19	—
Trifolium pratense	1	0.19	—
Festuca pratensis	1	0.22	—
Dactylis glomerata	2	0.32	0.27–0.33
Mean (weighted)		0.17	0.04–0.33

[a]Adapted from Rook and Campling (1962).

TABLE 9.6

Effect of Magnesium Concentration in Forage on the Level of Blood Magnesium

Species and cultivar	Forage Mg (g/kg DM)	Blood Mg (mg/liter)	Reference
Lolium perenne			
Grasslands Ruanui	1.3	13	Butler and Metson (1967)
Lolium perenne × *L. multiflorum*			
Grasslands Tama	1.4	17	
Grasslands Paroa	1.6	17	
Mixed pasture	1.9	21	
Bromus unioloides	1.1	18	Wilson and Grace (1978)
Lolium multiflorum			
Grasslands Tama	1.3	18	
Mixed pasture	1.7	23	
Lolium perenne			
Grasslands Ruanui	1.9	20	Grace and Wilson (1972)
Trifolium repens			
Grasslands Huia	2.8	18	

insufficient time for the establishment of a microbial population conducive to a high Mg absorption.

8. Forage Species

The availability of Mg in the grasses varies between both species and cultivars (Table 9.6). This difference in blood Mg is associated with the concentration of Mg in the grasses. However, there is an indication of lower absorption of Mg from the legume (Table 9.6), possibly due to a high protein concentration.

9. Forage Calcium and Phosphorus

No studies appear to have been conducted with forage on the effect of varying the level of calcium (Ca) or phosphorus (P) on the absorption of Mg. With semipurified diets fed to sheep, trebling the level of Ca from 1.4 to 4.2 g/kg only depressed the Mg absorption coefficient from an initial value of 0.51 down to 0.46. A similarly small depression was found when P concentration in the diet was increased from 1.2 to 3.6 g/kg (Chicco *et al.*, 1973b).

E. Requirement for Magnesium

The quantity of Mg required in the diet depends on the absolute requirements of the animal and the efficiency of absorption of Mg from the

TABLE 9.7

Magnesium Requirements of Beef Cattle Fed ad Libitum Expressed as Dietary Concentrations (g/kg DM)[a]

Liveweight (kg)	Liveweight gain (kg/day)		
	0	0.5	1.0
200	0.9	1.1	1.1
400	1.2	1.2	1.1
600	1.3	1.2	1.2

[a]Derived from ARC (1980).

digestive tract. However, to be of any value in the interpretation of forage analytical data, dietary requirements must be expressed as concentration of Mg in the forage DM and not in terms of the quantity of Mg required each day.

The dietary requirements in terms of forage composition can be derived from the quantitative values published by ARC (1980) and estimates of the ad libitum forage intake required to achieve a specified level of production. The Mg concentrations in the forage needed to meet the requirements for four classes of ruminant production are shown in Tables 9.7, 9.8, 9.9, and 9.10. These tables show that grazing animals will rarely require forage containing more than 2 g Mg/kg DM. Very often lower concentrations of Mg are adequate, particularly for nonlactating animals. However, as noted earlier, the absorption of Mg is affected by K and ammonia in the rumen and if these are high, then insufficient Mg may be absorbed from a forage with an apparently adequate level of Mg.

TABLE 9.8

Magnesium Requirements of Lactating Cows Fed ad Libitum, Expressed as Dietary Concentrations (g/kg DM)[a]

Liveweight (kg)	Breed	Milk yield (kg/day)			
		0	10	20	30
400	Jersey	1.2	1.3	1.7	—
500	Ayrshire	1.3	1.4	1.7	2.1
600	Friesian	1.3	1.5	1.8	2.1

[a]Derived from ARC (1980).

TABLE 9.9

Magnesium Requirements of Castrated Lambs Fed ad Libitum, Expressed as Dietary Concentrations (g/kg DM)[a]

	Liveweight gain (kg/day)			
Liveweight	0	0.1	0.2	0.3
20	0.8	1.1	1.5	—
30	0.9	0.9	1.0	1.3
40	0.9	0.9	1.0	1.1

[a]Derived from ARC (1980).

TABLE 9.10

Magnesium Requirements of Lactating Ewes Fed ad Libitum, Expressed as Dietary Concentrations (g/kg DM)[a]

Liveweight (kg)	Milk yield (kg/day) 0	1.0	2.0	3.0
40	0.9	1.2	1.7	—
75	1.1	1.3	1.5	1.8

[a]Derived from ARC (1980).

II. MAGNESIUM IN FORAGE

A. Mean Concentration

The mean Mg concentration in a range of 930 forage samples was 2.3 g/kg DM with a range from 1.0 to more than 9.0 g/kg DM (Fig. 9.2). The samples were divided into four groups: temperate and tropical grasses and temperate and tropical legumes. These had mean Mg concentrations of 1.8, 3.6, 2.6, and 2.8 g/kg, respectively (Fig. 9.3). Of practical importance is that 65% of temperate grass samples contained less than 2 g Mg/kg and may not have met the Mg requirements of lactating ruminants. Conversely, only 14% of the tropical grasses sampled contained less than 2 g Mg/kg DM.

B. Species Differences

Temperate legumes generally contain more Mg than grasses (Table 9.11) but there are large differences between the various studies and no

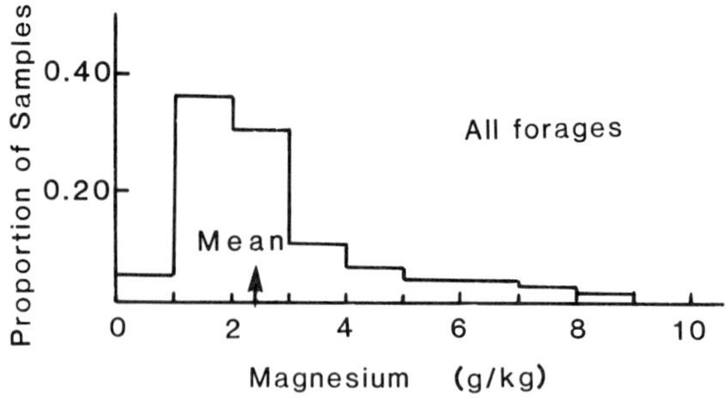

Fig. 9.2 Magnesium concentration in a range of forages (data from the world literature).

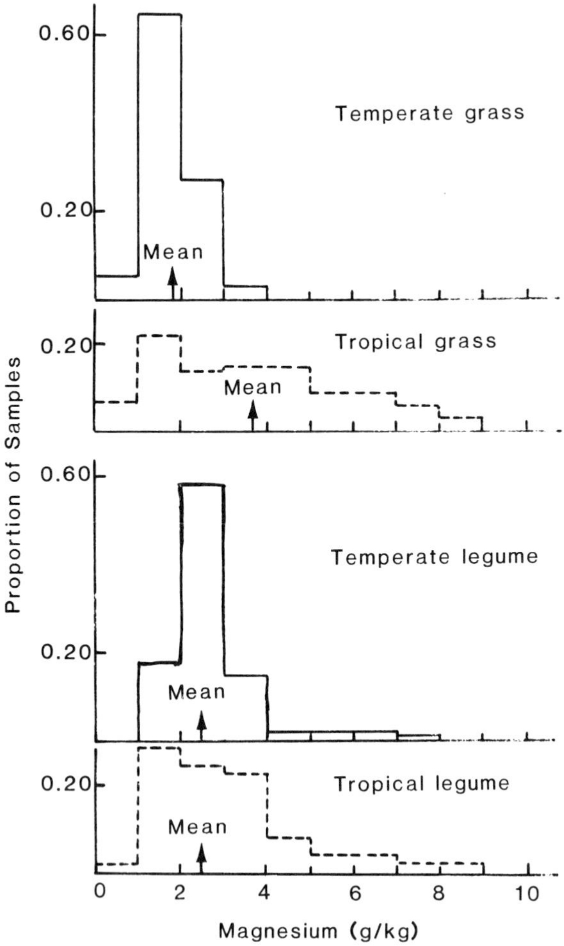

Fig. 9.3 Magnesium concentration in temperate and tropical grasses and legumes (data from the world literature).

TABLE 9.11

Magnesium Concentrations in Temperate Forage Species[a]

Grasses	Mg (g/kg)	Legumes	Mg (g/kg)
Agrostis sp.	1.3–1.4	*Medicago lupulina*	7.6
Bromus inermis	2.3–3.1	*Lotus corniculatus*	6.3–6.5
Cynosurus cristatus	1.8	*Medicago sativa*	1.5–5.5
Dactylis glomerata	1.7–4.2	*Onobrychis viciifolia*	1.7–7.9
Festuca arundinacea	3.5–4.0	*Trifolium hybridum*	3.7–6.2
Festuca elatior	1.7–5.0	*Trifolium pratense*	2.1–5.0
Festuca pratensis	1.6–2.6	*Trifolium repens*	1.8–4.8
Festuca rubra	1.5–1.6		
Holcus lanatus	1.6–1.8		
Lolium multiflorum	1.8		
Lolium perenne	1.2–6.9		
Phalaris arundinacea	5.0–5.5		
Phleum pratense	1.3–5.0		
Poa trivialis	1.6–1.8		
Mean	2.7		4.8

[a]Derived from Thomas *et al.* (1952a), Walshe and Conway (1960), Todd (1961b), Whitehead and Jones (1969), Fleming and Murphy (1968), Gross and Jung (1978).

general conclusions can be drawn on the relative Mg concentration of different species of grasses or legumes. In mixed pastures Mg concentration is positively correlated with the proportion of legume in the sward (Hutton *et al.*, 1965). In one study, *T. repens* contained 3.2 g Mg/kg compared with 1.5 g Mg/kg in a companion *D. glomerata* (Bartlett *et al.*, 1954). Pasture herbs in other plant families generally contain higher concentrations of Mg than grasses or legumes (Grunes *et al.*, 1970).

C. Stage of Growth

As *P. pratense* matures there is a large decrease in Mg concentration (Fig. 9.4), possibly due to a lower concentration of Mg in the leaf than in the stem (Pritchard *et al.*, 1964). This decrease occurs even when the grass is fertilized with Mg (Reid *et al.*, 1984). However, only small changes in Mg concentration have been found in other studies with *D. glomerata* (Reid *et al.*, 1970), *L. perenne* (Thomas *et al.*, 1952a; Wilson and McCarrick, 1967; Fleming and Murphy, 1968), *P. pratense* (Wilson and McCarrick, 1967; Fleming and Murphy, 1968), and four temperate legumes (Whitehead and Jones, 1969).

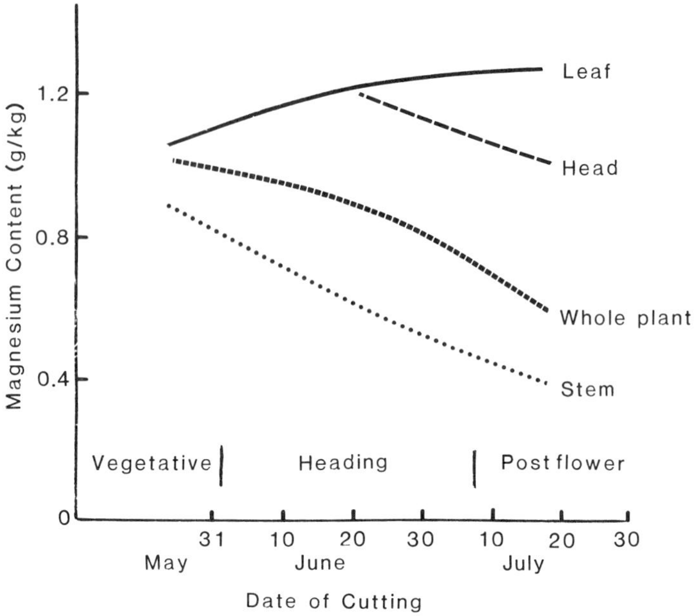

Fig. 9.4 Effect of stage of growth on the magnesium concentration of *Phleum pratense*. From Pritchard *et al.* (1964.)

D. Soil Fertility

Magnesium concentration of forages is affected by soil type (Cooper *et al.*, 1947; Metson, 1974; Mayland and Grunes, 1979). Soils developed from acidic igneous rocks are generally low in Mg, while high levels of Mg are found in soils derived from more basic rocks. Soils developed on sedimentary rocks, especially on sandstone and shales, are generally low in Mg (Grunes *et al.*, 1970; Metson, 1974). In New Zealand, hypomagnesemic tetany is particularly prevalent on yellow-brown loam soils; in one study by Young *et al.*, (1981) 70% of the herds had low serum Mg concentrations and 10% of the blood samples contained less than 10 mg Mg/liter. The level of available Mg in soils also depends on the fertilizers applied (McIntosh *et al.*, 1973) and previous cropping history (Metson, 1974).

Magnesium concentration in *Cynodon aethiopicus, D. glomerata, Agropyron desertorum,* and mixed temperate pastures was increased by 0.2–0.7 g Mg/kg following the application of 56–580 kg N/ha (Stewart and Holmes, 1953; Kemp, 1960; Stillings *et al.*, 1964; Reid *et al.*, 1967b; Rudert and Oliver, 1978). There was no interaction among fertilizers N, K, or P except when soil Mg had been depleted by frequent cutting and removal of the forage (Stewart and Holmes, 1953).

Fertilizer K (291 kg/ha/yr) reduced Mg level in a mixed pasture from 3.1 to 2.4 g/kg (Stewart and Holmes, 1953). In another study annual applications of 211 kg K/ha reduced Mg concentration in *L. perenne* from 1.6 to 1.3 g/kg (Bartlett *et al.*, 1954). This change was accompanied by low serum Mg in grazing cows, with 75% of the herd below 10 mg/liter and clinical symptoms of hypomagnesemic tetany in 25% of the animals. Similar depressions in forage Mg were observed in extensive studies in The Netherlands (Kemp, 1960), but with small applications of K there was little or no change in the Mg concentration (Walshe and Conway, 1960; McConaghy *et al.*, 1963).

Fertilizer Ca has caused a small depression in the Mg concentration of forages (Mayland and Grunes, 1979), but most of the evidence is inconclusive, possibly because the uptake of Mg by the plant is affected not only by the levels of Ca and Mg in the soil but also by the concentrations of other cations and by the soil pH.

E. Climate

1. Seasonal Changes

The Mg concentration in regrowths of forage varies throughout the year, with lowest values in the spring (Ferguson, 1932; Cooper *et al.*, 1947; Stewart and Holmes, 1953; Stewart and Reith, 1956; Griffiths, 1959; Todd, 1961b; Fleming and Murphy, 1968; McIntosh *et al.*, 1973). With frequently cut mixed temperate pastures Mg concentration remains relatively constant in spring and early summer but is higher in autumn and winter (Fig. 9.5). In Ireland and New Zealand, autumn-grown pasture contained 33–40% more Mg than pasture grown in the spring (Walshe and Conway, 1960; McNaught *et al.*, 1973). Although the Mg concentration of grasses *(D. glomerata* and *L. perenne)* increased in the autumn, the legumes *(T. pratense* and *T. repens)* remained relatively constant throughout the year (Todd, 1961b; Jones, 1963b).

2. Light Intensity

Reducing light intensity by 71% increased the Mg concentration of *Cynodon dactylon* from 2.0 g/kg for the control to 2.7 g/kg for shaded forage when the plots were not fertilized with N. However, when the grass received adequate levels of fertilizer N, shading had little effect on Mg concentration (Burton *et al.*, 1959).

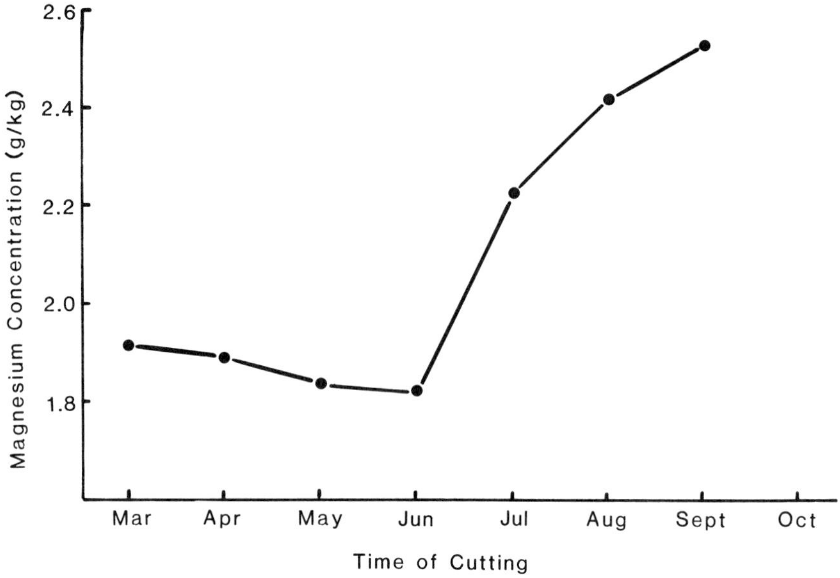

Fig. 9.5 Seasonal variation in the magnesium concentration of forage regrowths. From Jones (1963b.)

3. TEMPERATURE

Magnesium concentration is positively related to the temperature at which the forage is grown. At low temperatures and in waterlogged soils uptake of Mg is generally low (Elkins and Hoveland, 1977). Growing at low temperatures reduced the Mg concentration in *A. desertorum* (Grunes, 1967), *L. perenne* (Dijkshoorn and Hart, 1957; Grunes, 1967), *Lotus corniculatus*, *Melilotus officinalis*, and *Trifolium hybridum* (Smith, 1970b). This positive relation between Mg concentration and temperature probably accounts for the low incidence of hypomagnesemic tetany in cattle grazing temperate pastures in summer (Kemp and Hart, 1957) and all tropical pastures (Minson and Norton, 1982).

III. DIAGNOSIS OF MAGNESIUM DEFICIENCY

A. Clinical Signs

An acute deficiency of Mg can lead to the development of the nutritional disease hypomagnesemic tetany. The clinical signs of the disease have been described by Underwood (1981):

> The initial signs are those of nervous apprehension with ears pricked, head held high and staring eyes. At this stage the animal's movements are stiff and stilted, it staggers when walking and there is twitching of the muscles, especially of the face and ears. Within a few hours or days, extreme excitement and violent convulsions develop; the animal lies flat on its side; the forelegs "pedal" periodically and the jaws work, making the teeth grate. If treatment is not given at this stage death usually occurs during or after one of the convulsions or the animal may pass into a coma and die.

These clinical signs will disappear within minutes following a subcutaneous injection of 400 ml of a 25% solution of Mg sulfate (Allen and Davies, 1981) provided treatment has not been delayed and irreversible pathological changes have not occurred (Littledike and Cox, 1979).

B. Forage Analysis

Analysis of forages can provide a guide to their potential for supplying the animal with Mg. If the forage contains more than about 2 g/kg it will meet the recommended requirements (Tables 9.7–9.10) and no deficiency is likely to occur. In spring-grown forage low serum Mg levels have not been observed where the Mg concentration in the forage exceeds 2 g/kg (Smyth *et al.*, 1958; Kemp *et al.*, 1960), but in autumn-grown forage hypomagnesemic tetany has occurred at this level and 2.5 g Mg/kg is probably a more realistic critical figure for diagnostic purposes (Todd, 1966).

The absorption of Mg is depressed by high levels of potassium especially in the presence of low levels of Mg. Kemp and Hart (1957) took this into account when they predicted the probability of hypomagnesemic tetany from the K:(Ca + Mg) ratio in the forage, expressed in milliequivalents. Few cases of tetany (0.77%) occur when the ratio is less than 2.2, but when the ratio exceeds 3.0 the incidence of tetany among individual grazing cows in a herd may exceed 15% (Fig. 9.6). The ratio is a useful indicator for grasses but is not suitable for legumes due to their high Ca concentration (Gross and Jung, 1978). Other anion/cation indices have been suggested as indicators of potential hypomagnesemic tetany (Rahman *et al.*, 1960; Rook and Wood, 1960) but these do not appear to have been widely used.

None of these ratio methods takes into account the effect of CP on Mg absorption. This shortcoming was overcome in The Netherlands by using a nomogram which predicted serum Mg from the Mg, K, and CP concentrations in the forage (Fig. 9.7).

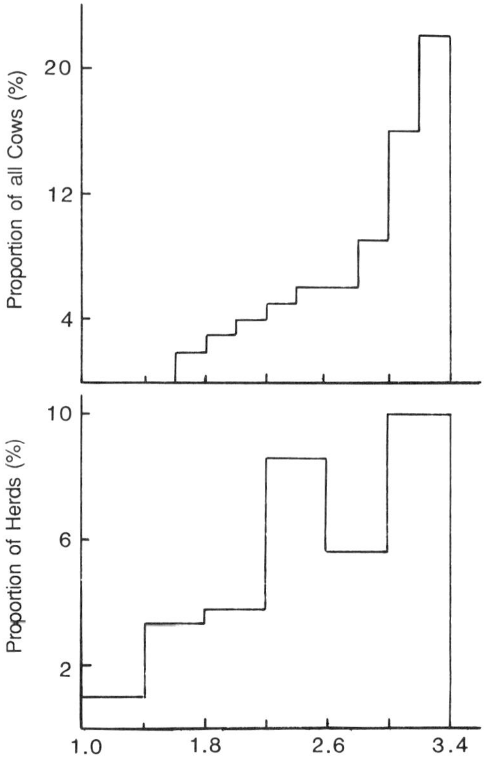

Fig. 9.6 Incidence of hypomagnesemic tetany in individual cows and herds in relation to the K:(Ca + Mg) ratio in forage (mEq). From Butler (1963.)

C. Urine Analysis

When animals are put out to grass in the spring there is a rapid fall in excretion of Mg in the urine (Rook and Balch, 1958; Kemp, 1968; Field, 1961), indicating that the quantity of Mg absorbed from the diet is less than is required by the animal. The concentration of Mg in the serum is positively correlated with the output of Mg in the urine (Kemp *et al.*, 1960; Rook and Storry, 1962) and a semiquantitative test for urine Mg, based on visual color estimation, has been developed for use in the field (Simesen, 1977). The renal threshold for Mg is about 20 mg/liter in sheep and 18 mg/liter in cattle (Gardner, 1973), and animals can be classified into three groups on the basis of urine Mg concentration: normal, >100 mg/liter; borderline, 50–100 mg/liter; and low Mg status, <50 mg/liter of urine (Simesen, 1977).

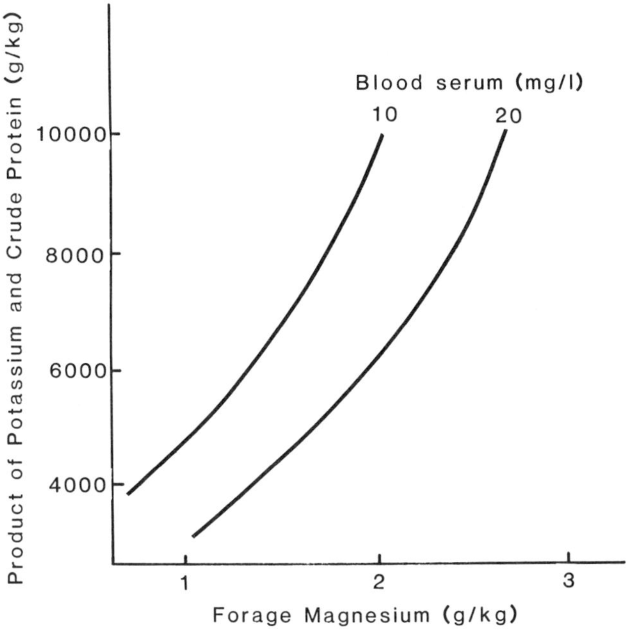

Fig. 9.7 Nomogram for predicting blood magnesium from forage composition. From Committee on Mineral Nutrition (1973.)

IV. PREVENTING MAGNESIUM DEFICIENCY

A. Supplementary Magnesium

It is generally accepted that none of the methods at present used to prevent Mg deficiency is applicable to all management systems. The method chosen will depend on the class of animal, the grazing intensity, and local economic and climatic factors. Seven contrasting methods have been used to supplement ruminants with Mg.

1. Grain Mixtures

In The Netherlands a daily allowance of 1 kg pelleted concentrates containing 50 g Mg oxide has proved a practical method for supplying Mg to dairy cows (Kemp, 1971). This is also the main method of preventing hypomagnesemia in the United Kingdom. Although widely used, Mg oxide is considered to be unpalatable and Mg phosphate has been recommended as an alternative, feeding at a rate of 57 g/day (Ritchie and Fishwick, 1977).

2. Treatment of Hay and Silage

In winter, extra Mg can be given to both beef and dairy cattle by feeding hay treated with Mg oxide. The supplement is applied at a rate of 50–56 g/animal/day to the cut aside of baled hay as a watery molasses solution (Young, 1967; Grace, 1983a). This was sufficient to maintain the blood Mg of all animals above 18 mg/liter compared with the unsupplemented group, in which half the blood samples contained less than 18 mg/liter and a quarter were below the critical level of 10 mg/liter (Young, 1967). Silage can also be used as the carrier for Mg supplements.

3. Drench

With intensively managed animals Mg can be administered daily as a suspension of Mg oxide in water. This method is widely used to supplement lactating cows in New Zealand. The rates used have varied: 56 g/day (Allcroft, 1953), 42 g/day (Young and Rys, 1977), and 18 g/day (Young *et al.*, 1979) supplying 34, 25, and 12 g Mg/day, respectively. The apparent absorption of Mg oxide used in the drench is increased by reducing particle size: fine (under 75μm), medium (150–250 μm), and coarse (500–1000 μm) particle sizes have apparent absorption coefficients of 0.48, 0.28, and 0.08, respectively (Wilson and Ritchie, 1981). Magnesium oxide can be mixed with a pluronic if bloat is also to be controlled. A stable suspension suitable for administering 45 g of drench with an automatic drench gun can be made with 2.5 kg Mg oxide, 1 liter pluronic, and 1 liter water (Young *et al.*, 1979).

Magnesium sulfate, at a rate of 100 g/day, has been used as a drench for cattle. Where cows are simultaneously dosed with a pluronic and Mg, Mg chloride (118 g Mg/kg) has been recommended (88 g/day) (Grace, 1983a) starting 2 to 3 weeks before calving (Young *et al.*, 1979). Neither Mg sulfate nor Mg chloride is palatable and acceptance of the drench can be improved by adding about 5% molasses for the first few weeks (Grace, 1983a).

Another potential supplement is Mg carbonate. Laboratory-grade Mg carbonate has a high availability but when supplied as the mineral magnesite, Mg availability is low (Ammerman *et al.*, 1972). This difference is apparently associated with a difference in the physical properties of the two forms; laboratory-grade Mg carbonate is fluffy and indefinite in structure compared with the crystalline structure of magnesite.

4. Capsules

In studies with sheep, Mg has been supplied daily in gelatin capsules containing 0.96 mg of Mg as the oxide (Ammerman *et al.*, 1971), but this procedure is only suitable for experimental work. To overcome the need for daily dosing of cattle, Laby (1973) produced a variable-geometry capsule which released Mg into the rumen at an average rate of 2 g/day. This improved milk-fat production more than drenching with 10 g/day Mg (Table 9.3). The high efficiency of the sustained release device would possibly be associated with the maintenance of a higher concentration of soluble Mg close to the absorptive surfaces of the rumen wall. Sustained release capsules of this type would be of great value for supplementing beef cattle grazing at low stocking rates.

5. Pasture Dusting

The intake of Mg can be increased by dusting the pasture with finely ground calcined magnesite, a method first used in Northern Ireland by Todd (1962). When applied at a rate of 63 kg Mg/ha just before grazing, the Mg concentration in the forage was increased and blood Mg levels restored to normal levels. In later studies the application rate was reduced to 31 kg/Mg/ha. This increased the Mg content of the pasture from 1.6 to 3.1 g/kg and maintained the serum Mg level of grazing cows above 18 mg/liter (Todd and Morrison, 1964). A similar level of calcined magnesite applied to pastures in The Netherlands maintained safe levels of Mg in the forage for 1 week (Kemp, 1971). In New Zealand, calcined magnesite was applied to the pasture at weekly intervals at the rate of 500 g/cow. Application is made in the early morning when adhesion of the dust is improved by the presence of dew (Young, 1967; Grace, 1983a). This raised the Mg level of the forage from 1.8 to 4.4 g Mg/kg immediately after dusting, but this level dropped to 2.2 g Mg/kg after 7 days (Young, 1967). If rainfall exceeds 40–50 mm within 2–3 days of dusting a further application is recommended (Grace, 1983a). The method is suitable for intensively grazed pastures but is not ideal for low-stocked pastures where animals are eating only a small proportion of the forage offered.

6. Water Treatment

Where animals have access to a controllable source of drinking water, Mg may be supplied through the watering system (Cunningham, 1934b). Magnesium chloride or Mg sulfate added to the water trough at the rate of 50 or 60 g/cow/day, respectively, equivalent to 6 g elemental Mg, was sufficient to reduce clinical signs of hypomagnesemic tetany from 14–25% to less than 3% of all cows in a herd (Young, 1975). For a bulk water supply, Young (1975) recommended 3 g Mg chloride or 3.3 g Mg sulfate/liter on the assumption that water intake is 26 liters/cow/day. Higher levels of Mg chloride are required in wet weather, when less water is drunk.

7. Magnesium Licks

Magnesium can be supplied in the form of solid licks but consumption is generally too variable to ensure a constant supply of Mg (Cunningham, 1934b; Grace, 1983a). Normal levels of serum Mg have been maintained by feeding cows with 30 g Mg/day in a mixture containing equal quantities of Mg oxide and molasses (Todd *et al.*, 1966).

B. Fertilizer Magnesium

The Mg content of forage can be increased by the application of Mg fertilizers. Many sources of Mg are available (Metson, 1974) and these may be divided into two categories according to their solubility (Table 9.12).

1. Magnesium Carbonate

Dolomite was the first fertilizer used to prevent hypomagnesemia (Cunningham, 1936). When applied at rates between 100 and 540 kg Mg/ha the Mg concentration of temperate forage is increased by 0.2–1.2 g/kg DM (Table 9.13). These increases in Mg concentration in the forage were associated with a rise in the level of exchangeable Mg in the soil (Jones, 1963b; Simpson, 1964). Dolomite has a low solubility and maximum increase in Mg occurred 3 yr after the fertilizer was applied (Fig. 9.8).

TABLE 9.12

Magnesium in Different Fertilizers[a]

Name	Composition	Mg (g/kg)
Slightly soluble forms		
Magnesium carbonate		
Dolomite	$MgCO_3 \cdot CaCO_3$	100
Magnesite	$MgCO_3$	260
Magnesium oxide and hydroxide		
Dolomite (selectively calcined)	$MgO \cdot CaCO_3$	160
Dolomite (hydrated)	$MgO \cdot Ca(OH)_2$	170
Brucite	$Mg(OH)_2$	360
Calcined magnesite	MgO	560
Very soluble forms		
Magnesium sulphate		
Kieserite	$MgSO_4 \cdot H_2O$	160
Epsom salts	$MgSO_4 \cdot 7H_2O$	90
Kainite	$MgSO_4 \cdot KCl, 3H_2O$	70
Langbeinite	$2MgSO_4 \cdot K_2SO_4$	110

[a] From Metson (1974).

TABLE 9.13

Effect of Different Forms and Rates of Fertilizer Magnesium on the Magnesium Concentration in Temperate Forage

Fertilizer	Rate applied (kg Mg/ha)	Increase in Mg (g/kg DM)	Reference
Magnesium carbonate	100	0.2	Cunningham (1936)
	230	0.3	Simpson (1964)
	460	0.4	
	496	0.8	Jones (1963b)
	540	1.2	Stewart and Reith (1956)
Magnesium oxide	330	1.3	Griffiths (1959)
	670	0.9	Birch and Wolton (1961)
	670	1.2	Parr and Allcroft (1957)
	1600	1.6	Bartlett *et al.* (1954)
	1660	2.0	Griffiths (1959)
Magnesium sulfate	38	0.4	Griffiths (1959)
	48	0.6	Walshe and Conway (1960)
	60	0.2	Birch and Wolton (1961)
	62	0.3	Jones (1963b)
	120	1.2	Walshe and Conway (1960)
	390	0.8	Reid *et al.*(1984)

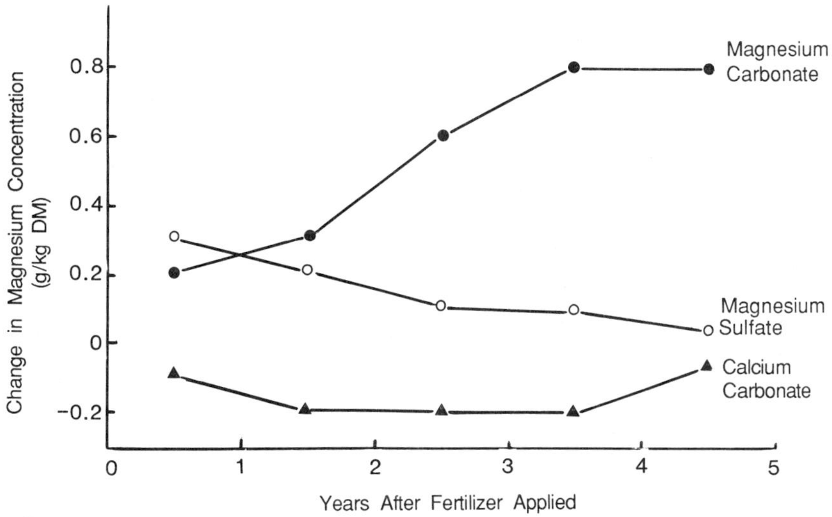

Fig. 9.8 Initial and residual effect of different sources of magnesium. From Jones (1963b).

2. Magnesium Oxide and Hydroxide

The solubility of calcined magnesite (Mg oxide) and hydrate forms are much greater than that of Mg limestone and applying between 330 and 1600 kg Mg/ha increased the concentration of Mg in forage by 0.9–2.0 g/kg DM (Table 9.13). When Mg oxide was applied to a mixed pasture the increase in Mg concentration in *T. repens* were twice that found in *L. perenne* (Bartlett *et al.*, 1954). On light soil with a pH of 5.8, 320 kg/ha Mg as calcined magnesite protected grazing cows against hypomagnesemic tetany for 3 years and provided marginal protection in the fourth year, but on a heavy loam with a pH of 6.6 the response only lasted a year (Todd, 1965). In recent New Zealand studies, the application of 60 kg Mg/ha proved more effective in raising serum Mg and preventing hypomagnesemic tetany than drenching with 10 g Mg/day (O'Connor *et al.*, 1987).

3. Magnesium Sulfate

All forms of Mg sulfate are very soluble and when applied at rates between 38 and 120 kg Mg/ha the Mg concentration of forage is increased by 0.4–1.2 g/kg DM (Table 9.13). The largest increase in Mg concentration occurs in the year the fertilizer is applied, with smaller residual effects in the following two years (Fig. 9.8).

C. Other Management Practices

1. Rectal Infusion

Magnesium chloride can be absorbed from the rectum (Meyer and Busse, 1975) and the quantity absorbed is sufficient to raise plasma Mg level to the same level as can be achieved by drenching (Wilson, 1980). The Mg may be administered as a solution containing 300 g/liter of magnesium chloride in two portions at milking times using an automatic drenching gun. However the method causes irritation and is not recommended for routine use (Wilson, 1980).

2. Subcutaneous Injection

In experimental work Mg has been supplied as a daily subcutaneous injection. To provide 2 g Mg/day, 100 ml of a sterile solution was used containing 200 g Mg sulfate dissolved in 1 liter water (Wilson, 1980).

3. Pasture Species Naturally Rich in Magnesium

Legumes generally contain a higher concentration of Mg than grasses (Table 9.11), so encouraging a higher proportion of legume may be ex-

TABLE 9.14

Narrow-Sense Heritability Estimates for Magnesium in Five Grasses

Grass	Heritability (%)	Reference
Lolium perenne	86	Cooper (1973)
Festuca arundinacea	76	Sleper *et al.* (1977)
Phalaris arundinacea	66	Hovin *et al.* (1978)
Setaria sphacelata	40	Bray and Hacker (1981)
Dactylis glomerata	81–86	Currier *et al.* (1983)

pected to reduce hypomagnesemia. However, the higher Mg concentration of the legumes could be partly offset by the lower absorption of the Mg (Table 9.6).

4. Fertilizer Sodium

In Section I,D it was shown that the absorption of Mg is reduced if ruminants are deficient in Na. Applying fertilizer Na (79 kg/ha/yr) halved the incidence of hypomagnesemic tetany in beef cows but the treatment was less effective than applying Mg oxide (Smith *et al.*, 1983).

5. Breeding Forages for Increased Magnesium Concentration

Breeding forages to prevent hypomagnesemia is complex since low serum Mg is affected by many factors in addition to low forage Mg. High Mg will reduce the risk of hypomagnesemia provided none of the other factors affecting Mg absorption is adversely changed during the selection process. Studies have been made of the inheritance of Mg concentration in temperate and tropical grasses (Table 9.14). These estimates appear to be sufficiently high to enable lines of higher Mg concentrations to be bred. With *D. glomerata* it has been suggested that breeding could improve Mg intake in both nonflooded and flooded soil conditions (Currier *et al.*, 1983). In *Medicago sativa* low- and high-Mg lines have been identified which contained 1.8 and 3.0 g Mg/kg (Hill and Jung, 1975).

V. CONCLUSION

Magnesium is an essential element required by ruminants to prevent hypomagnesemic tetany and maintain maximum milk production. Magnesium deficiency is associated with heavily fertilized temperate grasses in cold and some warm environments. There are no reports of hypomagnesemic tetany occurring in ruminants grazing tropical forages.

To maintain normal Mg levels in the blood, Mg must be absorbed continuously from the digestive tract, because ruminants have no large reserve of Mg that can be rapidly mobilized. Magnesium is rapidly released from forage during digestion in the rumen, but transfer through the rumen wall into the bloodstream is reduced by the presence of other elements and compounds in the forage. These anatagonists include high concentrations in the forage of K, CP, and higher fatty acids. Absorption of Mg can also be depressed by rapid changes in composition of the diet but there was no consistent difference between forage species. Absorption of Mg is often improved by high levels of sodium and readily available carbohydrates.

Temperate grasses generally contain less Mg than temperate legumes and tropical grasses. There are no consistent differences in Mg concentration between either grasses or legume species, but large genetic differences in Mg content have been reported within species of both grasses and legumes. Leaves contain more Mg than stem but increasing maturity has no consistent effect on Mg concentration in forage. Magnesium concentration in forage varies with available soil Mg, is increased by fertilizer N, but is depressed by fertilizer K and low temperature.

Magnesium deficiency can be positively diagnosed by low levels of Mg in the plasma and a positive response of animals with hypomagnesemic tetany to a subcutaneous injection of magnesium sulfate. The ratio of K to (Ca + Mg) in the forage can be used to indicate potential Mg deficiency. A low urine Mg concentration indicates insufficient Mg is being absorbed to meet ruminant requirements.

Magnesium deficiency can be reduced by feeding supplementary Mg, applying fertilizer Mg and Na, and avoiding excess fertilizer K and N.

10

Sodium

I. SODIUM IN RUMINANT NUTRITION

A. Function of Sodium

Sodium (Na) is the principal cation in the extracellular fluid of ruminants. It plays a major role in the regulation of osmotic pressure, acid–base balance, maintenance of membrane potentials, and transmission of nerve impulses (Hays and Swenson, 1970) and is required for the growth of rumen bacteria (Caldwell and Hudson, 1974). Cattle and sheep contain 1–2 g Na/kg empty body weight (ARC, 1980) plus a large pool of Na in the rumen which can be used when the diet is low in Na (Buchan *et al.*, 1986). A deficiency of Na reduces Mg absorption (Martens *et al.*, 1987) and increases the incidence of hypomagnesemia (Chapter 9).

Sodium concentration in the blood is constant (Manston and Rowlands, 1973) and is not affected by differences in dietary Na (Smith and Aines, 1959; Morris and Gartner, 1971; Morris and Murphy, 1972). When the intake of dietary Na is low, Na homeostasis in the extracellular fluids is achieved through increased production of aldosterone by the adrenal cortex (Blair-West *et al.*, 1970), which reduces Na loss in the urine (Kemp, 1968; Morris and Gartner, 1971; Leche, 1977a) and feces (Jones *et al.*, 1967b; Morris and Gartner, 1971). The lower fecal loss of Na is achieved by substituting K for Na in the parotid saliva (Bailey and Balch, 1961; Jones *et al.*, 1967b; Morris and Gartner, 1971).

The Na concentration in milk is constant at all levels of dietary Na (Murphy and Plasto, 1973; Leche, 1977a) because aldosterone has no influence on milk composition (Peaker and Linzell, 1973). This lack of ability of animals to reduce the Na content of their milk leaves them more susceptible to Na deficiency during lactation. Deaths of lactating cattle from Na deficiency have been reported in Papua New Guinea where the disease is known as "nutritional lactation stress" (Leche, 1977a).

B. Voluntary Intake of Forage

It is generally accepted that appetite is depressed by a deficiency of Na (McClymont *et al.*, 1957; Underwood, 1971; Morris and Murphy, 1972). Voluntary intake of *Medicago sativa* by weaners and yearling cattle was depressed by 3 and 8%, respectively, by Na deficiency (Joyce and Brunswick, 1975).

In arid and semiarid regions some forages accumulate up to 82 g Na/kg DM and when eaten by stock voluntary intake will be depressed if insufficient drinking water of a low Na content is available to prevent toxic levels of Na accumulating in the body (Wilson, 1966a). In arid regions drinking water often contains concentrations of soluble salts, including salts of Na, which may be high enough to cause deleterious effects on livestock (Shirley, 1985).

C. Effect on Production

Feeding an Na supplement improves the growth rate of cattle and sheep and the milk production of cattle eating some forages (Table 10.1). Responses to Na supplementation are found when animals with a high requirement for Na are fed forage low in Na for such a long time that the large Na reserves in the rumen are depleted. There appear to be no published reports of a Na response in nonlactating ruminants and animals that are not growing. This is to be expected since the Na reserves in the rumen are sufficient to meet the Na requirments of these animals for many months and possibly more than a year. In one study with lactating cows a low-Na diet was fed for 10 months before appetite and body weight was depressed (Smith and Aines, 1959).

D. Efficiency of Absorption

Sodium compounds are very soluble in water so the potential absorption of forage Na is very high. However Na is only absorbed from the lower gut if required (ARC, 1980), and apparent absorption coefficients measured in pens with mature dry stock (Table 10.2) are unlikely to reflect the maximum availability of Na in forages. Absorption and excretion of Na in urine are under the control of the hormone aldosterone (Bott *et al.*, 1964).

E. Dietary Requirements

The concentration of Na required in forage by different classes of ruminants can be derived from the quantitative values published by the ARC

TABLE 10.1

Depression in Ruminant Production Caused by Low Sodium in the Forage

Product	Diet	Production (kg/day)			Reference
		Sodium supplement	No sodium supplement	Depression (%)	
Cattle					
Beef growth	Hay (0.1 g Na/kg)	0.411	0.317	23	Horrocks (1964)
Beef carcass	*Medicago sativa*	0.220	0.178	19	Joyce and Brunswick (1975)
Beef growth	*M. sativa*	0.414	0.338	18	Joyce and Brunswick (1975)
	Savanna	0.555	0.135	76	Winter and McLean (1988)
	Savanna	0.374	0.301	20	Rhodes (1956)
	Savanna	0.292	0.236	19	Walker (1957)
	Sorghum bicolor var. sudanensis	0.781	0.659	16	Archer and Wheeler (1978)
	Stylosanthes guianensis	0.168	0.039	77	Hunter *et al.* (1979b)
	Temperate pasture	0.54	0.34	37	Towers *et al.* (1984)
Milk	*M. sativa*	14.54	12.96	11	Joyce and Brunswick (1975)
	Grass/*Neonotonia wightii*	12.00	10.80	10	Davison *et al.* (1980)
Lactating cows					
Growth dams	Native pastures	0.35	0.14	60	Murphy and Plasto (1973)
Growth calves		0.77	0.62	19	
Lactating cows					
Growth dams	Native pasture	0.847	−0.319	138	Leche (1977a)
Growth calves		0.681	0.486	29	
Sheep					
Lamb growth	*Sorghum almum*	0.080	0.049	39	Said *et al.* (1977) Wheeler *et al.* (1978)
	Medicago sativa	0.166	0.099	40	Joyce and Brunswick (1975)
		0.264	0.241	9	Jagusch *et al.* (1977b)
	M. sativa/Lolium perenne	0.154	0.096	38	Joyce and Brunswick (1975)
Wool	*M. sativa*	1.47	1.23	16	Joyce and Brunswick (1975)
	M. sativa/L. perenne	1.38	1.20	13	Joyce and Brunswick (1975)

TABLE 10.2

Apparent Absorption of Sodium from Forages

Species	Absorption coefficient	Reference
Brachiaria decumbens	0.25–0.67	Perdomo *et al.* (1977)
Cynodon dactylon	0.63–0.70	Perdomo *et al.* (1977)
Digitaria decumbens	0.68–0.80	Perdomo *et al.* (1977)
Lolium perenne	0.61–0.98	Joyce and Rattray (1970a)
Temperate pasture	0.21–0.85	Hutton *et al.* (1967)
	0.85	Kemp (1964)
	0.11–0.94	Lomba *et al.* (1970)
Trifolium repens	0.85–0.91	Joyce and Rattray(1970a), Grace *et al.* (1974)

(1980) and estimates of forage intakes (Tables 10.3 to 10.6). These tables show that 1.5 g Na/kg DM should meet the requirements for most forms of production. However, these recommended requirements may sometimes be higher than necessary. For example, Table 10.6 shows that the Na requirement of lactating ewes is up to 1.5 g/kg DM, but studies with ewes fed different levels of Na showed that 0.9 g/kg was sufficient to maintain normal Na levels in the saliva (Morris and Peterson, 1975). Conversely, the requirement for weaner cattle (Table 10.3) appears to be too low; a 59% improvement in growth occurred when an Na supplement was fed to animals grazing forage containing 1.0 g Na/kg DM (Towers *et al.*, 1984). Ruminants normally have large reserves of Na which can be used to augment a low level of Na in the diet. This can lead to an underestimation of the Na concentration required in forages unless the production studies are of long duration.

TABLE 10.3

Sodium Requirement of Beef Cattle Fed ad Libitum Expressed as Dietary Concentrations (g/kg DM)[a]

Livewieght (kg)	Rate of daily gain (kg/day)		
	0	0.5	1.0
200	0.4	0.5	0.6
400	0.5	0.5	0.6
600	0.6	0.6	0.6

[a]Adapted from ARC (1990).

TABLE 10.4

Sodium Requirements of Lactating Cows Fed ad Libitum (g/kg DM)[a]

Liveweight (kg)	Breed	Milk yield (kg/day) 0	10	20	30
400	Jersey	0.4	0.8	1.1	—
500	Ayrshire	0.5	0.8	1.1	1.5
600	Friesian	0.5	0.9	1.2	1.5

[a]Adapted from ARC (1980).

II. SODIUM IN FORAGE

A. Mean Concentration

The Na concentration in forage dry matter ranges from 0.05 to 21.3 g/kg DM with a mean of 2.2 g/kg DM for 671 values cited in the scientific literature. The frequency distribution (Fig. 10.1) is skewed with 50% of all samples containing more than the 1.5 g Na/kg required by ruminants

TABLE 10.5

Sodium Requirements of Castrated Lambs Fed ad Libitum (g/kg DM)[a]

Liveweight (kg)	Liveweight gain (kg/day) 0	0.1	0.2	0.3
20	1.1	1.3	1.5	—
30	1.3	1.1	1.1	1.4
40	1.3	1.1	1.1	1.1

[a]Adapted from ARC (1980).

TABLE 10.6

Sodium Requirements of Lactating Ewes Fed ad Libitum (g/kg DM)[a]

Liveweight (kg)	Milk yield (kg/day) 0	1.0	2.0	3.0
40	1.5	1.2	1.4	—
75	1.7	1.3	1.4	1.5

[a]Adapted from ARC (1980).

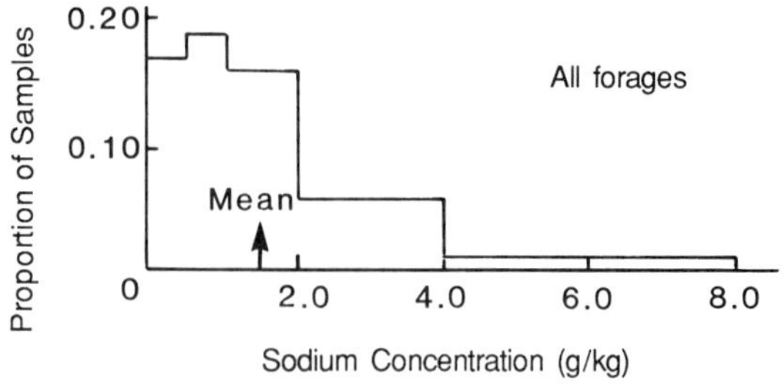

Fig. 10.1 Frequency distribution of sodium concentration in a wide range of forages (data from the world literature).

with the highest requirement for Na. Some ruminants only require 0.4 g Na/kg and 15% of forage samples failed to reach this level.

Part of the large variation in Na concentration is associated with forage type; grasses tend to contain more Na than legumes and within legumes the temperate species contain more Na than the tropical species (Fig. 10.2). Half the samples of tropical legumes contained less than 0.4 g Na/

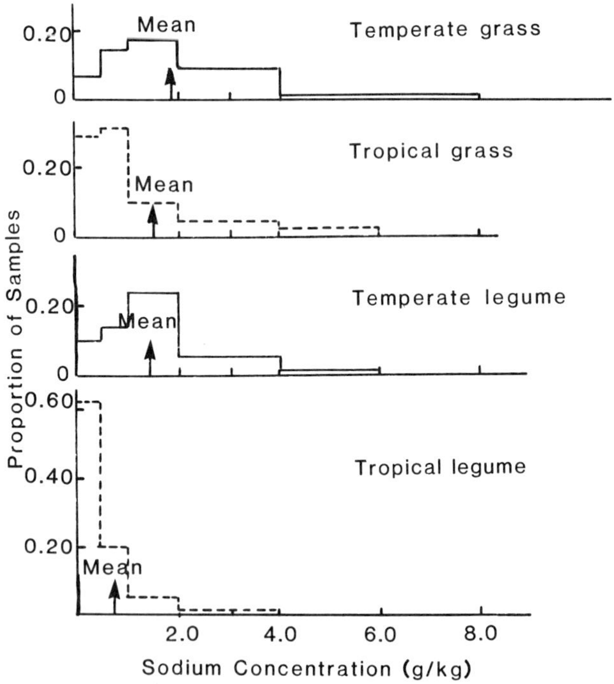

Fig. 10.2 Frequency distribution of sodium concentration in temperate and tropical grasses and legumes (data from the world literature).

TABLE 10.7

Effect of Fertilizer Potassium on the Sodium Concentration in Forage (g/kg DM)[a]

	KCl applied (kg/ha)			
	0	125	250	500
Dactylis glomerata	9.1	1.6	1.1	0.4
Paspalum dilatatum	1.2	0.6	0.5	0.4

[a]Adapted from McNaught (1959).

kg and so were unable to meet the requirements of any form of ruminant production without some form of supplementation.

B. Species Differences

There are two classes of forage plants, the Na accumulators and the nonaccumulators (Youssef, 1988). Sodium-accumulating forages contain >2 g Na/kg provided the soil contains sufficient Na and soil K is not in excess. The nonaccumulating forages always contain <2 g Na/kg even when grown on soils high in Na and the concentration is almost independent of the level of soil K. This difference between Na-accumulating and nonaccumulating forages is demonstrated in Table 10.7. It shows the range of Na that can occur in an Na-accumulating forage such as *Dactylis glomerata* and the way fertilizer K can depress it to a low level. This is an extreme case caused by the high rates of KCl applied, and under normal conditions Na-accumulating forages generally contain >2 g Na/kg, sufficient to meet the Na requirements of all classes of ruminants. Table 10.8 lists forage species in terms of mean Na concentration. In both the temperate and the tropical grasses there are species with either high or low Na concentration, but all legumes appear to be nonaccumulators of Na.

C. Within-Species Variation

Large varietal differences in Na concentration occur in *Dactylis glomerata* and *Lolium perenne,* smaller variation in *Festuca arundinacea,* and very little variation in *Festuca pratensis* (Fig. 10.3). Most *D. glomerata* cultivars contains sufficient Na to meet the requirements of all forms of livestock production. There is sufficient Na variation in *F. pratensis* to allow genotypes to be selected with levels of Na adequate for all forms of ruminant production. In the case of temperate grasses, De Loose

TABLE 10.8

Mean Sodium Concentration in Different Forage Species

Species	Sodium concentration (g/kg DM) Mean	Range	Reference
Grasses			
Sorghum almum	0.1	Trace–0.1	Said *et al.* (1977)
Sorghum bicolor var. *sudanensis*	0.1	Trace–0.1	Archer and Wheeler (1978)
Agrostis tenuis	0.2	0.1–0.4	Sherrell (1978)
Themeda triandra	0.2	0.1–0.2	Du Toit *et al.* (1934)
Cynodon dactylon	0.3	0.1–0.9	Du Toit *et al.* (1934), Jones (1963a), Long *et al.* (1970), Perdomo *et al.* (1977), Gonzalez and Everitt (1982)
Heteropogon contortus	0.3	0.1–1.3	Playne (1970a), Playne and Haydock (1972), Murphy and Plasto (1973), Playne *et al.*(1978)
Pennisetum clandestinum	0.3	—	Kaiser (1975), Sherrell (1978)
Pennisetum purpureum	0.3	0.2–0.4	Long *et al.* (1970)
Saccharum officinarum	0.3		Devendra (1979)
Zea mays	0.3	0.2–0.3	Devendra (1979), Ando *et al.* (1985)
Cenchrus ciliaris	0.5	0.1–3.1	Du Toit *et al.* (1934), Playne (1970a), Gonzalez and Everitt (1982)
Phleum pratense	0.5	0.1–1.9	Coppenet and Calvez (1962), Griffith *et al.* (1965), Griffith and Walters (1966), Patil and Jones (1970)
Festuca pratensis	0.6	0.2–1.8	Coppenet and Calvez (1962), Griffith *et al.* (1965)
Paspalum dilatatum	0.6	0.3–1.2	McNaught (1959), Sherrell (1978)
Brachiaria decumbens	0.8	0.2–5.2	Long *et al.* (1970), Perdomo *et al.* (1977), Poland and Schnabel (1980), Youssef (1988)
Digitaria polevansii	1.1	0.6–2.1	Minson (1984)
Digitaria milanjiana	1.1	0.6–1.6	Minson (1984)
Lolium multiflorum	1.1	0.2–3.5	Griffith *et al.* (1965), Joyce and Brunswick (1975)
Bromus unioloides	1.4	0.8–1.9	Rumball *et al.* (1972)
Digitaria smutsii	1.4	0.5–2.7	Minson (1984)
Chloris gayana	2.0	0.1–7.0	Jones (1964), Long *et al.* (1970), Playne (1970a)
Lolium perenne × *L. multiflorum*	2.3	1.6–5.1	Butler *et al.* (1962), Coppenet and Calvez (1962), Grace *et al.* (1974)

TABLE 10.8

(continued)

Species	Sodium concentration (g/kg DM) Mean	Range	Reference
Panicum maximum	2.3	0.3–8.0	Long *et al.* (1970), Minson (1975), Perdomo *et al.* (1977), Youssef (1988)
Festuca arundinacea	2.9	0.6–6.9	Coppenet and Calvez (1962), Griffith *et al.* (1965), Patil and Jones (1970)
Phalaris aquatica	3.6	—	Langlands (1966)
Lolium perenne	4.2	1.1–8.7	McNaught (1959), Butler *et al.* (1962), Coppenet and Calvez (1962), Griffith *et al.* (1965), Griffith and Walters (1966), Jones *et al.* (1967b), Patil and Jones (1970), Grace *et al.* (1974), Moseley and Jones (1974), Joblin and Keogh (1979)
Digitaria pentzii	4.5	2.9–5.8	Minson (1984)
Dactylis glomerata	5.1	0.4–10.1	McNaught (1959), Coppenet and Calvez (1962), Griffith *et al.* (1965), Patil and Jones (1970)
Digitaria decumbens	5.7	3.2–8.2	Perdomo *et al.* (1977), Poland and Schnabel (1980), Minson (1984)
Panicum coloratum	8.1	3.1–16.5	Minson (1975)
Setaria sphacelata	8.6	0.5–18.0	Hacker (1974b)
Cenchrus setiger	15.3	3.2–17.4	Playne (1970a)
Legumes			
Neonotonia wightii	0.2	0.1–0.2	Davison *et al.* (1980)
Macroptilium atropurpureum	0.2	0.2–0.4	Playne (1970a)
Stylosanthes guianensis	0.2	0.1–0.3	Hunter *et al.* (1979b)
Stylosanthes humilis	0.4	0.1–1.5	Playne (1969a, 1970a), Playne and Haydock (1972), Playne *et al.* (1978)
Medicago sativa	0.4	0.2–0.8	Jagusch *et al.* (1977a), Schultz *et al.* (1979), Sherrell (1978)
Trifolium pratense	0.9	0.2–1.6	De Groot (1963), Davies *et al.* (1966, 1968)
Vigna unguiculata	0.9	—	Devendra (1979)
Trifolium repens	1.1	—	Sherrell (1978)
Trifolium subterraneum	1.5	—	Sherrell (1978)

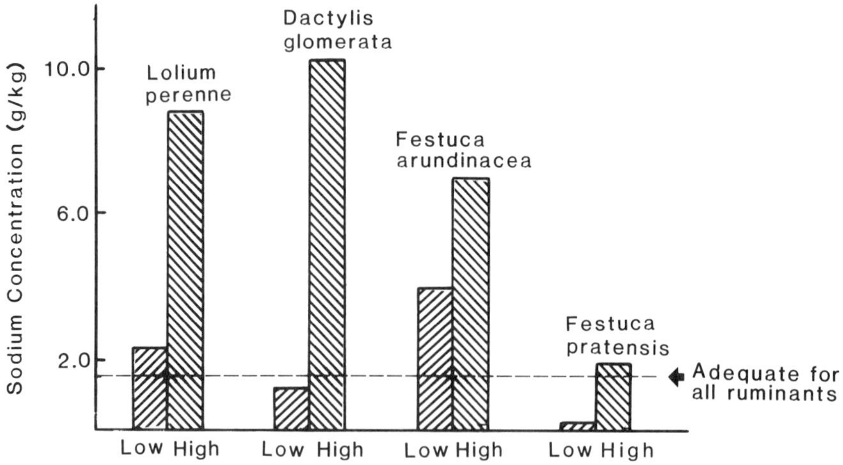

Fig. 10.3 Sodium of four forage species showing values for lowest and highest genotype. Adapted from Griffith *et al.* (1965).

and Baert (1966) concluded that "while a certain gain in Na content may be expected by selecting those timothy clones with the highest Na uptake, the values will never attain those for perennial and italian ryegrass and cocksfoot. There will always be a difference in uptake by a factor of 3 to 5."

Tropical grasses exhibit greater genetic variation in Na concentration than temperate forages (Table 10.9). Not only is there a larger variation, but it occurs in species where some ecotypes or cultivars are very deficient in Na, e.g., *Chloris gayana, Digitaria milanjiana,* and *Setaria sphacelata*. Variation in Na concentration of *D. milanjiana* is associated with climatic and edaphic differences between the areas where they grow in Africa (Hacker *et al.,* 1985). Lines low in Na came from arid regions, while Na-accumulating plants were collected from coastal areas with a high rainfall. It was postulated that these low Na concentrations are associated with genetic adaptation to soils low in K and that breeding grasses for a higher Na concentration may reduce their drought tolerance.

D. Plant Parts

In the classic studies of Du Toit *et al.* (1934) the Na concentration was determined in the leaves, stems, and inflorescences of four tropical grasses each cut at regrowth periods from 1 to 6 months. There was no consistent difference in Na concentration between leaf and stem fractions

TABLE 10.9

Genetic Differences in Sodium within Species, Grown in the Field

Species	Variety or line	Sodium concentration (g/kg) Mean	Range	Reference
Bromus unioloides	Chapel Hill	1.0	—	Rumball *et al.* (1972)
	N.Z. Commercial	1.5	—	
	Priebes	1.9	—	
Chloris gayana	Giant	0.3	0.1–0.5	Jones (1964)
	Katambora	3.1	2.1–7.0	
Digitaria milanjiana	Low-Na ecotypes	0.4	0.2–0.8	Hacker *et al.* (1985)
	High-Na ecotypes	7.2	3.0–12.5	
Lolium perenne	Melle	2.0	1.2–2.7	Coppenet and Calvez (1962)
	Primevere	3.2	2.6–4.4	
	Sceempter	2.0	1.5–2.6	Griffith and Walters (1966)
	S.23	3.5	1.7–5.5	
	S.24	5.0	2.5–6.5	
	S.321	6.1	3.6–7.5	
L. perenne × *L. multiflorum*	Clone G	1.6	—	Butler *et al.* (1962)
	Clone F	5.1	—	
Panicum coloratum	Burnett	5.4	4.8–6.5	Minson (1975)
	Kabulabula	13.3	9.6–16.5	
Phleum pratense	S.51	0.8	0.4–1.3	Griffith and Walters (1966)
	Omnia	1.3	0.5–1.9	
Setaria sphacelata	Nandi	0.3	—	Hacker and Jones (1969)
	Kazungula	3.1	—	
S. sphacelata var. *stolonifera*	From Nigeria	1.4	0.6–2.2	Hacker (1974a)
	From S. Africa	15.9	14.2–18.0	
Trifolium pratense	Line AA14	0.6	0.5–1.0	Davies *et al.* (1968)
	S.123	1.3	0.7–1.6	
Trifolium repens	S.100	2.9	2.7–3.2	Davies *et al.* (1968)
	Line AC9	3.4	2.5–3.9	

but the inflorescence contained approximately twice as much Na (Table 10.10). No difference in Na concentration has been found between the leaf and stem fractions of *Digitaria decumbens* and *C. gayana* (Poppi *et al.*, 1981a), but with *L. perenne* (Laredo and Minson, 1975a) and *Lablab purpureus* (Hendricksen and Minson, 1980) the Na concentration in the leaf was approximately double that in the stem.

TABLE 10.10

Sodium Concentration in Different Parts of Four Tropical Grasses[a]

Forage	Sodium concentration (g/kg) Leaves	Stems	Inflorescence
Bothriochloa insculpta	0.22	0.27	0.41
Eragrostis superba	0.23	0.19	0.55
Hyparrhenia hirta	0.10	0.15	0.24
Cenchrus ciliaris	0.33	0.25	0.40
Mean	0.22	0.22	0.40

[a]Adapted from Du Toit *et al.* (1934).

E. Stage of Growth

As forage regrowths increase in age there is usually a reduction in the concentration of Na (Fig. 10.4). In a recent study with different *Digitaria* species increasing maturity depressed Na concentration of both high- and low-Na-accumulating species (Table 10.11).

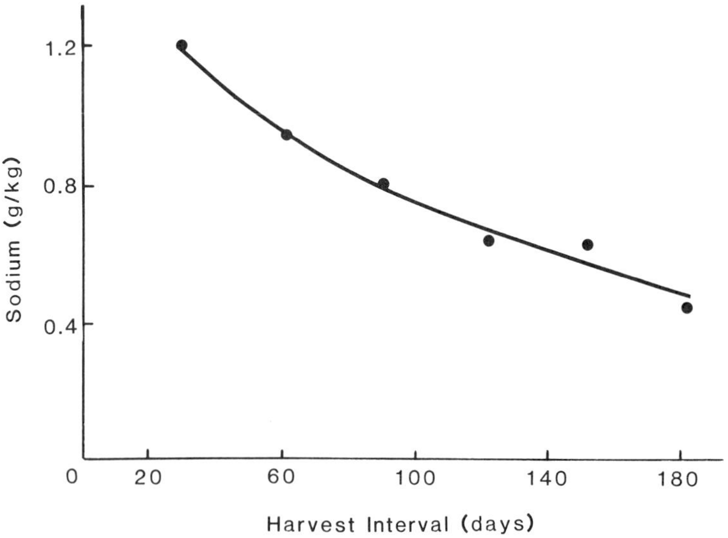

Fig. 10.4 The effect of harvest frequency on sodium concentration; mean of eleven grasses. Adapted from Du Toit *et al.* (1934).

TABLE 10.11

Effect of Age of Regrowth on the Sodium Concentration (g/kg) in *Digitaria* Species[a]

Forage	Age of regrowths (days)		
	28	70	98
D. polevansii	1.0	0.8	0.6
D. decumbens	5.9	5.1	3.6

[a]From D.J. Minson, unpublished data.

Increasing maturity depressed Na concentration in *D. glomerata, Festuca elatoir, F. pratensis, L. perenne, L. multiflorum* (Griffith *et al.*, 1965), and *Phleum pratense* (Pritchard *et al.*, 1964; Griffith *et al.*, 1965). Most of these changes were relatively small with minimum values of 33 to 72% of the maximum concentration. Similar changes have also been reported for *Heteropogon contortus* and *Stylosanthes humilis* (Playne and Haydock, 1972). However, no differences in Na were found with maturity in *Bromus inermis* (Pritchard *et al.*, 1964), *D. decumbens* (Poppi *et al.*, 1981a), or *L. purpureus* (Hendricksen and Minson, 1985a).

F. Location

Studies have been made of the effect of geographic distribution within New Zealand on Na concentration in *Anthoxanthum odoratum* (Wells, 1962) and grasslegume pastures (Smith and Middleton, 1978). The Na concentrations were high and generally adequate for all classes of stock in areas close to the coast but were low and often deficient in inland pastures. Sea spray increased the Na concentration of grasslegume pastures up to 20 km from the coast (Fig. 10.5).

G. Soil Fertility

In forages that accumulate Na, the concentration varies with the level of both the available Na and the available K in the soil (Reith *et al.*, 1964; Henkens, 1965; Kemp, 1971). Increasing the level of soil K can depress Na concentrations in the temperate forage (Hemingway, 1961a,b; Reith *et al.*, 1964) to a point where it is insufficient to meet the requirements of most forms of ruminant production. Conversely, the concentration in Na-accumulating forages is increased by a high soil Na. In a study with *L. perenne* and *Trifolium repens*, changing the fertilizer applied from KCl to

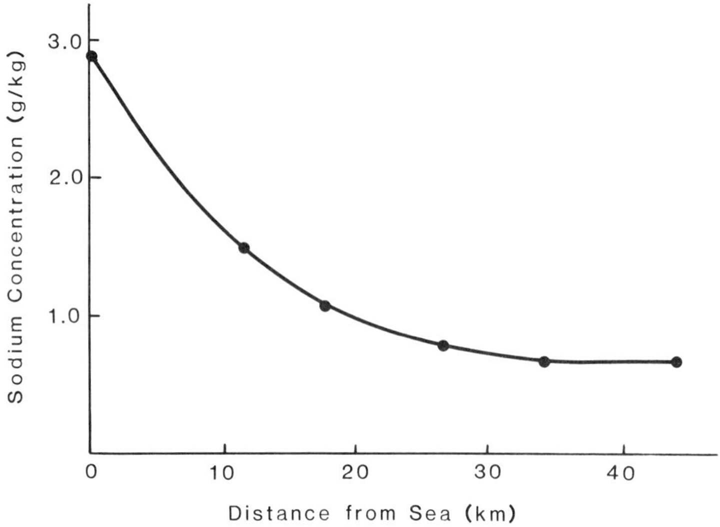

Fig. 10.5 The sodium concentration in forage as affected by distance from the sea. Adapted from Smith and Middleton (1978).

NaCl (250 kg/ha) increased the Na concentration of the grass from 2.4 to 7.9 g/kg DM and that of the legume from 2.7 to 8.2 g/kg DM (McNaught and Karlovsky, 1964).

At the other end of the spectrum, forages that do not accumulate Na are unaffected by large changes in soil Na and K. Among these forages are the tropical legumes *Desmodium intortum, D. uncinatum, Centrosema pubescens, Macroptilium atropurpureum,* and *S. humilis* (Andrew and Robins, 1969a,b), the tropical grasses *Brachiaria decumbens* (Poland and Schnabel, 1980) and *H. contortus* (Smith, 1981), and the temperate grass *P. pratense* (Hasler, 1962; Griffith and Walters, 1966; Jarvis, 1982).

High soil N increases Na concentration in the forage if the N fertilizer is applied *in the absence of K fertilizer* (Stewart and Holmes, 1953; Hemingway, 1961a,b; Reith *et al.*, 1964) (Table 10.12). With *L. perenne,* sulfate of ammonia increases the Na concentration, especially if the fertilizer is frequently applied (Hemingway, 1961a,b; Griffith *et al.*, 1965). This has been attributed to a change in the ratio of available K to Na in the soil (Henkens, 1965). Only small changes in Na concentration have been found when *T. repens* is fertilized with sulfate of ammonia (Hemingway, 1961a,b).

TABLE 10.12

Effect of Nitrogen Fertilizer on the Sodium Concentration in Mixed Pasture in the Absence of Potassium Fertilizer[a]

Nitrogen applied (kg/ha)	Sodium concentration (g/kg DM)
0	3.9
195	6.3
390	6.9

[a]Adapted from Reith *et al.* (1964).

III. DIAGNOSIS OF SODIUM DEFICIENCY

A. Forage Species

As previously discussed there are very large differences in Na concentration between and within forage species (Tables 10.8 and 10.9). Cultivars containing more than 1.5 g Na/kg supply sufficient Na for all forms of animal production and no Na deficiency would be anticipated. However, in some circumstances these normally high Na cultivars may have too low an Na content for animal production, for example, where the ratio of available soil K to Na is high. This can occur when excess K fertilizer is applied or Na in the soil has been leached out by heavy rain, as can occur in the humid tropics.

If animals are fed exclusively on species that contain less than 1.5 g Na/kg they may eventually become Na deficient *provided* there is no other source of Na in the form of saline soil, saline drinking water, or salt blocks. The time taken for animals to become Na deficient may be only a few weeks in the case of dairy cows (Paterson and Crichton, 1960), but with fattening beef cattle the loss of Na is so slow that it may take a year before production is affected.

B. Forage Analysis

The Na concentration in forage can be readily determined by either flame photometry or atomic absorption spectrophotometry (AOAC, 1980). Samples of forage similar to that being eaten by the animal can be plucked by hand but plastic gloves should be worn to avoid contaminating the sample with sweat. Esophageally fistulated animals cannot be used to collect the sample for Na analysis due to the high level of Na in the saliva.

C. Animal Behavior

The main signs of Na deficiency in ruminants are languor, debility, and pica (Aines and Smith, 1957; Smith and Aines, 1959; Kemp and Geurink, 1966; Denton, 1969; Morris, 1980). The most striking sign is a salt hunger characterized by licking or chewing wood and soil, licking the sweat on other animals, and drinking the urine of other animals during urination.

D. Tissue Analysis

1. Saliva

Saliva analysis is the most reliable method of detecting Na deficiency. A deficiency leads to increased secretion of aldosterone by the adrenal cortex and a decrease in the Na concentration in the parotid saliva (Denton, 1956, 1957; Kay, 1960; Bailey and Balch, 1961; Kemp and Geurink, 1966; Gartner and Murphy, 1974). The fall in the Na concentration is balanced by a rise in the K concentration and analysis of parotid saliva for Na:K ratio provides a sensitive index of whether the animal is actively conserving Na (Kemp and Hartmans, 1968; Kemp, 1971; Morris and Gartner, 1971; Morris and Murphy, 1972; Morris, 1980). The levels of Na and K in saliva of cattle differing in Na status are shown in Table 10.13. No response to feeding Na supplements has been observed when the Na:K ratio in the saliva exceeds 2.0 (Hennessy and Sundstrom, 1975; Murray *et al.*, 1976; Leche, 1977b; Winks and O'Rourke, 1977). Samples of parotid saliva may be collected from cattle using a sponge (Kemp and Geurink, 1966) or with a suction system (Murphy and Connell, 1970).

2. Rumen Fluid

Sodium deficiency reduces the quantity of Na entering the rumen in both the forage and the saliva. This leads to a change in the Na:K ratio. With Na-deficient cattle and sheep the Na:K ratio in the rumen was 0.6 and 0.2, respectively (Kemp and Geurink, 1966; Martens *et al.*, 1987), compared with a normal value of 1.0 (Martens *et al.*, 1987). Samples of rumen contents may be collected by stomach tube or when the animals are slaughtered.

3. Urine

Sodium homeostasis is achieved primarily by regulation of urinary Na excretion (Morris, 1980; Martens *et al.*, 1987). However, when animals are Na deficient, urine volume is increased (Whitlock *et al.*, 1975), further reducing the concentration of Na. It has been suggested that cattle are

TABLE 10.13

Sodium Status of Cattle and Related Sodium and Potassium Concentrations in the Saliva (mEq/liter)[a]

Sodium status	Clinical signs	Saliva composition		
		Na	K	Na:K
Sufficient	None	>130	<13	>10.0
Insufficient	None	87–130	13–38	2.3–10.0
Insufficient	Sometimes	43–87	38–64	0.7–2.3
Severe deficiency	Always	<43	>64	<0.7

[a]Adapted from Kemp and Hartmans (1968).

probably Na deficient when the Na concentration in the urine falls below 69 mg/liter (Morris, 1980).

4. BLOOD

The Na concentration in blood is relatively constant, varying less than any other component (Manston and Rowlands, 1973). In cases of gross Na deficiency, small depressions have been found in plasma Na with a complementary rise in plasma K (Hennessy and McClymont, 1970; Morris and Murphy, 1972), but in other studies Na deficiency had no effect on blood composition (Morris and Gartner, 1971; Murphy and Plasto, 1973; Leche, 1977b). Morris (1980) concluded that blood analysis was of little value in diagnosing Na deficiency.

5. FECES

Feces are easily collected and exhibit large variations in Na:K ratio comparable to the variations found in saliva of cattle fed different levels of Na. The Na:K ratio in the feces of deficient cattle was only 0.05 compared with 0.27 to 0.34 for cattle fed adequate levels of Na and with normal Na:K ratios in the saliva (Morris and Gartner, 1971). Further work is requried to establish the value of the Na:K ratio in the feces as a method of diagnosing potential Na deficiency.

IV. PREVENTION OF SODIUM DEFICIENCY

A. Supplementary Sodium

Aines and Smith (1957) showed that when milk production was increased by feeding NaCl, the beneficial effect was caused by the Na ion and not by the Cl ion; this was confirmed in studies with fattening sheep and cattle (McClymont *et al.*, 1957; Joyce and Brunswick, 1975).

In experimental studies NaCl has been given as a drench at a rate of 1.2 g Na/day for sheep, and 14 g Na/day for dairy cattle (Joyce and Brunswick, 1975), or 11–22 g Na/day for beef cattle (Walker, 1957). With housed stock the NaCl can be mixed with the entire diet at the rate of 1 g Na/kg (McClymont *et al.*, 1957) or sprinkled on top of the forage (Joyce and Brunswick, 1975). Alternatively dairy cattle have been fed 16 g Na/day as NaCl in a grain supplement (Davison *et al.*, 1980).

Under grazing conditions supplementary Na is usually supplied in the form of a salt block, but consumption can be very variable (McDowell, 1985). More uniform intake can be achieved by using equipment which adds NaCl to the drinking water. This method ensures that all stock receive the supplement, provided there is no other source of drinking water, but can be wasteful if the Na is not required. Furthermore, the method may prove fatal if excess Na is added (Shirley, 1985). With salt blocks the intake by stock is to some extent naturally controlled by their need for Na.

B. Fertilizer Sodium

The Na concentration in Na-accumulating forages is related to the ratio of Na to K in the soil and can be increased by applying fertilizers containing Na (Henkens, 1965). With sodium-deficient pastures grown on humic sandy soil Lehr *et al.* (1963) recommended that the first nitrogen dressing in spring should be given either wholly or partly in the form of $NaNO_3$ and that NaCl could be used on farms with predominantly organic manuring. The method is not suitable for clay soils due to a breakdown of soil structure.

The Na concentration in forage species that do not accumulate Na is relatively independent of the level of soil Na. With this group of forages, applying Na fertilizers will not increase the Na concentration in the forage or prevent a deficiency in the animal.

V. CONCLUSION

Sodium deficiency in ruminants can occur when growing or lactating animals are eating forage with a low Na concentration, have no alternative source of Na, and have exhausted their body reserves of Na. No production response has been reported when Na supplements were fed to ruminants that were neither growing nor lactating.

Forages may be divided into those that are Na accumulators or those that are nonaccumulators. The accumulators usually contain sufficient Na for all types of production whereas the nonaccumulators generally contain insufficient Na for any forms of production. All legumes appear to be nonaccumulators of Na but in the grasses there are both accumulators and nonaccumulators; in some species there are both Na-accumulating and nonaccumulating ecotypes.

Sodium deficiency is characterized by licking or chewing of wood and soil and licking the sweat on other animals and is evident when stock eat species which are nonaccumulators of Na. Analysis of parotid saliva, rumen fluid, and possibly feces for Na:K ratio can provide positive diagnosis of an Na deficiency. Sodium deficiency can be prevented by providing supplementary Na or by sowing Na-accumulating forage cultivars.

11

Copper

I. COPPER IN RUMINANT NUTRITION

A. Function of Copper

Copper (Cu) is an essential element required by ruminants and deficiencies of the element occur in grazing animals in many parts of the world under a wide range of soil and climatic conditions (Underwood, 1981). In the tropics, 34 countries of Latin America, Africa, and Asia have reported various levels of Cu deficiency—more than any other essential element with the exception of phosphorus (McDowell, 1985). Copper is required in many enzyme systems, and this is reflected in the metabolic and clinical symptoms (Section III,A) associated with Cu deficiency (Table 11.1).

B. Effect on Production

1. Sheep

Copper deficiency in ewes can reduce growth rate and can cause the fatal disease swayback or enzootic ataxia in their lambs. The Scottish Blackface is very susceptible to Cu deficiency and in one study on improved upland pasture 33% of lambs died, 13% from swayback and 20% from other causes. Treating the lambs with Cu prevented swayback and reduced the nonswayback losses from 20 to 11% (Woolliams *et al.*, 1968a). Copper deficiency reduced the growth rate of surviving lambs by 15% (Woolliams *et al.*, 1986b). Wool production by Cu-deficient sheep can be reduced 30% (Fig. 11.1) and the wool has a low quality. Wool from Cu-deficient sheep is discolored and has little crimp, a condition described as "steely" (Lee, 1951) and "stringy" (Underwood, 1981).

2. Cattle

There are many reports of growth rate of cattle being improved by feeding a Cu supplement (Table 11.2). Many of these increases occur in cattle grazing temperate pastures which have high levels of Cu and the

TABLE 11.1

Manifestations of Copper Deficiency in Sheep and Cattle[a]

Clinical effect	Metabolic defect	Probable enzyme involved
Pigment loss from hair and wool	Melanin formation	Tyrosinase
Steely wool and hair	Keratinization	—
Nervous disorders	Demyelination, catecholamine deficiency	(Cytochrome oxidase?)
Skeletal defects	Cross-link formation in collagen and elastin	Lysyloxidase
Anemia	Iron metabolism	—
Scours	—	—
Depressed growth	—	—
Poor reproduction	—	—
Lowered resistance to infections	—	—

[a]Adapted from Underwood (1977), Walker (1980), and Grace (1983a).

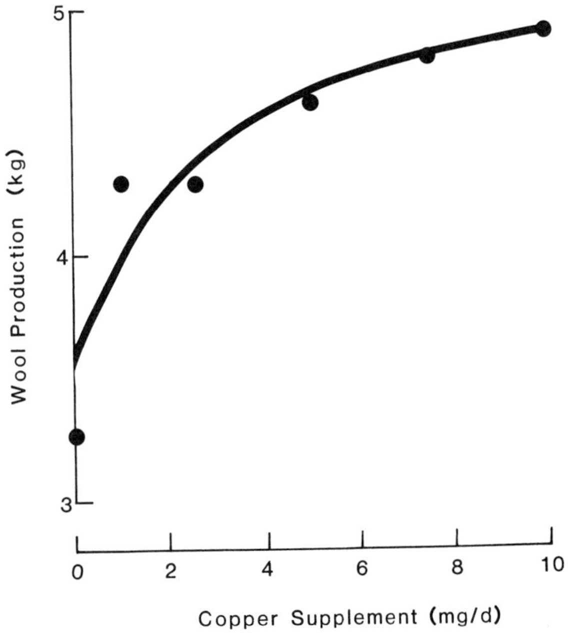

Fig. 11.1 Effect of supplementary copper on wool production. Adapted from Lee (1951).

TABLE 11.2

Increase in Growth of Cattle Following Copper Supplementation

Forage	Forage composition: Cu (mg/kg)	Mo (mg/kg)	S (g/kg)	Increase in growth (%)	Reference
Avena sativa	4	3.6	3.8	73	Smith and Thompson (1980)
Brassica oleracea	5	—	8	48	Barry *et al.* (1981)
Digitaria decumbens with *Lotononis bainesii*	—	—	—	10	Gartner *et al.* (1968)
Festuca arundinacea with					
Medicago sativa	—	—	—	20	Gomm *et al.* (1982)
Trifolium repens	—	—	—	12	
Temperate pasture	—	—	—	40	Field (1957)
	6	7.0	—	23	Drysdale and Lillie (1977)
	10	1.2	1.1	14	MacPherson *et al.* (1978)
	15	7.5	—	17[a]	Rogers and Poole (1978)
	—	—	—	13[b]	MacPherson *et al.* (1979)
	11	—	—	50	Poole *et al.* (1974b)
Tropical pasture	11	0.7	2.9	25	Donaldson *et al.* (1964)
	7	0.7	1.5	10	Alexander *et al.* (1967)
	4	1.4	1.3	13	Murphy *et al.* (1981)
	6	7.0	—	24	Thornton *et al.* (1972)
	5	5.0	0.7	8	Bingley and Anderson (1972)

[a]14 out of 92 herds.
[b]3 out of 7 comparisons.

deficiency in these animals is caused by a high level of molybdenum (Mo) in the forage (Section I,C). Copper deficiency is usually limited to cattle in their first year (Poole *et al.*, 1974b), but if Mo concentrations in forage are high then symptoms of Cu deficiency can occur in animals up to 3 yr of age (Field, 1957).

The liveweight gain of pregnant cows has been depressed 30 kg during the term of the pregnancy by a deficiency of Cu (Alexander *et al.*, 1967). Calves from deficient cows have lower Cu reserves, grow more slowly unless supplemented with Cu (Table 11.3), and have a lower feed conversion efficiency (Suttle and Angus, 1976). Feeding a grain diet low in Cu reduced voluntary intake by 31% (Phillippo *et al.*, 1987a).

3. Fertility

Cattle grazing pasture containing less than 3 mg Cu/kg DM show signs of Cu deficiency and have a high incidence of infertility (Bennetts *et al.*,

TABLE 11.3

Effect of Injecting Cows with Copper during the 3 Months before Calving Compared with Injecting the Calves with Copper on the Weight Gain of Calves in Their First 18 Weeks[a] (kg)

	Treatment applied to calves		
Treatment applied to cows	No copper	Copper injection	Increase (kg)
No copper	157	185	28
Copper injection	180	191	11
Increase	23[b]	6	

[a]From MacPherson *et al.* (1979).
[b]Significant at $p < 0.05$.

1948) caused by delays in both estrus (O'Moore, 1960b) and conception (Donaldson *et al.*, 1964). Infertility in cattle is also associated with Cu deficiency caused by high levels of Mo in the diet (Phillippo *et al.*, 1987b).

C. Efficiency of Absorption

Absorption and retention of Cu depends on the level of Mo and sulfur (S) in the diet (Dick, 1956; Hogan *et al.*, 1971; Suttle, 1975, 1988; ARC, 1980). Dick *et al.* (1975) concluded "that the Cu combined with thiomolybdate is not available for absorption and that much of the thiomolybdate that is not combined with Cu in the rumen is absorbed into the blood stream and mobilizes tissue Cu," leading to low concentrations of Cu in the liver. Increases in dietary S and Mo cause large reductions in Cu absorption and equations have been derived for predicting the true availability of Cu in forages containing different levels of Mo and S (Fig. 11.2). When both Mo and S are high, availability of Cu is less than 0.01, compared with over 0.04 for diets low in the two antagonists. Absorption of Cu is also reduced by cadmium (Mills and Dalgarno, 1972), iron (Campbell *et al.*, 1974; Humphries *et al.*, 1983), and zinc (Dynna and Havre, 1963), but no equations have been published for these antagonists. Grazing animals usually ingest soil, which reduces the availability of Cu in forage, the effect being most marked when soils high in Fe are combined with forages high in S (Suttle *et al.*, 1984).

The availability of Cu in fresh forage is very much less than in dried forage of similar composition (Suttle, 1980) (Table 11.4). This difference in Cu availability is found in both sheep (Table 11.4) and cattle (Hartmans and Bosman, 1970) and is probably caused by a higher level of sulfide in

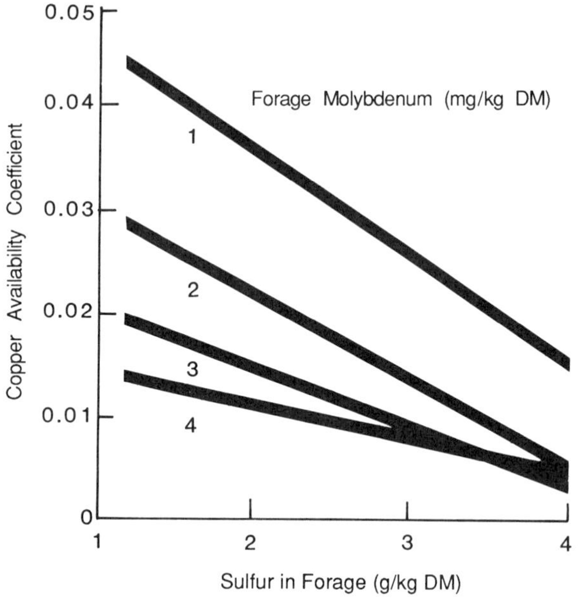

Fig. 11.2 True availability of dietary copper to sheep fed forages containing different levels of molybdenum and sulfur. Adapted from Suttle (1986).

the rumen of animals fed fresh forage. When fed mature grass cattle retain more Cu than when fed immature grass, as mature grass contains less S (Hartmans and Bosman, 1970). Treatment of silage with formaldehyde reduces sufide concentration in the rumen and increases the availability of Cu (Suttle, 1986).

TABLE 11.4

Availability Coefficient of Copper in Contrasting Temperate Forages[a]

Forage	Availability coefficient (mg/kg DM)	Forage composition: Cu (mg/kg DM)	Mo (g/kg DM)	S (g/kg DM)
Summer pasture	0.023	7–8	0.6–3.5	2–4
Autumn pasture	0.012	6–10	0.5–6.1	3–5
Hay	0.072	2–9	0.3–4.3	1–4
Silage	0.049	6–10	0.3–1.3	2–5
Brassica oleracea var. *acephala*	0.128	3–6	0.4–1.1	7–12

[a]From Suttle (1981).

D. Requirement for Copper

The requirements of ruminants for Cu can be calculated from the net requirements for maintenance and production and the availability of Cu in the forage (ARC, 1980). If the availability of Cu is assumed to be 0.04 then the concentration in forage of Cu required by different classes of stock varies from 6 to 13 mg/kg DM (Table 11.5). These values should be adjusted when dietary Mo and S concentrations are known, appropriate coefficients being selected from Fig. 11.2. It is probable that the absorption coefficient for Cu will be 0.03 or less when dietary cadmium concentration is >3 mg/kg DM, iron is >1 g/kg DM, or zinc is >200 mg/kg DM (ARC, 1980).

Adult cattle are extremely tolerant of excess Cu (Cunningham, 1946; Chapman *et al.*, 1962), but chronic Cu toxicity can occur in sheep reared

TABLE 11.5

Requirement of Ruminants for Copper in Forages[a]

Ruminant	Liveweight (kg)	Production (kg/day)		Requirements[b] (mg Cu/kg DM)
Sheep				
		Growth		
Growing lambs	40	0.075		6
		0.150		6
		0.300		9
Adult	50	0		7
Pregnant ewe	75	Milk		8
Lactating ewe	75	1		6
		2		6
		3		8
Cattle		Growth		
Growing bullocks	100	0.5		10
		1.0		10
	200	0.5		9
		1.0		10
	300	0.5		9
		1.0		10
Adult	500	0	70	13
Pregnant cows	500	Milk	91	12
Lactating cows	500	10	90	10
		20	110	8
		30	130	8

[a]Derived from ARC (1980).

[b]These values may be too high where forage is low in Mo, and Cu toxicity may occur in some breeds of sheep (ARC, 1980). Absorption efficiency of copper assumed to be 0.04.

indoors (ARC, 1980; Ivan *et al.*, 1986) and in sheep grazing forage with a low Mo content (Dick, 1969). Dick (1969) reported that *Trifolium subterraneum*

> has the peculiarity that it does not take up Mo from the soil when grown under warm environmental conditions. Thus, if there is an early autumn, it will take up normal amounts of Cu but practically no Mo which may be as low as 0.01 mg/kg. It would seem that Cu retention by the animal is limited by the "normal" Mo levels of 1–4 mg/kg in the plant but not by the sub-normal levels of less than 1 mg/kg. Consequently Cu accumulates in the liver to values of over 1000 mg/kg and this may ultimately lead to a haemolytic crisis and death of the animal with jaundice.

This Cu toxicity can be prevented by increasing the intake of Mo (Dick and Bull, 1945).

II. COPPER IN FORAGE

The Cu concentrations in 1278 forage samples, cited in the world literature, are shown in Fig. 11.3. The mean concentration is 6.1 mg/kg DM, with approximately half of all samples containing less than the minimum quantity of Cu recommended for ruminants (Table 11.6). This proportion is very similar to that reported for forage grown in British Columbia, where 54% of samples contained <5.7 mg/kg DM (Miltimore and Mason, 1971). Latin American forages generally contain more Cu and only 23%

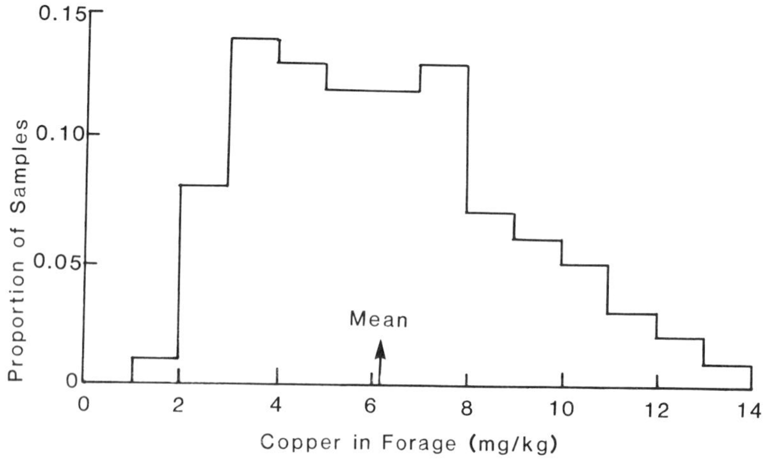

Fig. 11.3 Frequency distribution of copper in a wide range of forages (data from the world literature).

of all samples contained less than 5.0 mg Cu/kg DM (McDowell *et al.*, 1977). Higher mean concentrations of Cu have been found in forages grown in Kenya and Scotland (Howard *et al.*, 1962; Hemingway *et al.*, 1968). In Western Australia, forage collected from areas where Cu-deficiency symptoms were commonly observed in sheep and cattle contained, on average, 2.6 mg Cu/kg DM compared with 7.7 mg/kg DM for forage from areas where ruminants showed no symptoms of Cu deficiency (Beck, 1962). Concentrations of Mo in these two groups of forages were similar (0.54 and 0.78 mg/kg DM).

A. Species Differences

1. Legume versus Grass

Temperate legumes generally contain more Cu than temperate grasses (7.8 versus 4.7 mg/kg DM) and are more variable (Fig. 11.4). When a temperate legume is grown in association with a grass the legume generally has the higher Cu concentration (Adams and Elphick, 1956; Mitchell *et al.*, 1957b). However there are no differences between species when the soil is low in Cu and levels of Cu in both legumes and grasses are low (Beck, 1962). Tropical legumes generally contain less Cu than tropical grasses (3.9 versus 7.8 mg/kg DM), with a wider distribution of Cu levels in the grasses (Fig. 11.4)

2. Individual Species

There are large differences in Cu concentration between species but many of these are associated with differences between experiments (Table 11.6). When the same species were compared in different studies the ranking order was not the same. For example, *Trifolium pratense* had a higher Cu content than *Medicago sativa* and *Trifolium repens* in two studies (Loper and Smith, 1961; Fleming, 1965) but there was no difference in a third study (Whitehead and Jones, 1969).

B. Genotype Differences

Large differences in Cu concentration occur between cultivars of *T. subterraneum*, and the ranking order is not affected by the level of soil Cu (Table 11.7). With temperate grasses large variation exists in the Cu concentration in cultivars of *Lolium multiflorum* (84%), *Festuca arundinacea* (57%) (Montalvo-Hernandez *et al.*, 1984), and *Lolium perenne* (24 and 25%) (Forbes and Gelman, 1981; Montalvo-Hernandez *et al.*, 1984). Variable results have been found in studies of differences between cultivars of *Dactylis glomerata* with no difference in one comparison (Forbes

TABLE 11.6

Copper Concentration in Forages

Species	Copper content (mg/kg DM)		Reference
	Mean	Range	
Avena sativa	3.0	1.7–6.4	Piper (1942), Gladstones (1962), Burridge (1970), Miltimore and Mason (1971), Gladstones *et al.* (1975)
Brachiaria decumbens	10.1	2.9–25.0	Perdomo *et al.* (1977), Poland and Schnabel (1980)
Centrosema pubescens	4.1	2.0–5.9	Andrew and Thorne (1962)
Cynodon dactylon	8.0	4.3–12.2	Kappel *et al.* (1985)
Dactylis glomerata	9.4	1.9–54.0	Thomas *et al.* (1952a), Coppenet and Calvez (1962), Forbes and Gelman (1981), Whitehead (1966), Reid *et al.* (1967b), Patil and Jones (1970), Montalvo-Hernandez *et al.* (1984)
Digitaria decumbens	10.0	5.3–19.0	Perdomo *et al.* (1977), Poland and Schnabel (1980)
Festuca arundinacea	7.5	2.2–19.0	Thomas *et al.* (1952a), Coppenet and Calvez (1962), Reid *et al.* (1967a), Patil and Jones (1970), Montalvo-Hernandez *et al.* (1984)
Festuca pratensis	7.4	4.0–23.0	Thomas *et al.* (1952a), Coppenet and Calvez (1962)
Festuca rubra	7.8	3.9–14.5	Thomas *et al.* (1952a), Lessard *et al.* (1970)
Leucaena leucocephala	5.0	3.1–6.9	Kabaija and Smith (1988)
Lolium perenne	7.1	2.5–20.0	Thomas *et al.* (1952a), Coppenet and Calvez (1962), Whitehead (1966), Burridge (1970), Patil and Jones (1970), Forbes and Gelman (1981), Sherrell and Rawnsley (1982), Montalvo-Hernandez *et al.* (1984)
Lolium rigidum	3.9	2.2–7.5	Gladstones (1962), Gladstones *et al.* (1975), Reddy *et al.* (1981b)
Lotus corniculatus	8.3	9.4–11.3	Lessard *et al.* (1970), Forbes and Gelman (1981)
Medicago sativa	5.9	1.7–8.4	Andrew and Thorne (1962), Whitehead and Jones (1969)
Onobrychis viciifolia	7.0	5.9–8.2	Whitehead and Jones (1969)
Panicum maximum	9.2	5.8–13.0	Perdomo *et al.* (1977), Kabaija and Smith (1988)
Phalaris arundinacea	9.1	7.4–10.8	Lessard *et al.* (1970)

TABLE 11.6

(continued)

Species	Copper content (mg/kg DM)		Reference
	Mean	Range	
Phleum pratense	7.0	5.0–9.4	Thomas *et al.* (1952a), Patil and Jones (1970), Lessard *et al.* (1970), Coppenet and Calvez (1962)
Trifolium alexandrinum	11.4	2.9–21.0	Andrew and Thorne (1962), Gupta *et al.* (1979)
Trifolium fragiferum	4.6	2.1–7.9	Andrew and Thorne (1962)
Trifolium pratense	7.4	5.8–9.1	Whitehead and Jones (1969)
Trifolium repens	7.7	3.3–17.5	Andrew and Thorne (1962), Whitehead and Jones (1969), Metson *et al.* (1979), Forbes and Gelman (1981), Sherrell and Rawnsley (1982)
Trifolium subterraneum	10.5	2.6–19.4	Dick *et al.* (1953) Gladstones (1962), Gladstones *et al.* (1975), Reddy *et al.* (1981a,b)
Zea mays	5.9	2.5–13.9	Kappel *et al.* (1985)

and Gelman, 1981), but in another study one cultivar had twice the Cu concentration of another cultivar (Montalvo-Hernandez *et al.*, 1984). In *Phalaris arundinacea* Cu is significantly heritable in the broad sense (Hovin *et al.*, 1978).

C. Plant Parts

The leaves of temperate grasses contain, on average, 35% more Cu than the stem fraction (Table 11.8). The extent of the difference depends on the age of the plant part; immature leaf and stem of *D. glomerata* contain similar concentrations of Cu but as they mature the Cu level of the stem falls while that of the leaf remains constant (Davey and Mitchell, 1968). Flowering heads of grasses contain similar levels of Cu to the leaf fraction (Table 11.8). Leaflets of temperate legumes have a higher Cu concentration than the stem (Table 11.8) but there is no difference in Cu concentration between leaf and stem of the tropical legume *Lablab purpureus* (Hendricksen and Minson, 1980).

When pastures are leniently grazed, animals selectively graze the leaf

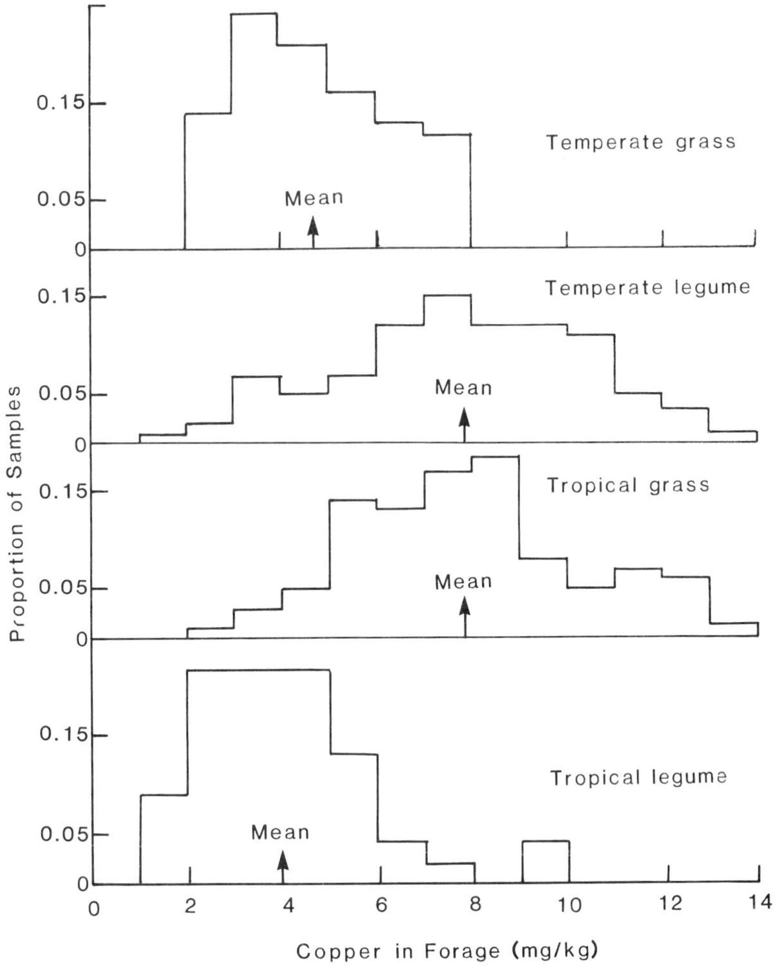

Fig. 11.4 Frequency distribution of copper in temperate and tropical grasses and legumes (data from the world literature).

fraction (Chapter 3). Since the leaf contains 30% more Cu than the stem fraction, selective grazing will improve the intake of total Cu. However, this advantage may be offset by the higher S concentration in the leaf, which will reduce the availability of the Cu (Section I,C). Whether the quantity of Cu absorbed is increased or decreased by selective grazing will also depend on the quantity of Fe in any soil ingested (Section I,C).

TABLE 11.7

Copper Concentration in Three Cultivars of *Trifolium subterraneum* Grown at Four Levels of Soil Copper[a]

Fertilizer copper (kg/ha)	Dwalganup	Cultivar Mt. Barker (mg/kg)	Yarloop	Mean
0	5.6	8.2	9.7	7.8
1.4	6.9	8.6	10.4	8.6
2.9	6.8	7.7	10.8	8.4
5.7	8.7	8.7	12.7	10.0
Mean	7.0	8.3	10.9	8.7

[a]From Gladstones (1962).

D. Stage of Growth

As forage matures there is a decrease in the concentration of Cu, probably due to a decrease in the proportion of leaf. Young *Avena sativa* plants contained 9.4 mg Cu/kg DM and this fell to 3.2 mg Cu/kg DM when the plant reached the milk-ripe stage (Piper, 1942), a result confirmed in later studies (Burridge, 1970; McDonald and Wilson, 1980). Similar changes have been reported for many other grasses (Beeson and Mac-

TABLE 11.8

Copper Concentrations in Plant Parts[a]

Species	Plant part (mg/kg) Leaf	Stem	Flowering head
Grasses			
Dactylis glomerata	7.1	5.4	7.2
Festuca pratensis	4.9	3.5	4.6
Lolium perenne	5.0	4.0	6.5
Phleum pratense	4.6	3.2	6.5
Legumes			
Lotus corniculatus	9.8	7.6[b]	—
Medicago sativa	10.5	7.9[b]	—
Trifolium repens	10.5	6.9[b]	—
	17.0[b]	14.0	11.0
Mean	8.7	6.6	

[a]Data from Beeson and MacDonald (1951) and Fleming (1968).
[b]Including petiole.

Donald, 1951; Thomas *et al.*, 1952a; Gomide *et al.*, 1969b; Lessard *et al.*, 1970; Perdomo *et al.*, 1977; Poland and Schnabel, 1980; Reddy *et al.*, 1981b; Kabaija and Smith, 1988) and legumes (Beeson and MacDonald, 1951; Thomas *et al.*, 1952a; Whitehead and Jones, 1969; Lessard *et al.*, 1970; Gladstones *et al.*, 1975; Reddy *et al.*, 1981b; Kabaija and Smith, 1988).

Copper concentration in forages falls with increasing age but the effect varies with season of the year. In early summer the Cu levels of regrowths fall rapidly, but later in the season the decrease is much slower and in *Festuca rubra* a small rise in Cu level has been reported (Lessard *et al.*, 1970). This seasonal difference is probably due to the presence of stem in the spring cuts and only vegetative growth in the autumn harvests.

E. Soil Fertility

1. Soil Type

Forage will contain low levels of Cu if insufficient available Cu is maintained in the soil solution for uptake by the plant. Copper-deficient soils can be divided into two groups depending on the cause of the low copper availability: (1) soils with low total Cu, and (2) soils with normal levels of total Cu but with a high proportion of the Cu in an unavailable form (Alloway and Tills, 1984).

Three groups of soils with low total Cu contents have been identified:

1. Young sandy-textured soils with little clay or silt.
2. Coarse-textured ferrallitic and ferruginous soils composed of hydrous oxides of iron and aluminium and sometimes kaolinite.
3. Calcareous soils developed on chalk, crystalline limestone, and calcite-cemented sandstone.

When soils contain normal levels of Cu, availability may be reduced by the formation of Cu complexes with organic matter. This occurs in peat soils and in mineral soils with >10% organic matter (Alloway and Tills, 1984). In many countries it has been possible to relate Cu levels in forage to soil type (Lee, 1951; Cunningham, 1960; Alloway and Tills, 1984; Cornforth, 1984) or to the parent material (Fraser, 1984; Berrow and Ure, 1985).

Areas where forages may be low in Cu or high in Mo can be identified from geochemical reconnaissance data (Thornton *et al.*, 1969) and from soil type (Cunningham, 1960). In the United States areas high in Mo have been located in parts of Florida, California, Colorado, Idaho, Montana, Nevada, Oregon, Washington, and Wyoming (Kubota, 1975). High soil

Mo can also result from excessive use of molybdenized superphosphate (Cunningham, 1960).

2. Soil Copper

The level of Cu in forage varies with the concentration of available Cu in the soil (Wells, 1957). For example, *T. subterraneum* grown on a lateritic podsolic soil or calcareous sand containing 2.1 or 0.5 mg extractable Cu/kg soil contained 12.9 and 6.0 mg Cu/kg DM, respectively (Reddy *et al.*, 1981b). Applying Cu sulfate increased available soil Cu and the Cu in *A. sativa* (Piper, 1942), *Centrosema pubescens* (Andrew and Thorne, 1962), *D. glomerata* (Forbes and Gelman, 1981), *Indigofera spicata* (Andrew and Thorne, 1962), *L. perenne* (Forbes and Gelman, 1981; Sherrell and Rawnsley, 1982), *Lolium rigidum* (Gladstones, 1962), *M. sativa, Macroptilium lathyroides, Stylosanthes guianensis* (Andrew and Thorne, 1962), *T. repens* (Andrew and Thorne, 1962; Forbes and Gelman, 1981; Sherrell and Rawnsley, 1982), and *T. subterraneum* (Gladstones, 1962; Reddy *et al.*, 1981a).

3. Soil Nitrogen

With swards of *D. glomerata* and *L. perenne* that were regularly cut, fertilizer soil N reduced Cu levels in the forage, possibly due to a depletion of soil Cu (Whitehead, 1966). However, when available soil Cu was high, increasing the level of soil N had no effect on the Cu contents of *Cynodon plectostachyus* (Rudert and Oliver, 1978), *D. glomerata,* and *F. arundinacea* (Reid *et al.*, 1967a). Forage grown on urine patches has the same Cu concentration as forage cut from the interexcreta zone (Joblin and Keogh, 1979).

4. Soil Phosphorus

Applying superphosphate to a lateritic podsolic soil can stimulate the growth of *T. subterraneum,* dilute the Cu available to the legume, and halve the concentration of Cu in the forage (Reddy *et al.*, 1981a). However, where soil Cu is high, forage Cu is unaffected by the level of soil P (Table 11.9).

F. Climate

Season of the year appears to have a small but inconsistent effect on the Cu concentration in forage. With regularly cut or grazed temperate forage, lowest levels of Cu generally occur in the summer (Bingley and Anderson, 1972; Reay and Marsh, 1976; Metson *et al.*, 1979; Langlands *et al.*, 1981). *Festuca arundinacea* appears to be an exception, with high-

TABLE 11.9

Interaction between Fertilizer Phosphorus and Copper on the Concentration of Copper in *Trifolium subterraneum* Grown in Pots[a]

Fertilizer (kg/ha)		Yield	Copper in forage	Total forage copper
Phosphorus	Copper	(g/pot)	(mg/kg)	(mg/pot)
0	0	6.2	11.4	7.1
135	0	15.5	4.9	7.6
0	15	6.4	18.0	11.5
135	15	14.1	18.5	26.1

[a]From Reddy *et al.* (1981a).

est levels of Cu in midsummer (Reid *et al.*, 1967b). Seasonal changes in Cu concentration can be caused by differences in temperature. Increasing soil temperature from 12 to 20°C raised the Cu concentration in *T. subterraneum* by between 20 and 93%, depending on soil type (Reddy *et al.*, 1981b).

Irrigation and waterlogging has little effect on Cu concentration in forage (Adams and Honeysett, 1964; Whitehead, 1966) but increases the concentration of Mo in both the soil solution and the forage (Mitchell *et al.*, 1957a; Kubota *et al.*, 1963). It appears that high soil water can cause Cu deficiency in ruminants by increasing the concentration of Mo in the forage, which is antagonistic to the absorption of Cu (Section I,C).

III. DIAGNOSIS OF COPPER DEFICIENCY

The survival and production of ruminants may be reduced by a deficiency of Cu and many different approaches have been used to diagnose a deficiency or potential deficiency of this element.

A. Clinical Signs

A change in pigmentation of the hair or wool is a characteristic of Cu deficiency. The black coat of Angus and Friesian cattle turns brown and the deep red of Hereford cattle changes to a sandy yellow or a washed out appearance (Davies, 1983). Depigmentation of the black hairs around the eyes gives cattle the appearance of wearing spectacles, a typical symptom of Cu deficiency (Hartmans, 1962). A "greying" of wool in black-wooled sheep is a sensitive and early sign of Cu deficiency (Lee, 1951). The crimp in the wool becomes progressively less distinct until the fibers are almost straight.

Swayback of lambs is perhaps the most striking clinical sign of Cu deficiency in sheep, causing death or low production due to lesions of the brain stem and spinal cord (Smith *et al.*, 1977a; Howell *et al.*, 1981). This disease occurs in lambs born to Cu-deficient ewes and is evident from birth to about 4 months of age (Bennetts and Chapman, 1937). The movements of the hind legs are uncoordinated, producing a staggering gait and swaying of the hind quarters. In cattle, Cu deficiency is occasionally characterized by sudden death, almost invariably without warning signs, a symptom described as "falling disease" (Bennetts *et al.*, 1941). Other more common clinical symptoms of Cu deficiency are less useful for diagnostic purposes and include slow growth, loss of weight, anemia, fragile long bones, and heart lesions (Cunha *et al.*, 1964; Underwood, 1977). Diarrhea has often been reported as a clinical symptom of Cu deficiency in cattle, particularly where Cu deficiency is induced by high Mo concentrations in the forage as in the "teart" pastures of Somerset in England and the "peat scours" in New Zealand (Davies, 1983).

B. Soil Analysis

Copper deficiency has been reported in ruminants grazing forage grown on consolidated coastal sands, leached sandy soils, marine silts, and some river silts (Grace, 1983a). These deficiencies are associated with a primary deficiency of Cu in the soil, but Cu deficiencies due to excess Mo also occur on organic soils.

The Cu and Mo content of soil, and to a certain extent forage, is influenced by the nature of the bedrock from which the soil parent material is derived. Geochemical reconnaissance by sampling of stream sediments can aid detecting and mapping of areas of potential Cu deficiency (Webb *et al.*, 1968; Thornton and Webb, 1970). In County Limerick, Ireland, geochemical reconnaissance delineated an area of excess Mo where the cattle were unthrifty and had persistent scours consistent with an induced Cu deficiency (Thornton *et al.*, 1966). There is a distinct tendency for Cu-deficient cattle to be found in areas where river sediments contain more than 2 μg Mo/g, although some herds with low blood Cu levels are found in areas where the sediment is 1 μg/Mo/g or less (Thompson and Todd, 1976). Where sediment Cu levels are greater than 56 μg/g, virtually no hypocupremia is observed.

C. Forage Analysis

The quantity of Cu absorbed by the animal will be affected by the levels of Cu, Mo, and S in the diet (Fig. 11.2). Forage should be analyzed for all three elements so allowance can be made for any depression in Cu availability caused by high levels of Mo and S. Forage analysis can be used to provide a measure of potential of the forage to supply Cu, but if available Cu is low it may be many months before the animal responds to supplementary Cu because animals can maintain normal blood Cu levels by drawing on Cu stored in the liver.

In Western Australia, sheep with deficiency symptoms were usually grazing pastures containing <2 mg Cu/kg, while safe pastures contained 7–12 mg Cu/kg DM (Beck, 1941). "Falling disease" was only encountered in the most acutely deficient areas, where Cu concentration in the forage was below 3 mg/kg, generally on the order of 2 mg/kg (Bennetts *et al.*, 1948). In many areas of the world, Cu deficiency occurs in animals fed forage containing more than 7 mg Cu/kg DM due to the presence of Mo and other antagonists of Cu absorption (Table 11.2).

D. Animal Tissue Analysis

1. Liver Copper

The liver is the main storage organ for surplus Cu (Ammerman, 1970; Grace, 1983b) and the concentration of Cu provides a useful measure of the reserves which can be used when insufficient Cu is available in the diet. The concentration of Cu in the liver is a measure of the quantity of Cu previously absorbed from the diet (Allcroft and Uvarov, 1959; Suttle, 1976).

Liver samples may be obtained from slaughtered animals or from the live animal by biopsy (Dick, 1944, 1952; Loosmore and Allcroft, 1951). The Cu level in the liver can vary within wide limits without affecting blood Cu. It is not until liver Cu falls below 20–40 mg Cu/kg DM (Fig. 11.5) that blood Cu is reduced and animals become responsive to Cu supplements. Great care must be exercised when interpreting liver Cu analysis since there are many reports of animals with low levels of Cu in the liver failing to respond to supplementary Cu (Sutherland, 1956; Todd *et al.*, 1967; Rogers and Poole, 1978; Davies, 1980; Gartner *et al.*, 1980; Murphy *et al.*, 1981).

2. Blood

The normal range of Cu in plasma of ruminants is 0.6 to 1.1 mg/liter (Fig. 11.5); below this level a response may be found to Cu supplements.

In lambs, plasma Cu has to fall below 0.2 mg/liter before survival and growth are affected, a level approaching the limits of detection for many analytical methods (Suttle, 1986). In dairy cows, clinical symptoms occur when blood Cu levels fall below 0.3 mg/liter, a value that has been recommended for diagnostic purposes (Smith and Coup, 1973). This value is lower than the 0.5 mg/liter which is generally accepted as indicative of deficiency in sheep and cattle (McDowell *et al.*, 1983). This difference in the recommended critical value may be caused by differences in the Mo content of the forage. When the forage contains high levels of Mo (5–20 mg/kg DM) growth response to supplementary Cu can occur even when plasma Cu is 0.7 mg/liter (Hogan *et al.*, 1971).

The estimation of Cu requires rigorous standards of cleanliness at all stages from the collection of the blood sample to laboratory analysis. This high standard is not required if blood Cu is measured as ceruloplasmin, a protein containing Cu as an integral part of the molecule (Todd, 1970; Thompson and Todd, 1976). Ceruloplasmin oxidase activity is positively correlated ($r = 0.83$ and 0.92) with the Cu level in the serum of cattle and sheep, respectively (Blakley and Hamilton, 1984), but has the serious limitation of a variable interval between the onset of low values and the development of pathological changes (Mills, 1987).

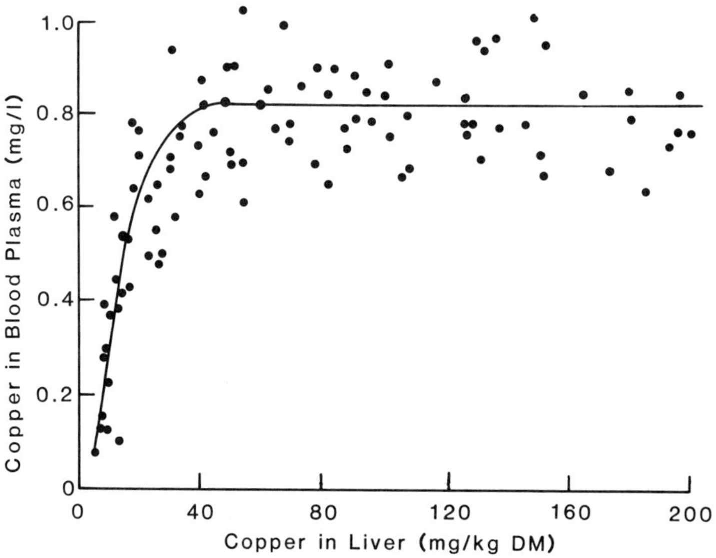

Fig. 11.5 Relation between blood plasma copper concentration and concentration of copper in the liver. Adapted from Hartmans (1969).

Superoxide dismutase (ESOD) is a Cu-containing enzyme, present in the erythrocytes (Underwood, 1977), which is correlated ($r = 0.88$) with serum Cu concentration (Andrewartha and Caple, 1980). However, when there is a change in diet it is 90 days before ESOD reaches a steady level, compared with a delay of only 1 day for serum Cu (Andrewartha and Caple, 1980). Analysis for ESOD has little to offer where a high level of dietary Mo (more than 10 mg/kg DM) rapidly induces clinical "teart-type" symptoms (Suttle and McMurray, 1983). In this situation, plasma and liver Cu are considered more useful diagnostic criteria.

3. Hair and Fleece

Studies in the 1970s showed that Cu concentration in hair of cattle is correlated with Cu levels in the plasma (Suttle and Angus, 1976; Kellaway *et al.*, 1978). Subsequent studies showed that low levels of Cu in "hair or fleece may have diagnostic significance, indicating a more prolonged or intense deficiency of Cu and a higher probability of clinical and production responses to Cu therapy than low plasma or liver Cu values" (Suttle and McMurray, 1983). It was suggested that an improved diagnosis of subclinical Cu deficiency should be based on combined use of ESOD, hair or fleece Cu, and plasma Cu concentrations (Table 11.10). This method of diagnosing Cu deficiency may not be suitable where forages are high in Mo (Suttle and McMurray, 1983).

4. Milk

Copper concentration in milk can be used for diagnosing Cu deficiency. Milk from cattle grazing pastures in Cu-deficient districts of Western Australia contained 0.01–0.02 mg Cu/liter compared with 0.05–0.20 mg Cu/liter in milk from districts where cattle were healthy (Bennetts *et al.*, 1941).

IV. PREVENTION OF COPPER DEFICIENCY

A. Supplements

Deficiencies of Cu can be prevented by supplying Cu in the form of feed supplements, by dosing or drenching the animal at intervals, or by an injection of organic complexes of Cu.

1. Mineral Supplements

Mineral supplements containing 1–2 g/kg copper sulfate are generally consumed by grazing animals in amounts sufficient to maintain adequate

TABLE 11.10

Scheme for the Combined Use of Three Criteria of Copper Status in the Diagnosis of Subclinical and Clinical Copper Deficiency[a]

Erythrocyte superoxide dismutase (mg/g hemoglobin)		Hair/fleece copper (mg/kg)	Plasma copper (mg/liter)	Diagnosis
Adult	Growing			
			Cattle	
>0.3	>0.7	>100	<9.4	Deficiency of limited duration. Production response unlikely.
<0.3	<0.5	<63	<9.4	Prolonged deficiency. Production response likely.
<0.3	<0.7	<63	>9.4	Infection or other stress increasing plasma Cu *or* recent recovery in Cu status.
			Sheep	
>0.3	>0.5	>40	<9.4	Deficiency of limited duration. Production response unlikely.
<0.2	<0.3	<30	<9.4	Prolonged deficiency. Production response likely.
<0.3	<0.5	<30	>9.4	Infection or other stress increasing plasma Cu *or* recent recovery in Cu status.

[a]From Suttle and McMurray (1983).

and safe levels of Cu (McDowell *et al.*, 1983). With sheep, a salt lick containing 5 g/kg copper sulfate removed all symptoms of Cu deficiency (Bennetts and Chapman, 1937). When cattle were offered a salt mixture containing 5 g/kg copper sulfate, liver copper was increased from 111 to 328 mg/kg, with no difference between copper sulfate and a chelated Cu supplement (Miltimore *et al.*, 1978). It should be noted that the inclusion of Cu in salt licks is at present illegal in Europe due to concern relating to Cu toxicity.

Attempts have been made to provide additional Cu through the drinking water. Using a metering device that adds 2 mg Cu/liter water, plasma Cu levels can be maintained at satisfactory levels provided there is no other source of water available (MacPherson, 1981). When animals are eating hay, Cu can be applied direct to the bales at 20 mg/kg DM (Suttle, 1983).

2. Copper Needles

Only small quantities of Cu are required each day and systems which slowly release Cu in the abomasum have been developed. In 1963 it was found that copper oxide (CuO) needles were retained in the digestive tract of cattle, only 4% being excreted in the four days following dosing (Chapman and Bell, 1963). Copper oxide needles have a high specific gravity (6.3) and when administered in a gelatin capsule into the reticulorumen are quickly carried over into the abomasum, where they are retained for over a month in sheep (Dewey, 1977) and 3 to 4 months in cattle (Costigan and Ellis, 1980). The CuO is slightly soluble in the acidic environment of the abomasum and the extent of accumulation of Cu in the liver is related to the quantity of CuO retained in the abomasum (Ellis, 1980). A single 10-g dose of CuO increased liver Cu from 54 to 1183 mg/kg in sheep (Dewey, 1977), while in cattle 40 g of CuO increased liver Cu from 5 to 342 mg/kg (Suttle, 1979).

For calves, heifers, and cows, 4-, 10-, and 20-g CuO needles proved effective in maintaining normal levels of Cu for 4 to 6 months (MacPherson, 1983). Needles should be administered to cows 3 months before parturition to ensure that calves are born with adequate Cu reserves (MacPherson, 1984). A dose of 0.045 g CuO/kg liveweight induced the maximum response in Merino sheep and Hereford cattle (Langlands *et al.*, 1986). No clinical signs of Cu toxicity were observed in Cu-sensitive sheep when the dose rate was 0.1 g/kg liveweight (Suttle, 1987a).

3. Soluble-Glass Boluses

It is possible to make a Cu-containing glass which releases Cu into the rumen at a controlled rate (Knott *et al.*, 1985). A controlled-release glass (CRG) bolus formulated for cattle weighs 75 g, contains 178 g Cu/kg, and is 2.5 cm in diameter and 5 cm in length. The sheep bolus weighs approximately 17 g and is 1.5 cm in diameter and 3.5 cm in length. The density of all Cu CRG boluses approaches 2.8. Copper is released from the CRG bolus at a minimum rate of 2.2 mg Cu/day in grazing sheep and 8.0 mg Cu/day in cattle, equivalent to approximately 50 and 15% of daily requirement for Cu (Allen *et al.*, 1985).

4. Parenteral Injections

Subcutaneous and intramuscular injections of solutions or suspensions of Cu sulfate, oxide, glycinate, methionate, oxyquinoline sulfonate, and the mixed Cu–Ca complex with ethylenediaminetetraacetate (EDTA) have been used to increase blood Cu concentration (Lamand, 1978; Ellis, 1980). Positive responses in cattle have been obtained with injectable copper glycinate in many countries, including Australia, New Zealand, the United Kingdom, and the U.S.A. (Ammerman and Miller, 1972; Smith and Thompson, 1978). Pregnant cows should be injected 1–4 months before calving to ensure a high Cu level in the liver of the calf (Allcroft and Uvarov, 1959). Inflammation and abscess formation often occur at the site of injection causing damage to skins and hides and the rejection of some parts of the carcass at the abattoirs (Ellis, 1980). The severity of the reaction varies between different compounds and between preparations of the same compound (Boila *et al.*, 1984).

B. Fertilizers

Copper deficiency in ruminants can usually be prevented by applying fertilizer Cu to the forage. In New Zealand, Cu sulfate ($CuSO_4{\cdot}5H_2O$) is applied at 5–10 kg/ha every 3 or 4 yr (Grace, 1983a), but on soils with a high organic matter, larger quantities are required to raise Cu concentration of the forage to the recommended level and these can be uneconomical.

In contrast, on calcareous soils in Western Australia one application of Cu sulfate is effective in preventing Cu deficiency for 12–20 yr (Gartrell, 1979). It is recommended that Cu-fertilized pastures should not be grazed for 3 weeks after the application or until heavy rain has fallen if Cu toxicity is to be prevented (Pryor, 1959).

C. Plant Breeding

Copper levels in forages are genetically controlled and large differences have been found between cultivars (Section II,B). Plant breeding has not been used to increase the Cu content of forages but should be considered.

V. CONCLUSION

Ruminants can be Cu deficient when eating fresh forage due to either a primary deficiency of Cu in the forage or the presence in the diet of antagonists to Cu absorption. The main antagonists are Mo and S in the forage and Fe from contamination of the diet with soil. Copper deficiency is associated with soils low in Cu and poorly drained soils with a high organic matter. Absorption of Cu from young fresh forage is low but is increased by drying. Copper deficiency causes infertility, swayback in lambs, low production of meat and wool, low wool quality, and reduced pigmentation of wool and hair.

Copper levels are higher in temperate legumes than in temperate grasses, but in tropical forages grasses have a higher level of Cu than legumes. There are no consistent differences in Cu concentration between forage species but there are large differences between cultivars. Leaf contains more Cu than stem, a difference that may be important when pastures are selectively grazed. Copper concentration is decreased by plant maturity, by high soil N and P if available soil Cu is low, and by low soil temperature. Waterlogging has no effect on total Cu but increases the concentration of Mo and hence reduces the quantity of Cu absorbed by ruminants.

Geochemical reconnaissance by stream sediment analysis has been used to detect areas of potential primary or induced Cu deficiency. Other diagnostic analyses used include forage Cu, Mo, and S; liver Cu; blood Cu; caeruloplasmin; superoxide dismutase; and the Cu concentration in hair, wool, and milk.

Copper deficiency can be prevented by feeding Cu supplements, offering salt licks containing Cu, or adding Cu to the drinking water. Other methods used include Cu oxide needles, glass boluses which slowly release Cu into the rumen, and parenteral injections of Cu compounds.

12

Iodine

I. IODINE IN RUMINANT NUTRITION

A. Function of Iodine

The sole function of iodine (I) is for the synthesis of the hormones thyroxine (T_4) and triiodothyronine (T_3), which affect a variety of physiological processes including lipid, carbohydrate, and nitrogen metabolism, regulation of energy metabolism, growth and development, reproductive performance (Underwood, 1977), and brain development (Potter *et al.*, 1981). Iodine deficiency causes large losses in newborn ruminants in cold weather because thyroid hormone is required for the mobilization of the brown fat energy reserves (Caple *et al.*, 1980a).

The main clinical sign of iodine deficiency is an enlargement of the thyroid gland. This can occur when there is a simple iodine deficiency but can also be caused by the presence in forage of goitrogenic substances. There are two main types of goitrogen, those like thiocyanates known as anionic goitrogens, and the organic or thiouracil-type goitrogens (Mason and Wilkinson, 1973). Anionic goitrogens such as the cyanogenetic glucosides in *Cynodon aethiopicus, Panicum coloratum, Paspalum dilatatum* (Rodel, 1972), and *Trifolium repens* (Corkill, 1952) owe their potency to the conversion of the HCN into thiocyanate in the tissues. Thiocyanate acts by inhibiting the selective concentration of iodine by the thyroid but its action can be overcome if the diet contains sufficient I. By contrast, goiter caused by organic goitrogens like DHP produced in the rumen of animals eating *Leucaena leucocephala* (Jones and Hegarty, 1984) cannot be prevented by supplementary I (Jones *et al.*, 1978).

There are many reports of goiter in lambs born to ewes grazing *Brassica oleracea* var. *acephala*. The goitrogen involved is of the anionic type and the goiter is prevented by supplementing with I (Sinclair and Andrews, 1954; David, 1976).

B. Effect on Production

1. Reproduction and Lamb Survival

Iodine plays a major role in the reproduction of ruminants. This effect is illustrated in Table 12.1 for goats and sheep fed a semisynthetic diet containing 0.04 mg I/kg DM. Iodine deficiency depressed conception, increased the number of abortions or stillbirths, and increased the incidence of congenital goiter and neonatal deaths. Similar effects have been reported for ewes fed a range of I-deficient forages, with lower birth weights, a sevenfold increase in thyroid size, and a halving of lamb survival (Table 12.2). Ewes grazing temperate forage on some farms in New Zealand produced 12% more lambs when injected with I bound in poppyseed oil. Iodine injections also improved lamb survival between docking and weaning by an average 12% and the weaning percentage of fawns from 65 to 85 (McGowan, 1983).

2. Growth

The growth rate of weaned lambs has been depressed by I deficiency in New Zealand. In two farm studies the growth rate of two-tooth ewes was 20 and 37% less than when the ewes were treated with I (McGowan, 1983), but in another part of the country the difference was only 4% (Grace *et al.*, 1973). A deficiency of I in ewes eating *Medicago sativa* reduced the growth rate of their lambs from 165 to 143 g/day (Knights *et al.*, 1979). In other studies, differences were found in the growth rate of lambs in one year (Flux *et al.*, 1963) but not in another season (Flux *et al.*, 1960). Mild goiter in lambs produced by ewes grazing a crop of *Brassica oleracea* var. *acephala* had no effect on growth rate (Sinclair and Andrews, 1954, 1959).

TABLE 12.1

Reproductive Performance of Goats and Sheep with Insufficient and Sufficient Iodine Supply[a]

	Goats		Sheep	
	Control	Iodine	Control	Iodine
Success of first insemination (%)	27	73	—	—
Final conception rate (%)	79	83	87	100
Abortions or stillbirths (%)	47	0	15	0
Lambs				
With goiter (%)	—	—	100	0
Survived first week (%)	—	—	50	72

[a]Adapted from Groppel *et al.* (1985).

TABLE 12.2

Effect of Iodine Deficiency in Ewes on Birth Weight, Thyroid Size, and Survival

Species	Birth weight (kg)		Thyroid (gm)		Survival (%)		Reference
	Control	Iodine	Control	Iodine	Control	Iodine	
Brassica oleracea var. *acephala*	4.0	4.8	[a]	—	83	87	Sinclair and Andrews (1954)
	3.5	4.2	17	3	22	82	Sinclair and Andrews (1958)
	—	—	10	2	41	90	Sinclair and Andrews (1961)
	—	—	23	6	—	—	David (1976)
Cynodon aethiopicus	2.7	3.6	36	2	30	75	Rudert and O'Donovan (1974)
Medicago sativa	3.5	3.9	—	—	64	84	Knights *et al.* (1970)
Mean	3.4	4.1	21	3	48	84	

[a]80% of lambs had enlarged thyroids.

3. Milk Production

Milk-fat production is depressed by a deficiency of I (McGowan, 1983). Iodine-deficient cows injected with I bound to poppy-seed oil, a nonirritant carrier, produced on average 4.5% more milk fat, with a maximum difference of 8.2%.

C. Requirements for Iodine

The published estimates of I requirements differ considerably, but most suggest that even during pregnancy and lactation dietary concentrations of 0.5 mg/kg DM are adequate for cattle and sheep provided goitrogens are absent from the diet (ARC, 1980). A lower level of I appears to be required in summer and it has been suggested that 0.15 g/kg DM might be adequate for grazing ruminants (ARC, 1980). In an experiment with goats, a semisynthetic diet containing 0.13 mg I/kg DM proved adequate for growth and reproduction when fed for 1 yr (Groppel *et al.*, 1985). With white-tail deer, a diet containing 0.26 mg I/kg DM proved adequate for normal growth and reproduction when fed for 2 yr (Watkins *et al.*, 1981). Barry *et al.*, (1983) considered that I requirements should be based on the ability to maintain normal levels of serum T_3. On this basis a dietary concentration of 0.27 mg I/kg DM was adequate for growing lambs and 0.18 mg I/kg DM was adequate for growing cattle. Where goitrogens are present in the diet the requirement for I is increased and a concentration of 2 mg/kg DM has been recommended when substantial quantities of goitrogens are suspected (ARC, 1980).

II. IODINE IN FORAGE

A. Variation in Concentration

The mean concentration of I in a wide range of forages grown in different environments was 0.26 mg/kg DM, with values varying from 0.05 to 1.90 mg/kg DM (Table 12.3). The majority of this variation was due to differences between environments with low values (mean 0.09 mg I/kg DM) recorded for forages grown on sandy soil in The Netherlands (Hartmans, 1974) compared with values four times higher (mean 0.42 mg I/kg DM) for forage grown in New Zealand (Johnson and Butler, 1957). When grown under the same conditions there were differences between forage species (Johnson and Butler, 1957; Alderman and Jones, 1967; Hartmans, 1974) but the ranking order for species varied between studies, possibly

TABLE 12.3

Iodine Concentration in Forages

Species	Iodine (mg/kg DM)		Reference
	Mean	Range	
Agrostis tenuis	0.16	0.11–0.30	Johnson and Butler (1957), Hartmans (1974)
Arrhenatherum elatius	0.07	—	Hartmans (1974)
Brassica oleracea var. *acephala*	0.12	0.02–0.27	David (1976), Barry *et al.* (1981, 1983)
Bromus unioloides	0.09	0.07–0.12	Rumball *et al.* (1972)
Cymbopogon plurinodis	0.22	0.08–0.36	Blom (1934)
Cynosurus cristatus	0.10	—	Hartmans (1974)
Dactylis glomerata	0.19	0.08–0.59	Johnson and Butler (1957), Alderman and Jones (1967), Hartmans (1974), Horn *et al.* (1974)
Festuca pratensis	0.15	0.07–0.45	Alderman and Jones (1967), Hartmans (1974)
Festuca rubra	0.06	—	Hartmans (1974)
Holcus lanatus	0.07	0.05–0.05	Johnson and Butler (1957)
Lolium multiflorum	0.39	0.07–1.22	Johnson and Butler (1957), Alderman and Jones (1967), Hartmans (1974)
Lolium perenne	0.50	0.05–1.86	Johnson and Butler (1957), Alderman and Jones (1967), Hartmans (1974), Statham and Bray (1975)
Lolium perenne × *L. multiflorum*	0.31	0.16–0.52	Johnson and Butler (1957)
Medicago sativa	0.56	—	Simpson (1931)
Panicum maximum	0.20	0.14–0.31	Blom (1934)
Paspalum dilatatum	1.42	1.28–1.70	Johnson and Butler (1957)
Phleum pratense	0.20	0.07–0.76	Alderman and Jones (1967), Hartmans (1974)
Poa annua	0.11	0.08–0.19	Johnson and Butler (1957)
Poa pratensis	0.08	—	Hartmans (1974)
Poa trivialis	0.12	0.08–0.28	Johnson and Butler (1957), Hartmans (1974)
Rhynchelythrum repens	0.24	0.09–0.39	Blom (1934)
Themeda triandra	0.24	0.14–0.51	Blom (1934)
Trifolium pratense	0.37	0.11–0.62	Simpson (1930, 1931), Johnson and Butler (1957)
Trifolium repens	0.62	0.17–1.90	Newton and Toth (1951), Johnson and Butler (1957), Alderman and Jones (1967), Hartmans (1974)
Trifolium subterraneum	0.20	0.11–0.44	Johnson and Butler (1957), Statham and Bray (1975)
Temperate pasture	0.24	0.11–0.58	Simpson (1930), Alderman and Stranks (1967)
Urochloa pullulans	0.19	0.06–0.33	Blom (1934)
Mean	0.26	0.05–1.90	

due to the use of different cultivars in each comparison. Large genetic differences in I concentration were found in *Lolium perenne* (Butler and Glenday, 1962; Alderman and Jones, 1967), *Bromus unioloides* (Rumball *et al.*, 1972), and *T. repens* (Alderman and Jones, 1967). Some of the other factors controlling I concentration will now be considered.

B. Plant Parts

Very little has been published on the distribution of I in forages. The only data are for *Bothriochloa insculpta, Eragrostis superba,* and *Hyparrhenia hirta* cut at monthly intervals throughout the year (Blom, 1934). Mean I concentrations were 0.10 and 0.24 mg/lg DM for stems and leaves, respectively.

C. Stage of Growth

In common with most elements, I concentration tends to decrease with age. The I concentration is high in young leafy *L. perenne* and rapidly declines as the forage matures (Fig. 12.1).

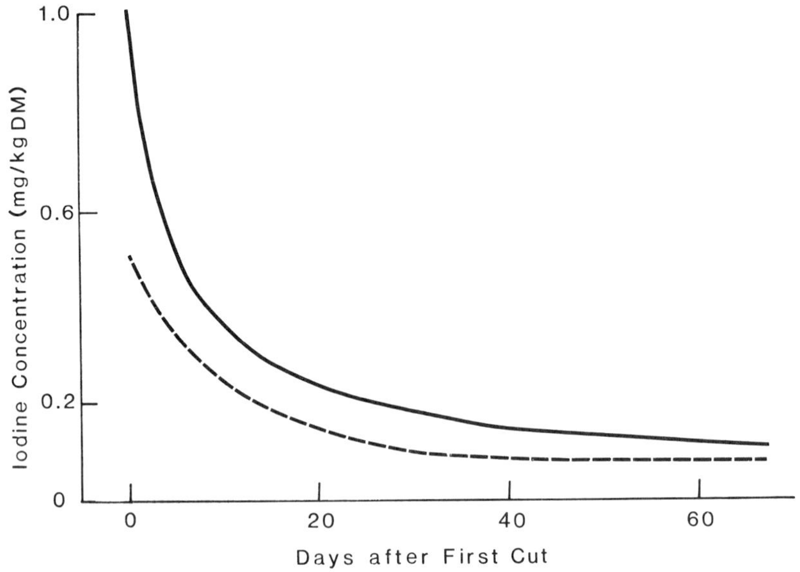

Fig. 12.1 Effect of fertilizer nitrogen and stage of growth on the iodine concentration in *Lolium perenne*. (———) 26 kg N/ha; (- - - - -) 104 kg N/ha. Adapted from Alderman and Jones (1967).

D. Soil Fertility

The I concentration of forages depends on the I available in the soil. Highest total I was found in soils of the younger marine clays and peats and lowest values in sandy soils and river clays low in organic matter. For example, *L. perenne* and *Trifolium subterraneum* contained 55% more I when grown on fine sandy light clay than they did when grown on a sandy loam (Table 12.4), a difference that had a significant effect on the incidence of goiter and neonatal deaths.

The application of fertilizer I increased the concentration of I in *L. perenne* from 0.08 to 1.41 mg/kg DM (Hartmans, 1974). The applied I was rapidly immobilized or lost from the soil, and the second cut contained only 0.21 mg I/kg DM.

The effect of fertilizer N on the I concentration of *Dactylis glomerata, L. perenne,* and *Phleum pratense* has been studied in regrowths cut at monthly intervals (Alderman and Jones, 1967). Fertilizer N reduced the mean I concentration from 0.41 to 0.27 mg/kg DM with no difference between species. In another study fertilizer N applied in the spring to *L. perenne* depressed the I concentration in the young forage but had a smaller effect on mature forage (Fig. 12.1). Fertilizer N not only de-

TABLE 12.4

Effect of Soil Type on the Iodine Concentration of Forage and the Incidence of Thyroid Enlargement and Neonatal Death of Lambs[a]

	Soil type	
	Sand	Clay
Soil iodine (mg/kg DM)		
Top 1.5 cm	1.08	1.18
Remaining A horizon	0.25	0.88
Soil organic matter (g/kg DM)	17	50
Forage iodine (mg/kg DM)		
Lolium perenne	0.18	0.29
Trifolium subterraneum	0.16	0.23
Lambs		
Born per ewe	1.25	1.42
Enlarged thyroids (%)	72	2
Died within 4 days (%)	22	12
Survived per ewe	0.98	1.25

[a]Adapted from Statham and Bray (1975).

TABLE 12.5

Cyanogenic Glucoside Concentration (Expressed as Hydrocyanic Acid) of Four Grasses Fertilized with Different Quantities of Nitrogen[a]

	Fertilizer nitrogen (kg/ha/yr)		
Species	0	450	1110
Cynodon aethiopicus	96	195	ND
Panicum coloratum	ND[b]	37	63
Paspalum dilatatum	ND	33	43
Paspalum notatum	ND	18	39

[a]Adapted from Rodel (1972) and Rudert and Oliver (1978).
[b]ND, not determined.

pressed the I concentration but it increased the requirement for I by raising the level of cyanogenic glucosides in the forage (Table 12.5).

E. Climate

In early New Zealand studies, the highest I concentrations occurred in forage samples collected in autumn and winter. It was recognized that this may have been due to contamination of the samples with soil which generally had a higher I content than the forage (Simpson, 1930). In subsequent work no consistent seasonal changes in I concentration have been reported (Johnson and Butler, 1957; Hartmans, 1974; Statham and Bray, 1975).

III. DIAGNOSIS OF IODINE DEFICIENCY

A. Clinical Signs

The main clinical sign of I deficiency is the enlargement of the thyroid gland to form a goiter and the occurrence of stillborn or weak offspring which are often edematous and with a poor growth of hair (Russell and Duncan, 1956). Goiter is not always apparent in live ruminants, particularly sheep, and post-mortem determinations of thyroid weight and other tests of thyroid function may be required for diagnosis of I deficiency (ARC, 1980). High losses within a few days of birth are also characteristic of I deficiency (Table 12.2).

B. Soil Analysis

The quantity and availability of I in the soil can be determined by soil extraction techniques (Whitehead, 1973). However no attempt has been made to relate these analyses to clinical cases of I deficiency and to establish critical values for extractable soil iodine.

C. Forage Analysis

Plucked samples of the forage eaten can be analyzed for I and the results compared with the requirements (Section I,C). The recommended levels of I in the diet have a wide margin of safety and in many cases forage analysis can give a false indication of I deficiency.

D. Tissue Analysis

1. Blood

Determination of serum or plasma T_4 and T_3 concentrations by radioimmunoassay has been used to assess the I status of grazing cattle and sheep. Depressions in both body growth and wool production were found to be better correlated with serum T_3 than with serum T_4 concentration (Barry *et al.*, 1981, 1983). This is consistent with T_3 being the active form of the hormone, T_4 functioning as a prohormone.

For maximum resistance against hypothermia, plasma T_4 of newborn lambs should exceed 50 nmol/liter (Caple and Nugent, 1982) and benefits of I supplementation of ewes are likely when the level of T_4 in the lambs falls below 40 nmol/liter (Caple *et al.*, 1985). With cattle fed *L. leucocephala* clinical signs were observed when T_4 levels fell below 25–30 nmol/liter (Jones and Winter, 1982). Growth rate of the cattle was positively correlated with level of serum T_4 up to 80 nmol/liter.

2. Milk

In 1934 it was found that the I concentration of milk was related to I level in the diet and could be increased 17 times by feeding potassium iodide (Blom, 1934). With 18 dairy herds grazing pasture containing 0.14 to 0.58 mg I/kg DM, the intake of I was closely correlated (r = 0.98) with the I level in a sample of milk from the whole herd (Alderman and Stranks, 1967). An I concentration of 15 μg/liter in the milk corresponded to a forage I level of 0.5 mg/kg. Iodophors which contain I are used to sterilize milking equipment and extreme care must be taken that equipment is free of I when collecting milk samples for I analysis.

The histological status of the thyroid of lambs has been related to the I concentration in the milk of ewes collected 10 days after birth (Mason, 1976). In pen studies, all lambs had goiter when the milk contained 20 μg/liter or less I and no goiter when the level exceeded 97 μg/liter (Fig. 12.2). Under field conditions 0 to 4% of lambs had palpable thyroid enlargement when the milk contained 98 to 111 μg I/liter, but 23 to 69% of all lambs were affected when the milk contained 23 to 64 μg I/liter (Mason, 1976). For practical purposes a milk I concentration of less than 80 μg/liter has been accepted as indicating an inadequate I intake for ewes (Mason, 1967) and goats (Caple *et al.*, 1985).

3. Feces

The I concentration in the feces can be used to predict the I content in the forage (Whitehead, 1973). Lambs fed *M. sativa* containing 0.2 mg I/kg DM produced feces with the same I concentration, while for ewes fed hay and a commercial concentrate with 2.5 mg I/kg DM the feces contained 1.5 mg I/kg DM. Fecal I concentration has been used to study the relative intake of soil I by ewes grazing a pasture at different intensities (Healy *et al.*, 1972).

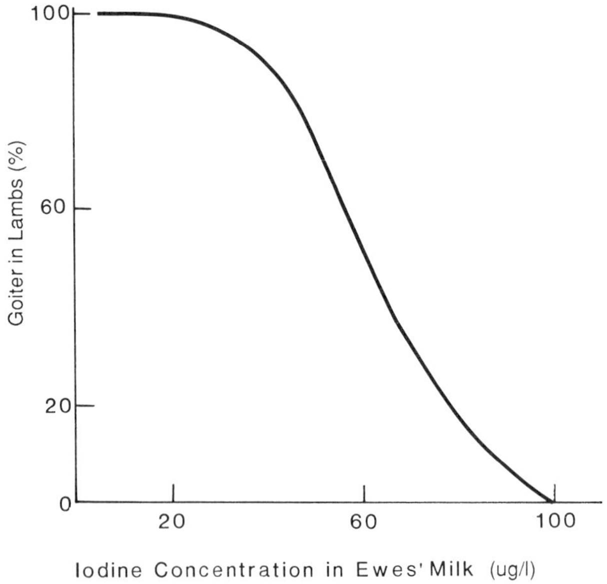

Fig. 12.2 Relation between the incidence of goiter in lambs and iodine concentration in ewes' milk 10 days after lambing. Adapted from Mason (1976).

IV. PREVENTING IODINE DEFICIENCY

A. Supplementary Iodine

In many parts of the world iodized salt, usually in the form of licks, has eliminated I deficiency in ruminants. The I is usually supplied as potassium iodate, which is readily available to sheep and cattle and much more stable than potassium iodide (Davidson *et al.*, 1951). The recommended level of I in salt supplements is 8 mg/kg (McDowell, 1985).

Iodine has been administered successfully as an intramuscular injection of poppy-seed oil containing 40% by weight of bound I (Myers and Ross, 1959; Sinclair and Andrews, 1961; Statham and Koen, 1982). Single injections of 1 ml of iodized poppy-seed oil given to pregnant ewes 7 to 9 weeks before parturition prevented severe goiter and reduced neonatal mortality in the lambs (Sinclair and Andrews, 1961). A single injection of pregnant ewes in an area of low I availability prevented goiter in lambs for two seasons (Statham and Koen, 1982) and raised milk I concentration for at least 16 months (Azuolas and Caple, 1984). Injections given to ewes 4 weeks before lambing only partially prevented thyroid enlargement and a high mortality rate of the lambs (Sinclair and Andrews, 1961). The effects of iodized poppy-seed oil on other forms of ruminant production have been considered in Section I,C.

A more recent development has been the use of intraruminal devices (IRD) for the controlled release of I. Elemental I (1.4 g) sealed within a polyethylene capsule attached to a 6 × 6 cm square of rubber is retained in the rumen (Mason and Laby, 1978). Alternatively, a plastic wing structure may be used (Siebert and Hunter, 1982). The I was released at a rate of 107 μg/day into the rumen by diffusion through the polyethylene (Mason and Laby, 1978), each capsule lasting 3 to 5 yr (Laby, 1980). These IRDs have reduced the incidence of enlarged thyroids in lambs from 78 to 0% (Mason and Laby, 1978) and lamb mortality from 12.6 to 8.6% (Ellis and Coverdale, 1982).

B. Fertilizer Iodine

The I concentration of forage is related to the level of I in the soil and can be increased by applying potassium iodide. This immediately increases the I concentration in the forage but the residual effect is small and the fertilizer has little effect of the I concentration in forage regrowths (Hartmans, 1974). It appears that the I in potassium iodide is rapidly lost by volatilization and some more stable compound of I is required for use as a fertilizer.

C. Plant Breeding

The I content of the leaves of different clones of *L. perenne* grown on the same soil ranged from 0.18 to 2.47 mg/kg (Butler and Glenday, 1962). This difference was partly genetic, I concentration being a strongly inherited characteristic (Section II, A). No attempt appears to have been made to select forage cultivars for high I concentration. The concentration of cyanogenetic glucosides in *T. repens* is genetically controlled and cultivars with low levels of goitrogen could be produced if required (Corkill, 1952).

D. Other Management Practices

The topsoil often contains more I than the forage grown on that soil (Table 12.4). When pastures are short, grazing animals will ingest considerable quantities of soil and this can be sufficient to prevent I deficiency. In New Zealand studies, ewes which grazed leniently produced lambs of which 69% had enlarged thyroids compared with only 4% when the ewes were forced to graze tightly. The feces of the leniently grazed ewes contained 4% soil and 0.4 mg I/kg DM, compared with 33% soil and 1.3 mg I/kg DM in feces from the tightly grazed sheep (Healy *et al.*, 1972). It has been suggested that difference in the intake of soil is probably an important contributor to the large year-to-year variation in the occurrence of I deficiency on farms (Healy *et al.*, 1972; Statham and Bray, 1975).

The I deficiency associated with the browse legume *L. leucocephala* is caused by a goitrogen (DHP) produced by the rumen microflora from the amino acid mimosine present in the forage (Jones and Hegarty, 1984). Thyroid enlargement in steers grazing *L. leucocephala* is the same for the cultivars Peru and Hawaii (Holmes, 1979). Jones (1981) recognized that ruminants grazing this forage in Hawaii did not suffer from an induced I deficiency and that this was due to the presence of rumen bacteria capable of detoxifying DHP (Jones and Megarrity, 1983). It was subsequently shown that these bacteria could be successfully transferred to ruminants in areas when I deficiency was caused by *L. leucocephala* (Jones and Lowry, 1984; Jones and Megarrity, 1986). Once the detoxifying bacteria have been established in one animal they spread rapidly to other ruminants in the vicinity. This has greatly increased the potential value of *L. leucocephala* as a browse shrub.

V. CONCLUSION

Iodine is an essential element required to prevent goiter, stillbirths, high rates of neonatal death, infertility, and reduced production. Iodine deficiency in ruminants is caused by a low level of I in the forage or by the presence of a goitrogen which interferes with I metabolism and increases the requirements for I.

The I concentration in forage varies with forage species, genotype, and soil fertility and is often below recommended requirements for ruminants. A deficiency of I is usually identified by an enlarged thyroid gland, but greater use in diagnostic studies could be made of the I concentration in feces and milk.

Iodine deficiency has been traditionally overcome by including iodate in a salt lick, but other methods now used include intramuscular injection of iodized poppy-seed oil and controlled-release intraruminal devices. Where I deficiency is caused by the goitrogen DHP (a rumen metabolite produced from *L. leucophala*), the deficiency can be prevented by inoculating the rumen with bacteria that metabolize DHP. Fertilizer I and plant breeding have not been used for overcoming I deficiency.

13

Zinc

I. ZINC IN RUMINANT NUTRITION

A. Function of Zinc

Zinc (Zn) was first shown to be essential for rats in 1937 but it was not until 1958 in Finland that a Zn deficiency was recorded in ruminants (Mills, 1978). Production and health of ruminants fed forages has been improved by feeding Zn in Guyana (Legg and Sears, 1960), Norway (Dynna and Havre, 1963; Haaranen, 1963), Australia (Masters and Fels, 1980), and the United States (Mayland *et al.*, 1980). In New Zealand interest has largely focused on the use of large doses of Zn to control facial eczema, a disease caused by the pasture-borne fungus *Pithomyces chartarum* (Smith *et al.*, 1977b, 1978). Large doses of Zn also have a protective effect against Cu toxicosis (Bremner *et al.*, 1976).

Zinc is widely distributed throughout the body with high concentrations (100–300 mg/kg) found in the skin, wool, hair, and horn (Towers and Grace, 1983). It is a constituent of many metalloenzymes that are involved in arachidonic acid and prostaglandin metabolism, water and cation balance, and membrane peroxidation (O'Dell, 1981). When ruminants are fed semisynthetic diets deficient in Zn, protein utilization is impaired (Somers and Underwood, 1969) and voluntary intake is depressed 39% (Underwood and Somers, 1969) but there is no effect on dry-matter digestibility (Miller *et al.*, 1966; Somers and Underwood, 1969). Testicular development and spermatogenosis are dependent on adequate Zn in the diet (Miller *et al.*, 1964b; Underwood and Somers, 1969; Smith and Gawthorne, 1975).

B. Effect on Production

1. Growth

When the Zn level in a semisynthetic diet was reduced from 32.4 to 2.4 mg/kg the growth rate of ram lambs was reduced from 118 to 27 g/

day (Underwood and Somers, 1969). This reduction in growth was mainly caused by the lower intake of the Zn-deficient diet with only 19% of the difference accounted for by a reduction in efficiency of feed utilization. Retardation of growth and severe epidermal lesions in young ruminants occurred whenever semisynthetic diets contained less than 8 mg Zn/kg (Mills, 1978).

Although many reviews cite cases of Zn deficiency in ruminants these usually refer to studies where animals suffering from parakeratosis have been cured by supplying Zn (Legg and Sears, 1960; Dynna and Havre, 1963; Spais and Papasteriadis, 1974; McDowell, 1985). Parakeratosis and other diseases responsive to feeding Zn can be caused by many factors other than a deficiency of Zn in the diet (Mills, 1978), and there is very little evidence that forage contains insufficient Zn for the normal growth of ruminants (McDonald, 1968a).

Trials were conducted on 21 farms in northeastern Scotland with beef cattle fed typical winter rations containing 13–32 mg Zn/kg. This level proved adequate for bulls and steers on all farms, but with heifers Zn supplement improved the weight gain on three farms, a response that was not related to the level of Zn in either the forage or the plasma (Price and Humphries, 1980). The only reported effect of Zn on growth of grazing animals is a 6% lower weight gain of unsupplemented calves when both calves and dams grazed *Agropyron desertorum* and *Bromus tectorum* containing between 7 and 17 mg Zn/kg (Mayland *et al.*, 1980). The unsupplemented cows grew 17% more slowly but the difference was not statistically significant.

2. Milk

Milk production may be reduced by low concentrations of Zn in forage (Voelker *et al.*, 1969). When cows were fed *Zea mays* silage containing 12.5 mg Zn/kg and a fixed quantity of concentrates they produced 9% less milk than cows fed additional Zn in the concentrates.

3. Reproduction

Testicular development was impaired and there was a complete cessation of spermatogenesis in ram lambs fed a semisynthetic diet containing 2.4 mg Zn/kg DM (Underwood and Somers, 1969). When the level of Zn was raised to 17.4 mg/kg DM testicular development and function remained lower than in similar lambs fed a diet containing 32.4 mg Zn/kg DM.

In South Australia, ewes grazing forage containing between 12 and 21 mg Zn/kg were slow to conceive and produced fewer lambs than when provided with a Zn supplement (Table 13.1). This may have been due to

TABLE 13.1

Effect of Zinc Deficiency on Cumulative Lambing of up to Six Services[a]

Supplement	Cumulative lambing to service (%)		
	1 Estrus cycle	4 Estrus cycles	6 Estrus cycles
None	30	51	72
Zinc	46	67	94
Difference	16	16	22

[a]From Egan (1972).

increased embryonic mortality (Hidiroglou, 1979b). In other studies Zn supplementation of ewes throughout mating and pregnancy increased the number of lambs born by 14% (Masters and Fels, 1980).

C. Requirement for Zinc

The requirements of ruminants for Zn have also been estimated by the factorial method, which uses data on the Zn retained in various tissues, endogenous losses of Zn in the feces and urine, and the apparent availability of Zn in the diet (Tables 13.2 and 13.3). The ARC (1980) requirements tend to be higher than those of Towers and Grace (1983), a difference mainly caused by the use of a higher coefficient of absorption (0.3) for dietary Zn when calculating the New Zealand Zn requirements (Towers and Grace, 1983). The use of the higher figure is probably justified since the absorption of Zn increases as the requirement for Zn rises (Miller *et al.*, 1970) and the level of Zn in the diet falls (Davies *et al.*, 1981). These calculated requirements for Zn are generally less than the 30–40 mg Zn/kg recommended in the United States (NRC, 1978, 1984).

The absorption of Zn is probably suppressed by high levels of Ca in the diet. Feed conversion efficiency was reduced by 9% when high levels of Ca were added to a pelleted mixed diet low in Zn but had no effect if the diet was high in Zn (Fontenot *et al.*, 1964). Raising the level of Ca in a Zn-deficient diet also increases the incidence of hyperkerotosis (Mills and Dalgarno, 1967). Forage legumes generally contain high levels of Ca (Chapter 7) and it is possible that the Zn requirement of ruminants eating legumes is higher than for grass diets.

Production trials have indicated that the requirement for Zn is usually less than the values shown in Tables 13.2 and 13.3. With semisynthetic

TABLE 13.2

Factorial Estimates of the Zinc Requirements of Sheep

	Liveweight (kg)	Production level	Dietary requirement (mg Zn/kg DM)	
			ARC (1980)	New Zealand[a]
		Growth (g/day)		
Lambs	10	150	24	18
	20	150	32–36	17
	30	150	29–49	17
	40	75	27–48	19
		150	51	18
		300	30–48	17
Ewes				
Maintenance	50	—	20–28	—
	55	—	—	23
Late pregnancy	55	Single	—	21
		Twins	—	23
	75	Twins	20–24	—
		Milk (kg/day)		
Lactation	55	1	—	24
		2	—	26
	75	1	24–31	—
		2	25–32	—
		3	25–39	—

[a]Towers and Grace (1983).

diets, 8.6 mg Zn/kg DM was sufficient for growth and sexual maturity of bull calves (Miller *et al.*, 1963), 7–8 mg Zn/kg for maintaining maximum growth in young sheep and cattle (Mills *et al.*, 1967), and 9 mg Zn/kg to maintain ewes throughout pregnancy (Masters, 1981). Conversely, ram lambs required a diet with more than 17 mg Zn/kg DM for maximum testicular development and spermatogenesis (Underwood and Somers, 1969).

Zinc is relatively nontoxic to ruminants and under normal circumstances Zn toxicity is unlikely. However the use of large quantities of Zn salts to protect grazing ruminants against facial eczema increases the risk of Zn toxicity. Reduced intake and growth rate have been found in lambs and calves fed diets containing 900–1500 mg Zn/kg (Ott *et al.*, 1966a,b), while a single dose of zinc oxide (120 mg Zn/kg body weight) given to lactating cows to control facial eczema depressed milk production by 1.5 kg/day for 2 days (Smith *et al.*, 1984).

TABLE 13.3

Factorial Estimates of the Zinc Requirements of Cattle

	Liveweight (kg)	Production level	Dietary requirement (mg Zn/kg DM)	
			ARC (1980)	New Zealand[a]
		Growth (kg/day)		
Growing cattle	100	0.5	14–28	18
		1.0	21–34	24
	200	0.5	12–20	16
		1.0	17–28	19
	300	0.5	16–28	15
		1.0	22–35	16
		Milk (kg/day)		
Adult				
Maintenance	380	—	—	15
	500	—	15–25	—
Late pregnancy	380	—	—	14
	500	—	13–21	—
Lactation	380	10	—	22
		20	—	24
		30	—	24
	450	10	—	25
	500	10	18–25	—
		20	19–27	—
		30	25–31	—

[a]Towers and Grace (1983)

II. ZINC IN FORAGE

A. Variation in Concentration

The Zn concentration in forages varies from 7 to over 100 mg/kg DM with a mean of 36 mg/kg DM for 719 samples described in the world literature (Table 13.4). Of these samples 11% contained less than 20 mg Zn/kg DM and were potentially deficient for some forms of production (Tables 13.2 and 13.3). The Pennsylvania State Forage Testing Service analyzed nearly 17,000 samples over a period of 5 yr (Adams, 1975); the Zn concentration of these forages varied from 3 to 300 mg/kg DM with a mean of 29 mg/kg DM. A slightly lower mean Zn concentration (27 mg/kg DM) was found for a range of forages cut at various stages of growth in Louisiana (Kappel *et al.*, 1985). Samples of forage analyzed by the British Columbia Feed Analysis Service had an even lower mean Zn content of 22 mg/kg DM (Miltimore *et al.*, 1970). These differences in mean values for Zn

TABLE 13.4

Concentration of Zinc in Forages[a]

Forages	Number of samples	Mean (mg Zn/kg)	Samples containing less than 20 mg Zn/kg DM (%)
Temperate grasses	370	34	16
Temperate legumes	184	38	7
Tropical grasses	145	36	6
Tropical legumes	20	40	5
All samples	719	36	11

[a]Data from the world literature.

concentration may be caused by differences in species, maturity, or level of Zn in the soil. These factors will be considered in the following sections.

B. Species Differences

1. Temperate versus Tropical

Temperate forages tend to contain slightly less Zn than tropical forages but the difference is only 2 mg/kg (Table 13.4). There were important differences in the distribution of the values, with twice as many samples of temperate forages containing less than 20 mg Zn/kg compared with tropical forages (Table 13.4).

2. Grasses versus Legumes

Grasses, both temperate and tropical, tend to contain slightly less Zn than legumes (Table 13.4), a difference that has been confirmed when grasses and clovers have been grown together (Metson *et al.*, 1979). In a comparison of a range of temperate grasses and legumes grown at different levels of soil Zn, mean Zn concentration in the grasses and legumes was 25 and 47 mg/kg DM, respectively (Gladstones and Loneragan, 1967).

3. Difference between Grasses

When grown at the same site there are often large differences in Zn concentration between grass species, but the ranking order is not consistent between different studies. For example, in one study (Perdomo *et al.*, 1977) *Panicum maximum* contained less Zn than *Digitaria decumbens*, 24 versus 37 mg/kg DM, while in another study (Gomide *et al.*, 1969b) *P. maximum* had the higher concentration of Zn, 35 versus 26 mg/kg DM.

Avena sativa is usually low in Zn, with a mean content of 14 mg/kg DM, compared with 20 mg/kg DM and 38 mg/kg for *Hordeum sativum* and *Secale cereale,* respectively (Gladstones and Loneragan, 1967). Differences in Zn concentrate have been found between genotypes of *Lolium perenne* (Butler *et al.,* 1962) and *Bromus unioloides* (Rumball *et al.,* 1972). In *L. perenne* the maximum difference in level of Zn was 21% but in *B. unioloides* the difference was 136%. Among five species of *Digitaria* there was only a 10% difference in Zn concentration (Minson, 1984).

4. Difference between Legumes

The Zn content of four temperate legumes was similar at a range of harvest dates (Whitehead and Jones, 1969). *Medicago sativa* had a mean Zn content of 24 mg/kg DM; *Onobrychis viciifolia,* 28 mg/kg DM; *Trifolium pratense,* 24 mg/kg DM; and *T. repens,* 25 mg/kg DM. Similar results were reported for *M. sativa, T. pratense,* and *T. repens* (Loper and Smith, 1961). Other studies have demonstrated large differences in Zn content between legume species at both low and high levels of soil Zn (Gladstones and Loneragan, 1967). *Trifolium hirtum* had a mean Zn content of 58 mg/kg DM, compared with 38 and 31 mg/kg DM in *Trifolium subterraneum* and *Medicago truncatula,* respectively.

C. Plant Parts

The most recently formed leaves of *Dactylis glomerata* have the highest concentration of Zn and leaf blades generally have more Zn than either the leaf sheath or the stem, which have similar levels of Zn (Davey and Mitchell, 1968). In a comparison of *D. glomerata, Festuca pratensis, L. perenne,* and *Phleum pratense* the mean Zn content of the leaf blades was 20 mg/kg DM compared to 15 mg/kg for the stem (Fleming, 1963). Flowering heads of the grasses contained 36 mg Zn/kg DM. There was no difference between the Zn content of the flowering head and the leaf plus petiole of *T. pratense,* 40 versus 42 mg/kg DM, but the stem was very much lower at 12 mg/kg DM (Fleming, 1963). Smaller differences were found between the leaf and the stem of *Lablab purpureus,* with mean Zn contents of 34 and 27 mg/kg DM, respectively (Hendricksen and Minson, 1985a).

D. Stage of Growth

As forages mature there is an increase in the proportion of stem, and a fall in Zn concentration would be expected. In a study involving 24 different forages including *Lupinus sp., T. subterraneum,* and *A. sativa*

TABLE 13.5

Effect of Time of Harvest on the Zinc Concentration (mg/kg) in 24 Forages[a]

Fertilizer zinc (kg/ha)	Time of harvest (days)			
	88	114	150	Mean
0	32	24	18	25
2.3	54	37	27	39
7.2	71	54	40	55
Mean	52	38	28	

[a]From Gladstones and Loneragan (1967).

Zn concentration fell with increasing maturity, a difference that was not affected by level of soil Zn (Table 13.5). This simple generalization has not been found in other studies. Very young growths of *M. sativa* and *O. viciifolia* had higher concentrations of Zn, but there was no difference of Zn concentration between cuts at later stages of maturity (Whitehead and Jones, 1969). A similar effect was found in six tropical grasses (Gomide *et al.*, 1969b), but with a mixed temperate pasture a high value occurred in the first spring but in the following season Zn concentration remained constant (Fleming and Murphy, 1968). Likewise, no changes in Zn concentration have been found when the following forages mature: *Cynodon dactylon* (Miller *et al.*, 1964a), *Festuca arundinacea* (Reid *et al.*, 1967a), *L. purpureus* (Hendricksen and Minson, 1985a), and *T. repens* (Whitehead and Jones, 1969).

E. Soil Fertility

1. Soil Zinc

The Zn concentration in *Z. mays* varied between 7 and 38 mg/kg DM when grown on different soils (Brown *et al.*, 1962) and similar variations have been found with other forages (Adams, 1975; Metson *et al.*, 1979; Tejada *et al.*, 1987). These differences are partly due to differences in the level of available Zn in the soil, and raising soil Zn increases the concentration of Zn in the forage (Brown *et al.*, 1962; Winter and Jones, 1977) (Table 13.5).

2. Soil Nitrogen

Very large applications of fertilizer N will increase the Zn concentration in forage. When fertilizer N was applied at 1600 kg/ha/yr, Zn concentration in *C. dactylon* was increased by 12 mg/kg DM (Miller *et al.*, 1964a). Similar increases in Zn concentration were found in *L. perenne* where fertilizer N was applied at over 1000 kg/ha/yr (Hodgson and Spedding, 1966; Large and Spedding, 1966). This rise in Zn concentration may be due to a reduction in soil pH associated with the high levels of fertilizer N (Miller *et al.*, 1964a).

The Zn concentration in forages is not affected by more normal levels of soil nitrogen. Applying fertilizer N (up to 504 kg/ha/yr) had no consistent effect on Zn concentration in *Cynodon aethiopicus* (Rudert and Oliver, 1978), *D. glomerata* (Reid *et al.*, 1967b), and *F. arundinacea* (Reid *et al.*, 1967a). Urine patches had a high concentration of soil N but this was insufficient to alter the Zn concentration, which was the same in forage grown on urine patches as in the interexcreta zone (Joblin and Keogh, 1979).

3. Soil pH

High levels of lime (40 ton/ha) have raised soil pH from 4.9 to 6.8 and depressed mean Zn concentration in *C. dactylon* from 37 to 28 mg/kg DM (Miller *et al.*, 1964a). Application of smaller quantities of lime (8 ton/ha) had no effect on the Zn concentration although it increased the pH to 5.5.

4. Soil Phosphorus and Potassium

Very little information has been published on the effect these nutrients have on the Zn concentration in forages. The only study appears to be with *F. arundinacea*, in which high levels of P and K fertilizer had little or no effect on the Zn concentration in the forage (Reid *et al.*, 1967a).

F. Climate

There is no consistency between experiments on variation in Zn concentration in forage throughout the year. No seasonal variation in Zn concentration could be found in *C. dactylon* (Miller *et al.*, 1964a), *L. perenne* (Reay and Marsh, 1976), *Paspalum notatum* (Kappel *et al.*, 1985), *T. pratense* (Reay and Marsh, 1976), and temperate pastures (Egan, 1972; Towers, 1977; Mayland *et al.*, 1980; Kappel *et al.*, 1985). However, seasonal variations in Zn concentration have been reported in *A. sativa* (McDonald and Wilson, 1980), *C. dactylon* (Kappel *et al.*, 1985), and mixed temperate species (Fleming, 1970; Masters and Somers, 1980). In these forages the maximum Zn concentration sometimes occurred in the winter, while in other studies it was found in midsummer.

Tropical forages grown in the rainy and dry seasons in Guatemala showed no consistent differences in Zn concentration (Tejada *et al.*, 1987). Waterlogging of soils has no consistent effect on the Zn concentration of *D. glomerata, L. perenne, T. pratense,* or mixed temperate species (Mitchell *et al.,* 1957a).

III. DIAGNOSIS OF ZINC DEFICIENCY

A. Clinical Signs

Ruminants fed Zn-deficient semisynthetic diets display a range of clinical symptoms. In lambs the first visible symptoms are the loosening, rubbing, and eating of the fleece followed by swelling of the area between the hocks and the hoof and lesions of the skin just above the hoof and around the eyes. The skin areas devoid of wool become slightly red and wrinkled, with encrusted areas developing on the lower surfaces of the body. In the most acute cases of Zn deficiency there is an excessive quantity of frothy saliva when chewing; shivering; swollen joints; and an arching of the back, with feet very close together (Ott *et al.,* 1964; Mills *et al.,* 1967; Miller, 1970; McDowell, 1985). Zinc-deficient calves are listless, salivate excessively, have little vitality, and develop scabby lesions on the skin above the hooves, with soft, edematous swelling of the feet in front of the fetlocks. These symptoms are followed by scabbing of the skin on the lower extremities, alopecia, and dermatitis of the entire body, especially severe on the nose, ears, eyes, neck, scrotum, heel of the mandible, and inside the legs. There is an enlargement of the hock bones and an appearance of fidgeting and stepping movement of the hind legs (Ott *et al.,* 1965; Mills *et al.,* 1967; Miller, 1970; McDowell, 1985). Similar symptoms have been found in Zn-deficient goats (Miller *et al.,* 1964b). These visual symptoms are accompanied by histological changes in many tissues (Ott *et al.,* 1965).

B. Forage Analysis

Samples of forage which are selectively grazed may be collected by hand for Zn analysis. Esophageally fistulated animals should not be used for sampling, Zn in the saliva inflating Zn concentration in the sample by 70% (Little, 1972). If a laboratory mill with tool steel knives and a bronze sieve are used the level of Zn can be overestimated by 60% (Ammerman *et al.,* 1970). Interpretation of the results of Zn analysis are difficult because no firm critical values for Zn have been established (Tables 13.2 and 13.3) and, even under controlled indoor feeding, variable responses to feeding Zn have been obtained with different rations containing similar levels of Zn (Perry *et al.,* 1968; Beeson *et al.,* 1977).

C. Tissue Analysis

1. Blood

Plasma Zn is often used in the diagnosis of Zn deficiency but interpreting the results can be difficult. Depletion and repletion studies with lambs and calves fed semisynthetic diets have shown that plasma Zn responded within a week to changes in Zn intake and that growth is likely to be impaired only when plasma Zn remains consistently below 0.4 mg/liter (Mills *et al.*, 1967), compared with normal values of 0.8 to 1.2 mg/liter (Mills *et al.*, 1967; Mayland *et al.*, 1980; Price and Humphries, 1980; Towers *et al.*, 1981). Once plasma Zn falls below 0.4 mg/liter clinical signs usually develop rapidly in young growing stock, but at other stages of development or when growth is limited by a deficiency of other nutrients there may be no visual signs of Zn deficiency (Mills, 1987). A critical value of <0.6–0.8 mg/liter has been recommended for determining the possibility of Zn deficiencies in large numbers of ruminants (McDowell, 1985). In Greece 1.5% of cattle had a plasma Zn concentration between 0.4 and 0.6 mg/liter and generally were affected with characteristic lesions. Mild signs of Zn deficiency were observed in cattle with plasma Zn levels of 0.6 to 1.0 mg/liter, and no symptoms of deficiency were observed where the level exceeded 1.0 mg/liter (Spais and Papasteriadis, 1974). The interpretation of plasma Zn is difficult but this may possibly be overcome by a technique that monitors the level of a Zn protein, metallothionein, in blood and body tissues (Mills, 1987).

2. Liver

The concentration of Zn in the liver is depressed when lambs and calves are fed Zn-deficient semisynthetic diets (Ott *et al.*, 1964, 1965; Miller, 1969), with a suggested critical value of 84 mg/kg of liver dry matter (McDowell *et al.*, 1984). This critical value is not affected by variations in the Mn content of the diet (Watson *et al.*, 1973).

3. Hair and Wool

Feeding a Zn-deficient diet reduced the Zn concentration in the wool of lambs by 17% (Ott *et al.*, 1964) and in the hair of calves by 44% (Ott *et al.*, 1965). However, the Zn concentration in wool and hair tends to be a reflection of the long-term Zn nutrition of ruminants (Beeson *et al.*, 1977; Mayland *et al.*, 1980; Combs *et al.*, 1982) and analysis may be of limited value in assessing the current Zn status in ruminants because it fails to reflect the severity and duration of Zn deficiency as manifested by impaired growth or clinical symptoms (Combs, 1987).

4. Milk

Milk usually contains about 4 mg Zn/liter, but cows fed forage low in Zn produced milk containing only 0.8 to 1.8 mg/liter (Miller, 1970). The Zn content of milk and blood are probably closely correlated and further work may show milk Zn to be a useful diagnostic technique, particularly for surveys of Zn status. Excess Zn in the diet has only a small effect on the Zn concentration in milk (Miller *et al.*, 1965b).

5. Feces

Dietary Zn that is in excess of demand is excreted in the feces, with only small quantities lost in the urine (Grace, 1975; Suttle *et al.*, 1982). Concentration of Zn in the feces should provide a simple indirect method of assessing the level of Zn in the diet and possibly provide a useful diagnostic criterion. For example, when a Zn supplement was given to calves the Zn concentration in the feces was increased from 12 to 166 mg/kg (Miller, 1969). In another study supplementary Zn increased fecal Zn concentration by a factor of 4 (Mayland *et al.*, 1980). No information has been published on the relation between Zn concentration in the feces and level of production.

IV. PREVENTION OF ZINC DEFICIENCY

A. Supplementary Zinc

Mineral mixtures containing 10–20 g Zn/kg can be used to prevent Zn deficiency (McDowell, 1985). Other methods studied include rumen pellets which slowly release Zn into the rumen for up to 10 weeks (Masters and Moir, 1980) and adding Zn sulfate to the drinking water (Wright *et al.*, 1978). An intramuscular injection into the neck of the ewes of 200 mg Zn oxide suspended in 4 ml of olive oil (Lamand, 1978) raised serum Zn for at least 3 months (Mahmoud *et al.*, 1985). A longer-term response can be achieved with an injection of metallic zinc (Lamand *et al.*, 1980).

B. Fertilizer Zinc

The Zn concentration of forage may be increased by the application of fertilizer Zn in the form of Zn sulfate (Brown *et al.*, 1962; Winter and Jones, 1977), Zn oxide (Gladstones and Loneragan, 1967), and Zn carbonate (Reuter, 1975). With Zn-deficient soils in Western Australia an initial application of 3 kg/ha of Zn oxide has been recommended (Gartrell and

Glencross, 1968), but on soil with a high pH, 4 kg Zn/ha had little effect on the Zn concentration of *M. sativa* (Reid *et al.*, 1987). The benefits of applying Zn fertilizer last many years, with little loss of Zn through leaching (Brown *et al.*, 1962; Reuter, 1975).

V. CONCLUSION

Zinc is required by ruminants to prevent the skin disease parakeratosis and to achieve maximum food intake, growth rate, testicular development, and spermatogenesis in rams and conception in ewes. The recommended requirements of ruminants for dietary Zn have been reduced to 16–26 mg/kg DM following the results of absorption studies with diets deficient in Zn and yield responses to Zn supplements. Zinc deficiency rarely occurs in ruminants fed forage.

Temperate grasses contain slightly less Zn than legumes and all tropical forages. Large differences in Zn concentration occur between different species and genotypes. Leaf blades contain more Zn than the leaf sheath and the stem of grasses.

Zinc concentration in forage is not affected consistently by forage maturity or changes in climate but is increased by the application of fertilizer Zn, N, and a decrease in soil pH.

Potential Zn deficiency can be determined from clinical signs and analysis for Zn of forage, blood, liver, hair, wool, milk, and possibly feces. Zinc deficiency in ruminants may be prevented by applying Zn fertilizer, feeding a mineral mixture containing Zn, using sustained-release rumen pellet, administering an intramuscular injection of Zn, or adding Zn to the drinking water.

14

Manganese

I. MANGANESE IN RUMINANT NUTRITION

A. Function of Manganese

In 1931 it was shown that manganese (Mn) is an essential element required to maintain normal growth, bone development, and reproduction in mice (Hidiroglou, 1979a). It was not until 1951 that Mn was found to be an indispensable element in the nutrition of ruminant fed semisynthetic diets (Bentley and Phillips, 1951), but occurrence of Mn deficiency in grazing ruminants has not been clearly established (McDonald, 1968b; ARC, 1980; McDowell, 1985).

Manganese is an essential co-factor for many enzymes with a wide range of activities (Smith and Gawthorne, 1975; Grace, 1983a) including the synthesis of mucopolysaccharides required for the organic matrix of bones and teeth, the synthesis of the steroid hormones, and gluconeogenesis and glucose utilization. Manganese is also required by rumen microorganisms (Martinez and Church, 1970). Manganese is absorbed from the small and large intestines at a low and relatively constant rate. Excess Mn is excreted in the feces (Bremner and Davies, 1980) and only a small proportion of dietary Mn appears in the urine (Grace, 1975).

B. Effect on Production

1. Growth

When goats were fed a diet containing 1.9 mg Mn/kg DM, growth was depressed and mortality higher than when fed a diet containing 5.5 mg Mn/kg DM, which appeared adequate for normal growth (Hennig *et al.*, 1972). Kids produced by goats fed a diet containing 6 or 20 mg Mn/kg DM were 20% lighter at birth than kids from the control group fed a diet containing 100 mg Mn/kg (Anke and Groppel, 1970). For lambs, a diet containing 0.8 mg Mn/kg DM was inadequate; voluntary intake was depressed and there were signs of joint pain with poor locomotion and balance (Lassiter and Morton, 1968). There appear to be no corresponding data for forages fed to other domestic ruminants.

2. Reproduction

Manganese is required for normal reproduction and a deficiency can cause delayed or irregular estrus and low conception rate in ruminants (ARC, 1980). This effect is illustrated for goats in Table 14.1. Similar results have been reported for ewes fed a grain-based diet containing 8 mg Mn/kg (Hidiroglou *et al.*, 1978). These required 2.5 services to conceive compared with 1.5 services for ewes fed a Mn-supplemented diet. A similar result was reported for ewes grazing a pasture grown on shallow calcareous soil and containing 30–40 mg Mn/kg DM, but this result could not be repeated the following year (Egan, 1972). Conception rate in cows was not adversely affected when fed a diet containing 7–10 mg Mn/kg DM (Bentley and Phillips, 1951; Rojas *et al.*, 1965). In some herds calcium phosphate appeared just as effective as manganese sulfate in restoring normal fertility (Wilson, 1966b). It was suggested that Mn deficiency was possibly induced by a high Ca and low P content of the diet.

C. Requirements for Manganese

Some uncertainty exists on the dietary Mn requirements of ruminants. Diets containing 10 mg Mn/kg DM appear adequate for growth (Bentley and Phillips, 1951), but 20–25 mg/kg DM may be needed to permit optimal skeletal development and reproduction (ARC, 1980). In North America the recommended Mn allowances for dairy cattle and beef cattle are 40 mg/kg DM and 20–50 mg/kg DM, respectively (NRC, 1978, 1984). Most studies have shown that Mn can be fed in large quantities with no adverse

TABLE 14.1

Fecundity and Mortality of Goats Fed Diets with Different Levels of Manganese[a]

	Manganese in diet	
	6–20 mg/kg	100 mg/kg
Goats pregnant (%)	93	100
Inseminations per pregnancy	1.42	1.07
Abortions (%)	23	0
Birth weight of kids (kg)	2.6	3.3
Mortality of dams (%)	43	18

[a]From Anke and Groppel (1970).

effect on the growth of cattle and sheep until Mn concentration in the diet exceeds 1000 and 2000 mg/kg DM, respectively (Cunningham *et al.*, 1966). However, with grazing lambs growth rate was depressed 40% when the Mn concentration was increased from 160 to 400 mg/kg DM (Grace, 1973).

II. MANGANESE IN FORAGE

A. Variation in Concentration

The concentration of Mn in forages varies between 1 and 2670 mg/kg DM (Adams, 1975), with very large differences between the values from different studies (Table 14.2). On a worldwide basis the mean concentration of Mn in forage is 86 mg/kg DM, with only 3% of samples containing less than the 20 mg/kg DM recommended by the ARC (1980) and 27% containing less than the 40 mg/kg recommended by the NRC (1978). Similar values have been reported for forages in Latin America and Scotland, but forage grown in New Zealand contains a very much higher level of Mn (Table 14.2). In North America, forages analyzed in Pennsylvania and British Columbia had mean concentrations of 47 and 60 mg Mn/kg DM,

TABLE 14.2

Concentration and Distribution of Manganese in Forages from Different Sources

Origin	Number of samples	Mean (mg/kg)	Percentage of samples <20 mg/kg	Percentage of samples <40 mg/kg	Reference
Ontario[a]	290	22	—	93	Buchanan-Smith *et al.* (1974)
South Australia[b]	89	38	24	61	Piper and Walkley (1943)
Pennsylvania	16844	47	—	—	Adams (1975)
British Columbia	634	60	—	49	Miltimore *et al.* (1970)
Scotland	278	80	8	27	Hemingway *et al.* (1968)
Latin America	293	—	5	20	McDowell *et al.* (1977)
Belgium	>2000	98	—	15	Devuyst *et al.* (1963)
New Zealand	3411	166	<1	2	Smith and Edmeades (1983)
World	843	86	3	27	Many publications

[a]*Zea mays* only.
[b]*Avena sativa* only.

but when the analysis was restricted to *Zea mays* silage produced in Ontario the mean Mn concentration was only 22 mg/kg DM (Table 14.2). Further evidence of the importance of location was found with *Avena sativa* grown at 89 sites in South Australia. When cut at the early flowering stage of growth Mn concentration varied from 4 to 93 mg/kg DM (Piper and Walkley, 1943). The cause of these large differences in Mn content of forage will now be considered.

B. Species Differences

1. Grass versus Legume

Grasses and legumes grown on the same site generally contain similar concentrations of Mn when the level of Mn in the grass is less than about 60 mg Mn/kg DM. However, grasses contain more Mn than legumes when the concentration of Mn in the grasses exceeds 60 mg Mn/kg DM (Table 14.3).

2. Grass Species

Large differences have been found between different species of temperate grass when grown and cut at the same site (Table 14.4). Despite considerable difference in the accumulation of Mn by grasses when grown at various sites, the ranking order of the grass species is similar in the five studies. *Zea mays* had a much lower Mn concentration than other forages analyzed at feed testing centers in North America (Miltimore *et*

TABLE 14.3

Mean Manganese Concentration of Grass and Legume When Grown on the Same Soil

Manganese in forage (mg/kg)			
Grass	Legume	Difference	Reference
18	14	+4	Baker and Chard (1961)
26	27	−1	Gladstones (1962)
28	46	−18	Thomas *et al.* (1952a)
30	30	0	McNaught (1970)
31	30	+1	Hemingway (1962)
43	45	−2	Beeson and MacDonald (1951)
65	60	+5	Kabaija and Smith (1988)
76	44	+32	Adams (1975)
76	45	+31	Reay and Marsh (1976)
92	32	+60	Miltmore *et al.* (1970)
100	40	+60	Mitchell *et al.* (1957a)
125	47	+78	Bolin (1934)
148	78	+70	Metson *et al.* (1979)

TABLE 14.4

Manganese Concentration in Temperate Grasses Grown at Five Sites (mg/kg)

Species	Sites[a] 1	2	3	4	5
Lolium perenne	22	77	158	—	—
Festuca arundinacea	23	—	134	—	288
Poa pratensis	—	—	—	78	149
Festuca pratensis	29	—	112	151	—
Phleum pratense	30	—	80	114	192
Bromus inermis	—	—	—	155	261
Agrostis gigantea	—	—	—	192	816
Dactylis glomerata	46	125	178	208	564
Mean	30	101	132	150	378

[a]1. Thomas *et al.* (1952a); 2. Mitchell *et al.* (1957a); 3. Coppenet and Calvez (1962); 4. Bolin (1934); 5. Beeson *et al.* (1947).

al., 1970; Adams, 1975) and *Zea mays* silage contained only half as much Mn as grass silages. Differences in Mn concentration have also been reported among various species of tropical grasses (Table 14.5).

3. Legume Species

Variation in Mn concentration in temperate legumes is, in general, less than in grasses (Table 14.6), although species of *Lupinus* may have values

TABLE 14.5

Manganese Concentration in Tropical Grasses Grown at Three Sites (mg/kg)

Species	Site[a] 1	2	3
Zea mays	—	52	—
Brachiaria mutica	—	—	96
Paspalum dilatatum	—	—	156
Cynodon dactylon	64	111	177
Paspalum notatum	—	140	166
Pennisetum clandestinum	96	—	—
Axonopus compressus	—	—	195
Pennisetum purpureum	123	—	—
Panicum maximum	196	—	—
Digitaria decumbens	248	—	—

[a]1. Gomide *et al.* (1969b); 2. Kappel *et al.* (1985); 3. Beeson *et al.* (1947).

TABLE 14.6

Manganese Concentration in Some Temperate Legumes Grown at Three Sites (mg/kg)

Species	Site[a] 1	2	3
Medicago sativa	32	37	42
Onobrychis viciifolia	—	44	52
Trifolium repens	56	—	49

[a]1. Beeson and MacDonald (1951); 2. Thomas *et al.* (1952a); 3. Whitehead and Jones (1969).

as high as 290 mg Mn/kg DM (Gladstones, 1962). This high concentration was attributed to deep rooting of *Lupinus* spp. in sandy soils (Gladstones and Loneragan, 1970).

C. Plant Parts

There appears to be no consistent relationship between the Mn content of the leaf and stem fractions. In one study similar Mn concentrations were found in the leaf and stem of *Dactylis glomerata* (Fleming, 1963), but in a later study the leaf blade contained twice as much Mn as the leaf sheath and stem (Davey and Mitchell, 1968). Higher Mn concentrations have been found in the leaf than in the stem fraction of *Lotus corniculatus, Medicago sativa, Trifolium repens* (Beeson and MacDonald, 1951), *Phleum pratense, Trifolium pratense* (Fleming, 1963), and *Lablab purpureus* (Hendricksen and Minson, 1985a), but in *Festuca pratensis* and *Lolium perenne* the highest levels of Mn were found in the stem (Fleming, 1963).

D. Stage of Growth

There is conflicting evidence concerning the changes in Mn concentration as forages mature. In most studies Mn concentration remained relatively constant as the forage matured (Reid *et al.*, 1967a; Fleming and Murphy, 1968; Gomide *et al.*, 1969b; Whitehead and Jones, 1969), but small rises (Beeson and MacDonald, 1951) and falls (Thomas *et al.*, 1952a) have been reported for a range of legumes. This variation cannot be attributed to forage species, level of Mn in the plant, or experimental site and is probably caused by differences in availability of soil Mn at the various stages of plant maturity.

E. Soil Fertility

Of the trace elements required by ruminants, Mn is second only to Fe in abundance in the earth's crust. Under most soil conditions it is present as insoluble oxides (Norrish, 1975), with availability inversely related to soil pH (Piper, 1931; Stewart and McConaghy, 1963; Lutz *et al.*, 1972; Kabata-Pendias and Pendias, 1984). At a pH of 5.1 *Z. mays* contained 298 mg Mn/kg DM, but this declined to 70 mg/kg DM when the pH was increased to 6.1 (Lutz *et al.*, 1972). In another study liming acid soils caused a reduction in the concentration of forage Mn from 702 to 127 mg/kg DM (Whitehead, 1966). The availability of Mn is also depressed by draining the soil (Mitchell *et al.*, 1957a). Forage grown on soil with a high water table contained 106 mg Mn/kg DM, compared with 63 mg/kg DM in forage grown on well-drained soil.

The effect of fertilizer N on Mn content of forage depends on the secondary effects on soil pH. Ammonium sulfate reduces soil pH and has increased the Mn concentration in a mixed pasture from 31 to 63 mg/kg DM (Hemingway, 1962) and in six tropical grasses from 121 to 164 mg/kg DM (Gomide *et al.*, 1969b). Urine increases soil pH and had a large effect on Mn concentration of temperate forage, depressing it to 104 from 176 mg/kg DM for forage grown in the interexcreta zone (Joblin and Keogh, 1979). Ammonium nitrate has little effect on soil pH and had no effect on the Mn concentration of *D. glomerata* (Reid *et al.*, 1967b) and *Cynodon aethiopicus* (Rudert and Oliver, 1978).

F. Climate

Seasonal changes in Mn concentration of forage have been reported, but there is no consistency in the direction of the changes. The mean Mn concentration of regularly cut temperate grasses increased from 25 mg/kg DM in spring to 189 mg/kg DM in autumn (Coppenet and Calvez, 1962), but in another study Mn concentration steadily declined during the same period (Fleming, 1968). *Avena sativa* grown in the winter contained 30% less Mn than that grown in the summer, and in a controlled temperature study forage grown at 19 and 29°C contained 72 and 94 mg Mn/kg DM, respectively (Munns *et al.*, 1963). No consistent seasonal variation in Mn concentration was found in other experiments (Bolin, 1934; Hemingway, 1962; Egan, 1972; Reay and Marsh, 1976).

III. DIAGNOSIS OF MANGANESE DEFICIENCY

A. Clinical Signs

The characteristic signs of Mn deficiency in calves are twisted legs with enlarged knees and joints, stiffness, and general weakness (Rojas *et al.*, 1965). An Mn deficiency apparently leads to differential loss of the female fetus and an increase in the proportion of male calves that are born (Hidiroglou, 1979a). In lambs there is evidence of joint pains, reluctance to move, and a tendency to "rabbit hops" when forced (Lassiter and Morton, 1968). Kids born to goats fed a Mn-deficient diet had abnormally short forelegs and there was extensive damage caused by kneeling on the forefoot joints. In some kids paralysis appeared in the first day of life (Anke *et al.*, 1973).

B. Soil Analysis

Solubility of soil Mn in dilute acid extracts has been used to identify areas where ruminants may be grazing forage low in Mn. For optimum fertility in cows it has been recommended that available soil Mn should be >3 mg/kg DM (Wilson, 1966b) and 5 mg/kg DM has been recommended as the critical level (Tejada *et al.*, 1987). However, Mn deficiency in grazing ruminants has not been clearly established.

C. Forage Analysis

Samples of the forage selectively grazed may be collected from animals fitted with esophageal fistulae. Saliva contains very little Mn and the Mn content of the bolus is similar to that of the forage eaten (Little, 1975). Care must be taken when preparing the sample for analysis. Grinding in a mill with tool steel knives increased the Mn content of samples by 16% (Ammerman *et al.*, 1970). Results from forage analysis can be compared with the requirements for different forms of production presented in Section I,C. However, it has been recommended that an Mn deficiency should only be suspected if reproductive problems or joint and bone abnormalities are associated with a low forage Mn (Grace, 1983a).

D. Tissue Analysis

1. Blood

The Mn level in blood is normally about 0.03 mg/liter but is depressed when Mn-deficient diets are fed (Rojas *et al.*, 1965). Conversely, diets containing 4031 mg Mn/kg raised the Mn in serum to 0.161 mg/liter (Black *et al.*, 1985).

2. Liver

The Mn content of the liver is correlated with Mn concentration in the diet (Hidiroglou, 1979a); increasing the quantity of Mn in the diet from 31 to 4031 mg/kg increased Mn concentration in the liver from 9 to 232 mg/kg (Black *et al.*, 1985). The critical level is generally accepted as 6 mg/kg liver (McDowell, 1985).

3. Hair

Attempts have been made to use the Mn level of hair to assess the Mn status of cattle, but there are many problems in interpreting the results (Hidiroglou, 1979a). It has been suggested that color of the hair, time of the year, method of preparing the sample for analysis, as well as analytical procedures should be taken into account when using the Mn level in hair as an index of the Mn status of ruminants (Hidiroglou and Spurr, 1975).

4. Feces

Manganese in excess of requirements is mainly excreted in the feces (Bremner and Davies, 1980; Grace and Gooden 1980), so fecal analysis should provide a useful way of assessing the Mn status of ruminants. No critical value has been published for fecal Mn.

IV. PREVENTION OF MANGANESE DEFICIENCY

A. Supplementary Manganese

A daily dose of 4 g Mn sulfate containing 1 g Mn has been used to prevent Mn deficiency and restore fertility in cows (Wilson, 1966b).

B. Fertilizer Manganese

Low Mn concentrations in forage usually occur only on neutral or alkaline soils and can be increased by applying 40–100 kg/ha Mn sulfate (Cooke, 1972). Most of the applied Mn is fixed by the soil unless the pH is reduced by applying ammonium sulfate.

V. CONCLUSION

Manganese is an essential element required for normal growth, bone development, and reproduction, but there is no unequivocal evidence that

Mn deficiency occurs in grazing ruminants. Ruminants require a diet containing 10–20 mg Mn/kg DM and very few samples of forage contain <20 mg Mn/kg DM. Manganese concentration varies between forage species and is increased in soils by low pH, waterlogging, and high temperatures.

Potential Mn deficiency can be determined by analysis of the soil, forage, blood, liver, and possibly feces. A deficiency of Mn can be prevented by applying Mn fertilizer or feeding an Mn supplement.

15

Selenium

I. SELENIUM IN RUMINANT NUTRITION

A. Function of Selenium

Selenium (Se) and vitamin E are essential nutrients closely and mutually involved in a variety of metabolic processes (McMurray *et al.*, 1983; Robinson and Thomson, 1983). The metabolic and clinical effects of Se and vitamin E in ruminants reflect their similar but independent roles in protecting tissue membranes against damage arising from the end products of some oxidative processes. Selenium is required for the formation of the enzyme gluthathione peroxidase (GSH-Px), which destroys potentially toxic peroxides, while vitamin E is believed to act as a "scavenger" of any peroxide that escapes destruction (McMurray and McEldowney, 1977; ARC, 1980).

A deficiency of Se will cause nutritional muscular dystrophy (NMD) or white-muscle disease in young ruminants (Muth *et al.*, 1958, 1959; Muth, 1963; Drake *et al.*, 1960a; Allaway *et al.*, 1966). Lambs and calves are most frequently affected (Gardiner, 1969), and in the congenital form death occurs before birth or within the first 24 hr (ARC, 1980). Clinical symptoms are described in Section III,A.

A deficiency of Se can depress voluntary intake by 30–40% (Whanger *et al.*, 1970; Kuchel and Godwin, 1976), causing subclinical ill thrift with poor production and low reproductive performance. These aspects of Se deficiency will be considered in the following section.

B. Effect on Production

1. Growth

A deficiency of Se can reduce the growth rate of sheep and cattle, with young animals particularly affected (Blaxter, 1963). Examples are shown in Table 15.1. The reduction in growth rate is variable, probably due to differences in the Se content of forage or to reserves of Se in the liver at birth. Depressions in growth associated with Se deficiency are not caused by a deficiency of vitamin E, and growth rate cannot be improved by supplementary vitamin E (Oldfield *et al.*, 1960; Buchanan-Smith *et al.*, 1969; Johnson *et al.*, 1981).

Where forage is deficient in both selenium and cobalt (Co) the depression in growth rate is greater than it is for either element on its own (Table 15.2). Although there appears to be an interaction between the two elements there is no evidence that one element may "spare" or otherwise modify the effect of the other (McLean *et al.*, 1959). A similar effect on liveweight has been observed where forage is deficient in both Se and Cu

TABLE 15.1

Effect of Supplementary Selenium on the Death Rate and Growth of Lambs and Calves Grazing Selenium-Deficient Forage

Forage	Reduction in death rate (% of all animals)	Increase in growth of survivors (%)	Reference
Sheep			
Temperate pasture	19	65	Drake *et al.* (1959)
	—	23	McLean *et al.* (1959)
	12	57	Drake *et al.* (1960b)
	17	166	Jolly (1960)
	—	18	Setchell *et al.* (1962)
	—	58	Skerman (1962)
	22	18	Andrews *et al.* (1964)
	—	60	Grant (1965)
	—	20	Hartley (1967)
	—	33	Hartley (1967)
	—	30	Andrews *et al.* (1974)
	—	17[a]	Grace *et al.* (1974)
	—	18	Andrews *et al.* (1976)
	—	67	Kuchel and Godwin (1976)
	—	86[b]	Paynter (1979)
	—	9	Paynter *et al.* (1979)
	—	22	Watkinson (1983)
	—	18	Dove *et al.* (1986)
Trifolium repens	—	77	Oldfield *et al.* (1960)
Cattle			
Agropyron repens/ *Festuca arundinacea*	—	67	Hathaway *et al.* (1979)
Temperate forage	—	43	Hartley (1961)
	—	43	Johnson *et al.* (1981)
	11	3	Spears *et al.* (1986)

[a]Between weaning and mating.
[b]Wethers.

TABLE 15.2

Effect of Supplementary Selenium on Growth of Lambs in the Presence of a Cobalt Deficiency (g/day)

	Treatment				
Study	Control	Selenium	Cobalt	Selenium + cobalt	Reference
1	12	16	25	42	Drake *et al.* (1959)
2	26	30	69	82	Andrews *et al.* (1964)
3	105	179	167	197	McLean *et al.* (1959)
Mean	48	74	87	107	

(Thomson and Lawson, 1970). To determine the true effect on production of Se deficiency it is essential that all other nutrients are supplied in sufficient quantities.

2. Milk Production

Production of milk and butterfat is depressed by a deficiency of Se. Cows in 12 herds with low blood Se (<12 μg/liter) were given a subcutaneous injection of Se every 2 months. Mean production of milk and butterfat was increased by 4.6 and 2.3%, respectively, with the largest response occurring in herds where the control group had the lowest blood Se (Fraser *et al.*, 1987a). In another Se-deficient herd, a subcutaneous injection of 500 mg of barium selenate increased production of milk and butterfat by 7.2 and 5.4%, respectively (Tasker *et al.*, 1987).

3. Wool Production

A deficiency of Se depressed wool production by lambs by 35% (Hartley, 1961). By providing additional Se, wool production has been increased in both ewes (Slen *et al.*, 1961; Quarterman *et al.*, 1966; Hill *et al.*, 1969; Wilkins *et al.*, 1982) and lambs (Hill *et al.*, 1969; Gabbedy, 1971; Dove *et al.*, 1986), a response that is often associated with a higher liveweight of the Se-supplemented animals (Hill *et al.*, 1969). Selenium deficiency also reduces staple length (Slen *et al.*, 1961; Hill *et al.*, 1969) and fiber diameter of wool (Slen *et al.*, 1961; Gabbedy, 1971).

4. Reproduction

New Zealand studies showed that in areas where NMD in lambs is severe, a high proportion of the ewes are barren (Grant *et al.*, 1960) and feeding Se reduced embryonic mortality from 26 to 3% (Hartley, 1963). In the South Island of New Zealand Se deficiency decreased the proportion of ewes that conceived by 9 and 15% when they were grazing *Medicago sativa* and *Lolium perenne*, respectively, but had no effect on the proportion of multiple births (Scales, 1974). In other studies Se deficiency reduced lambing from 95 to 86% in ewes grazing a pasture containing *Trifolium repens* (Wilkins *et al.*, 1982) and from 64 to 46% when grazing *Trifolium subterranum* (Godwin *et al.*, 1970). Selenium deficiency reduced conception at first service in dairy cows from 68 to 51% (Fraser *et al.*, 1987b).

Selenium deficiency causes a retention of the placenta in cows, and in a trial in Ohio Se supplementation reduced the incidence from 38% to zero (Julien *et al.*, 1976). Similar improvements have been achieved with a combined injection of vitamin E and Se (Trinder *et al.*, 1973).

C. Requirement for Selenium

The ARC (1980), which reviewed the literature on the Se requirements of ruminants, concluded that 0.03 mg Se/kg DM should be regarded as inadequate and that Se in the range 0.03 to 0.05 mg/kg DM may lead to subclinical muscle damage and should therefore be regarded as marginal. The requirement for Se is related to growth rate (Grace and Watkinson, 1985), so any factor which increases intake and digestion of forage will increase the absolute requirement for Se and, to a lesser extent, the critical Se concentration in the forage. Outbreaks of NMD in Western Australia were associated with forage Se levels of 0.05 mg/kg DM or less, with most cases occurring when the levels were below 0.03 mg/kg DM (Gardiner *et al.*, 1962) although, with some synthetic diets, peak levels of Se in blood are only achieved when the diet contains more than 0.10 mg Se/kg DM (Millar, 1983). Part of this variation may be caused by differences in the level of vitamin E in the diet (Judson and Obst, 1975; ARC, 1980). In the United States it is recommended that diets for growing and finishing steers and heifers should contain 0.10 mg Se/kg DM, while those for breeding bulls and pregnant and lactating cows should contain 0.05 to 0.10 mg Se/kg DM (NRC, 1984). It was also suggested that 2 mg/kg DM is the maximum tolerable Se level for all ruminants' diets; above 2 mg/kg DM Se may result in toxicity. In 1987, the U.S. FDA allowed up to 0.3 ppm of selenium in complete feeds for cattle and sheep. In salt–mineral mixtures for free-choice feeding, the FDA allows up to 120 ppm of selenium for beef cattle (at a rate not to exceed an intake of 3 mg of selenium for beef cattle per day) and up to 90 ppm for sheep (at a rate not to exceed an intake of 0.7 mg of selenium per head daily).

Indirect evidence from several studies indicates that the requirements for Se may be affected by the levels of other elements in the diet (Judson and Obst, 1975). However, absorption of Se is not affected by additional Cu (Cadwallader *et al.*, 1980;) or Ca (Alfaro *et al.*, 1987). Additional sulfur slightly depresses the absorption of Se (Pope *et al.*, 1979) but has no effect on weight gain (Stadtmore and Reid, 1981).

II. SELENIUM IN FORAGE

A. Variation in Concentration

Very large differences occur in the Se concentration in plants. Selenium accumulator plants grown on seleniferous soils may contain >300 mg Se/kg DM (Johnson, 1975) and these concentrations are lethal to ruminants (Rosenfeld and Beath, 1964). Areas of the United States where these occur have been delineated by Ammerman and Miller (1975). Species normally sown for forage purposes contain less Se than accumlator plants but can still contain toxic levels if grown on soils high in Se (Hamilton and Beath, 1963). For example, *Phleum pratense* and *M. sativa* grown on seleniferous soil may contain as much as 33 and 40 mg Se/kg DM, respectively.

It is now recognized that Se deficiency is more widespread and of greater economic significance than Se toxicity. In the United States there are large areas in the Northwest, Northeast, and Southeast, including many of the states adjoining the Great Lakes, which produce forage containing <0.05 mg Se/kg DM (Ammerman and Miller, 1975). Forages grown in Ontario are also low in Se, with 54% of all samples containing <0.05 mg Se/kg DM (Young *et al.*, 1977). In Uganda 20% of forage samples analyzed contained <0.05 mg Se/kg (Long and Marshall, 1973), while in Western Australia many samples of forage contained only 0.02 mg Se/kg DM (Gardiner *et al.*, 1962; Gardiner and Gorman, 1963).

B. Species Differences

1. Grass versus Legume

In New Zealand studies the Se concentration in grasses was higher than that in *T. repens* (Table 15.3). This difference occurred at all levels of soil Se and of fertilizer Se (Table 15.4). The tropical legumes *Centrosema pubescens* and *Stylosanthes guianensis* also contained less Se than associated grasses, especially when the level of Se was above average (Long and Marshall, 1973).

No difference in Se concentration was found between *T. subterraneum*

TABLE 15.3

Selenium Concentration of Forage Species in a Mixed Sward Grown on a Selenium-Deficient Soil[a]

Species	Selenium (mg/kg DM)	Concentration relative to *Trifolium repens*
Anthoxanthum odoratum	0.049	3.8
Danthonia spp.	0.033	2.5
Holcus lanatus	0.031	2.4
Agrostis tenuis	0.027	2.1
Lolium perenne	0.020	1.5
Trifolium repens	0.013	1.0

[a]From Davies and Watkinson (1966).

and associated grasses in Australian studies (Caple *et al.*, 1980b), between *Phleum pratense* and *Trifolium pratense* when grown on a range of Se-deficient soils in Ontario (Gupta and Winter, 1975), or between *M. sativa* and grasses in Denmark (Gissel-Nielsen, 1975). In another study *M. sativa* contained about twice as much Se as *T. pratense, Dactylis glomerata,* and *Bromus inermis* (Ehlig *et al.*, 1968).

2. Grass Species

Agrostis tenuis contains more Se than *D. glomerata* and *L. perenne* (Table 15.4), a ranking order confirmed in other studies with forage grown on sand or soil (Peterson and Spedding, 1963). A similar ranking order of Se concentration was found when a mixed pasture grown on unfertilized soil was separated into the major species (Table 15.3).

TABLE 15.4

Selenium Concentration (mg/kg) of Second-Cut Forage Grown on Soils Fertilized with Selenium[a]

	Selenium applied (g/ha)				
Species	0	140	280	560	1120
Agrostis tenuis	0.035	0.27	0.58	1.10	1.9
Dactylis glomerata	0.030	0.19	0.42	0.70	1.5
Lolium perenne	0.020	0.12	0.26	0.57	1.1
Trifolium repens	0.017	0.07	0.13	0.19	0.5

[a]From Davies and Watkinson (1966).

C. Plant Parts

Very little appears to be known about the concentration of Se in different parts of forage plants. In *D. glomerata* the concentration of Se in the spikelets was double that in the leaf fraction, with intermediate values for the leaf sheath and stem (Davey and Mitchell, 1968).

D. Stage of Growth

Very little data apear to have been published on the Se level in forages at different stages of maturity. In *M. sativa* stage of growth had little effect on Se concentration (Ehlig *et al.*, 1968).

E. Soil Fertility

The Se concentration in forages is related to the Se content of the soil and the parent material from which it is derived (Kubota *et al.*, 1967; Andrews *et al.*, 1968; Doyle and Fletcher, 1977). For example, when *P. pratense* and *T. pratense* were grown on soils containing between 0.09 and 0.60 mg Se/kg there was a significant correlation (r = 0.4) between forage and soil Se (Gupta and Winter, 1975).

Soil type also affects the Se content of forage. When grown on mineral upland soils forage has a lower Se concentration than when grown on organic moorland soils, possibly due to a greater availability of Se in organic combination (Thornton *et al.*, 1985). Sandy soils tend to produce forage with a lower Se concentration than heavier soils in the same district (Gardiner and Gorman, 1963).

The incidence of NMD is closely associated in Australia and New Zealand with lush pastures and the regular application of superphosphate (Gardiner, 1969). This is probably caused by the higher proportion of legumes, which often have a lower Se content than grasses (Section II,B), and by the fertilizer stimulating growth, leading to a dilution in the forage of the available Se (Westerman and Robbins, 1974).

The Se concentration in *L. perenne* growing on urine patches is 40% lower than for forage grown in the interexcreta zone (Joblin and Pritchard, 1983). This is probably due to the increased growth caused by the urea, diluting a limited supply of Se.

F. Climate

It is generally accepted that NMD is more frequent after cold and wet summers than after hot and dry ones. This difference has been associated with the low Se content of pasture in a cold summer. *Avena sativa* forage grown in the field at a mean daily temperature of 14°C contained 0.012 mg Se/kg DM, compared with 0.042 mg/kg DM in a year when the mean temperature was 19°C (Lindberg and Lannek, 1970). An alternative explanation is the high leaching of Se that would occur in cold wet seasons, Se level of plants varying inversely with rainfall (Gardiner and Gorman, 1963). In Western Australia average Se level in forages was 0.26 mg/kg DM in areas with an annual rainfall of 250–375 mm but only 0.036 mg/kg DM where rainfall exceeded 750 mm.

Forage grown in Denmark exhibited large seasonal variations in Se concentration (Gissel-Nielsen, 1975). Maximum value was 0.20 mg/kg DM in March, reducing to a minimum of 0.02 mg/kg DM in the period June to September. In the same summer period, native pastures in Norway also had a constant Se concentration (Garmo *et al.*, 1986).

III. DIAGNOSING A SELENIUM DEFICIENCY

A. Clinical Signs

Nutritional muscular dystrophy in lambs and calves is the most characteristic sign of Se deficiency. In lambs the disease has been described by Rosenfeld and Beath (1964) as a

> progressive stiffness, resulting in complete disuse of skeletal muscles. The skeletal muscles of the neck or the extremities may be involved, and the involvement of the muscles is bilateral. The bilateral involvement of skeletal muscles impairs both voluntary movement and locomotion and results in muscular weakness. In animals with impaired locomotion secondary factors such as starvation, exposure, and bacterial infections are the contributing cause of death. Severely affected cardiac muscle damage, usually in calves, give rise to general weakness and death within a few hours after symptoms appear or even without appreciable symptoms.

Spontaneous recovery may occur without any apparent change in feeding and management "provided there is no severe cardiac damage. The incidence in lambs is from the first week to at least 12 weeks after birth with the majority of cases occurring at about 3 weeks" (Dodd, 1954). With older sheep and cattle myoglobinurea may occur as a direct result of muscle damage (Judson and Obst, 1975). Gross changes seen at post-mortem are symmetrical whitish-yellow discoloration of the large leg and back muscles and also of the intercostal muscles and the diaphragm (Lannek and Lindberg, 1975).

B. Soil Analysis

Soil collected from farms in New Zealand was analyzed for Se and the results related to the response of lambs to an Se supplement (Watkinson, 1962). It was concluded that an Se deficiency might be encountered when the Se content of the soil was less than 0.45 mg/kg. The Se content of the diet depends not only on soil Se but also on the species of forage (Section II,A). It is therefore not possible to establish a simple critical value for soil Se which can be used for diagnosis in all situations.

C. Forage Analysis

Samples of forage for Se analysis are usually collected by hand plucking. Esophagelly fistulated sheep have been used to collect samples of selectively grazed forage but there was poor agreement between the Se concentration in forage offered, whole extrusa or squeezed extrusa (Langlands *et al.*, 1980b). Some of the Se in forage is present as volatile compounds, but if forage is dried in an oven at 70°C only between 0.5 and 3.0% of Se is lost from the sample (Asher and Grundon, 1970). In other studies losses reported have also been low and it appeared "unnecessary to treat herbage samples to be analysed for Se differently, with regards to drying, from normal herbage samples" (Collier and Parker-Sutton, 1976). Provided the sample contains >0.05 mg Se/kg a response to additional Se is unlikely (Section I,C).

D. Tissue Analysis

1. Blood

Selenium deficiency can best be diagnosed by determining the Se level in whole blood, plasma, or erythrocytes (Millar, 1983). Based on many field studies with lambs, values for blood Se have been divided into four diagnostic categories (Table 15.5). Blood Se has to fall below about 0.12 mg/liter before lactating cows respond to an Se supplement (Fraser *et al.*, 1987a; Tasker *et al.*, 1987).

The concentration of Se in whole blood is correlated with the activity of the enzyme GSH-Px (Oh *et al.*, 1976; Langlands *et al.*, 1980a). The analysis of GSH-Px is now the main method of estimating blood Se (Millar, 1983) because it has advantages over the direct measurement of Se. GSH-Px analysis is rapid and less expensive than direct measurement of Se and a simple spot test has been developed.

Another enzyme that is sometimes measured in Se studies is creatine

TABLE 15.5

Relation between Blood Selenium Concentration and Diagnostic Category[a]

Blood selenium (mg/liter)	Diagnostic category
<0.05	Deficient
0.051–0.075	Low marginal
0.076–0.10	Marginal
>0.10	Adequate

[a]From Anderson *et al.* (1979).

phosphokinase. The activity of this enzyme rises by a factor of 10 in sheep fed diets deficient in Se and is a sensitive indicator of the presence of subclinical dystrophy (McMurray and McEldowney, 1977; Paynter, 1979). However, the activity of this enzyme only rises at an advanced stage of Se deficiency and is also increased by other forms of muscle damage. In consequence, a rise in creatine phosphokinase is not specific to Se deficiency.

2. LIVER

Selenium is stored in the liver and the concentration of Se can be used as a measure of the Se status. Lambs responsive to supplementary Se had an Se concentration in fresh liver of 0.12 mg/kg (Andrews *et al.*, 1964), and a concentration of 0.21 mg/kg is near the minimal requirement (Allaway *et al.*, 1966).

3. WOOL AND HAIR

The concentration of Se in wool is increased when sheep graze forage high in Se for long periods (Leonard and Burns, 1955). Fiber analysis may provide a useful method of assessing the long-term Se nutrition of ruminants once equations have been derived relating the Se content of hair or wool to the Se content of forage or production response to Se supplements.

4. MILK

The concentration of Se in milk is related to the Se concentration in the diet and blood of cattle (Little *et al.*, 1979) and sheep (Hunter *et al.*, 1979a). Milk composition responded very rapidly to changes in dietary Se (Little *et al.*, 1979) and should prove useful in diagnostic studies.

5. Feces

Over half the Se in the diet is excreted in the feces (White and Somers, 1978; Pope *et al.*, 1979), so there is probably a high correlation between the level of Se in forage and feces. No studies have been undertaken to determine this relationship. Feces samples are derived from forage selected by ruminants and so should provide a better diagnostic assessment than can be obtained by forage analysis.

IV. CORRECTING A SELENIUM DEFICIENCY

A. Supplementary Selenium

When forage contains insufficient Se the requirement can be met by oral or parenteral administration of either sodium or barium selenate (Hartley, 1967; Millar, 1983; Metherell *et al.*, 1987) or sodium selenite (Hartley, 1967; Millar, 1983). Oral administration is satisfactory for young stock but is less convenient with mature animals that are infrequently yarded.

1. Oral Administration

Problems of barren ewes and congenital NMD can be prevented by dosing with 5 mg Se (as sodium selenite or sodium selenate) 1 month before mating and 1 month before lambing (Hartley, 1961). Lambs which are unthrifty or with delayed NMD respond to 1 mg Se at tailing and further 5-mg Se doses at two or three monthly intervals (Hartley, 1961). Calves with NMD or which are unthrifty are usually given 10 mg Se at marking (Hartley, 1961) and then 20–30 mg at 2- to 3-month intervals (Millar, 1983).

To avoid regular dosing, a heavy pellet which released Se in the rumen for at least 35 weeks was produced (Kuchel and Buckley, 1969; Handreck and Godwin, 1970). Pellets now commercially available contain 5% coarse-grained elemental selenium dispersed in an iron matrix (Hudson *et al.*, 1981). The rate of release and absorption of Se from these bullets can be increased by also including a Zn bullet in the rumen (Peter *et al.*, 1981). The effectiveness of the bullet may be reduced because calcium phosphate in the rumen forms a coating on the bullet which prevents the release of Se. This can be overcome by including a 10-g grub screw (Kuchel and Godwin, 1976), but these can cause damage to machinery at the slaughterhouse.

Another form of slow-release device is the glass bolus containing 0.31% Se. When pregnant cows were given two 120-g boluses, blood GSH-Px was increased for at least 10 months, with peak levels five times that of controls (Hidiroglou *et al.*, 1985, 1987). Selenium concentration in their milk was also increased, as was the GSH-Px level in the blood of their calves.

The traditional method of feeding trace elements has been in the form of licks or salt blocks. Paulson *et al.* (1968) showed that including Se in trace-mineralized salt at rates of 26 or 132 mg/kg improved growth, but 264 g/kg depressed weight gain. In other studies 30 and 65 mg Se/kg proved to be a safe and effective way of raising the GSH-Px and Se in blood of ewes and lambs (Ullrey *et al.*, 1978). With grazing yearling heifers, mineral salt mixtures containing 25, 50, and 100 mg Se/kg increased growth rate from 0.46 kg/day for controls to 0.69, 0.68, and 0.78 kg/day, respectively (Hathaway *et al.*, 1979). No toxic effects were found and it was concluded that there is little animal or human health hazard associated with Se supplements where the total Se in the diet is <0.1 mg/kg DM (Ullrey *et al.*, 1977). Other aspects of possible Se toxicity are reviewed by Gardiner (1969).

2. Subcutaneous Injection

Injection of Se has been used where a slow release of Se is required and stock are not frequently handled. Materials are administered by subcutaneous injections in the shoulder or neck region at a rate of approximately 5 mg Se per 100 kg body weight (Johnson *et al.*, 1981). For unthrifty calves the recommendation is 10 mg Se at 3 months repeated every 3 months or as necessary (Hartley, 1961); with cattle weighing between 160 and 250 kg this treatment increased growth rate by 43% (Johnson *et al.*, 1981). For dairy cattle an injectable form of barium selenate has been used, each cow receiving 500 mg Se about 2 weeks before the breeding season (Tasker *et al.*, 1987). This treatment increased conception rate and milk production in cows grazing Se-deficient pasture (Section I,B).

For sheep a controlled-release glass pellet has been produced which is implanted intramuscularly into the neck (Allen *et al.*, 1981). Each pellet (3.2 mm long × 3.0 mm in diameter) weighs approximately 110 mg, contains 44% Se, and maintains an adequate supply of Se for up to 7 months.

B. Fertilizer Selenium

The Se content of forage can be increased by applying Se salts to the soil (Grant, 1965; Allaway *et al.*, 1966). When applied as fertilizer sodium selenate proved to be 5 to 15 times more effective than sodium selenite in raising the Se concentration of forage (Grant, 1965). Peak concentrations of Se were reached within a few days of applying sodium selenate but then declined logarithmically, with little residual benefit after a year. Recovery of Se in the forage was about 35% from selenate but only 1–2% when selenite was applied (Allaway *et al.*, 1966; Gissel-Nielsen and Bisbjerg, 1970).

The current practice in New Zealand is to apply sodium selenate every year at 10 g Se/ha, a rate that is 5% of the quantity considered toxic and does not affect the level of Se in the meat of grazing animals (Watkinson, 1983). It is difficult to apply 10 g Se/ha evenly, and granules containing 1% Se have been commercially produced. These are applied to pasture, either alone or mixed with fertilizer, at a rate not exceeding 1 kg of pellets per hectare (Watkinson, 1983; Halpin *et al.*, 1985). The cost of supplying Se through fertilizer can be reduced by putting the required Se on only one-sixth or less of the pasture (Hupkens Van Der Elsk and Watkinson, 1977; Watkinson, 1983). This method is very effective, the animals neither rejecting nor preferentially grazing the fertilized strips.

V. CONCLUSION

Selenium is a trace element required for the formation of the enzyme gluthathione peroxidase. A deficiency of this enzyme causes nutritional muscular dystrophy in young ruminants. Selenium deficiency has also been associated with a depression of appetite, reduced rates of growth, milk production, wool production, and conception, and an increase in embryonic mortality.

The Se concentration in forage is related to the level of available soil Se. Legumes tend to contain less Se than grasses. All forages can absorb toxic levels of Se when grown on seleniferous soils, but very high levels are associated with Se accumulator plants.

Selenium deficiency can be diagnosed from clinical signs and by analysis of soil, forage, blood, and liver. Other potential criteria are Se concentrations in milk, feces, hair, and wool.

Selenium deficiency can be prevented by fertilizing with very small quantities of sodium selenate, by oral administration of Se either in a soluble form or as a slow-release bullet, by subcutaneous injection of Se, and by including it in a mineral supplement or in trace-mineralized salt at proper levels.

16

Cobalt

I. COBALT IN RUMINANT NUTRITION

A. Function of Cobalt

Cobalt (Co) is not essential for growth of forages (Underwood, 1981) but is required in small quantities by ruminants. In 1935 it was found that Co prevented the nutritional diseases coast disease and enzootic marasmus in grazing sheep and cattle in Australia (Lines, 1935; Marston, 1935; Underwood and Filmer, 1935; Filmer and Underwood, 1937). Other wasting diseases, since found to be responsive to Co, include bush sickness in New Zealand, salt sickness in Florida, pine in Scotland and England, and nakuruitis in Kenya (Marston, 1952). The most severely deficient areas of the United States include portions of New England and the lower Atlantic Plain (Kubota, 1968). Cobalt deficiency has also been reported in Brazil and Norway (Ammerman, 1970). Long before anything definite was known about its cause, each disease was recognized as being associated with particular areas, and prevention and cure were effected by transferring stock to healthy areas (Beeson, 1950; Gardiner, 1977).

Cobalt deficiency is characterized by anemia and progressive loss of appetite leading to death in a state resembling hunger edema (Lee, 1975). In 1948 it was shown that Co was a constituent of vitamin B_{12}, the antipernicious anemia factor, and when Co-deficient lambs were given vitamin B_{12} there was a complete recovery of health (Smith *et al.*, 1951). In grazing studies with lambs, similar improvements in growth were achieved with vitamin B_{12} injections and supplementary Co (Anderson and Andrews, 1952). Cobalt is required for the synthesis of vitamin B_{12} by rumen bacteria (Marston, 1959) and most of the effects of Co deficiency in ruminants can be attributed to reduced activity of two enzymes, methylmalonyl-CoA mutase and 5-methyl-tetrahydrofolate: homocysteine methyltransferase (Smith and Gawthorne, 1975). Cobalt deficiency also decreases the concentration of hemoglobin in blood and the level of glucose, total protein, urea, and cholesterol in serum (Hannam *et al.*, 1980). Disease resistance is reduced by a deficiency of Co (MacPherson *et al.*,

1987). Cobalt is also required for the detoxification of the neurotoxin compound in *Phalaris* which causes the disease "staggers" in grazing ruminants, although the manner in which Co acts remains unknown (Lee *et al.*, 1957; Lee, 1975). Cobalt has no effect on the digestion of fiber (Marston, 1952).

B. Effect on Production

Cobalt deficiency adversely affects all forms of production in those well-defined areas where the soil is deficient in Co (Stewart *et al.*, 1946). Only ruminants are susceptible to Co deficiency. When ewes in pens were fed Co-deficient forage, hemoglobin level was reduced by 70%, appetite by 45%, and body weight to a point where death would have occurred if supplementary Co had not been fed (Lines, 1935). Marston (1959) has stated that the "failure of appetite is the most prominent and certainly the most basic feature of the ovine syndrome of vitamin B_{12} deficiency."

Cobalt deficiency in temperate and tropical forage reduces the growth rate of sheep and cattle (Table 16.1 and 16.2). In some studies the forage was so low in Co that all the sheep died (Lines, 1935). Cobalt deficiency can reduce the wool production of lambs by 9% (Hannam *et al.*, 1980) and the milk production of ewes by up to 86% (Filmer and Underwood, 1937). With pregnant cows, Co deficiency has no adverse effect on the weight of cows or on the birth weight of calves but reduces the Co reserves of the calf and reduces their subsequent rate of growth (Skerman

TABLE 16.1

Effect of Cobalt Supplement on the Growth of Cattle Grazing Cobalt-Deficient Forage

Species	Forage cobalt (mg/kg)	Liveweight gain (kg/day)		Reference
		No cobalt	Cobalt	
Cynodon plectostachyus	—	0.40	0.51	Howard (1970)
Panicum maximum/ *Stylosanthes guianensis*	0.01–0.03	−0.06	0.42	Winter *et al.* (1977a)
Tropical Forage	—	0.14	−0.07	Chapman and Kidder (1964)
Temperate Forage	—	0.55	1.03	Duncan *et al.* (1986)
	—	0.71	0.85	Fearn (1961)
	—	0.54	0.64	Skerman *et al.* (1969)
	0.009–0.014	Died	0.38	Mannetje *et al.* (1976)
	0.01–0.07	0.08	0.24	Nicol and Smith (1981)

TABLE 16.2

Effect of Cobalt Supplement or Vitamin B_{12} Injection on the Growth of Sheep Grazing Cobalt-Deficient Forage[a]

Species	Liveweight gain (kg/day)		Reference
	No cobalt	Cobalt	
Digitaria decumbens	0.144	0.243	Norton and Hales (1976)
Temperate pasture	0.005	0.086	Andrews and Anderson (1955)
	0.052	0.085	Andrews (1965)
	Died	0.112	Andrews *et al.* (1966)
	−0.019	0.067	Filmer and Underwood (1937)
	0.036[b]	0.057	Hannam *et al.* (1980)
	Died	0.015	Marston *et al.* (1948a)
	0.114	0.131	McDonald (1942)
	0.048	0.070	Millar and Lorentz (1979)
	0.008	0.053	O'Moore and Smyth (1958)
	0.118	0.201	Poole *et al.* (1974a)
	0.047	0.057	Skerman (1959)
	0.084	0.194	Stewart *et al.* (1946)
	Died	0.113	Underwood and Filmer (1935)

[a]Cobalt concentration of the forage offered was published for three studies: Andrews (1965), 0.07–0.08 mg/kg; O'Moore and Smyth (1958) and Poole *et al.* (1974a), 0.04 mg/kg.
[b]50% died.

and O'Halloran, 1962). Field experience indicates that Co deficiency impairs breeding performance in both sheep and cattle (Hidiroglou, 1979b).

C. Requirement for Cobalt

The Co requirement of sheep is the quantity of Co required to achieve maximum growth and maintain normal levels of vitamin B_{12} in the liver (ARC, 1980). Andrews (1965) concluded that forage Co concentrations of 0.11 mg/kg DM or higher are probably adequate for sheep, a value endorsed by ARC (1980). When forage contains 0.08 to 0.10 mg Co/kg DM there are usually no clinical signs of Co deficiency (Table 16.3), but liver storage of vitamin B_{12} may be depressed (ARC, 1980). It seems very doubtful that any estimate of minimal level will have general applicability since lambs are more sensitive than sheep and rapidly growing lambs need more Co than those growing at a slower rate (Andrews *et al.*, 1958).

There is little experimental evidence on the Co requirement of cattle but it appears to be lower than that for sheep. Forage containing 0.04 mg Co/kg DM was considered borderline for cattle (Smith and Loosli, 1957),

TABLE 16.3

Relation between Cobalt Concentration in New Zealand Forage and Stock Health[a]

Forage cobalt (mg/kg DM)	Effect on livestock	
	Sheep	Cattle
0.01	Serious sickness	Serious sickness
0.04	Serious sickness	Some sickness
0.04–0.07	Some sickness	Fairly healthy
0.07–0.08[b]	Healthy	Healthy

[a]From Wunsch (1939).

[b]For lambs 0.11 mg Co/kg DM has been recommended (Andrews *et al.*, 1958).

but in other studies this level proved to be deficient rather than borderline (MacPherson *et al.*, 1973). In order to avoid the risk of subclinical Co deficiencies in cattle a level of Co in the diet of 0.06 or 0.07 mg/kg DM was recommended. In Peru, cattle grazing forage containing 0.04–0.07 mg Co/kg responded to Co supplementation, but supplementary Co had no effect when the Co concentration was >0.10 mg/kg (Beeson and Guillermo-Gomez, 1970).

II. COBALT IN FORAGE

A. Species Differences

The Co concentration of forages grown in different parts of the world varies from <0.01 to 1.26 mg/kg DM (Beeson, 1950). Most of this variation is associated with differences in soil Co (Section II,E), but Co concentration is also affected by forage species, stage of growth, and season.

1. Legume versus Grass

The Co concentration of forage legumes is generally higher than that of grasses (Table 16.4). In other studies *Medicago sativa* and *Trifolium* species contained 0.30–0.40 mg Co/kg DM, while grasses growing on the same soil contained an average of only 0.08 mg/kg DM (Latteur, 1962). In a survey of Co levels in forages grown in Virginia, legumes contained between 0.06 and 0.48 mg Co/kg compared with 0.02–0.24 mg/kg in the grasses (Price and Hardison, 1963).

Although legumes generally contain more Co than associated grasses this only occurs when the soil contains relatively high levels of Co. In

TABLE 16.4

Cobalt Concentration of Forages Grown on the Same Soil[a]

Species	Cobalt concentration (mg/kg DM)
Phleum pratense	0.08
Dactylis glomerata	0.11
Festuca pratensis	0.12
Lolium perenne × *L. multiflorum*	0.13
Lolium perenne	0.16
Trifolium pratense	0.22
Trifolium repens	0.24

[a]From Andrews (1966).

Florida legumes and grasses had similar levels of Co when grown on Co-deficient Everglade peaty muck soils: *Avena sativa*, 0.05 mg/kg DM; *Lolium perenne*, 0.07 mg/kg DM: *M. sativa*, 0.07 mg/kg DM; *Trifolium repens*, 0.06 mg/kg DM (Kretschmer *et al.*, 1954). *Trifolium repens* and grasses also had similar levels of Co when collected from fields in New Zealand on which Co deficiency had occurred in sheep (Andrews, 1966). Similar Co concentrations were found in *T. repens*, *Dactylis glomerata*, and *L. perenne* (Forbes and Gelman, 1981) and in tropical legumes and grasses (Bryan *et al.*, 1960). *Stylosanthes guianensis* sometimes has a higher Co concentration than *Panicum maximum* (Mannetje *et al.*, 1976), but in other studies the legume had a lower Co concentration than the grasses (Winter *et al.*, 1977a).

2. Variation between Legumes

When grown on soil with relatively high levels of Co only small differences in Co concentration have been found between *M. sativa*, *Onobrychis viciifolia*, *Trifolium pratense*, and *T. repens* (Whitehead and Jones, 1969) and between *T. pratense*, *T. repens*, and *M. sativa* (Andrews, 1966).

3. Variation between Grasses

When grown and managed in the same way *L. perenne* generally contains more Co than *Phleum pratense* (Table 16.4). In another comparison, grasses were ranked in the following order: *Axonopus compressus* and *Poa pratensis* contained 0.13 mg Co/kg DM; *Agropyron repens* and *Bromus inermis*, 0.09 mg Co/kg DM; *Agrostis gigantea*, *Dactylis glomerata*, *Paspalum notatum*, and *Sorghum halepense*, 0.08 mg Co/kg DM; and

Brachiaria mutica, Cynodon dactylon, and *Paspalum dilatatum*, 0.07 mg Co/kg DM (Beeson *et al.*, 1947).

B. Genotype Differences

Very little appears to have been published on the genetic differences in Co concentration of forage species. In a study with cultivars of *D. glomerata, L. perenne*, and *T. repens* grown on Co-deficient soil, there was no significant difference in Co concentration between cultivars of the different species nor between forage species (Forbes and Gelman, 1981).

C. Plant Parts

Cobalt is concentrated in the margins of the leaflets of *M. sativa* and *Trifolium subterraneum* and at the base and extreme tips of the leaf blades of *Bromus catharticus* and *Phalaris aquatica* (Handreck and Riceman, 1969). The leaf blades of *D. glomerata* contain more Co than the leaf sheath and stem: 15, 7, and 10 mg/kg DM, respectively (Davey and Mitchell, 1968). However, when the concentration of Co is low there is no consistent difference in Co concentration of leaf and stem of *Lotus corniculatus, M. sativa, T. repens* (Beeson and MacDonald, 1951), *D. glomerata, Festuca pratensis, L. perenne*, or *P. pratense* (Fleming, 1963).

D. Stage of Growth

Increasing maturity of the forage has no consistent effect on Co concentration. Falls in Co concentration have been recorded for *L. perenne* (Fleming and Murphy, 1968), *O. viciifolia*, and *T. pratense* (Whitehead and Jones, 1969), but rises in concentration have occurred in *L. corniculatus, M. sativa, P. pratense*, and *T. repens* (Beeson and MacDonald, 1951). No consistent changes in Co concentration were found in studies with *M. sativa, T. repens* (Whitehead and Jones, 1969), and *T. subterraneum* (Rossiter *et al.*, 1948).

E. Soil Fertility

Cobalt deficiency in ruminants is associated with soils low in available Co. In general, unfertilized sands and fine sands have a much lower Co content and are more often Co deficient than are sandy loams, but there are many exceptions and the reader is referred to the reviews by Beeson (1950), Norrish (1975), and Reid and Horvath (1980).

TABLE 16.5

Relation between Soil and Forage Cobalt Concentration in New Zealand[a]

Cobalt in soil (mg/kg)		Cobalt in forage (mg/kg)	
Range	Mean	Mean	Range
0.0–2.0	1.2	0.04	0.03–0.07
2.1–5.0	3.4	0.06	0.03–0.12
5.1–20.0	9.2	0.08	0.03–0.21

[a]From Stanton and Kidson (1939).

In New Zealand, soils have been ranked into three classes with increasing level of Co (Table 16.5). The Co concentration in forage grown on these soils was positively correlated with the Co concentration of the soil but there were large differences within each group; some species grown on soils high in Co had low levels of Co which were similar to levels in forages grown on low-Co soils. This variation is possibly caused by differences in soil manganese and level of soil moisture (Norrish, 1975).

The uptake of Co by plants is dependent on the manganese (Mn) concentration in the soil (Adams *et al.*, 1969; Norrish, 1975). High levels of soil Mn depress the uptake of Co by forages, and when the level of Mn exceeds 1 g/kg the fixation of Co in the soil is so high that fertilizer Co has little effect on the Co concentration of forages.

Under waterlogged conditions the soil minerals which contain Co become unstable, releasing trapped cobalt into the soil solution. Forage grown on poorly drained soil contained seven times as much Co as forage grown on well-drained soil (Table 16.6), a result confirmed in controlled pot studies (Adams and Honeysett, 1964). In another study, high soil moisture increased the concentration of Co in the soil solution and doubled the uptake of Co by *Trifolium hybridum* (Kubota *et al.*, 1963). The Co concentration in forage is increased by high soil mositure provided there is no change in yield; if yield is increased by the additional moisture there may be no change in Co concentration (Table 16.7). Except where soil Mn is high, the available Co in soils can be increased by applying cobalt sulfate (Table 16.8).

Fertilizer nitrogen, phosphorus, and potassium appear to have no effect on the Co concentration of *A. sativa* (Wright and Lawton, 1954), *D. glomerata*, or *Festuca arundinacea* (Reid *et al.*, 1967a). Applying dolomi-

TABLE 16.6

Cobalt Concentration in Forage Grown on Well-Drained and Poorly Drained Portions of Adjoining Fields[a]

Species	Cobalt concentration (mg/kg)	
	Well drained	Poorly drained
Dactylis glomerata	0.13	0.73
Lolium perenne	0.14	1.07
Trifolium pratense	0.16	1.30
Mixed pasture	0.15	1.13
Mean	0.14	1.08

[a]From Mitchell *et al.* (1957a).

tic limestone depresses the Co concentration in *A. sativa* when the Co concentration is high but has little or no effect when Co levels are <0.10 mg/kg (Wright and Lawton, 1954).

F. Seasonal Variation

In continuously grazed temperate pasture, there is a tendency for Co concentration to increase in late autumn and winter, when growth is re-

TABLE 16.7

Effect of Soil Moisture on the Cobalt Uptake and Concentration in *Trifolium hybridum*[a]

Soil	Treatment	Total yield (g)	Cobalt		
			Soil solution (Relative)	Total uptake (Relative)	Concentration (mg/kg DM)
Gloucester fine sandy loam	Dry	36	0.06	2.0	0.06
	Wet	47	0.34	4.8	0.10
Gloucester fine sandy loam[b]	Dry	55	0.15	37.9	0.74
	Wet	58	1.17	64.7	1.17
Moltsville fine gravelly loamy sand	Dry	68	0.07	7.8	0.12
	Wet	110	0.13	14.0	0.12
Ophir fine gravelly loam	Dry	46	0.08	6.9	0.16
	Wet	101	0.24	15.9	0.16

[a]From Kubota *et al.* (1963).
[b]Received cobalt fertilizer.

TABLE 16.8

Cobalt Concentration in Forages Grown on an Old Red Sandstone Soil with and without Cobalt Sulfate[a]

Species	Cobalt concentration (mg/kg)	
	Untreated	Cobalt fertilized[b]
Phleum pratense	0.03	0.32
Mixed forage	0.06	0.63
Dactylis glomerata	0.06	0.64
Lolium perenne	0.06	0.64
Cynosurus cristatus	0.06	0.66
Trifolium pratense	0.05	0.98
Mean	0.05	0.64

[a]From Mitchell *et al.* (1957a)
[b]1.7 kg/ha cobalt sulfate.

tarded, and to be low in spring and summer, when growth is at its maximum (McNaught and Paul, 1939). However, in other studies there was no seasonal variation in Co concentration (Andrews, 1966; Fleming, 1970).

III. DIAGNOSIS OF COBALT DEFICIENCY

A. Clinical Signs

A striking feature of Co deficiency in ruminants is the loss of appetite, causing visual symptoms similar to those of starvation. There is a rapid loss of weight, dry coat, tight skin, muscular atrophy, severe anemia, and finally death (Marston, 1952). A watery ocular discharge also occurs (Robertson, 1971). Young growing sheep are most sensitive to Co deficiency, followed by mature sheep, calves between 6 and 18 months old, and finally mature cattle (Andrews, 1956). Subclinical forms of Co deficiency provide no visual signs that may be used for diagnosis other than a nonspecific state of unthriftiness. In areas which are only mildly to moderately deficient in Co only some lambs in a flock become unthrifty, and then not until they are 3–6 months of age. The number of deaths and degree of unthriftiness vary between years (Lee, 1951; Andrews, 1956).

Cobalt deficiency is often confused with other causes of unthriftiness such as worm infestation, chronic starvation, and ill-thrift of unknown causes. Analysis of soil, forage, or animal tissues may be a useful guide to Co deficiency but conclusive diagnosis can only be achieved by measuring the response to supplementary Co administered by mouth or by injection of vitamin B_{12}.

B. Soil Analysis

Various chemical extraction methods have been used to determine the Co available to plants in contrast to total Co in the soil. When acetic acid (2.5%) is used as the extractant, the critical concentration of Co in the soil below which deficiency disease may be experienced is about 0.25 mg/kg (Gardiner, 1977). Most of the Co in soil is immobilized by Mn minerals, so total soil Co is of little use in diagnosing areas of potential Co deficiency (Gardiner, 1977).

C. Forage Analysis

Absorption and utilization of forage Co by ruminants is not seriously affected by other elements, so analysis of samples of forage for Co can be a useful diagnostic technique provided the samples represent the diet of the animal over a period of several months (Clark and Millar, 1983). Special care is required to prevent the Co concentration of the forage from being inflated by inclusion of soil or Co from equipment used to cut or grind the samples. The results of the analysis can be compared with the recommended levels shown in Table 16.3. This will provide a guide to a possible Co deficiency and whether a response to Co supplementation might be obtained.

D. Tissue Analysis

1. Liver

The liver is the main storage site of vitamin B_{12} and analysis of liver samples provides a measure of B_{12} reserves and the likelihood of responsiveness to feeding a Co supplement. When the B_{12} level in fresh samples of liver exceeds 220 nmol/kg, sheep and cattle are unlikely to respond to Co supplements, but below this level ruminants are deficient or marginally deficient (Table 16.9).

The level of Co in liver is low in deficient animals (Underwood and Harvey, 1938; Marston *et al.*, 1948b). A concentration of Co in liver below 0.04–0.06 mg/kg DM has been suggsted as indicating Co deficiency in sheep and cattle (McNaught, 1948). Liver analysis for Co is still used to diagnose Co deficiency (Rentsch, 1980; McDowell *et al.*, 1983).

TABLE 16.9

Diagnostic Levels of Vitamin B_{12} in Liver and Serum[a]

Tissue	Species	Cobalt status: Deficient	Marginal	Adequate
Liver vitamin B_{12} (nmol/kg fresh)	Sheep	<110	110–220	>220
	Cattle	<75	75–220	>220
Serum vitamin B_{12} (pmol/liter)	Sheep	<185	185–370	>370
	Cattle	ND[b]	ND	ND

[a]From Clark and Millar (1983).
[b]ND, not yet determined.

2. Serum

Serum vitamin B_{12} concentration is controlled by both the synthesis of vitamin B_{12} in the rumen and the release of vitamin B_{12} from the liver. The small quantities of vitamin B_{12} present in serum were measured by microbiological methods but are now determined by radioassay techniques (Millar and Penrose, 1980). The relation between the level of serum vitamin B_{12} and response to supplementary Co is shown in Table 16.9.

3. Urinary Methylmalonic Acid

Propionic acid is metabolized via methylmalonic acid (MMA) to succinic acid and in healthy animals only small quantities of MMA are excreted in the urine (Millar and Lorentz, 1979). When ruminants are deficient in vitamin B_{12} the conversion to succinic acid is interrupted and there is a large (5- to 12-fold) increase in the concentration of MMA in the blood and excretion of MMA in the urine (Gawthorne, 1968; Hogan *et al.*, 1973; Millar and Lorentz, 1979). It has been suggested that urinary MMA concentrations greater than 30 mg/liter urine in 10 randomly selected sheep indicate Co deficiency in a flock (Millar and Lorentz, 1979). The rise in MMA concentration only occurs when there is a severe vitamin B_{12} deficiency and forage intake has been depressed by 75% (Gawthorne, 1968). The method is not considered to be as useful in the diagnosis of Co deficiency as serum or liver analysis for vitamin B_{12} (Clark and Millar, 1983). Other aspects of the use of metabolic indices for diagnosing Co deficiency are considered in the recent review by Mills (1987).

4. Milk

The vitamin B_{12} concentration of milk is related to the Co level in the diet (Smith and Loosli, 1957). Supplementing animals grazing marginally Co-deficient pasture with Co increased the vitamin B_{12} concentration in ewes' milk by a factor of four (Hart and Andrews, 1959) and the milk of cows by up to 71% (Skerman and O'Halloran, 1962). However, no diagnostic levels for vitamin B_{12} in milk appear to have been published.

5. Feces

Feces samples are easily collected by no attempt has been made to use Co or vitamin B_{12} concentration in the feces for diagnostic purposes. It has been shown that supplementary Co increases the level of fecal excretion of vitamin B_{12} (Jones and Anthony, 1965), so fecal analysis appears to warrant further study as a diagnostic method.

IV. PREVENTION OF COBALT DEFICIENCY

A. Supplementary Cobalt

1. Oral Administration

Oral dosing of 35 and 175 mg of cobalt sulfate (7 and 35 mg Co) at weekly intervals has been recommended for lambs and cattle, respectively, grazing Co-deficient forage (Clark and Millar, 1983). Drenching at weekly intervals is not a practical treatment for whole flocks or herds but can be a valuable technique in the diagnosis of Co deficiency. Dosing of lambs at monthly intervals reduces mortality but the growth rate is less than with weekly dosing (Andrews *et al.*, 1966).

The introduction of the Co "bullet" has overcome the need for weekly dosing. Cobalt bullets have a high specific gravity and are retained in the rumen for 6 months or more (Dewey *et al.*, 1958). The release of Co from the bullet is often reduced by a coating of calcium phosphate which tends to form within the rumen (Poole and Connolly, 1967), but this can be prevented by including a large grub screw as a grinder (Dewey *et al.*, 1969). Unfortunately, the grinder can cause damage to machinery at slaughterhouses and some countries have banned the use of Co bullets. Slow-release bullets are now made for use with both sheep and cattle.

The Co needs of cattle in South Florida are met by including 0.03% Co in a salt mixture that is consumed by cattle at a rate of about 17 kg/yr (Chapman and Kidder, 1964). Another method of providing Co is through the water supply, using a metering device which continuously adds Co (MacPherson, 1981).

2. Vitamin B_{12} Injection

Injection of vitamin B_{12} is commonly used in the diagnosis of Co deficiency but it also has a role in areas of marginal Co deficiency where lambs may respond to a boost in vitamin B_{12} (Clark and Millar, 1983). Single injections of 1 mg of vitamin B_{12} meet the requirements of lambs for 14 weeks (Hannam *et al.*, 1980), while 2 mg of vitamin B_{12} is effective in calves for 7–9 months (Judson *et al.*, 1981). However, the period of protection will depend on the Co level in the diet.

B. Fertilizer Cobalt

New Zealand scientists pioneered the use of fertilizer containing added Co for preventing Co deficiency. Two years after the discovery of the nutritional importance of Co they showed that the Co concentration of forage could be increased by applying cobalt chloride (Askew and Dixon, 1937), and 2 yr later they reported that when applied at 0.28, 0.56, and 2.20 kg/ha the Co concentration in temperate forage was increased from 0.04 to 0.08, 0.16, and 0.37 mg/kg, respectively (cited by Stewart *et al.*, 1941). Similar benefits from the use of very small quantities of Co compounds have been found in other countries, including Scotland (Mitchell *et al.*, 1941; Stewart *et al.*, 1941, 1942, 1946), Australia (Rossiter *et al.*, 1948), and the U.S. (Kretschmer *et al.*, 1954). It has been suggested that for moderately acid soils containing less than about 100 mg/kg total Mn, 1.1 kg/ha/yr of cobalt sulfate should result in adequate Co in the forage, but where total Mn exceeds 1 g/kg of soil, the problem of low forage Co can not be solved by applying fertilizer (Anonymous, 1969). Where soils are high in Mn, Co may be supplied as a foliar spray at 0.14 kg/ha/yr of cobalt sulfate or by direct supplementation of the stock.

The duration of the response to fertilizer Co appears to depend on the soil type (Adams *et al.*, 1969) and management. The benefits of Co fertilizer last a shorter time on soils that have a high Mn content (Norrish, 1975; McLaren *et al.*, 1979) and those that have been limed (Mitchell, 1945).

Only small quantities of fertilizer Co are required and carriers are used to improve distribution. Carriers include various types of superphosphate, serpentine, beach sand, and pumice sand (Maunsell and Simpson, 1944). In New Zealand, cobaltized superphosphate manufactured in the North Island contains 0.13% cobalt sulfate and the annual application of 250 kg/ha provides the required quantity of Co (Clark and Millar, 1983).

For unproductive pastures, fertilizer Co is not an economic method of overcoming Co deficiency (Lee, 1950, 1975).

C. Grazing Management

The concentration of Co in soil is higher than in forage (Stewart *et al.*, 1946), a difference that is very large where soils are high in Mn (Adams *et al.*, 1969). Thus, soil ingested by grazing animals can be a source of supplementary Co (Healy *et al.*, 1970; MacPherson *et al.*, 1978). When lambs graze either short or long temperate pastures more soil is eaten with the short pasture and this leads to a higher concentration of Co in the liver (Andrews *et al.*, 1958) and a higher growth rate (McQueen, 1984).

V. CONCLUSION

Cobalt is an element required in trace quantity by ruminants for the synthesis of vitamin B_{12}. A deficiency of cobalt leads to a loss of appetite, low growth rates, and death. Concentration of Co in forage depends on the level of available Co in the soil, with low concentrations in forage grown on some unfertilized sands. High levels of soil Mn depress Co uptake by plants, but waterlogged conditions increase the concentration of Co in forage. Legumes contain more Co than grasses when grown on soils with adequate Co.

Clinical signs of Co deficiency are similar to those of starvation. Analysis of soil for available Co and forage for Co can be a useful guide in diagnosing Co deficiency. Positive identification can be achieved by analysing serum for vitamin B_{12} or methylmalonic acid. The use of vitamin B_{12} concentrations in feces and milk for diagnosing Co deficiency should be examined.

Cobalt deficiency can be prevented by applying Co fertilizer provided the soil does not contain excess Mn. Supplementary Co can be given as a drench or a slow-release bullet in the rumen. Injections of vitamin B_{12} and systems of grazing management that ensure that animals eat forage contaminated with soil have also been used to prevent Co or vitamin B_{12} deficiency in ruminants. Cobalt may also be provided in a salt–mineral mixture.

Glossary of Forage Names

Acacia aneura	Mulga
Agropyron cristatum	Crested wheatgrass
Agropyron desertorum	See *Agropyron cristatum*
Agropyron gerardi	No common name
Agropyron intermedium	See *Elymus hispidus*
Agropyron repens	Quack grass; couch grass
Agrostis gigantea	Red top
Agrostis stolonifera	Creeping bent grass
Agrostis tenuis	Fine bent grass; browntop; common bent; New Zealand bent
Amphilophis insculpta	See *Bothriochloa insculpta*
Andropogon caricosus	Cuban grass; Nadi bluegrass; India bluegrass
Andropogon elongatum	Tall wheatgrass
Andropogon gayanus	Gamba grass
Anthoxanthum odoratum	Sweet vernal grass
Arachis glabrata	Perennial peanut
Arctotheca calendula	Capeweed
Arrhenatherum elatius	Tall oat grass
Artemisia nova	Black sagebush
Artemisia tridentata	Big sagebush
Astrebla spp.	Mitchell grass
Avena sativa	Oats
Axonopus compressus	Broad-leaved carpet grass; mat grass
Bothriochloa insculpta	Pitted bluegrass
Bothriochloa intermedia	Forest bluegrass
Bothriochloa spp.	Bluegrass
Bouteloua gracilis	Blue grama
Brachiaria brizantha	Signal grass
Brachiaria decumbens	Signal grass; Kenya sheep grass; Surinam grass

Brachiaria mutica	Para grass; giant couch; California grass
Brachiaria ruziziensis	Congo grass
Brassica napus	Rape
Brassica oleracea var. *acephala*	Marrow stem kale; thousand-headed kale
Bromus catharticus	See *Bromus unioloides*
Bromus inermis	Brome grass; smooth brome grass
Bromus tectorum	Dropping brome; downy brome
Bromus unioloides	Prairie grass
Calluna vulgaris	Ling; heather
Cenchrus ciliaris	Buffel grass; African foxtail; blue buffalo grass
Cenchrus setiger	Birdwood grass; Anjon
Cenchrus setigerus	See *Cenchrus setiger*
Centrosema pubescens	Centro; butterfly pea
Chloris barbata	Purpletop chloris
Chloris gayana	Rhodes grass
Chrysopogon fallax	Golden-beard grass
Cyamopsis tetragonoloba	Cluster bean; guar
Cymbopogon plurinodis	No common name
Cynodon aethiopicus	Giant star grass
Cynodon dactylon	Coastal Bermuda grass; Bermuda grass; couch grass; Suwanee Bermuda grass; star grass
Cynodon plectostachyus	Giant star grass; Naivasha star grass; star grass
Cynosurus cristatus	Crested dogstail
Dactylis glomerata	Cocksfoot; orchard grass
Danthonia spp.	Wallaby grass
Desmodium intortum	Green leaf desmodium
Desmodium uncinatum	Silver leaf desmodium
Digitaria decumbens	Pangola grass
Digitaria milanjiana	Woolly finger grass
Digitaria pentzii	Woolly finger grass
Digitaria polevansii	No common name
Digitaria setivalva	Woolly finger grass
Digitaria smutsii	Tall finger grass
Elymus hispidus	Intermediate wheatgrass
Elymus junceus	Russian wild ryegrass
Eragrostis superba	Love grass
Eriochloa polystachya	Malojilla

Festuca arundinacea	Altar fescue; tall fescue
Festuca elatior	Tall fescue
Festuca pratensis	Meadow fescue
Festuca rubra	Red fescue
Glycine javanica	See *Neonotonia wightii*
Glycine max	Soybean
Helianthus annuus	Sunflower
Hemarthria altissima	Limpo grass
Heteropogon contortus	Black speargrass; Pili or tanglehead grass
Holcus lanatus	Yorkshire fog; fog
Holcus mollis	Creeping soft grass
Hordeum leporinum	Barley grass
Hordeum murinum	Wall barley grass
Hordeum satium	See *Hordeum vulgare*
Hordeum vulgare	Barley
Hyparrhenia dissoluta	Yellow thatching grass
Hyparrhenia hirta	Coolatai grass
Hyparrhenia rufa	Jaragua
Imperata cylindrica	Cotton grass; Alang-alang; blady grass
Indigofera spicata	Creeping indigo
Ischaemum aristatum	Tocogras
Ischaemum timorense	Loekoentoegras
Iseilema spp.	Flinders grass
Lablab purpureus	Lablab; hyacinth bean; Egyptian bean
Lespedeza cuneata	See *Lespedeza juncea*
Lespedeza juncea	Sericea; Sericea lespedeza
Leucaena leucocephala	Leucaena
Lolium hybridum	See *Lolium perenne* × *L. multiflorum*
Lolium multiflorum	Italian ryegrass
Lolium perenne × *L. multiflorum*	Short rotation ryegrass; hybrid ryegrass
Lolium perenne	Perennial ryegrass
Lolium rigidum	Wimmera ryegrass
Lotononis bainesii	Lotononis
Lotus corniculatus	Birdsfoot trefoil
Lotus pedunculatus	Greater lotus
Lotus spp.	Trefoil
Lupinus spp.	Lupin
Macroptilium atropurpureum	Siratro
Macroptilium lathyroides	Phasey bean

Medicago denticulata	See *Medicago polymorpha*
Medicago falcata	Yellow flower lucerne
Medicago lupulina	Trefoil; black medic; hop clover
Medicago media	Variegated lucerne
Medicago polymorpha	Burr medic
Medicago sativa	Alfalfa; lucerne
Medicago tribuloides	See *Medicago truncatula*
Medicago truncatula	Barrel medic
Melilotus officinalis	Yellow sweet clover
Melinis minutiflora	Molasses grass
Neonotonia wightii	Glycine; Soja perene
Onobrychis viciifolia	Sainfoin
Panicum antidotale	Blue panic; giant panic grass
Panicum coloratum	Colored Guinea; small buffalo grass
Panicum maximum	Guinea grass; sempre verde grass; Tanganyika grass
Panicum virgatum	Switchgrass
Paspalum dilatatum	Dallas grass
Paspalum notatum	Bahia grass
Paspalum plicatulum	Plicatulum
Pennisetum americanum	See *Pennisetum glaucum*
Pennisetum ciliare	See *Cenchrus ciliaris*
Pennisetum clandestinum	Kikuyu grass
Pennisetum glaucum	Pearl millet
Pennisetum purpureum	Elephant grass; napier grass; Uganda grass; Merker grass
Pennisetum purpureum var. *merkeri*	See *Pennisetum purpureum*
Pennisetum purpureum × *P. typhoides*	Pusa giant napier grass; Bana grass
Pennisetum typhoides	See *Pennisetum glaucum*
Phalaris aquatica	Phalaris
Phalaris arundinacea	Reed canary grass
Phalaris tuberosa	See *Phalaris aquatica*
Phaseolus lathyroides	See *Macroptilium lathyroides*
Phleum pratense	Timothy
Pisum sativum	Field pea
Poa annua	Annual poa
Poa compressa	Canadian blugrass
Poa pratensis	Bluegrass, Kentucky bluegrass; smooth-stalked meadow grass
Poa trivialis	Rough-stalked meadow grass

Rhynchelytrum repens	Natal grass
Rhynchelytrum roseum	See *Rhynchelytrum repens*
Saccharum officinarum	Sugar cane
Secale cereale	Rye
Setaria italica	Foxtail millet; Boer millet
Setaria sphacelata	Setaria
Setaria sphacelata var. *aurea*	Setaria
Setaria sphacelata var. *splendida*	Setaria
Setaria splendida	See *Setaria sphacelata* var. *splendida*
Setaria trinervia	See *Setaria sphacelata* var. *aurea*
Sorghastrum nutans	Indiangrass
Sorghum almum	Perennial sorghum
Sorghum bicolor	Sorghum
Sorghum bicolor var. *sudanensis*	Sudan grass
Sorghum halepense	Johnson grass
Sorghum plumosum	Plume sorghum
Sorghum vulgare	See *Sorghum bicolor*
Stenotaphrum secundatum	Saint Augustine grass; buffalo grass
Stylosanthes bojeri	See *Stylosanthes guianensis*
Stylosanthes fruticosa	African stylo
Stylosanthes gracilis	See *Stylosanthes guianensis*
Stylosanthes guianensis	Schofield stylo; stylo; Brazilian lucerne; tropical lucerne
Stylosanthes hamata	Caribbean stylo
Stylosanthes humilis	Townsville stylo; annual stylo
Stylosanthes scabra	Shrubby stylo
Stylosanthes subsericea	No common name
Stylosanthes viscosa	Sticky stylo
Themeda australis	See *Themeda triandra*
Themeda triandra	Kangaroo grass; red oat grass
Trifolium alexandrinum	Berseem clover; Egyptian clover
Trifolium fragiferum	Strawberry clover
Trifolium hirtum	Rose clover
Trifolium hybridum	Alsike clover
Trifolium pratense	Red clover
Trifolium repens	White clover; Ladino clover
Trifolium resupinatum	Persian clover
Trifolium subterraneum	Subterranean clover
Tripsacum laxum	Guatemala grass; Honduras grass
Triticum aestivum	Wheat

Triticum vulgare	See *Triticum aestivum*
Urochloa mosambicensis	Sabi grass
Urochloa pullulans	Witgrass
Vicia villosa	Wollypod vetch
Vigna sinensis	See *Vigna unguiculata*
Vigna unguiculata	Cowpea
Vulpia spp.	Silver grass
Zea mays	Maize; Indian corn
Zea mays var. *saccharata*	Sweet maize

References

Abdalla, H. O., Fox, D. G., and Seaney, R. R. (1988). *J. Anim. Sci.* **66,** 2325–2329.

Abrams, S. M., Hartadi, H., Chaves, C. M., Moore, J. E., and Ocumpaugh, W. R. (1983). *Proc. Int. Grassl. Congr., 14th* pp. 508–511.

Abrams, S. M., Harpster, H. M., Wangsness, P. J., Shenk, J. S., Keck, E., and Rosenberger, J. L. (1987). *J. Dairy Sci.* **70,** 1235–1240.

Abruna, F., Vincente-Chandler, J., and Pearson, R. W. (1964). *Soil Sci. Soc. Am. Proc.* **28,** 657–661.

Adams, R. S. (1975). *J. Dairy Sci.* **58,** 1538–1548.

Adams, D. C. (1985). *J. Anim. Sci.* **61,** 1037–1042.

Adams, A. F. R., and Elphick, B. L. (1956). *N.Z. J. Sci. Technol., Sect. A* **38,** 346–358.

Adams, S. N., and Honeysett, J. L. (1964). *Aust. J. Agric. Res.* **15,** 357–367.

Adams, C. A., and Sheard, R. W. (1966). *Can. J. Plant Sci.* **46,** 671–680.

Adams, S. N., Honeysett, J. L., Tiller, K. G., and Norrish, K. (1969). *Aust. J. Soil Res.* **7,** 29–42.

Adamson, A. H., and Terry, G. R. (1980). *J. Sci. Food Agric.* **31,** 854–856.

Adegbola, A. A., and Paladines, O. (1977). *J. Sci. Food Agric.* **28,** 775–786.

Aerts, J. V., De Brabander, D. L., Cottyn, B. G., and Buysse, F. X. (1977). *Anim. Feed Sci. Technol.* **2,** 337–349.

Aii, T., and Stobbs, T. H. (1980). *Anim. Feed Sci. Technol.* **5,** 183–192.

Aines, P. D., and Smith, S. E. (1957). *J. Dairy Sci.* **40,** 682–688.

Aitchison, E. M., Gill, M., Dhanoa, M. S., and Osbourn, D. F. (1986). *Br. J. Nutr.* **56,** 463–476.

Akin, D. E. (1982). *In* "Nutritional Limits to Animal Production from Pastures" (J. B. Hacker, ed.), pp. 201–223. Commonw. Agric. Bur., Farnham Royal, England.

Akin, D. E., and Hogan, J. P. (1983). *Crop Sci.* **23,** 851–858.

Akin, D. E., Hanna, W. W., and Rigsby, L. L. (1986). *Agron. J.* **78,** 827–832.

Alawa, J. P., and Owen, E. (1984). *Anim. Feed Sci. Technol.* **11,** 149–157.

Alberda, T. (1965). *Neth. J. Agric. Sci.* **13,** 335–360.

Alder, F. E. (1968). *J. Br. Grassl. Soc.* **23,** 310–316.

Alder, F. E., and Cooper, E. M. (1967). *J. Agric. Sci.* **68,** 331–346.

Alder, F. E., and Minson, D. J. (1963). *J. Agric. Sci.* **60,** 359–369.

Alderman, G., and Jones, D. I. H. (1967). *J. Sci. Food Agric.* **18,** 197–199.

Alderman, G., and Stranks, M. H. (1967). *J. Sci. Food Agric.* **18,** 151–152.

Alexander, G. I. (1972). *World Anim. Rev.* **4,** 11–14.

Alexander, R. H., and McGowan, M. (1966). *J. Br. Grassl. Soc.* **21,** 140–147.

Alexander, G. I., Harvey, J. M., Lee, J. H., and Stubbs, W. C. (1967). *Aust. J. Agric. Res.* **18,** 169–181.

Alexander, G. I., Daly, J. J., and Burns, M. A. (1970). *Proc. Int. Grassl. Congr., 11th* pp. 793–796.

Alfaro, E., Neathery, M. W., Miller, W. J., Gentry, R. P., Crowe, C. T., Fielding, A. S., Etheridge, R. E., Pugh, D. G., and Blackmon, D. M. (1987). *J. Dairy Sci.* **70,** 831–836.
Allan, P. J., Litschner, J. C., and Swain, A. J. (1972). *Proc. Aust. Soc. Anim. Prod.* **9,** 271–275.
Allaway, W. H., Moore, D. P., Oldfield, J. E., and Muth, O. H. (1966). *J. Nutr.* **88,** 411–418.
Allcroft, R. (1953). *Proc. Int. Vet. Congr.* **1,** 573–578.
Allcroft, W. M., and Green, H. H. (1938). *J. Comp. Pathol.* **51,** 176–191.
Allcroft, R., and Uvarov, O. (1959). *Vet. Rec.* **71,** 797–810.
Allden, W. G. (1962). *Proc. Aust. Soc. Anim. Prod.* **4,** 163–166.
Allden, W. G. (1969). *Aust. J. Agric. Res.* **20,** 499–512.
Allden, W. G. (1970). *Nutr. Abstr. Rev.* **40,** 1167–1184.
Allden, W. G., and Whittaker, I. A. (1970). *Aust. J. Agric. Res.* **21,** 755–766.
Allden, W. G., and Young, R. S. (1964). *Aust. J. Agric. Res.* **15,** 989–1000.
Allen, W. M., and Davies, D. C. (1981). *Br. Vet. J.* **137,** 436–441.
Allen, W. M., Moore, P. R., and Sansom, B. F. (1981). *In* "Trace Element Metabolism in Man and Animals" (J. M. Howell, J. M. Gawthorne, and C. L. White, eds.), pp. 195–198. Aust. Acad. Sci. Canberra, Australia.
Allen, W. M., Drake, C. F., and Tripp, M. (1985). *In* "Trace Elements in Man and Animals" (C. F. Mills, I. Bremner, and J. K. Chesters, eds.), pp. 719–722. Commonw. Agric. Bur., Farnham Royal, England.
Allinson, D. W., and Osbourn, D. F. (1970). *J. Agric. Sci.* **74,** 23–36.
Allinson, D. W., Elliott, F. C., and Tesar, M. B. (1969). *Crop Sci.* **9,** 634–637.
Allison, M. J. (1970). *In* "Physiology of Digestion and Metabolism in the Ruminant" (A. T. Phillipson, ed.), pp. 456–473. Oriel, Newcastle-upon-Tyne, England.
Allison, M. J. (1983). *J. Sci. Food Agric.* **34,** 175–180.
Allison, M. J., Bryant, M. P., Katz, I., and Keeney, M. (1962). *J. Bacteriol.* **83,** 1084–1093.
Alloway, B. J., and Tills, A. R. (1984). *Outlook Agric.* **13,** 32–42.
Alwash, A. H., and Thomas, P. C. (1971). *J. Sci. Food Agric.* **22,** 611–615.
Alwash, A. H., and Thomas, P. C. (1974). *J. Sci. Food Agric.* **25,** 139–147.
Ammerman, C. B. (1970). *J. Dairy Sci.* **53,** 1097–1107.
Ammerman, C. B., and Miller, S. M. (1972). *J. Anim. Sci.* **35,** 681–694.
Ammerman, C. B., and Miller, S. M. (1975). *J. Dairy Sci.* **58,** 1561–1577.
Ammerman, C. B., Forbes, R. M., Garrigus, U. S., Newmann, A. L., Norton, H. W., and Hatfield, E. E. (1957). *J. Anim. Sci.* **16,** 796–810.
Ammerman, C. B., Martin, F. G., and Arrington, L. R. (1970). *J. Dairy Sci.* **53,** 1514–1515.
Ammerman, C. B., Chicco, C. F., Moore, J. E., Van Walleghem, P. A., and Arrington, L. R. (1971). *J. Dairy Sci.* **54,** 1288–1293.
Ammerman, C. G., Chicco, C. F., Loggins, P. E., and Arrington, L. R. (1972). *J. Anim. Sci.* **34,** 122–126.
Amos, H. E., Evans, J., and Burdick, D. (1976). *J. Anim. Sci.* **42,** 276.
Amos, M. E., Windham, W. R., and Evans, J. J. (1984). *J. Anim. Sci.* **58,** 987–995.
Anderson, R. (1982). *Grass Forage Sci.* **37,** 139–145.
Anderson, J. P., and Andrews, E. D. (1952). *Nature (London)* **170,** 807.
Anderson, D. C., and Ralston, A. T. (1973). *J. Anim. Sci.* **37,** 148–152.
Anderson, D. M., Christensen, D. A., and Christison, G. I. (1977). *Can. J. Anim. Sci.* **57,** 609–611.
Anderson, P. H., Berrett, S., and Patterson, D. S. P. (1979). *Vet. Rec.* **104,** 235–238.
Anderson, S. J., Klopfenstein, T. J., and Wilkerson, V. A. (1988). *J. Anim. Sci.* **66,** 237–242.

Ando, T., Nishihara, N., and Ogawa, S. (1972). *Nippon Sochi Gakkaishi* **18,** 226–235.
Ando, T., Masaoka, Y., and Matsumoto, K. (1985). *Soil Sci. Plant Nutr. (Tokyo)* **31,** 601–610.
Andrew, C. S., and Norris, D. O. (1961). *Aust. J. Agric. Res.* **12,** 40–55.
Andrew, C. S., and Robins, M. F. (1969a). *Aust. J. Agric. Res.* **20,** 665–674.
Andrew, C. S., and Robins, M. F. (1969b). *Aust. J. Agric. Res.* **20,** 1009–1021.
Andrew, C. S., and Thorne, P. M. (1962). *Aust. J. Agric. Res.* **13,** 821–835.
Andrewartha, K. A., and Caple, I. W. (1980). *Res. Vet. Sci.* **28,** 101–104.
Andrews, E. D. (1956). *N.Z. J. Agric.* **92,** 239–244.
Andrews, E. D. (1965). *N.Z. J. Agric. Res.* **8,** 788–817.
Andrews, E. D. (1966). *N.Z. J. Agric. Res.* **9,** 829–838.
Andrews, E. D., and Anderson, J. P. (1955). *N.Z. Vet. J.* **3,** 78–79.
Andrews, A. C., and Croft, F. C. (1979). *Aust. J. Exp. Agric. Anim. Husb.* **19,** 444–447.
Andrews, E. D., Stephenson, B. J., Anderson, J. P., and Faithful, W. C. (1958). *N.Z. J. Agric. Res.* **1,** 125–139.
Andrews, E. D., Grant, A. B., and Stephenson, B. J. (1964). *N.Z. J. Agric. Res.* **7,** 17–27.
Andrews, E. D., Stephenson, B. J., Isaacs, C. E., and Register, R. H. (1966). *N.Z. Vet. J.* **14,** 191–196.
Andrews, E. D., Hartley, W. J., and Grant, A. B. (1968). *N.Z. Vet. J.* **16,** 3–17.
Andrews, E. D., Grant, A. B., and Brunswick, L. F. C. (1974). *N.Z. Vet. J.* **22,** 46–50.
Andrews, E. D., Hogan, K. G., and Sheppard, A. D. (1976). *N.Z. Vet. J.* **24,** 111–116.
Anke, M. (1967). *Arch. Tierernaehr.* **17,** 1–26.
Anke, M., and Groppel, B. (1970). *In* "Trace Element Metabolism in Animals" (C. F. Mills, ed.), pp. 133–136. Livingstone, Edinburgh, Scotland.
Anke, M., Groppel, B., Reissig, W., Ludke, H., Grun, M., and Dittrich, G. (1973). *Arch. Tierernaehr.* **23,** 197–211.
Annison, E. F., and Armstrong, D. G. (1970). *In* "Physiology of Digestion and Metabolism in the Ruminant" (A. T. Phillipson, ed.), pp. 422–437. Oriel, Newcastle-upon-Tyne, England.
Annison, E. F., Chalmers, M. I., Marshall, S. B. M., and Synge, R. L. M. (1954). *J. Agric. Sci.* **44,** 270–273.
Annison, E. F., Lewis, D., and Lindsay, D. B. (1959). *J. Agric. Sci.* **53,** 34–41.
Anonymous (1969). *Rural Res. CSIRO* **65.**
Anthony, W. B., and Reid, J. T. (1958). *J. Dairy Sci.* **41,** 1715–1722.
AOAC (1980). "Official Methods of Analysis of the Association of Official Analytical Chemists" (W. Horwitz, ed.). AOAC, Washington, D.C.
Appelman, H., and Dirven, J. G. P. (1962). *Surinaamse Landbouw.* **10,** 95–102.
ARC (1965). "The Nutrient Requirements of Farm Livestock. 2. Ruminants." Agric. Res. Counc., Her Majesty's Stationery Office, London.
ARC (1980). "The Nutrient Requirements of Ruminant Livestock." Commonw. Agric. Bur., Farnham Royal, England.
ARC (1984). "The Nutrient Requirements of Ruminant Liverstock, Supplement No. 1." Commonw. Agric. Bur., Farnham Royal, England.
Archer, K. A., and Wheeler, J. L. (1978). *Proc. Aust. Soc. Anim. Prod.* **12,** 172.
Armsby, H. P., and Moulton, C. R. (1925). "The Animal as a Converter of Matter and Energy." Chem. Catalog Co., New York.
Armstrong, D. G. (1964). *J. Agric. Sci.* **62,** 399–416.
Armstrong, D. G., and Hutton, K. (1975). *In* "Digestion and Metabolism in the Ruminant" (I. W. McDonald and A. C. I. Warner, eds.), pp. 432–437. Univ. of New England, Armidale, Australia.

Arnold, G. W. (1962). *Aust. J. Agric. Res.* **13,** 701–706.
Arnold, G. W. (1975). *Aust. J. Agric. Res.* **26,** 1017–1024.
Arnold, G. W. (1981). *In* "Grazing Animals" (F. H. W. Morley, ed.), pp. 79–104. Elsevier, Amsterdam.
Arnold, G. W., and Birrell, H. A. (1977). *Anim. Prod.* **24,** 343–353.
Arnold, G. W., and Dudzinski, M. L. (1966). *Proc. Int. Grassl. Congr., 10th* pp. 367–370.
Arnold, G. W., and Dudzinski, M. L. (1967). *Aust. J. Agric. Res.* **18,** 657–666.
Arora, S. K., and Das, B. (1974). *J. Assoc. Off. Agric. Chem.* **57,** 1224–1226.
Arriaga-Jordan, C. M., and Holmes, W. (1986). *J. Agric. Sci.* **106,** 581–592.
Asher, C. J., and Grundon, N. J. (1970). *Proc. Int. Grassl. Congr., 11th* pp. 329–332.
Askew, H. O., and Dixon, J. K. (1937). *N.Z. J. Sci. Technol.* **18,** 688–693.
Aufrere, J. (1982). *Ann. Zootech.* **31,** 111–130.
Austenson, H. M. (1963). *Agron. J.* **55,** 149–153.
Axelsson, J. (1938). *Biedermanns Zentralbl., Abt. B* **10,** 238–248.
Azuolas, J. K., and Caple, I. W., (1984). *Aust. Vet. J.* **61,** 223–227.
Baile, C. A., and Pfander, W. H. (1967). *J. Dairy Sci.* **50,** 77–80.
Bailey, C. B., (1962). *Can. J. Anim. Sci.* **42,** 49–54.
Bailey, C. B., and Balch, C. C. (1961). *Br. J. Nutr.* **15,** 383–402.
Baker, H. K., and Chard, J. R. A. (1961). *Grassl. Res. Inst. (Hurley), Annu. Rep. 1959–1960* pp. 110–114.
Baker, R. D., Alvarez, F., and Le Du, Y. L. P. (1981). *Grass Forage Sci.* **36,** 189–199.
Balasko, J. A. (1977). *Agron. J.* **69,** 425–428.
Balch, C. C., and Campling, R. C. (1962). *Nutr. Abstr. Rev.* **32,** 669–686.
Bales, G. L., Kellogg, D. W., and Urquhart, N. S. (1978). *J. Anim. Sci.* **47,** 561–568.
Barlow, C. (1965). *J. Agric. Sci.* **64,** 439–447.
Barnes, D. L. (1960). *Rhod. Agric. J.* **57,** 311–317.
Barnes, J. E., and Jephcott, B. R. (1955). *Aust. Vet. J.* **31,** 302–315.
Barnes, R. F., and Mott, G. O. (1970). *Agron. J.* **62,** 719–722.
Barnes, R. F., Muller, L. D., Bauman, L. F., and Colenbrander, V. F. (1971). *J. Anim. Sci.* **33,** 881–884.
Barrow, N. J., and Lambourne, L. J. (1962). *Aust. J. Agric. Res.* **13,** 461–471.
Barry, T. N. (1976). *J. Agric. Sci.* **86.** 379–392.
Barry, T. N. (1980). *N.Z. J. Agric. Res.* **23,** 427–431.
Barry, T. N. (1981). *Br. J. Nutr.* **46,** 521–532.
Barry, T. N., and Blaney, B. J. (1987). *In* "Nutrition of Herbivores" (J. H. Ternouth and J. B. Hacker, eds.), pp. 91–118. Academic Press, Orlando, Florida.
Barry, T. N., and Fennessy, P. F. (1972). *N.Z. J. Agric. Res.* **15,** 712–722.
Barry, T. N., and Fennessy, P. F. (1973). *N. Z. J. Agric. Res.* **16,** 59–63.
Barry, T. N., and Manley, T. R. (1984). *Br. J. Nutr.* **51,** 493–504.
Barry, T. N., Fennessy, P. F., and Duncan, S. J. (1973). *N.Z. J. Agric. Res.* **16,** 64–68.
Barry, T. N., Cook, J. E., and Wilkins, R. J. (1978). *J. Agric. Sci.* **91,** 701–715.
Barry, T. N., Reid, T. C., Millar, K. R., and Sadler, W. A. (1981). *J. Agric. Sci.* **96,** 269–282.
Barry, T. N., Duncan, S. J., Sadler, W. A., Millar, K. R., and Sheppard, A. D. (1983). *Br. J. Nutr.* **49.** 241–253.
Barry, T. N., Manley, T. R., and Duncan, S. J. (1986). *Br. J. Nutr.* **55,** 123–137.
Bartlett, S., Brown, B. B., Foot, A. S., Rowland, J., Allcroft, R., and Parr, W. H. (1954). *Br. Vet. J.* **110,** 3–19.
Barton, F. E., and Burdick, D. (1979). *J. Agric. Food Chem.* **27,** 1248–1252.

Bass, J. M., Fiskwick, G., Hemingway, R. G., Parkins, J. J., and Ritchie, N. S. (1981). *J. Agric. Sci.* **97,** 365–372.
Bath, I. H., and Rook, J. A. F. (1965). *J. Agric. Sci.* **64,** 67–75.
Bath, D. L., Gall, G. A. E., and Ronning, M. (1974). *J. Dairy Sci.* **57,** 198–204.
Bathurst, N. O., and Mitchell, K. J. (1958). *N.Z. J. Agric. Res.* **1,** 540–552.
Baumgardt, B. R., and Oh, H. K. (1964). *J. Dairy Sci.* **47,** 263–266.
Baumgardt, B. R., Taylor, M. W., and Cason, J. L. (1962). *J. Dairy Sci.* **45,** 62–68.
Beaty, E. R., McCreery, R. A., and Brooks, D. L. (1960). *Proc. Int. Grassl. Congr., 8th* pp. 708–710.
Beck, A. B. (1941). *J. Agric., West. Aust.* **18,** 285–300.
Beck, A. B. (1962). *Aust. J. Exp. Agric. Anim. Husb.* **2,** 40–45.
Beeson, K. C. (1946). *Bot. Rev.* **12,** 424–455.
Beeson, K. C. (1950). *Agric. Inf. Bull. (U.S., Dep. Agric.)* **7.**
Beeson, K. C., and Guillermo-Gomez, G. (1970). *Proc. Int. Grassl. Congr., 11th* pp. 89–92.
Beeson, K. C., and MacDonald, H. A. (1951). *Agron. J.* **43,** 589–593.
Beeson, K. C., Gray, L., and Adams, M. B. (1947). *J. Am. Soc. Agron.* **39,** 356–362.
Beeson, W. M., Perry, T. W., and Zurcher, T. D. (1977). *J. Anim. Sci.* **45,** 160–165.
Beever, D. E., and Thomson, D. J. (1981). *Grass Forage Sci.* **36,** 211–219.
Beever, D. E., Harrison, D. G., Thomson, D. J., and Osbourn, D. F. (1971a). *Proc. Nutr. Soc.* **30,** 15A.
Beever, D. E., Thomson, D. J., Pfeffer, E., and Armstrong, D. G. (1971b). *Br. J. Nutr.* **26,** 123–134.
Beever, D. E., Cammell, S. B., and Wallace, A. S. (1974). *Proc. Nutr. Soc.* **33,** 73A.
Beever, D. E., Thomson, D. J., and Cammell, S. B. (1976). *J. Agric. Sci.* **86,** 443–452.
Beever, D. E., Terry, R. A., Cammell, S. B., and Wallace, A. S. (1978). *J. Agric. Sci.* **90,** 463–470.
Beever, D. E., Osbourn, D. F., Cammell, S. B., and Terry, R. A. (1981). *Br. J. Nutr.* **46,** 357–370.
Beever, D. E., Cammell, S. B., Haines, M. J., Gale, D. L., and Thomas, C. (1984). *Anim. Prod.* **38,** 533.
Beever, D. E., Thomson, D. J., Ulyatt, M. J., Cammell, S. B., and Spooner, M. C. (1985). *Br. J. Nutr.* **54,** 763–775.
Beever, D. E., Losada, H. R., Gale, D. L., Spooner, M. C., and Dhanoa, M. S. (1987). *Br. J. Nutr.* **57,** 57–67.
Behaeghe, T., and Carlier, L. A. (1974). *Vaextodling* **28,** 52–66.
Bekker, J. G. (1932). *Rep. Dir. Vet. Serv. Anim. Ind., Onderstepoort* **18,** 751–797.
Belonje, P. C. (1978). *Onderstepoort J. Vet. Res.* **45,** 7–22.
Belonje, P. C., and Van den Berg, A. (1980). *Onderstepoort J. Vet. Res.* **47,** 169–172.
Belyea, R. L., Foster, M. B., and Zinn, G. M. (1983). *J. Dairy Sci.* **66,** 1277–1281.
Bengtsson, S., and Larsson, K. (1984). *J. Sci. Food Agric.* **35,** 951–958.
Bennetts, H. W., and Chapman, F. E. (1937). *Aust. Vet. J.* **13,** 138–149.
Bennetts, H. W., Beck, A. B., Harley, R. and Evans, S. T. (1941). *Aust. Vet. J.* **17,** 85–93.
Bennetts, H. W., Beck, A. B., and Harley, R. (1948). *Aust. Vet. J.* **24,** 237–244.
Bentley, O. G., and Phillips, P. H. (1951). *J. Dairy Sci.* **34,** 396–403.
Benzie, D., Boyne, A. W., Dalgarno, A. C., Duckworth, J., and Hill, R. (1959). *J. Agric. Sci.* **52,** 1–12.
Benzie, D., Boyne, A. W., Dalgarno, A. C., Duckworth, J., Hill, R., and Walker, D. M. (1960). *J. Agric. Sci.* **54,** 202–221.

Berger, L. L., Klopfenstein, T. J., and Britton, R. A. (1980). *J. Anim. Sci.* **50,** 745–749.
Berrow, M. L., and Ure, A. M. (1985). *In* "Trace Elements in Man and Animals" (C. F. Mills, I. Bremner, and J. K. Chesters, eds.), pp. 841–843. Commonw. Agric. Bur., Farnham Royal, England.
Bhattacharya, A. N., and Warner, R. G. (1968). *J. Dairy Sci.* **51,** 1091–1094.
Bines, J. A. (1976). *Livest. Prod. Sci.* **3,** 115–128.
Bines, J. A., Suzuki, S., and Balch, D. C. (1969). *Br. J. Nutr.* **23,** 695–704.
Bines, J. A., Napper, D. J., and Johnson, V. W. (1977). *Proc. Nutr. Soc.* **36,** 146A.
Bingley, J. B., and Anderson, N. (1972). *Aust. J. Agric. Res.* **23,** 885–904.
Binnie, R. C., Harrington, F. J., and Murdoch, J. C. (1974). *J. Br. Grassl. Soc.* **29,** 57–62.
Birch, J. A., and Wolton, K. M. (1961). *Vet. Rec.* **73,** 1169–1173.
Bird, P. R. (1974). *Aust. J. Agric. Res.* **25,** 631–642.
Bird, S. H., and Leng, R. A. (1984). *Proc. Aust. Soc. Anim. Prod.* **15,** 654.
Bird, S. H., and Leng, R. A. (1985). *Rev. Rural Sci.* **6,** 109–117.
Bisschop, J. H. R. (1964). Sci. Bull.—S. Afr., Dep. Agric. Tech. Serv. **365.**
Black, J. L., and Kenney, P. A. (1984). *Aust. J. Agric. Res.* **35,** 565–578.
Black, J. L., and Tribe, D. E. (1973). *Aust. J. Agric. Res.* **24,** 763–773.
Black, W. H., Tash, L. H., Jones, J. M., and Kleberg, R. J. (1943). *U.S. Dep. Agric., Tech. Bull.* **856.**
Black, W. H., Tash, L. H., Jones, J. M., and Kleberg, R. J. (1949). *U.S. Dep. Agric., Tech. Bull.* **981.**
Black, J. R., Robards, G. E., and Thomas, R. (1973). *Aust. J. Agric. Res.* **24,** 399–412.
Black, J. L., Faichney, G. J., Beever, D. E., and Howarth, B. R. (1982). *In* "Forage Protein in Ruminant Animal Production" (D. J. Thomson, D. E. Beever, and R. G. Gunn, eds.), Occas. Publ. 6, pp. 107–118. Br. Soc. Anim. Prod., London.
Black, J. R., Ammerman, C. B., and Henry, P. R. (1985). *J. Dairy Sci.* **68,** 433–436.
Blair-West, J. R., Coghlan, J. P., Denton, D. A., and Wright, R. D. (1970). *In* "Physiology of Digestion and Metabolism in the Ruminant" (A. T. Phillipson, ed.), pp. 350–361. Oriel, Newcastle-upon-Tyne, England.
Blakley, B. R., and Hamilton, D. L. (1984). *Can. J. Comp. Med.* **49,** 405–408.
Bland, B. F., and Dent, J. W. (1964). *J. Br. Grassl. Soc.* **19,** 306–315.
Blaney, B. J., Gartner, R. J., and Head, T. A. (1982). *J. Agric. Sci.* **99,** 533–539.
Blaser, R. E., Hammes, R. C., Bryant, H. T., Hardison, W. A., Fontenot, J. P., and Engel, R. W. (1960). *Proc. Int. Grassl. Congr., 8th* pp. 601–606.
Blaxter, K. L. (1960). *Proc. Int. Grassl. Congr., 8th* pp. 479–484.
Blaxter, K. L. (1962). "The Energy Metabolism of Ruminants." Hutchinson, London.
Blaxter, K. L. (1963). *Br. J. Nutr.* **17,** 105–115.
Blaxter, K. L., and Graham, N. M. (1956). *J. Agric. Sci.* **47,** 207–217.
Blaxter, K. L., and Wainman, F. W. (1964). *J. Agric. Sci.* **63,** 113–128.
Blaxter, K. L., and Wilson, R. S. (1963). *Anim. Prod.* **5,** 27–42.
Blaxter, K. L., Rook, J. A. F., and MacDonald, A. M. (1954). *J. Comp. Pathol.* **64,** 157–185.
Blaxter, K. L., Graham, N. M., and Wainman, F. W. (1956). *Br. J. Nutr.* **10,** 69–91.
Blaxter, K. L., Wainman, F. W., and Wilson, R. S. (1961). *Anim. Prod.* **3,** 51–61.
Blaxter, K. L., Wainman, F. W., and Davidson, J. L. (1966). *Anim. Prod.* **8,** 75–83.
Blaxter, K. L., Wainman, F. W., Dewey, P. J. S., Davidson, J. L., Denerley, H., and Gunn, J. B. (1971). *J. Agric. Sci.* **76,** 307–319.
Blom, I. J. B. (1934). *Onderstepoort J. Vet. Sci. Anim. Ind.* **11,** 139–150.
Blue, W. G., and Tergas, L. E. (1969). *Proc.—Soil Crop Sci. Soc. Fla.* **29,** 224–238.
Boda, J. M. (1956). *J. Dairy Sci.* **39,** 66–72.

Boda, J. M., and Cole, H. H. (1954). *J. Dairy Sci.* **37,** 360–372.

Bohman, V. R., Lesperance, A. L., Harding, G. D., and Grunes, D. L. (1969). *J. Anim. Sci.* **29,** 99–102.

Boila, R. J., Devlin, T. J., Drysdale, R. A., and Lillie, L. E. (1984). *Can. J. Anim. Sci.* **64,** 365–378.

Bolin, D. W. (1934). *J. Agric. Res.* **48,** 657–663.

Boniface, A. N., Murray, R. M., and Hogan, J. P. (1986). *Proc. Aust. Soc. Anim. Prod.* **16,** 151–154.

Borle, A. B. (1974). *Annu. Rev. Physiol.* **36,** 361–390.

Bosman, M. S. M. (1967). *Meded., Inst. Biol. Scheikd. Onderz. Landbouwgewassen, Wageningen* **349,** 97–100.

Bosman, M. S. M. (1970). *Meded., Inst. Biol. Scheikd. Onderz. Landbouwgewassen, Wageningen* **413,** 1–15.

Bott, E., Denton, D. A., Goding, J. R., and Sabine J. R. (1964). *Nature (London)* **202,** 461–463.

Bowman, J. G. P., and Asplund, J. M. (1988a). *Anim. Feed Sci. Technol.* **20,** 19–31.

Bowman, J. G. P., and Asplund, J. M. (1988b). *Anim. Feed Sci. Technol.* **20,** 33–44.

Braithwaite, G. D. (1976). *J. Dairy Res.* **43,** 501–520.

Braithwaite, G. D. (1978). *Br. J. Nutr.* **40,** 17–21.

Braithwaite, G. D. (1984). *J. Agric. Sci.* **102,** 135–139.

Braithwaite, G. D., and Riazuddin, S. H. (1971). *Br. J. Nutr.* **26,** 215–225.

Braver, E. D., and Eriksson, S. (1967). *Lantbrukshoegsk. Ann.* **33,** 751–765.

Bray, R. A., and Hacker, J. B. (1981). *Aust. J. Agric. Res.* **32,** 311–323.

Bray, R. A., and Pritchard, A. J. (1976). *Forage Res.* **2,** 1–7.

Breese, E. L. (1970). *Rep.—Welsh Plant Breed. Stn. (Aberystwyth,) Wales* pp. 33–37.

Breirem, K. (1944). *Kurgl. Lant. Tidskr.* **83,** 345–405.

Bremner, I., and Davies, N. T. (1980). *In* "Digestion Physiology and Metabolism in Ruminants" (Y. Ruckebusch and P. Thivend, eds.), pp. 409–427. MTP Press. Lancaster, England.

Bremner, I., Young, B. W., and Mills, C. F. (1976). *Br. J. Nutr.* **36,** 551–561.

Breves, G., Hoeller, H., and Lessmann, H. W. (1985). *Proc. Nutr. Soc.* **44,** 145A.

Brown, E. M. (1939). *Res. Bull.—MO., Agric. Exp. Stn.* **299,** 1–76.

Brown, A. L., Krantz, B. A., and Martin, P. E. (1962). *Soil Sci. Soc. Am. Proc.* **26,** 167–170.

Brown, R. H., Blaser, R. E., and Fontenot, J. P. (1968). *J. Anim. Sci.* **27,** 562–567.

Bryan, W. W., Thorne, P. M., and Andrew, C. S. (1960). *J. Aust. Inst. Agric. Sci.* **26,** 273–275.

Bryant, H. T., Blaser, R. E., Hammes, R. C., and Hardison, W. A. (1961). *J. Dairy Sci.* **44,** 1733–1741.

Bryant, H. T., Hammes, R. C., Blaser, R. E., and Fontenot, J. P. (1976). *J. Anim. Sci.* **42,** 554–559.

Buchan, W., Kay, R. N. B., Sasaki, Y., and Scott, D. (1986). *J. Physiol. (London)* **381,** 84P.

Buchanan-Smith, J. G., Nelson, E. C., Osburn, B. I., Wells, M. E., and Tillman, A. D. (1969). *J. Anim. Sci.* **29,** 808–815.

Buchanan-Smith, J. G., Evans, E., and Poluch, S. O. (1974). *Can. J. Anim. Sci.* **54,** 253–256.

Buchman, D. T., and Hemken, R. W. (1964). *J. Dairy Sci.* **47,** 861–864.

Buettner, M. R., Lechtenberg, V. L., Hendrix, K. S., and Hertel, J. M. (1982). *J. Anim. Sci.* **54,** 173–178.

Budhi, S. P. S., and Ternouth, J. H. (1988). *Proc. Nutr. Soc. Aust.* **13,** 110.
Bugge, G. (1978). *Z. Pflanzenzuecht.* **81,** 235–240.
Bull, L. S., and Tamplin, C. B. (1974) *J. Anim. Sci.* **39,** 234.
Burdick, D., and Sullivan, J. T. (1963). *J. Anim. Sci.* **22,** 444–447.
Burdick, D., Barton, F. E., and Nelson, B. D. (1981). *Agron. J.* **73,** 399–403.
Burlison, A. J., and Hodgson, J. (1985). *Anim. Prod.* **40,** 530–531.
Burridge, J. C. (1970). *In* "Trace Element Metabolism in Animals" (C. F. Mills, ed.), pp. 412–415. Livingstone, Edingburgh, Scotland.
Burroughs, W., Nelson, D. K., and Mertens, D. R. (1975). *J. Anim. Sci.* **41,** 933–944.
Burt, A. W. A., and Thomas, D. C. (1961). *Nature (London)* **192,** 1193.
Burton, G. W. (1974). *Proc. Int. Grassl. Congr., 12th* pp. 705–714.
Burton, G. W., and Monson, W. G. (1972). *Crop Sci.* **12,** 375–378.
Burton, G. W., Jackson, J. E., and Knox, F. E. (1959). *Agron. J.* **51,** 537–542.
Burton, G. W., Jackson, J. E., and Hart, R. H. (1963). *Agron. J.* **55,** 500–502.
Burton, G. W., Knox, F. E., and Beardsley, D. W. (1964). *Agron. J.* **56,** 160–161.
Burton, G. W., Hart, R. H., and Lowrey, R. S. (1967). *Crop Sci.* **7,** 329–332.
Burton, G. W., Gunnells, J. B., and Lowrey, R. S. (1968). *Crop Sci.* **8,** 431–434.
Burton, G. W., Monson, W. G., Johnson, J. C., Lowrey, R. S., Chapman, H. D., and Marchant, W. H. (1969). *Agron. J.* **61,** 607–612.
Butcher, J. E., Call, J. W., Blake, J. T., and Shupe, J. L. (1979). *J. Anim. Sci.* **49,** 35–39.
Butler, E. J. (1963). *J. Agric. Sci.* **60,** 329–340.
Butler, G. W., and Glenday, A. C. (1962). *Aust. J. Biol. Sci.* **15,** 183–187.
Butler, G. W., and Jones, D. I. H. (1973). *In* "Chemistry and Biochemistry of Herbage" (G. W. Butler and R. W. Bailey, eds.), Vol. 2, pp. 127–162. Academic Press, New York.
Butler, G. W., and Metson, A. J. (1967). *N.Z. Dairyfarming Annu.* pp. 142–153.
Butler, G. W., Barclay, P. C., and Glenday, A. C. (1962). *Plant Soil* **16,** 214–228.
Butris, G. Y., and Phillips, C. J. C. (1987). *Grass Forage Sci.* **42,** 259–264.
Butterworth, M. H. (1961). *Trop. Agric. (Trinidad)* **38,** 189–193.
Butterworth, M. H. (1965). *J. Agric. Sci.* **65,** 233–239.
Butterworth, M. H. (1966). *Turrialba* **16,** 253–256.
Buxton, D. R., Hornstein, J. S., Wedin, W. F., and Marten, G. C. (1985). *Crop Sci.* **25,** 273–279.
Byers, J. H. (1965). *J. Dairy Sci.* **48,** 206–208.
Bywater, A. C. (1984). *Agric. Syst.* **13,** 167–186.
Cadwallader, T. K., Hoekstra, W. G., and Pape, A. L. (1980). *J. Anim. Sci.* **51,** 351.
Caird, L., and Holmes, W. (1986). *J. Agric. Sci.* **107,** 43–54.
Calder, F. W., and MacLeod, L. B. (1968). *Can. J. Plant Sci.* **48,** 17–24.
Calder, F. W., Langille, J. E., and Nicholson, J. W. G. (1977). *Can. J. Anim. Sci.* **57,** 65–73.
Caldwell, D. R., and Hudson, R. F. (1974). *Appl. Microbiol.* **27,** 549–552.
Call, J. W., Butcher, J. E., Blake, J. T., Smart, R. A., and Shupe, J. L. (1978). *J. Anim. Sci.* **47,** 216–225.
Call, J. E., Butcher, J. W., Shape, J. L., Olsen, A. E., and Blake, J. T. (1986). *Proc. World Congr. Dis. Cattle, 14th* pp. 834–839.
Cameron, C. D. T. (1967). *Can. J. Anim. Sci.* **47,** 123–125.
Cammell, S. B. (1977). *Grassl. Res. Inst. (Hurley), Tech. Rep.* **24.**
Cammell, S. B., Thomson, D. J., Beever, D. E., Haines, M. J., Dhanoa, M. S., and Spooner, M. C. (1986). *Br. J. Nutr.* **55,** 669–680.
Campbell, C. M., Sherrod, L. B., and Ishizaki, S. M. (1969). *J. Anim. Sci.* **29,** 634–637.

Campbell, A. G., Coup, M. R., Bishop, W. H., and Wright, D. E. (1974). *N.Z. J. Agric. Res.* **17,** 393–399.
Campling, R. C. (1966). *Outlook Agric.* **2,** 74–79.
Campling, R. C., and Freer, M. (1966). *Br. J. Nutr.* **20,** 229–244.
Campling, R. C., Freer, M., and Balch, C. C. (1962). *Br. J. Nutr.* **16,** 115–124.
Campling, R. C., Freer, M., and Balch, C. C. (1963). *Br. J. Nutr.* **17,** 263–272.
Caple, I. W., and Nugent, G. F. (1982). *Proc. Aust. Soc. Anim. Prod.* **14,** 657.
Caple, I. W., Andrewartha, K. A., and Nugent, G. F. (1980a). *Victorian Vet. Proc.* **38,** 43–44.
Caple, I. W., Andrewartha, K. A., Edwards, S. J. A., and Halpin, C. G. (1980b). *Aust. Vet. J.* **56,** 160–167.
Caple, I. W., Azuolas, J. K., and Nugent, G. F. (1985). *In* "Trace Elements in Man and Animals" (C. F. Mills, I. Bremner, and J. K. Chesters, eds.), pp. 609–613. Commonw. Agric. Bur., Farnham Royal, England.
Carolin, R. C., Jacobs, S. W. L., and Vsk, M. (1973). *Bot. J. Linn. Soc.* **66,** 259–275.
Castle, M. E. (1972). *J. Br. Grassl. Soc.* **27,** 207–210.
Castle, M. E., and Watson, J. N. (1970). *J. Br. Grassl. Soc.* **25,** 278–284.
Castle, M. E., and Watson, J. N. (1984a). *Grass Forage Sci.* **39,** 93–99.
Castle, M. E., and Watson, J. N. (1984b). *Grass Forage Sci.* **39,** 187–193.
Castle, M. E., Drysdale, A. D., and Watson, J. N. (1962). *J. Dairy Res.* **29,** 199–206.
Castle, M. E., Retter, W. C., and Watson, J. N. (1979). *Grass Forage Sci.* **34,** 293–301.
Chacon, E., and Stobbs, T. H. (1976). *Aust. J. Agric. Res.* **27,** 709–727.
Chacon, E., Rodriguez-Carrasquel, S., and Chicco, C. F. (1971a). *Agron. Trop. (Maracay, Venez.)* **21,** 495–502.
Chacon, E., Rodriguez-Carrasquel, S., and Chicco, C. F. (1971b). *Agron. Trop. (Maracay, Venez.)* **21,** 503–509.
Chacon, E., Stobbs, T. H., and Sandland, R. L. (1976). *J. Br. Grassl. Soc.* **31,** 81–87.
Chai, K., Kennedy, P. M., Milligan, L. P., and Mathison, G. W. (1985). *Can. J. Anim. Sci.* **65,** 69–76.
Chalmers, M. I., and Synge, R. L. M. (1954). *J. Agric. Res.* **44,** 263–269.
Chalupa, W. (1972). *Fed. Proc., Fed. Am. Soc. Exp. Biol.* **31,** 1152–1164.
Chalupa, W. (1975). *J. Dairy Sci.* **58,** 1198–1218.
Chalupa, W. V., Cason, J. L., and Baumgardt, B. R. (1961). *J. Dairy Sci.* **44,** 874–878.
Chamberlain, D. G., and Thomas, P. C. (1979). *Proc. Nutr. Soc.* **38,** 138A.
Chamberlain, D. G., Thomas, P. C., and Wait, M. K. (1982). *Grass Forage Sci.* **37,** 159–164.
Chambers, A. R. M., Hodgson, J., and Milne, J. A. (1981). *Grass Forage Sci.* **36,** 97–105.
Chapman, H. L., and Bell, M. C. (1963). *J. Anim. Sci.* **22,** 82–85.
Chapman, H. L., and Kidder, R. W. (1964). *Bull.—Fla., Agric. Exp. Stn.* **674.**
Chapman, H. L., and Kretschmer, A. E. (1964). *Proc.—Soil Crop Sci. Soc. Fla.* **24,** 176–183.
Chapman, H. L., Nelson, S. L., Kidder, R. W., Sippel, W. L., and Kidder, C. W. (1962). *J. Anim. Sci.* **21,** 960–962.
Chenost, M. (1966). *Proc. Int. Grassl, Congr., 10th* pp. 406–411.
Chenost, M. (1975). *Ann. Zootech.* **24,** 327–349.
Chenost, M. (1985). *Ann. Zootech.* **34,** 205–227.
Chenost, M. (1986). *Ann. Zootech.* **35,** 1–20.
Chenost, M., and Dermarquilly, C. (1982). *In* "Herbage Intake Handbook" (J. D. Leaver, ed.), pp. 95–112. Br. Grassl. Soc., Hurley, Maidenhead, England.
Chenost, M., and Martin-Rosset, W. (1985). *Ann. Zootech.* **34,** 291–312.

Cherney, J. H., Axtell, J. D., Hassen, M. M., and Anliker, K. S. (1988). *Crop Sci.* **28,** 783–787.
Chestnutt, D. M. B. (1966). *Rec. Agric. Res.* **15,** 135–142.
Chicco, C. F., Ammerman, C. B., Moore, J. E., Van Walleghem, P. A., Arlington, L. R., and Shirley, R. L. (1965). *J. Anim. Sci.* **24,** 355–363.
Chicco, C. F., Ammerman, C. B., Feaster, J. P., and Demavant, B. G. (1973a). *J. Anim. Sci.* **36,** 986–993.
Chicco, C. F., Ammerman, C. B., and Loggins, P. E. (1973b). *J. Dairy Sci.* **56,** 822–824.
Chonan, N. (1978). *JARQ* **12,** 128–131.
Christie, B. R., and Mowat, D. N. (1968). *Can. J. Plant Sci.* **48,** 67–73.
Cipolloni, M. A., Schneider, B. H., Lucas, H. L., and Pavlech, H. M. (1951). *J. Anim. Sci.* **10,** 337–343.
Clancy, M. J., and Wilson, R. K. (1966). *Proc. Int. Grassl. Congr., 10th* pp. 445–452.
Clark, R. (1953). *Onderstepoort J. Vet. Res.* **26,** 137–140.
Clark, K. W. (1958). *Diss. Abstr.* **19,** 926.
Clark, J., and Beard, J. (1977). *Anim. Feed Sci. Technol.* **2,** 153–159.
Clark, D. A., and Brougham, R. W. (1979). *Proc. N.Z. Soc. Anim. Prod.* **39,** 265–274.
Clark, D. A., and Harris, P. S. (1985). *N.Z. J. Agric. Res.* **28,** 233–240.
Clark, R. G., and Millar, K. R. (1983). *In* "The Mineral Requirements of Grazing Ruminants" (N. D. Grace, ed.), pp. 27–37. N.Z. Soc. Anim. Prod., Wellington, New Zealand.
Clark, C. K., and Petersen, M. K. (1988). *J. Anim. Sci.* **66,** 743–749.
Clarke, E. A. (1959). *Proc. N.Z. Soc. Anim. Prod.* **19,** 91–97.
Clarke, T., Flinn, P. C., and McGowan, A. A. (1982). *Grass Forage Sci.* **37,** 147–150.
Clements, R. J. (1973). *Aust. J. Agric. Res.* **24,** 35–45.
Clements, R. J., Oram, R. N., and Scowcroft, W. R. (1970). *Aust. J. Agric. Res.* **21,** 661–675.
Coates, D. B. (1987). *In* "Herbivore Nutrition Research" (M. Rose, ed.), pp. 213–214. Aust. Soc. Anim. Prod., Melbourne, Australia.
Cochran, R. C., Adams, D. C., Wallace, J. D., and Galyean, M. L. (1986). *J. Anim. Sci.* 63, 1476–1483.
Cochrane, M. J. (1976). *Proc. Aust. Soc. Anim. Prod.* **11,** 497–500.
Cochrane, M. J., and Brown, D. C. (1974). *J. Aust. Inst. Agric. Sci.* **40,** 67–68.
Cochrane, M. J., and Radcliffe, J. C. (1977). *J. Aust. Inst. Agric. Sci.* **43,** 151–153.
Coelho da Silva, J. F., Seely, R. C., Thomson, D. J., Beever, D. E., and Armstrong, D. G. (1972a). *Br. J. Nutr.* **28,** 43–61.
Coelho da Silva, J. F., Seely, R. C., Beever, D. E., Prescott, J. H. D., and Armstrong, D. G. (1972b). *Br. J. Nutr.* **28,** 357–371.
Cohen, R. D. H. (1972). *Aust. J. Exp. Agric. Anim. Husb.* **12,** 455–459.
Cohen, R. D. H. (1973a). *Aust. J. Exp. Agric. Anim. Husb.* **13,** 5–8.
Cohen, R. D. H. (1973b). *Aust. J. Exp. Agric. Anim. Husb.* **13,** 625–629.
Cohen, R. D. H. (1974). *Aust. J. Exp. Agric. Anim. Husb.* **14,** 709–715.
Cohen, R. D. H. (1975a). *Aust. Meat Res. Counc. Rev.* **23.**
Cohen, R. D. H. (1975b). *World Rev. Anim. Prod.* **11,** 27–43.
Cohen, R. D. H. (1980). *Livest. Prod. Sci.* **7,** 25–37.
Colburn, M. W., Evans, J. L., and Ramage, C. H. (1968). *J. Dairy Sci.* **51,** 1450–1457.
Cole, M., Seath, D. M., Lassiter, C. A., and Rust, J. (1957). *J. Dairy Sci.* **40,** 252–257.
Collier, R. E., and Parker-Sutton, J. (1976). *J. Sci. Food Agric.* **27,** 743–744.

Colovos, N. F., Keener, H. A., and Davis, H. A. (1961). *Bull.—N.H., Agric. Exp. Stn.* **742,** 11–19.
Colovos, N. F., Peterson, N. K., Blood, P. T., and Davis, H. A. (1966a). *Bull.—N.H., Agric. Exp. Stn.* **486.**
Colovos, N. F., Holter, J. B., Peterson, N. K., Blood, P. T., and Davis, H. A. (1966b). *Bull.—N.H., Agric. Exp. Stn.* **488.**
Colovos, N. F., Koes, R. M., Holter, J. B., Mitchell, J. R., and Davis, H. A. (1969). *Agron. J.* **61,** 503–505.
Combellas, J., and Gonzalez, E. (1972a). *Agron. Trop. (Maracay, Venez.)* **22,** 623–634.
Combellas, J., and Gonzalez, J. E. (1972b). *Agron. Trop. (Maracay, Venez.)* **22,** 635–641.
Combellas, J., and Hodgson, J. (1979). *Grass Forage Sci.* **34,** 209–214.
Combellas, J., Gonzalez, J. E., and Parra, R. (1971). *Agron. Trop. (Maracay, Venez.)* **21,** 483–494.
Combellas, J., Gonzalez, J. E., and Trujillo, A. (1972). *Agron. Trop. (Maracay, Venez.)* **22,** 231–238.
Combellas, J., Baker, R. D., and Hodgson, J. (1979). *Grass Forage Sci.* **34,** 303–310.
Combs, D. K. (1987). *J. Anim. Sci.* **65,** 1753–1758.
Combs, D. K., Goodrich, R. D., and Meiske, J. C. (1982). *J. Anim. Sci.* **54,** 391–398.
Committee on Mineral Nutrition (1973). "Tracing and Treating Mineral Disorders in Dairy Cattle." Cent. Agric. Publ. Doc., Wageningen, The Netherlands.
Cook, C. W., and Harris, L. E. (1968). *Bull.—Utah, Agric. Exp. Stn.* **475.**
Cook, C. W., Mattox, J. E., and Harris, L. E. (1961). *J. Anim. Sci.* **20,** 866–870.
Cooke, G. W. (1972). "Fertilising for Maximum Yield." Crosby Lockwood, London.
Coombe, J. B., Christian, K. R., and Holgate, M. D. (1971). *J. Agric. Sci.* **77,** 159–174.
Coop, I. E., and Hill, M. K. (1962). *J. Agric Sci.* **58,** 187–199.
Coop, R. L., Sykes, A. R., and Angus, K. W. (1977). *Res. Vet. Sci.* **23,** 76–83.
Cooper, J. P. (1962). *Rep.—Welsh Plant Breed. Stn. (Aberystwyth, Wales)* pp. 145–156.
Cooper, J. P. (1973). *In* "Chemistry and Biochemistry of Herbage" (G. W. Butler and R. W. Bailey, eds.), Vol. 2, pp. 379–417. Academic Press, New York.
Cooper, J. P., and Breese, E. L. (1980). *Proc. Nutr. Soc.* **39,** 281–286.
Cooper, H. P., Paden, W. R., and Garman, W. H. (1947). *Soil Sci.* **63,** 27–41.
Cooper, J. P., Tilley, J. M. A., Raymond, W. F., and Terry, R. A. (1962). *Nature (London)* **195,** 1276–1277.
Copland, J. W., ed. (1985). "Draught Animal Power for Production." Aust. Cent. Int. Agric. Res., Canberra, Australia.
Coppenet, M., and Calvez, J. (1962). *Ann. Agron.* **13,** 203–219.
Coppock, C. E., Everett, R. W., and Merrill, W. G. (1972). *J. Dairy Sci.* **55,** 245–256.
Corbett, J. L. (1978). *In* "Measurement of Grassland Vegetation and Animal Production" (L. Mannetje, ed.), pp. 163–231. Commonw. Agric. Bur., Farnham Royal, England.
Corbett, J. L., and Boyne, A. W. (1958). *J. Agric. Sci.* **51,** 95–107.
Corbett, J. L., Greenhalgh, J. F. D., and MacDonald, D. A. P. (1958). *Nature (London)* **182,** 1014–1016.
Corbett, J. L., Langlands, J. P., and Reid, G. W. (1963). *Anim. Prod.* **5,** 119–129.
Corbett, J. L., Langlands, J. P., McDonald, I., and Pullar, J. D. (1966). *Anim. Prod.* **8,** 13–27.
Corbett, J. L., Furnival, E. P., Inskip, M. W., and Pickering, F. S. (1982). *In* "Forage Protein in Ruminant Production" (D. J. Thomson, D. E. Beever, and R. G. Gunn, eds.), Occas. Publ. 6, pp. 141–143. Br. Soc. Anim. Prod., London.
Corkill, L. (1952). *N.Z. J. Sci. Technol.* **34,** 1–16.

Cornforth, I. S. (1984). *Proc. N.Z. Soc. Anim. Prod.* **44,** 135–137.
Corrall, A. J., Heard, A. J., Fenlon, J. S., Terry, C. P., and Lewis, G. C. (1977). *Grassl. Res. Inst. (Hurley), Tech. Rep.* **22.**
Corrall, A. J., Lavender, R. H., and Terry, C. P. (1979). *Grassl. Res. Inst. (Hurley), Tech. Rep.* **26.**
Costigan, P., and Ellis, K. J. (1980). *Proc. Aust. Soc. Anim. Prod.* **13,** 451.
Cote, M., Seoane, J. R., and Gervais, P. (1983). *Can. J. Anim. Sci.* **63,** 367–371.
Cottrill, B. R. (1982). *In* "Forage Protein in Ruminant Production" (D. J. Thomson, D. E. Beever, and R. G. Gunn, eds.), Occas. Publ. 6, pp. 121–128. Br. Soc. Anim. Prod. London.
Coulman, B. E., and Knowles, R. P. (1974). *Can. J. Plant Sci.* **54,** 651–657.
Coward-Lord, J., Arroyo-Aguilu, J. A., and Garcia-Molinari, O. (1974). *J. Agric. Univ. P.R.* **58,** 293–304.
Cox, C. P., Foot, A. S., Hosking, Z. D., Line, C., and Rowland, S. R. (1956). *J. Br. Grassl. Soc.* **11,** 107–118.
Crabtree, J. R., and Williams, G. L. (1971a). *Anim. Prod.* **13,** 71–82.
Crabtree, J. R., and Williams, G. L. (1971b). *Anim. Prod.* **13,** 83–92.
Craig, W. M., and Broderick, G. A. (1981). *J. Dairy Sci.* **64,** 769–774.
Craig, W. M., Ulloa, J. A., Watkins, K. L., and Nelson, B. D. (1988). *J. Anim. Sci.* **66,** 185–193.
Crampton, E. W., and Maynard, L. A. (1938). *J. Nutr.* **15,** 383–395.
Crampton, E. W., Campbell, J. A., and Lange, E. H. (1940). *Can. J. Agric. Sci.* **20,** 504–509.
Crampton, E. W., Donefer, E., and Lloyd, L. E. (1960). *J. Anim. Sci.* **19,** 538–544.
Craven, C. P. (1964). *Aust. Vet. J.* **40,** 127–130.
Crawford, R. J., Hoover, W. H., Sniffen, C. J., and Crocker, B. A. (1978). *J. Anim. Sci.* **46,** 1768–1775.
Crooker, B. A., Sniffen, C. J., Hoover, W. H., and Johnson, L. L. (1978). *J. Dairy Sci.* **61,** 437–447.
Cruickshank, G. J., Poppi, D. P., and Sykes, A. R. (1985). *Proc. N.Z. Soc. Anim. Prod.* **45,** 113–116.
Cunha, T. J., Shirley, R. L., Chapman, H. L., Ammerman, C. B., Davis, G. K., Kirk, W. G., and Hentges, J. F. (1964).*Bull—Fla., Agric. Exp. Stn.* **683.**
Cunningham, I. J. (1934a). *N.Z. J. Sci. Technol.* **16,** 81–87.
Cunningham, I. J. (1934b). *N.Z. J. Sci. Technol.* **16,** 414–422.
Cunningham, I. J. (1936). *N.Z. J. Sci. Technol.* **18,** 775–778.
Cunningham, I. J. (1946). *N.Z. J. Sci. Technol.* **27A,** 372–376.
Cunningham, I. J. (1960). *N.Z. J. Agric.* **100,** 419–428.
Cunningham, G. N., Wise, M. B., and Barrick, E. R. (1966). *J. Anim. Sci.* **25,** 532–538.
Curll, M. L. (1977a). *Aust. J. Agric. Res.* **28,** 991–1006.
Curll, M. L. (1977b). *Aust. J. Agric. Res.* **28,** 1007–1014.
Curll, M. L., Wilkins, R. J., Snaydon, R. W., and Shanmugalingham, V. S. (1985). *Grass Forage Sci.* **40,** 129–140.
Currier, C. G., Haaland, R. L., Hoveland, C. S., Elkins, C. B., and Odom, J. W. (1983). *Proc. Int. Grassl. Congr., 14th* pp. 127–129.
Dabo, S. M., Taliaferro, C. M., Coleman, S. W., Horn, F. P., and Claypool, P. L. (1988). *J. Range Manage.* **41,** 40–48.
Dafaal'a, B. F. M., and Kay, R. N. B. (1980). *Proc. Nutr. Soc.* **39,** 71A.
Daniel, H. A., and Harper, H. J. (1935). *J. Am. Soc. Agron.* **27,** 644–652.
Darcy, B. K., and Belyea, R. L. (1980). *J. Anim. Sci.* **51,** 798–803.

Das, B., Arora, S. K., and Luthra, Y. P. (1975). *J. Dairy Sci.* **58,** 1347–1351.
Davey, A. W. F., and Holmes, C. W. (1977). *Anim. Prod.* **24,** 355–362.
Davey, B. G., and Mitchell, R. L. (1968). *J. Sci. Food Agric.* **19,** 425–431.
David, J. S. E. (1976). *J. Comp. Pathol.* **86,** 235–241.
Davidson, W. M., Finlayson, M. M., and Watson, C. J. (1951). *Sci. Agric. (Ottawa)* **31,** 148–151.
Davies, H. L. (1962). *Proc. Aust. Soc. Anim. Prod.* **4,** 167–171.
Davies, H. L. (1980). *In* "CSIRO Symposium on the Importance of Copper in Biology and Medicine" (B. R. McAuslan, ed.), pp. 78–81. Commonw. Sci. Ind. Res. Org., Canberra, Australia.
Davies, H. L. (1983). *Aust. Meat Res. Counc. Rev.* **44.**
Davies, D. A., and Morgan, T. E. H. (1982). *J. Agric. Sci.* **99,** 153–161.
Davies, E. B., and Watkinson, J. H. (1966). *N.Z. J. Agric. Res.* **9,** 317–327.
Davies, W. E., Griffith, G., and Ellington, A. (1966). *J. Agric. Sci.* **66,** 351–357.
Davies, W. E., Thomas, T. A., and Young, N. R. (1968). *J. Agric. Sci.* **71,** 233–241.
Davies, H. L., Suttle, N. F., and Field, A. C. (1981). *In* "Trace Element Metabolism in Man and Animals" (J. M. Howell, J. M. Gawthorne, and C. L. White, eds.), pp. 96–100. Aust. Acad. Sci., Canberra, Australia.
Davis, P., and Weston, R. H. (1986). *Proc. Nutr. Soc. Aust.* **11,** 168–171.
Davis, L. E., Jordan, R. M., and Marten, G. C. (1968). *Agron. J.* **60,** 420–422.
Davison, T. M., Murphy, G. M., Maroske, M. M., and Arnold, G. (1980). *Aust. J. Exp. Agric. Anim. Husb.* **20,** 543–546.
Davison, T. M., Isles, D. H., and McGuigan, K. R. (1986). *Proc. Aust. Soc. Anim. Prod.* **16,** 179–182.
De Boever, J. L., Cottyn, B. G., Buysse, F. X., Wainman, F. W., and Vanacker, J. M. (1986). *Anim. Feed Sci. Technol.* **14,** 203–214.
De Boever, J. L., Cottyn, B. G., Andries, J. I., Buysse, F. X., and Vanacker, J. M. (1988). *Anim. Feed Sci. Technol.* **19,** 247–260.
De Groot, T. (1963). *J. Br. Grassl. Soc.* **18,** 112–118.
Dehority, B. A., and Johnson, R. R. (1961). *J. Dairy Sci.* **44,** 2242–2249.
Deinum, B. (1966a). *Meded. Landbouwhogesch. Wageningen* **66** (11), 1–91.
Deinum, B. (1966b). *Proc. Int. Grassl. Congr., 10th* pp. 415–418.
Deinum, B. (1973). *Vaextodling* **28,** 42–51.
Deinum, B. (1976). *Misc. Pap.–Landbouwhogesch. Wageningen* **12,** 29–41.
Deinum, B., and Bakker, J. J. (1981). *Neth. J. Agric. Sci.* **29,** 92–98.
Deinum, B., Van Es, A. J. H., and Van Soest, P. J. (1968). *Neth. J. Agric. Sci.* **16,** 217–223.
Deinum, B., De Beyer, J., Nordfeldt, P. H., Kornher, A., Ostgard, O., and Van Bogaert, G. (1981). *Neth. J. Agric. Sci.* **29,** 141–150.
Deinum, B., Steg, A., and Hof, G. (1984). *Anim. Feed Sci. Technol.* **10,** 301–313.
De Loose, R., and Baert, L. (1966). *Plant Soil* **24,** 343–350.
DeLuca, H. F. (1979). *Nutr. Rev.* **37,** 161–193.
Demarquilly, C. (1965). *Proc. Int. Grassl. Congr., 9th* pp. 877–885.
Demarquilly, C. (1970a). *Ann. Zootech.* **19,** 413–422.
Demarquilly, C. (1970b). *Ann. Zootech.* **19,** 423–437.
Demarquilly, C. (1973). *Ann. Zootech.* **22,** 1–35.
Demarquilly, C., and Chenost, M. (1969). *Ann. Zootech.* **18,** 419–436.
Demarquilly, C., and Dulphy, J. P. (1977). *Proc. Int. Meet. Anim. Prod. Temperate Pastures, Dublin* pp. 53–61.
Demarquilly, C., and Jarrige, R. (1964). *Ann. Zootech.* **13,** 301–340.

Demarquilly, C., and Jarrige, R. (1970). *Proc. Int. Grassl. Congr., 11th* pp. 733–737.
Demarquilly, C., and Weiss, P. H. (1970). *Inst. Natl. Rech. Agron.* **42,** 1–64.
Demeyer, D., Van Nevel, C. J., Teller, E., and Godeau, J. M. (1986). *Arch. Anim. Nutr.* **36,** 132–143.
Dent, J. W. (1963). *Agric. Prog.* **38,** 40–49.
Dent, J. W., and Aldrich, D. T. A. (1963). *J. Natl. Inst. Agric. Bot. (G.B.)* **9,** 261–281.
Dent, J. W., and Aldrich, D. T. A. (1968). *J. Br. Grassl. Soc.* **23,** 13–19.
Denton, D. A. (1956). *J. Physiol. (London)* **131,** 516–525.
Denton, D. A. (1957). *J. Physiol. (London)* **142,** 72–95.
Denton, D. A. (1969). *Nutr. Abstr. Rev.* **39,** 1043–1049.
Deswysen, A. G., and Ehrlein, H. J. (1981). *Br. J. Nutr.* **46,** 327–335.
Deswysen, A., Vanbelle, M., and Focant, M. (1978). *J. Br. Grassl. Soc.* **33,** 107–115.
Devendra, C. (1977). *Trop. Agric. (Trinidad)* **54,** 29–38.
Devendra, C. (1978). *World Rev. Anim. Prod.* **14,** 9–22.
Devendra, C. (1979). "Malaysian Feedstuffs." Malaysian Agric. Res. Dev. Inst., Selangor, Malaysia.
Devendra, C., ed. (1988). "Non-conventional Feed Resources and Fibrous Agricultural Residues." Int. Dev. Res. Cent. Indian Counc. Agric. Res., New Delhi, India.
Devuyst, A., Vanbelle, M., Arnould, R., Moreels, A., and Vervack, W. (1963). *Agricultura (Heverlee, Belg.)* **11,** 451–466.
Dewey, D. W. (1977). *Search* **8,** 326–327.
Dewey, D. W., Lee, H. J., and Marston, H. R. (1958). *Nature (London)* **181,** 1367–1371.
Dewey, D. W., Lee, H. J., and Marston, H. R. (1969). *Aust. J. Agric. Res.* **20,** 1109–1116.
Dhanoa, M. S. (1988). *Grass Forage Sci.* **43,** 441–444.
Dhanoa, M. S., and Deriaz, R. E. (1984). *Grass Forage Sci.* **39,** 17–25.
Dick, A. T. (1944). *Aust. Vet. J.* **20,** 298–303.
Dick, A. T. (1952). *Aust. Vet. J.* **28,** 234–235.
Dick, A. T. (1956). *Soil Sci.* **81,** 229–236.
Dick, A. T. (1969). *Outlook Agric.* **6,** 14–19.
Dick, A. T., and Bull, L. B. (1945). *Aust. Vet. J.* **21,** 70–76.
Dick, A. T., Moore, C. W. E., and Bingley, J. B. (1953). *Aust. J. Agric. Res.* **4,** 44–51.
Dick, A. T., Dewey, D. W., and Gawthorne, J. M. (1975). *J. Agric. Sci.* **85,** 567–568.
Dijkshoorn, W., and Hart, M. L. (1957). *Neth. J. Agric. Sci.* **5,** 18–36.
Dijkstra, N. D. (1954). *Neth. J. Agric. Sci.* **2,** 273–297.
Dijkstra, N. D. (1971). *Neth. J. Agric. Sci.* **19,** 257–263.
Dinius, D. A., Simpson, M. E., and Marsh, P. B. (1976). *J. Anim. Sci.* **42,** 229–234.
Dinius, D. A., Goering, H. K., Oltjen, R. R., and Cross, H. R. (1978). *J. Anim. Sci.* **46,** 761–768.
Dodd, D. C. (1954). *N.Z. J. Agric.* **89,** 369–370.
Donaldson, C. H., and Rootman, G. T. (1977). *Proc.–Grassl. Soc. South Afr.* **12,** 91–93.
Donaldson, C. H., and Rootman, G. T. (1980). *Agric. Res. Rep.—S. Afr., Dep. Agric. Tech. Serv.* pp. 64–66.
Donaldson, L. E., Harvey, J. M., Beattie, A. W., Alexander, G. I., and Burns, M. A. (1964). *Queensl. J. Agric. Sci.* **21,** 167–179.
Donefer, E., Lloyd, L. E., and Crampton, E. W. (1960). *J. Anim. Sci.* **19,** 1304–1305.
Donefer, E., Niemann, P. J., Crampton, E. W., and Lloyd, L. E. (1963). *J. Dairy Sci.* **46,** 965–970.
Donefer, E., Crampton, E. W., and Lloyd, L. E. (1966). *Proc. Int. Grassl. Congr., 10th* pp. 442–445.

Doney, J. M., Gunn, R. G., Peart, J. N., and Smith, W. F. (1981). *Anim. Prod.* **33,** 241–247.
Donker, J. D., Marten, G. C., Jordan, R. M., and Bhargava, P. K. (1975). *J. Anim. Sci.* **41,** 333.
Donnelly, E. D., and Anthony, W. B. (1969). *Crop Sci.* **9,** 361–362.
Donnelly, E. D., and Anthony, W. B. (1970). *Crop Sci.* **10,** 200–202.
Donnelly, E. D., and Anthony, W. B. (1973). *Agron. J.* **65,** 993–994.
Donnelly, E. D., Anthony, W. B., and Langford, J. W. (1971). *Agron. J.* **63,** 749–751.
Dougall, H. W. (1963). *East Afr. Agric. For. J.* **28,** 182–189.
Dougall, H. W., and Bogdan, A. V. (1966). *East Afr. Agric. For. J.* **32,** 45–49.
Dove, H., and McCormack, H. A. (1986). *Grass Forage Sci.* **41,** 129–136.
Dove, H., Axelsen, A., and Watt, R. (1986). *Proc. Aust. Soc. Anim. Prod.* **16,** 187–190.
Dowe, T. W., Matsushima, J., and Arthaud, U. H. (1957). *J. Anim. Sci.* **16,** 811–820.
Dowman, M. G., and Collins, F. C. (1977). *J. Sci. Food Agric.* **28,** 1071–1074.
Dowman, M. G., and Collins, F. C. (1982). *J. Sci. Food Agric.* **33,** 689–696.
Doyle, P. J., and Fletcher, W. K. (1977). *Can. J. Plant Sci.* **57,** 859–864.
Drake, C., Grant, A. B., and Hartley, W. J. (1959). *Proc. Ruakura Farmers' Conf.* **12,** 61–71.
Drake, C., Grant, A. B., and Hartley, W. J. (1960a). *N.Z. Vet. J.* **8,** 4–6.
Drake, C., Grant, A. B., and Hartley, W. J. (1960b). *N.Z. Vet. J.* **8,** 7–10.
Drysdale, R. A., and Lillie, L. E. (1977). *Can. J. Anim. Sci.* **57,** 842.
Duckworth, J. E., and Shirlaw, D. W. (1958). *Anim. Behav.* **6,** 147–154.
Duncan, I. F., Greentree, P. L., and Ellis, K. J. (1986). *Aust. Vet. J.* **63,** 127–128.
Durand, M. R. E. (1974). "Phosphorus Deficiency and Supplementation of Grazing Cattle in Queensland," Tech. Bull. 3. Beef Cattle Husb. Branch, Queensl. Dep. Primary Ind., Queensland, Australia.
Durand, M., Foret, R., Dumay, C., and Gueguen, L. (1976). *Ann. Zootech.* **25,** 119–134.
Du Toit, P. J., Malan, A. I., Louw, J. G., Holzapfel, C. R., and Roets, G. (1934). *Onderstepoort J. Vet. Sci. Anim. Ind.* **2,** 607–648.
Du Toit, P. J., Malan, A. I., Louw, J. G., Holzapfel, C. R., and Roets, G. (1935). *Onderstepoort J. Vet. Sci. Anim. Ind.* **5,** 201–214.
Du Toit, P. J., Louw, J. G., and Malan, A. I. (1940). *Onderstepoort J. Vet. Sci. Anim. Ind.* **14,** 123–327.
Dutton, J. E., and Fontenot, J. P. (1967). *J. Anim. Sci.* **26,** 1409–1414.
Dynna, O., and Havre, G. N. (1963). *Acta Vet. Scand.* **4,** 197–208.
Edmeades, D. C., Smart, C. E., and Wheeler, D. M. (1983). *N.Z. J. Agric. Res.* **26,** 473–481.
Edmeades, D. C., O'Connor, M. B., and Toxopeus, M. R. J. (1987). *Proc. Anim. Sci. Congr. Asian–Australas. Assoc. Anim. Prod. Soc., 4th* p. 425.
Edye, L. A., Ritson, J. B., Haydock, K. P., and Davies, J. G. (1971). *Aust. J. Agric. Res.* **22,** 963–977.
Edye, L. A., Ritson, J. B., and Haydock, K. P. (1972). *Aust. J. Exp. Agric. Anim. Husb.* **12,** 7–12.
Egan, A. R. (1972). *Aust. J. Exp. Agric. Anim. Husb.* **12,** 131–135.
Egan, J. K., and Doyle, P. T. (1985). *Aust. J. Agric. Res.* **36,** 483–495.
Ehlig, C. F., Allaway, W. H., Cary, E. E., and Kubota, J. (1968). *Agron. J.* **60,** 43–47.
Ekern, A., Blaxter, K. L., and Sawers, D. (1965). *Br. J. Nutr.* **19,** 417–434.
Elkins, C. B., and Hoveland, C. S. (1977). *Agron. J.* **69,** 626–628.
Elliott, R. C., and Fokkema, K. (1960). *Rhod. Agric. J.* **57,** 301–304.
Elliot, R. C., and Fokkema, K. (1961). *Rhod. Agric. J.* **58,** 49–57.

Elliot, R. C., Fokkema, K., and French, C. H. (1961). *Rhod. Agric. J.* **58,** 124–130.
Ellis, K. J. (1980). *In* "CSIRO Symposium on the Importance of Copper in Biology and Medicine" (B. R. McAuslan, ed.), pp. 73–77. Commonw. Sci. Ind. Res. Org., Canberra, Australia.
Ellis, K. J., and Coverdale, O. R. (1982). *Proc. Aust. Soc. Anim. Prod.* **14,** 660.
Ellis, K. J., Laby, R. H., and Burns, R. G. (1981). *Proc. Nutr. Soc. Aust.* **6,** 145.
Emerick, R. J., and Embry, L. B. (1963). *J. Anim. Sci.* **22,** 510–513.
Eng, P. K., Mannetje, L., and Chen, C. P. (1978). *Trop. Grassl.* **12,** 198–207.
Engels, E. A. (1981). *S. Afr. J. Anim. Sci.* **11,** 171–182.
England, P., and Gill, M. (1983). *Anim. Prod.* **36,** 73–77.
Erizian, C. (1932). *Z. Tierz. Zuechtungsbiol.* **25,** 443–459.
Ernst, A. J., Limpus, J. F., and O'Rourke, P. K. (1975). *Aust. J. Exp. Agric. Anim. Husb.* **15,** 451–455.
Ernst, P., Le Du, Y. L. P., and Carlier, L. (1980). *In* "The Role of Nitrogen in Intensive Grassland Production" (W. H. Prins and G. H. Arnold, eds.), pp. 119–127. Pudoc, Wageningen, The Netherlands.
Evans, P. S. (1964). *N.Z. J. Agric. Res.* **7,** 508–513.
Evans, P. S. (1967a). *J. Agric. Sci.* **69,** 171–174.
Evans, P. S. (1967b). *J. Agric. Sci.* **69,** 175–181.
Evans, T. R. (1970). *Proc. Int. Grassl. Congr., 11th* pp. 803–807.
Evans, T. R., and Bryan, W. W. (1973). *Aust. J. Exp. Agric. Anim. Husb.* **13,** 530–536.
Fagan, T. W. (1928). *Welsh J. Agric.* **4,** 92–102.
Fahey, G. C., and Jung, H. G. (1983). *J. Anim. Sci.* **57,** 220–225.
Faichney, G. J., and Teleki, E. (1988). *Proc. Nutr. Soc. Aust.* **13,** 114.
Farhan, S. M. A., and Thomas, P. C. (1978). *J. Br. Grassl. Soc.* **33,** 151–158.
Fearn, J. T. (1961). *Aust. J. Exp. Agric. Anim. Husb.* **1,** 95–98.
Fels, H. E., Moir, R. J., and Rossiter, R. C. (1959). *Aust. J. Agric. Res.* **10,** 237–247.
Ferguson, W. S. (1932). *J. Agric. Sci.* **22,** 251–256.
Ferguson, W. S. (1948). *J. Agric. Sci.* **38,** 33–36.
Ferguson, W. S., and Terry, R. A. (1954). *J. Sci. Food Agric.* **5,** 515–524.
Ferguson, W. S., and Terry, R. A. (1956). *J. Agric. Sci.* **48,** 149–152.
Ferguson, K. A., Hemsley, J. A., and Reis, P. J. (1967). *Aust. J. Sci.* **30,** 215–216.
Fick, K. R., Ammerman, C. B., McGowan, C. H., Loggins, P. E., and Cornell, J. A. (1973). *J. Anim. Sci.* **36,** 137–143.
Field, H. I. (1957). *Vet. Rec.* **69,** 788–795.
Field, A. C. (1961). *Br. J. Nutr.* **15,** 287–295.
Field, A. C. (1983). *Livest. Prod. Sci.* **10,** 327–338.
Field, A. C., Suttle, N. F., and Nisbet, D. I. (1975). *J. Agric. Sci.* **85,** 435–442.
Field, A. C., Coop, R. L., Dingwall, R. A., and Munro, C. S. (1982). *J. Agric. Sci,* **99,** 311–317.
Filmer, J. F., and Underwood, E. J. (1937). *Aust. Vet. J.* **13,** 57–64.
Fisher, M. J. (1969). *Aust. J. Exp. Agric. Anim. Husb.* **9,** 196–208.
Fisher, M. J. (1970). *Aust. J. Exp. Agric. Anim. Husb.* **10,** 716–724.
Fisher, L. J. (1978). *Can. J. Anim. Sci.* **58,** 313–317.
Fishwick, G. (1976). *N.Z. J. Agric. Res.* **19,** 307–309.
Fishwick, G., and Hemingway, R. G. (1973). *J. Agric. Sci.* **81,** 139–143.
Fleck, A. T., Lusby, K. S., Owens, F. M., and McCollum, F. T. (1988). *J. Anim. Sci.* **66,** 750–757.
Fleming, G. A. (1963). *J. Sci. Food Agric.* **14,** 203–208.

Fleming, G. A. (1965). *Outlook Agric.* **4,** 270–285.
Fleming, G. A. (1968). *Agric. Dig.* **14,** 28–32.
Fleming, G. A. (1970). *Agric. Dig.* **19,** 25–32.
Fleming, G. A. (1973). *In* "Chemistry and Biochemistry of Herbage" (G. W. Butler and E. W. Bailey, eds.), Vol. 1, pp. 529–566. Academic Press, New York.
Fleming, G. A., and Murphy, W. E. (1968). *J. Br. Grassl. Soc.* **23,** 174–185.
Flores, J. F., Stobbs, T. H., and Minson, D. J. (1979). *J. Agric. Sci,* **92,** 351–357.
Flores, D. A., Phillip, L. E., Veira, D. M., and Ivan, M. (1986). *Can. J. Anim. Sci.* **66,** 1019–1027.
Flux, D. S., Butler, G. W., Rai, A. L., and Brougham, R. W. (1960). *J. Agric. Sci.* **55,** 191–196.
Flux, D. S., Butler, G. W., and Glenday, A. C. (1963). *J. Agric. Sci.* **61,** 197–200.
Fontenot, J. P., and Blaser, R. E. (1965). *J. Anim. Sci.* **24,** 1202–1208.
Fontenot, J. P., Miller, R. F., and Price, N. O. (1964). *J. Anim. Sci.* **23,** 874–875.
Fontenot, J. P., Wise, M. B., and Webb, K. E. (1973). *Fed. Proc., Fed. Am. Soc. Exp. Biol.* **32,** 1925–1928.
Foot, J. Z. (1972). *Anim. Prod.* **14,** 131–134.
Forbes, J. M. (1970a). *Br. Vet. J.* **126,** 1–11.
Forbes, J. M. (1970b). *J. Anim. Sci.* **31,** 1222–1227.
Forbes, J. M. (1977a). *Anim. Prod.* **24,** 91–101.
Forbes, J. M. (1977b). *Anim. Prod.* **24,** 203–214.
Forbes, R. M., and Garrigus, W. P. (1950). *J. Anim. Sci.* **9,** 354–362.
Forbes, J. C., and Gelman, A. L. (1981). *Grass Forage Sci.* **36,** 25–30.
Forbes, T. J., and Jackson, N. (1971). *J. Br. Grassl. Soc.* **26,** 257–264.
Forbes, E. B., Fries, J. A., and Braman, W. W. (1925). *J. Agric. Res.* **31,** 987–995.
Forbes, T. J., Raven, A. M., and Robinson, K. L. (1966). *J. Br. Grassl. Soc.* **21,** 167–173.
Forbes, T. J., Raven, A. M., Irwin, J. H. D., and Robinson, K. L. (1967a). *J. Br. Grassl. Soc.* **22,** 158–164.
Forbes, J. M., Rees, J. K. S., and Boaz, T. G. (1967b). *Anim. Prod.* **9,** 399–408.
Ford, C. W. (1978). *Aust. J. Agric. Res.* **29,** 1157–1166.
Forde, B. J., Slack, C. R., Roughan, P. G., Haslemore, R. M., and McLeod, M. N. (1976). *N.Z. J. Agric. Res.* **19,** 489–498.
Forero, O., Owens, F. N., and Lusby, K. S. (1980). *J. Anim. Sci.* **50,** 532–538.
Franklin, M. C. (1950). "Diet and Dental Development in Sheep." Bull. 252, Commonw. Sci. Ind. Res. Org., Melbourne, Australia.
Franklin, M. C., and Johnstone, I. L. (1948). *In* "Studies on Dietary and Other Factors Affecting the Serum-Calcium Levels of Sheep" (M. C. Franklin, R. L. Reid, and I. L. Johnstone, eds.), Bull. 240, pp. 63–71. Commonw. Sci. Ind. Res. Org., Melbourne, Australia.
Franklin, M. C., and Reid, R. L. (1948). *In* "Studies on Dietary and Other Factors Affecting the Serum-Calcium Levels of Sheep" (M. C. Franklin, R. L. Reid, and I. L. Johnstone, eds.), Bull. 240, pp. 53–58. Commonw. Sci. Ind. Res. Org., Melbourne, Australia.
Fraser, A. J. (1984). *Proc. N.Z. Soc. Anim. Prod.* **44,** 125–133.
Fraser, A. J., Ryan, T. J., Sproule, R., Clark, R. G., Anderson, D., and Pederson, E. O. (1987a). *Proc. N.Z. Soc. Anim. Prod.* **47,** 61–64.
Fraser, A. J., Ryan, T. J., Clark, R. G., and Sproule, R. (1987b). *Proc. Anim. Sci. Congr. Asian–Australas. Assoc. Anim. Prod. Soc., 4th* p. 427.
Freer, M., and Campling, R. C. (1963). *Br. J. Nutr.* **17,** 79–88.
Freer, M., and Jones, D. B. (1984). *Aust. J. Exp. Agric. Anim. Husb.* **24,** 156–164.

Freer, M., Campling, R. C., and Balch, C. C. (1962). *Br. J. Nutr.* **16,** 279–295.
French, M. H. (1957). *Herb. Abstr.* **27,** 1–9.
French, M. H. (1961). *Turrialba* **11,** 78–84.
Fribourg, H. A., Edwards, N. C., and Barth, K. M. (1971). *Agron. J.* **63,** 786–788.
Fritz, J. O., Cantrell, R. P., Lechtenberg, V. L., Axtell, J. D., and Hertel, J. M. (1981). *Crop Sci.* **21,** 706–709.
Fukazawa, K., Revol, J. F., Jurasek, L., and Goring, D. A. I. (1982). *Wood Sci. Technol.* **16,** 279–285.
Fulkerson, R. S., Mowat, D. N., Tossell, W. E., and Winch, J. E. (1967). *Can. J. Plant Sci.* **47,** 683–690.
Funk, M. A., Galyean, M. L., and Branine, M. E. (1987). *J. Anim. Sci.* **65,** 1354–1361.
Gabbedy, B. J. (1971). *Aust. Vet. J.* **47,** 318–322.
Gabrielsen, B. C., Vogel, K. P., and Knudsen, D. (1988). *Crop Sci.* **28,** 44–47.
Gaillard, B. D. E. (1962). *J. Agric. Sci.* **59,** 369–373.
Gaillard, B. D. E. (1966). *Neth. J. Agric. Sci.* **14,** 215–223.
Gallup, W. D., and Briggs, H. M. (1948). *J. Anim. Sci.* **7,** 110–116.
Galt, H. D., and Theurer, B, (1976). *J. Anim. Sci.* **42,** 1272–1279.
Gardener, C. J., Megarrity, R. G., and McLeod, M. N. (1982). *Aust. J. Exp. Agric. Anim. Husb.* **22,** 391–401.
Gardiner, M. R. (1969). *Outlook Agric.* **6,** 19–28.
Gardiner, M. R. (1977). *Tech. Bull.—West. Aust., Dep. Agric.* **36.**
Gardiner, M. R., and Gorman, R. C. (1963). *Aust. J. Exp. Agric. Anim. Husb.* **3,** 284–289.
Gardiner, M. R., Armstrong, J., Fels, H., and Glencross, R. N. (1962). *Aust. J. Exp. Agric. Anim. Husb.* **2,** 261–269.
Gardner, J. A. A. (1973). *Res. Vet. Sci.* **15,** 149–157.
Garmo, T. J., Froslie, A., and Hoie, R. (1986). *Acta Agric. Scand.* **36,** 147–161.
Garrigus, W. P. (1934). *Am. Soc. Anim. Prod., Rec. Proc. Annu. Meet.* pp. 66–69.
Garstang, J. R., and Mudd, C. H. (1971). *J. Br. Grassl. Soc.* **26,** 194.
Garstang, J. R., Thomas, C., and Gill, M. (1979). *Anim. Prod.* **28,** 423.
Gartner, R. J. W., and Murphy, G. M. (1974). *Proc. Aust. Soc. Anim. Prod.* **10,** 95–98.
Gartner, R. J. W., Ryley, J. W., and Beattie, A. W. (1965). *Aust. J. Exp. Biol. Med. Sci.* **43,** 713–724.
Gartner, R. J. W., Young, J. G., and Pepper, P. M. (1968). *Aust. J. Exp. Agric. Anim. Husb.* **8,** 679–682.
Gartner, R. J. W., Callow, L. L., Grazien, C. K., and Pepper, P. (1969). *Res. Vet. Sci.* **10,** 7–12.
Gartner, R. J. W., Dimmock, C. K., Stokoe, J., and Laws, L. (1970). *Queensl. J. Agric. Sci.* **27,** 405–410.
Gartner, R. J. W., McLean, R. W., Little, D. A., and Winks, L. (1980). *Trop. Grassl.* **14,** 266–272.
Gartner, R. J. W., Murphy, G. M., and Hoey, W. A. (1982). *J. Agric. Sci.* **98,** 23–29.
Gartrell, J. W. (1979). *In* "Mineral Requirements of Sheep and Cattle" (D. P. Purser, ed.), pp. 28–41. Aust. Soc. Anim. Prod., Perth, Australia.
Gartrell, J. W., and Glencross, R. N. (1968). *West. Aust., Dep. Agric. J.* **9,** 517–521.
Garza, R. T., Barnes, R. F., Mott, G. O., and Rhykerd, C. L. (1965). *Agron. J.* **57,** 417–420.
Gates, R. N., Klopfenstein, T. J., Waller, S. S., Stroup, W. W., Britton, R. A., and Anderson, B. F. (1987). *J. Anim. Sci.* **64,** 1821–1834.
Gawthorne, J. M. (1968). *Aust. J. Biol. Sci.* **21,** 789–794.
Geenty, K. G., and Sykes, A. R. (1986). *J. Agric. Sci.* **106,** 351–367.

Gibb, M. J., and Treacher, T. T. (1976). *J. Agric. Sci.* **86,** 355–365.
Gibb, M. J., and Treacher, T. T. (1978). *J. Agric. Sci.* **90,** 139–147.
Gibb, M. J., Treacher, T. T., and Shanmugalingam, V. S. (1981). *Anim. Prod.* **33,** 223–232.
Giduck, S. A., and Fontenot, J. P. (1984). *Can. J. Anim. Sci., Suppl.* **64,** 217–218.
Giduck, S. A., and Fontenot, J. P. (1987). *J. Anim. Sci.* **65,** 1667–1673.
Gihad, E. A. (1976). *J. Anim. Sci.* **43,** 879–883.
Gill, M., and England, P. (1984). *Anim. Prod.* **39,** 31–36.
Gissel-Nielsen, G. (1975). *Acta Agric. Scand.* **25,** 216–220.
Gissel-Nielsen, G., and Bisbjerg, B. (1970). *Plant Soil* **32,** 382–396.
Gladstones, J. S. (1962). *Aust. J. Exp. Agric. Anim. Husb.* **2,** 213–220.
Gladstones, J. S., and Loneragan, J. F. (1967). *Aust. J. Agric. Res.* **18,** 427–446.
Gladstones, J. S., and Loneragan, J. F. (1970). *Proc. Int. Grassl. Congr., 11th* pp. 350–396.
Gladstones, J. S., Loneragan, J. F., and Simmons, W. J. (1975). *Aust. J. Agric. Res.* **26,** 113–126.
Glenn, S., Rieck, C. E., Ely, D. G., and Bush, L. P. (1980). *J. Agric. Food Chem.* **28,** 391–393.
Glenn, B. P., Bond, J., and Glenn, S. (1987). *J. Anim. Sci.* **65,** 797–807.
Godwin, K. O., Kuchel, R. E., and Buckley, R. A. (1970). *Aust. J. Exp. Agric. Anim. Husb.* **10,** 672–678.
Goering, H. K., Hemken, R. W., Clark, N. A., and Vandersall, J. H. (1969). *J. Anim. Sci.* **29,** 512–518.
Goering, H. K., Smith, L. W., Van Soest, P. J., and Gordon, C. H. (1973). *J. Dairy Sci.* **56,** 233–240.
Goering, H. K., Derbyshire, J. C., Gordon, C. H., and Waldo, D. R. (1976). *J. Anim. Sci.* **43,** 263–264.
Gohl, B. (1975). "Tropical Feeds." Food Agric. Org., Rome.
Goings, R. L., Jacobson, N. L., Beitz, D. C., Littledike, E. T., and Wiggers, K. D. (1974). *J. Dairy Sci.* **57,** 1184–1188.
Golding, E. J., Moore, J. E., Franke, D. E., and Ruelke, D. C. (1976). *J. Anim. Sci.* **42,** 717–723.
Goldman, A., Genizi, A., Yulzari, A., and Seligman, N. G. (1987). *Anim. Feed Sci. Technol.* **18,** 233–245.
Gomide, J. A., Noller, C. H., Mott, G. O., Conrad, J. H., and Hill, D. L. (1969a). *Agron. J.* **61,** 116–120.
Gomide, J. A., Noller, C. H., Mott, G. O., Conrad, J. H., and Hill, D. L. (1969b). *Agron. J.* **61,** 120–123.
Gomm, F. B., Weswig, P. H., and Raleigh, R. J. (1982). *J. Range Manage.* **35,** 515–518.
Gonzalez, C. L., and Everitt, J. H. (1982). *J. Range Manage.* **35,** 733–736.
Gordon, F. J. (1979). *Anim. Prod.* **28,** 183–189.
Gordon, F. J. (1982). *Grass Forage Sci.* **37,** 59–65.
Gordon, F. J., and Peoples, A. C. (1986). *Anim. Prod.* **43,** 355–366.
Gordon, C. H., Melin, C. G., Wiseman, H. G., and Irvin, H. M. (1958). *J. Dairy Sci.* **41,** 1738–1746.
Gorsline, G. W., Thomas, W. I., and Baker, D. E. (1964). *Crop Sci.* **4,** 207–210.
Goto, I., and Minson, D. J. (1977). *Anim. Feed Sci. Technol.* **2,** 247–253.
Grace, N. D. (1973). *N.Z. J. Agric. Res.* **16,** 177–180.
Grace, N. D. (1975). *Br. J. Nutr.* **34,** 73–82.
Grace, N. D. (1981). *Br. J. Nutr.* **45,** 367–374.
Grace, N. D. (1983a). *In* "The Mineral Requirements of Grazing Ruminants" (N. D. Grace, ed.), Occas. Publ. 9, pp. 56–66. N.Z. Soc. Anim. Prod., Wellington, New Zealand.

Grace, N. D. (1983b). *N.Z. J. Agric. Res.* **26,** 59–70.
Grace, N. D., and Body, D. R. (1979). *N.Z. J. Agric. Res.* **22,** 405–410.
Grace, N. D., and Gooden, J. M. (1980). *N.Z. J. Agric. Res.* **23,** 293–298.
Grace, N. D., and Healy, W. B. (1974). *N.Z. J. Agric. Res.* **17,** 73–78.
Grace, N. D., and Watkinson, J. H. (1985). *In* "Trace Elements in Man and Animals" (C. F. Mills, I. Bremner, and J. K. Chesters, eds.), pp. 490–493. Commonw. Agric. Bur., Farnham Royal, England.
Grace, N. D., and Wilson, G. F. (1972). *N.Z. J. Agric. Res.* **15,** 72–78.
Grace, N. D., Maunsell, L. A., and Scott, D. (1973). *N.Z. J. Exp. Agric.* **2,** 99–102.
Grace, N. D., Ulyatt, M. J., and MacRae, J. C. (1974). *J. Agric. Sci.* **82,** 321–330.
Graham, N. M. (1967). *Aust. J. Agric. Res.* **18,** 137–147.
Graham, N. M. (1969). *Aust. J. Agric. Res.* **20,** 365–373.
Grant, A. B. (1965). *N.Z. J. Agric. Res.* **8,** 681–690.
Grant, A. B., Hartley, W. J., and Drake, C. (1960). *N.Z. Vet. J.* **8,** 1–3.
Grant, R. J., Van Soest, P. J., McDowell, R. E., and Perez, C. B. (1974). *J. Anim. Sci.* **39,** 423–434.
Grassland Research Institute (1961). "Research Techniques in Use at the Grassland Research Institute," Bull. 45. Commonw. Bur. Pastures Field Crops, Hurley, Maidenhead, England.
Green, H. H. (1925). *Physiol. Rev.* **5,** 336–348.
Green, J. O., Corrall, A. J., and Terry, R. A. (1971). *Grassl. Res. Inst. (Hurley), Tech. Rep.* **8.**
Greene, L. W., Fontenot, J. P., and Webb, K. E. (1983). *J. Anim. Sci.* **56,** 1208–1213.
Greene, L. W., Schelling, G. T., and Byers, F. M. (1986). *J. Anim. Sci.* **63,** 1960–1967.
Greenhalgh, J. F. D., and Corbett, J. L. (1960). *J. Agric. Sci.* **55,** 371–376.
Greenhalgh, J. F. D., and Reid, G. W. (1973). *Anim. Prod.* **16,** 223–233.
Greenhalgh, J. F. D., and Reid, G. W. (1974). *Anim. Prod.* **19,** 77–86.
Greenhalgh, J. F. D., and Reid, G. W. (1975). *Proc. Nutr. Soc.* **34,** 74A.
Greenhalgh, J. F. D., and Wainman, F. W. (1972). *Proc. Br. Soc. Anim. Prod.* pp. 61–72.
Greenhalgh, J. F. D., McDonald, I., and Corbett, J. L. (1959). *Proc. Nutr. Soc.* **18,** 18.
Greenhalgh, J. F. D., Corbett, J. L., and McDonald, I. (1960). *J. Agric. Sci.* **55,** 377–386.
Greenhill, W. L., Couchman, J. F., and De Freitas, J. (1961). *J. Sci. Food Agric.* **12,** 293–297.
Griffith, G., and Walters, R. J. K. (1966). *J. Agric. Sci.* **67,** 81–89.
Griffith, G., Jones, D. I. H., and Walters, R. J. K. (1965). *J. Sci. Food Agric.* **16,** 94–98.
Griffiths, T. W. (1959). *Br. Grassl. Soc. J.* **14,** 199–205.
Grings, E. E., and Males, J. R. (1987). *J. Anim. Sci.* **65,** 821–829.
Groppel, B., Anke, M., and Kronemann, H. (1985). *In* "Trace Elements in Man and Animals" (C. F. Mills, I. Bremner, and J. K. Chesters, eds.), pp. 279–282. Commonw. Agric. Bur., Farnham Royal, England.
Gross, C. F., and Jung, G. A. (1978). *Agron. J.* **70,** 397–403.
Gross, C. F., and Jung, G. A. (1981). *Agron. J.* **73,** 629–634.
Grunes, D. L. (1967). *Proc. Cornell Conf. Feed Manufacturers, 1967* pp. 105–110.
Grunes, D. L., Stout, P. R., and Brownell, J. R. (1970). *Adv. Agron.* **22,** 331–374.
Grunes, D. L., Wilkinson, S. R., Joo, P. K., Jackson, W. A., and Patterson, R. P. (1985). *Proc. Int. Grassl. Congr., 15th* pp. 509–510.
Gueguen, L., and Demarquilly, C. (1965). *Proc. Int. Grassl. Congr., 9th* pp. 745–754.
Gueguen, L., and Fauconneau, G. (1960). *Proc. Int. Grassl. Congr., 8th* pp. 621–625.
Gueguen, L., Foret, R., Durand, M., Allex, M., and Camus, P. (1976). *Ann. Zootech.* **25,** 111–118.

Gunn, R. G. (1969). *J. Agric. Sci.* **72,** 371–378.
Gupta, V. P., and Sehgal, K. L. (1971). *Indian J. Genet. Plant Breed.* **31,** 416–419.
Gupta, U. C., and Winter, K. A. (1975). *Can. J. Soil Sci.* **55,** 161–166.
Gupta, P. C., Singh, R., and Pradhan, K. (1979). *Indian J. Anim. Sci.* **49,** 462–463.
Gutierrez, O., Geerken, C. M., Funes, F., and Diaz, A. (1980). *Cuban J. Agric. Sci.* **14,** 159–163.
Haag, J. R., Jones, I. R., and Brant, P. M. (1932). *J. Dairy Sci.* **15,** 23–28.
Haaland, G. L., Matsushima, J. K., Nockels, C. F., and Johnson, D. E. (1977). *J. Anim. Sci.* **46,** 826–831.
Haaranen, S. (1963). *Nord. Veterinaermed.* **15,** 536–542.
Habib, G., and Leng, R. A. (1986). *Proc. Aust. Soc. Anim. Prod.* **16,** 223–226.
Hacker, J. B. (1974a). *Trop. Grassl.* **8,** 145–154.
Hacker, J. B. (1974b). *Aust. J. Agric. Res.* **25,** 401–406.
Hacker, J. B. (1982). *In* "Nutritional Limits to Animal Production from Pastures" (J. B. Hacker, ed.), pp. 305–326. Commonw. Agric. Bur., Farnham Royal, England.
Hacker, J. B., and Jones, R. J. (1969). *Trop. Grassl.* **3,** 13–34.
Hacker, J. B., and Minson, D. J. (1972). *Aust. J. Agric. Res.* **23,** 959–967.
Hacker, J. B., Strickland, R. W., and Basford, K. E. (1985). *Aust. J. Agric. Res.* **36,** 201–212.
Hadjipanayiotou, M. (1982). *Grass Forage Sci.* **37,** 89–93.
Hadjipanayiotou, M. (1984). *Anim. Feed Sci. Technol.* **11,** 67–74.
Hadjipieris, G., Jones, J. G. W., and Holmes, W. (1965). *Anim. Prod.* **7,** 309–317.
Haenlein, G. F. W., Richards, C. R., Salsbury, R. L., Yoon, Y. M., and Mitchell, W. H. (1966). *Del., Agric. Exp. Stn., Bull.* **359.**
Haggar, R. J. (1970). *J. Agric. Sci.* **74,** 487–494.
Haggar, R. J., and Ahmed, M. B. (1970). *J. Agric. Sci.* **75,** 369–373.
Hall, O. G., Baxter, H. D., and Hobbs, C. S. (1961). *J. Anim. Sci.* **20,** 817–819.
Halpin, C., McDonald, J., Hanrahan, P., and Caple, I. (1985). *In* "Trace Elements in Man and Animals" (C. F. Mills, I. Bremner, and J. K. Chesters, eds.), pp. 746–748. Commonw. Agric. Bur., Farnham Royal, England.
Hamilton, J. W., and Beath, O. A. (1963). *J. Range Manage.* **16,** 261–265.
Hancock, J. (1952). *Proc. Int. Grassl. Congr., 6th* pp. 1399–1407.
Handreck, K. A., and Godwin, K. O. (1970). *Aust. J. Agric. Res.* **21,** 71–84.
Handreck, K. A., and Riceman, D. S. (1969). *Aust. J. Agric. Res.* **20,** 213–226.
Hanna, W. W., Monson, W. G., and Burton, G. W. (1973). *Crop Sci.* **13,** 98–102.
Hanna, W. W., Gaines, T. P., and Monson, W. G. (1979). *Agron. J.* **71,** 1027–1029.
Hannam, R. J., Judson, G. J., Reuter, D. J., McLaren, L. D., and McFarlane, J. D. (1980). *Aust. J. Agric. Res.* **31,** 347–355.
Hannaway, D. B., Claypool, D. W., Adams, H. P., Beuttner, M. R., Carter, G. R., Adams, F. W., Allison, L., and Vough, L. R. (1981). *Oreg., Agric. Exp. Stn., Bull.* **141.**
Hansard, S. L., Crowder, H. M., and Lyke, W. A. (1957). *J. Anim. Sci.* **16,** 437–443.
Harb, M. Y., and Campling, R. C. (1983). *Grass Forage Sci.* **38,** 115–119.
Hardison, W. A., and Reid, J. T. (1953). *J. Nutr.* **51,** 35–52.
Harkess, R. D., and Alexander, R. H. (1969). *J. Br. Grassl. Soc,* **24,** 282–289.
Harris, L. E., and Mitchell, H. H. (1941). *J. Nutr.* **22,** 167–182.
Harris, C. E., and Raymond, W. F. (1963). *J. Br. Grassl. Soc.* **18,** 204–212.
Harris, L. E., Cook, C. W., and Stoddart, L. A. (1956). *Bull.—Utah Agric. Exp. Stn.* **398.**
Harris, C. E., Raymond, W. F., and Wilson, R. F. (1966). *Proc. Int. Grassl. Congr., 10th* pp. 564–567.
Hart, L. I., and Andrews, E. D. (1959). *Nature (London)* **184,** 1242.

Hartley, R. D., and Jones, E. C. (1978). *J. Sci. Food Agric.* **29,** 777–789.
Hartley, W. J. (1961). *N.Z. J. Agric.* **103,** 475–483.
Hartley, W. J. (1963). *Proc. N.Z. Soc. Anim. Prod.* **23,** 20–27.
Hartley, W. J. (1967). *Int. Symp. Trace Elem., Oreg. State Univ., 1st* pp. 79–96.
Hartmans, J. (1962). *Inst. Biol. Scheikd. Onderz. Landbouwgewassen, Wageningen, Jaarb.* **193,** 157–166.
Hartmans, J. (1969). *Agric. Dig.* **18,** 42–48.
Hartmans, J. (1971). *In* "Potassium in Biochemistry and Physiology," pp. 207–211. Int. Potash Inst., Berne, Switzerland.
Hartmans, J. (1974). *Neth. J. Agric. Sci.* **22,** 195–206.
Hartmans, J. (1975). *Stikstof (Engl. Ed.)* **18,** 12–21.
Hartmans, J., and Bosman, M. S. M. (1970). *In* "Trace Element Metabolism in Animals" (C. F. Mills, ed.), pp. 362–366. Livingstone, Edinburgh, Scotland.
Harvey, J. M. (1952). *Queensl. J. Agric. Sci.* **9,** 169–184.
Hasler, A. (1962). *Schweiz. Landwirtsch. Forsch.* **1,** 60–73.
Hathaway, R. L., Allison, L., Oldfield, J. E., and Carter, G. E. (1979). *J. Anim. Sci.* **49,** 373–374.
Hawkins, G. E., Paar, G. E., and Little, J. A. (1964). *J. Diary Sci.* **47,** 865–870.
Hays, V. W., and Swenson, M. J. (1970). *In* "Dukes' Physiology of Domestic Animals" (M. J. Swenson, ed.), pp. 663–690. Cornell Univ. Press, Ithaca, New York.
Head, M. J., and Rook, J. A. F. (1957). *Proc. Nutr. Soc.* **16,** 25–30.
Healy, W. B. (1972). *Proc. N.Z. Grassl. Assoc.* **34,** 84–90.
Healy, W. B., McCabe, W. J., and Wilson, G. F. (1970). *N.Z. J. Agric. Res.* **13,** 503–521.
Healy, W. G., Crouchley, G., Gillett, R. L., Rankin, P. C., and Watts, H. M. (1972). *N.Z. J. Agric. Res.* **15,** 778–782.
Heaney, D. P. (1973). *Can. J. Anim. Sci.* **53,** 431–438.
Heaney, D. P. (1979). *In* "Standardization of Analytical Methodology for Feeds" (W. J. Pigden, C. C. Balch, and M. Graham, eds.), pp. 45–48. Int. Dev. Res. Cent., Ottawa, Ontario, Canada.
Heaney, D. P., and Pigden, W. J. (1963). *J. Anim. Sci.* **22,** 956–960.
Heaney, D. P., Pigden, W. J., Minson, D. J., and Pritchard, G. I. (1963). *J. Anim. Sci.* **22,** 752–757.
Heaney, D. P., Pigden, W. J., and Pritchard, G. I. (1966a). *J. Anim. Sci.* **25,** 142–149.
Heaney, D. P., Pigden, W. J., and Pritchard, G. I. (1966b). *Proc. Int. Grassl. Congr., 10th* pp. 379–384.
Heaney, D. P., Pritchard, G. I., and Pigden, W. J. (1968). *J. Anim. Sci.* **27,** 159–164.
Heaney, D. P., Pigden, W. J., and Minson, D. J. (1969). *In* "Experimental Methods for Evaluating Herbage" (J. B. Campbell, ed.), Ottawa Publ. 1315, pp. 185–199. Can. Dep. Agric., Ottawa, Ontario, Canada.
Heinrichs, D. H., Troelsen, J. E., and Warder, F. G. (1969). *Can. J. Plant Sci.* **49,** 293–305.
Hemingway, R. G. (1960). *J. Sci. Food Agric.* **11,** 355–362.
Hemingway, R. G. (1961a). *J. Br. Grassl. Soc.* **16,** 106–116.
Hemingway, R. G. (1961b). *J. Sci. Food Agric.* **12,** 398–406.
Hemingway, R. G. (1962). *J. Br. Grassl. Soc.* **17,** 182–187.
Hemingway, R. G., and Fishwick, G. (1975a). *Proc. Nutr. Soc.* **34,** 78–79A.
Hemingway, R. G., and Fishwick, G. (1975b). *J. Agric. Sci.* **84,** 381–382.
Hemingway, R. G., and Fishwick, G. (1976). *In* "Feed Energy Sources for Livestock" (H. Swan and D. Jones, eds.), pp. 95–115. Butterworths, London.
Hemingway, R. G., MacPherson, A., Duthie, A. K., and Brown, N. A. (1968). *J. Agric. Sci.* **71,** 53–59.

Hemsley, J. A., and Moir, R. J. (1963). *Aust. J. Agric. Res.* **14,** 509–517.
Hemsley, J. A., Hogan, J. P., and Weston, R. H. (1970). *Proc. Int. Grassl. Congr., 11th* pp. 703–706.
Hendricksen, R. E., and Minson, D. J. (1980). *J. Agric. Sci.* **95,** 547–554.
Hendricksen, R. E., and Minson, D. J. (1985a). *Trop. Grassl.* **19,** 81–87.
Hendricksen, R. E., and Minson, D. J. (1985b). *Herb. Abstr.* **55,** 215–228.
Hendricksen, R. E., Poppi, D. P., and Minson, D. J. (1981). *Aust. J. Agric. Res.* **32,** 389–398.
Henkens, C. H. (1965). *Neth. J. Agric. Sci.* **13,** 21–47.
Hennessy, D. W., and McClymont, G. C. (1970). *Proc. Aust. Soc. Anim. Prod.* **8,** 207–211.
Hennessy, D. W., and Sundstrom, B. (1975). *J. Aust. Inst. Agric. Sci.* **41,** 59–60.
Hennessy, D. W., Nolan, J. V., Norton, B. W., Ball, F. M., and Leng, R. A. (1978). *Aust. J. Exp. Agric. Anim. Husb.* **18,** 477–482.
Hennessy, D. W., Williamson, P. J., Nolan, J. V., Kempton, T. J., and Leng, R. A. (1983). *J. Agric. Sci.* **100,** 657–666.
Hennig, A., Anke, M., Groppel, B., Ludke, H., Reissig, W., Dittrich, G., and Grun, M. (1972). *Arch. Tierernaehr.* **22,** 601–614.
Henrici, M. (1934). *Sci. Bull.—S. Afr., Dep. Agric. Tech. Serv.* **134.**
Henry, D. P., Greenfield, P. F., Ouano, E. A., Thomson, R. H., Fleming, G., and Minson, D. J. (1978). *Nature (London)* **274,** 619–620.
Henzell, E. F. (1963). *Aust. J. Exp. Agric. Anim. Husb.* 3, 290–299.
Herrera, R. S. (1977). *Cuban J. Agric. Sci.* **11,** 331–345.
Herrera, R. S. (1979). *Cuban J. Agric. Sci.* **13,** 101–112.
Hershberger, T. V., Long, T. A., Hartsook, E. W., and Swift, R. W. (1959). *J. Anim. Sci.* **18,** 770–779.
Hewitt, C. (1974). *Acta Vet. Scand.* **50,** 152.
Hides, D. H., Hayward, M. V., and Lovatt, J. A. (1983). *Grass Forage Sci.* **38,** 33–38.
Hidiroglou, M. (1979a). *Can. J. Anim. Sci.* **59,** 217–236.
Hidiroglou, M. (1979b). *J. Dairy Sci.* **62,** 1195–1206.
Hidiroglou, M., and Spurr, D. T. (1975). *Can. J. Anim. Sci.* **55,** 31–38.
Hidiroglou, M., Dermine, P., Hamilton, H. A., and Troelsen, J. E. (1966). *Can. J. Plant Sci.* **46,** 101–109.
Hidiroglou, M., Ho, S. K., and Standish, J. F. (1978). *Can. J. Anim. Sci.* **58,** 35–41.
Hidiroglou, M., Proulx, J., and Jolette, J. (1985). *In* "Trace Elements in Man and Animals" (C. F. Mills, I. Bremner, and J. K. Chesters, eds.), pp. 744–746. Commonw. Agric. Bur., Farnham Royal, England.
Hidiroglou, M., Proulz, J., and Jolette, J. (1987). *J. Anim. Sci.* **65,** 815–820.
Hight, G. K., Sinclair, D. P., and Lancaster, R. J. (1968). *N.Z. J. Agric. Res.* **11,** 286–302.
Hignett, S. L., and Hignett, P. G. (1951). *Vet. Rec.* **63,** 603–609.
Hill, R. R., and Jung, G. A. (1975). *Crop Sci.* **15,** 652–657.
Hill, M. K., Walker, S. D., and Taylor, A. G. (1969). *N.Z. J. Agric. Res.* **12,** 261–270.
Hino, T., and Russell, J. B. (1987). *J. Anim. Sci.* **64,** 261–270.
Hobson, P. N., ed. (1988). "The Rumen Microbial Ecosystem." Elsevier, Amsterdam.
Hodge, R. W. (1973). *Aust. J. Agric. Res.* **24,** 237–243.
Hodge, R. W., Pearce, G. R., and Tribe, D. E. (1973). *Aust. J. Agric. Res.* **24,** 229–236.
Hodges, E. M., Kirk, W. G., Peacock, F. M., Jones, D. W., Davis, G. K., and Neller, J. R. (1964). *Bull.—Fla., Agric. Exp. Stn.* **686.**
Hodges, E. M., Kirk, W. G., Davis, G. K., Shirley, R. L., Peacock, F. M., Easley, J. F., Breland, H. L., and Martin, F. C. (1968). *Circ.—Fla., Agric. Exp. Stn.* **S190.**
Hodgson, J. (1968). *J. Agric. Sci.* **70,** 47–51.
Hodgson, J. (1975). *J. Br. Grassl. Soc.* **30,** 307–313.

Hodgson, J. (1985). *Proc. Nutr. Soc.* **44,** 339–346.
Hodgson, J., and Spedding, C. R. W. (1966). *J. Agric. Sci.* **67,** 155–167.
Hodgson, J., Rodriguez, J. M., and Fenlon, J. S. (1977). *J. Agric. Sci.* **89,** 743–750.
Hoehne, O. E., Clanton, D. C., and Streeter, C. L. (1967). *J. Anim. Sci.* **26,** 628–631.
Hoey, W. A., Murphy, G. M., and Gartner, R. J. W. (1982). *J. Agric. Sci.* **98,** 31–37.
Hogan, J. P. (1974). *J. Diary Sci.* **58,** 1164–1177.
Hogan, J. P. (1981). *In* "Forage Evaluation: Concepts and Techniques" (J. L. Wheeler and R. D. Mochrie, eds.), pp. 177–188. Commonw. Sci. Ind. Res. Org. and Am. Forage and Grassl. Counc., Melbourne, Australia.
Hogan, J. P. (1982). *In* "Nutritional Limits to Animal Production from Pastures" (J. B. Hacker, ed.), pp. 245–257. Commonw. Agric. Bur., Farnham Royal, England.
Hogan, J. P., and Weston, R. H. (1969). *Aust. J. Agric. Res.* **20,** 347–363.
Hogan, J. P., and Weston, R. H. (1970). *In* "Physiology of Digestion and Metabolism in the Ruminant" (A. T. Phillipson, ed.), pp. 474–485. Oriel, Newcastle-upon-Tyne, England.
Hogan, J. P., and Weston, R. H. (1981). *In* "Forage Evaluation: Concepts and Techniques" (J. L. Wheeler and R. D. Mochrie, eds.), pp. 75–88. Commonw. Sci. Ind. Res. Org. and Am. Forage Grassl. Counc., Melbourne, Australia.
Hogan, M. R., Henderson, B. W., Berousek, E. R., Wakefield, R. C., and Gilbert, R. W. (1967). *J. Diary Sci.* **50,** 87–89.
Hogan, K. G., Money, D. F. L., White, D. A., and Walker, R. (1971). *N.Z. J. Agric. Res.* **14,** 687–701.
Hogan, K. G., Lorentz, P. P., and Gibb, F. M. (1973). *N.Z. Vet. J.* **21,** 234–237.
Holder, J. M. (1962). *Proc. Aust. Soc. Anim. Prod.* **4,** 154–159.
Holechek, J. L., Galyean, M. L., Wallace, J. D., and Wofford, H. (1985). *Grass Forage Sci.* **40,** 489–492.
Holm, A. M., and Payne, A. L. (1980). *Aust. J. Exp. Agric. Anim. Husb.* **20,** 398–405.
Holmes, J. H. G. (1979). *Papua New Guinea Agric. J.* **30,** 65–69.
Holmes, J. H. G. (1981). *Trop. Anim. Health Prod.* **13,** 169–176.
Holmes, J. C., and Lang, R. W. (1963). *Anim. Prod.* **5,** 17–26.
Holmes, J. C., and Osmon, H. S. (1962). *Anim. Prod.* **2,** 131–139.
Holmes, J. H. G., Franklin, M. C., and Lambourne, L. J. (1966). *Proc. Aust. Soc. Anim. Prod.* **6,** 354–363.
Holroyd, R. G., O'Rourke, P. K., Clarke, M. R., and Loxton, I. D. (1983). *Aust. J. Exp. Agric. Anim. Husb.* **23,** 4–13.
Holt, E. C., and Conrad, B. E. (1986). *Agron. J.* **78,** 433–436.
Holt, E. C., and Dalrymple, R. L. (1979). *Agron. J.* **71,** 59–62.
Holter, J. A., and Reid, J. T. (1959). *J. Anim. Sci.* **18,** 1339–1349.
Homb, T. (1953). *Acta Agric. Scand.* **3,** 1–32.
Homb, T. (1984). *In* "Straw and Other Fibrous By-products as Feed" (F. Sundstol and E. Owen, eds.), pp. 106–126. Elsevier, Amsterdam.
Hoover, W. H. (1986). *J. Dairy Sci.* **69,** 2755–2766.
Horn, J. P., and Smith, R. H. (1978). *Br. J. Nutr.* **40,** 473–484.
Horn, F. P., Reid, R. L., and Jung, G. A. (1974). *J. Anim. Sci.* **38,** 968–974.
Horrocks, D. (1964). *J. Agric. Sci.* **63,** 373–375.
Horst, R. L. (1986). *J. Dairy Sci.* **69,** 604–616.
Horton, G. M. J. (1978). *Can. J. Anim. Sci.* **58,** 471–478.
Horton, G. M. J., and Steacy, G. M. (1979). *J. Anim. Sci.* **48,** 1239–1249.
House, W. A., and Van Campen, D. (1971). *J. Nutr.* **101,** 1483–1492.
Hovell, F. D. D., and Greenhalgh, J. F. D. (1978). *Br. J. Nutr.* **40,** 171–183.
Hovell, F. D. D., Greenhalgh, J. F. D., and Wainman, F. W. (1976). *Br. J. Nutr.* **35,** 343–363.

Hovell, F. D. D., Ngambi, J. W. W., Barber, W. P., and Kyle, D. J. (1986). *Anim. Prod.* **42,** 111–118.
Hovin, A. W., Stucker, R. E., and Marten, G. C. (1974). *Proc. Int. Grassl. Congr., 12th* **3,** 793–798.
Hovin, A. W., Marten, G. C., and Stucker, R. E. (1976). *Crop Sci.* **16,** 575–578.
Hovin, A. W., Tew, T. L., and Stucker, R. E. (1978). *Crop Sci.* **18,** 423–427.
Howard, D. A. (1970). *Vet. Rec.* **87,** 771–774.
Howard, D. A., Burdin, M. L., and Lampkin, G. H. (1962). *J. Agric. Sci.* **59,** 251–256.
Howell, J. M., Pass, D. A., and Terlecki, S. (1981). *In* "Trace Element Metabolism in Man and Animals" (J. M. Howell, J. M. Gawthorne, and C. L. White, eds.), pp. 298–301. Aust. Acad. Sci., Canberra, Australia.
Hubbard, W. A. (1952). *Proc. Int. Grassl. Congr., 6th* pp. 1343–1347.
Hubbert, F., Cheng, E., and Burroughs, W. (1958). *J. Anim. Sci.* **17,** 559–568.
Hudson, D. R., Hunter, R. A., and Peter, D. W. (1981). *Aust. J. Agric. Res.* **32,** 935–946.
Hume, I. D. (1970). *Aust. J. Agric. Res.* **21,** 297–304.
Hume, I. D., and Bird, P. R. (1970). *Aust. J. Agric. Res.* **21,** 315–322.
Hume, I. D., Somers, M., and McKeown, N. R. (1968). *Aust. J. Exp. Agric. Anim. Husb.* **8,** 295–300.
Humphries, W. R., Phillippo, M., Young, B. W., and Bremmer, I. (1983). *Br. J. Nutr.* **49,** 77–86.
Hungate, R. E. (1966) "The Rumen and Its Microbes." Academic Press, New York.
Hunt, C. W., Paterson, J. A., and Williams, J. E. (1985). *J. Anim. Sci.* **60,** 301–306.
Hunter, R. A., and Siebert, B. D. (1980). *Aust. J. Agric. Res.* **31,** 1037–1047.
Hunter, R. A., and Siebert, B. D. (1985a). *Br. J. Nutr.* **53,** 637–648.
Hunter, R. A., and Siebert, B. D. (1985b). *Br. J. Nutr.* **53,** 649–656.
Hunter, R. A., and Siebert, B. D. (1986a). *Aust. J. Agric. Res.* **37,** 549–560.
Hunter, R. A., and Siebert, B. D. (1986b). *Aust. J. Agric. Res.* **37,** 665–671.
Hunter, R. A., and Siebert, B. D. (1987). *Aust. J. Agric. Res.* **38,** 209–218.
Hunter, R. A., and Vercoe, J. E. (1984). *Outlook Agric.* **13,** 154–159.
Hunter, R. A., Peter, D. W., and Buscall, D. J. (1979a). *Proc. Nutr. Soc. Aust.* **4,** 147.
Hunter, R. A., Siebert, B. D., and Webb, C. D. (1979b). *Aust. J. Exp. Agric. Anim. Husb.* **19,** 517–521.
Hupkens Van Der Elsk, F. C. C., and Watkinson, J. H. (1977). *N.Z. J. Exp. Agric.* **5,** 79–84.
Hutchinson, K. J., Wilkins, R. J., and Osbourn, D. F. (1971). *J. Agric. Sci.* **77,** 545–547.
Hutton, J. B. (1961). *N.Z. J. Agric. Res.* **4,** 583–590.
Hutton, J. B. (1963). *Proc. N.Z. Soc. Anim. Prod.* **23,** 39–52.
Hutton, J. B., Jury, K. E., and Davies, E. B. (1965). *N.Z. J. Agric. Res.* **8,** 479–496.
Hutton, J. B., Jury, K. E., and Davies, E. B. (1967). *N.Z. J. Agric. Res.* **10,** 367–388.
Hutton, J. B., Jury, K. E., Hughes, J. M., Parker, O. F., and Lancaster, R. J. (1971). *N.Z. J. Agric. Res.* **14,** 393–405.
Hutton, J. B., Hughes, J. W., Bryant, A. M., Pluck, L. J., and Taylor, R. E. C. (1975). *N.Z. J. Agric. Res.* **18,** 37–43.
Imperial Chemical Industries (1966). "Jealott's Hill Research Station Guide to Field Experiments," p. 90. [Cited by Whitehead, D. C. (1966). "The Role of Nitrogen in Grassland Productivity." Commonw. Agric. Bur., Farnham Royal, England.]
Ingalls, J. R., Thomas, J. W., Tesar, M. B., and Carpenter, D. L. (1966). *J. Anim. Sci.* **25,** 283–289.
Innes, R. F. (1947). *Jam. Dep. Sci. Agric., Bull.* **35.**
Ishiguri, T. (1979). *Nippon Sochi Gakkaishi* **25,** 156–160.
Ishiguri, T. (1980). *Nippon Sochi Gakkaishi* **26,** 324–329.

Ishiguri, T. (1983). *Nippon Sochi Gakkaishi* **29,** 148–153.
Ishizaki, S. M., and Stanley, R. W. (1967). *Hawaii, Agric. Exp. Stn., Tech. Prog. Rep.* **154.**
Ivan, M., Ihnat, M., and Veira, D. M. (1983). *Anim. Feed Sci. Technol.* **9,** 131–142.
Ivan, M., Veira, D. M., and Kelleher, C. A. (1986). *Br. J. Nutr.* **55,** 361–367.
Jackson, N., and Forbes, T. J. (1970). *Anim. Prod.* **12,** 591–599.
Jacques, K. A., Cochran, R. C., Corah, L. R., Avery, T. B., Zoellner, K. O., and Higgingbotham, J. F. (1987). *J. Anim. Sci.* **65,** 777–785.
Jagusch, K. T., Gumbrell, R. C., Mobley, M. C., and Jay, N. P. (1977a). *N.Z. J. Exp. Agric.* **5,** 15–18.
Jagusch, K. T., Gumbrell, R. C., Mobley, M. C., and Jay, N. P. (1977b). *N.Z. J. Exp. Agric.* **5,** 19–22.
Jamieson, W. S., and Hodgson, J. (1979). *Grass Forage Sci.* **34,** 261–271.
Jarl, F. (1938). *Nord. Jordbrugsforsk.* **20,** 1–30.
Jarrige, R. (1960). *Proc. Int. Grassl. Congr., 8th* pp. 628–635.
Jarrige, R. (1980). *Ann. Zootech.* **29,** 299–323.
Jarrige, R., and Minson, D. J. (1964). *Ann. Zootech.* **13,** 117–150.
Jarrige, R., Thivend, P., and Demarquilly, C. (1970). *Proc. Int. Grassl. Congr., 11th* pp. 762–766.
Jarrige, R., Demarquilly, C., and Dulphy, J. p. (1982). *In* "Nutritional Limits to Animal Production from Pastures" (J. B. Hacker, ed.), pp. 363–387. Commonw. Agric. Bur., Farnham Royal, England.
Jarrige, R., Demarquilly, C., Dulphy, J. P., Hoden, A., Robelin, J., Beranger, C., Geay, Y., Journet, M., Malterrre, C., Micol, D., and Petit, M. (1986). *J. Anim. Sci.* **63,** 1737–1758.
Jarvis, S. C. (1982). *Ann. Bot. (London)* **49,** 199–206.
Jayasuriya, M. C. N., and Owen, E. (1975). *Anim. Prod.* **21,** 313–322.
Jewell, S. N., and Campling, R. C. (1986). *Anim. Feed Sci. Technol.* **14,** 81–93.
Joblin, K. N., and Keogh, R. G. (1979). *J. Agric. Sci.* **92,** 571–574.
Joblin, K. N., and Pritchard, M. W. (1983). *Plant Soil* **70,** 69–76.
John, A., and Ulyatt, M. J. (1987). *Proc. N.Z. Soc. Anim. Prod.* **47,** 13–16.
Johnson, C. M. (1975). *In* "Trace Elements in Soil–Plant–Animal Systems" (D. I. D. Nicholas and A. R. Egan, eds.), pp. 165–180. Academic Press, New York.
Johnson, R. R. (1976). *J. Anim. Sci.* **43,** 184–191.
Johnson, J. M., and Butler, G. W. (1957). *Physiol. Plant.* **10,** 100–111.
Johnson, J. C., and McCormick, W. C. (1976). *J. Anim. Sci.* **42,** 175–179.
Johnson, H. D., and Yeck, R. G. (1964). *Mo., Agric. Exp. Stn., Bull.* **865.**
Johnson, R. R., Dehority, B. A., and Parsons, J. L. (1965). *Proc. Int. Grassl. Congr., 9th* pp. 773–778.
Johnson, R. R., McClure, K. E., Johnson, L. J., Klosterman, E. W., and Triplett, G. (1966). *Agron. J.* **58,** 151–153.
Johnson, W. H., Norman, B. B., and Dunbar, J. R. (1981). *In* "Trace Element Metabolism in Man and Animals" (J. M. Howell, J. M. Gawthorne, and C. L. White, eds.), pp. 203–206. Aust. Acad. Sci., Canberra, Australia.
Johnston, M. J., and Waite, R. (1965). *J. Agric. Sci.* **64,** 211–219.
Jolliff, G. D., Garza, A., and Hertel, J. M. (1979). *Agron. J.* **71,** 91–94.
Jolly, R. D. (1960). *N.Z. Vet. J.* **8,** 11–12.
Jones, D. I. H. (1963a). *Rhod. J. Agric. Res.* **1,** 35–38.
Jones, E. (1963b). *J. Br. Grassl. Soc.* **18,** 131–138.
Jones, D. I. H. (1964). *Rhod. J. Agric. Res.* **2,** 57–59.
Jones, R. K. (1968). *Aust. J. Exp. Agric. Anim. Husb.* **8,** 521–527.

Jones, A. S. (1976a). *In* "Green Crop Fractionation" (R. J. Wilkins, ed.), Occas. Symp. 9, pp. 1–7. Br. Grassl. Soc., Hurley, Maidenhead, England.
Jones, D. I. H. (1976b). *Misc. Pap.—Landbouwhogesch. Wageningen* **12,** 67–77.
Jones, R. J. (1981). *Aust. Vet. J.* **57,** 55.
Jones, O., and Anthony, W. B. (1965). *J. Anim. Sci.* **24,** 285.
Jones, R. J., and Ford, C. W. (1972a). *Trop. Grassl.* **6,** 201–204.
Jones, R. J., and Ford, C. W. (1972b). *Aust. J. Exp. Agric. Anim. Husb.* **12,** 400–406.
Jones, L. H. P., and Handreck, K. A. (1965). *J. Agric. Sci.* **65,** 129–134.
Jones, D. I. H., and Hayward, M. V. (1975). *J. Sci. Food Agric.* **26,** 711–718.
Jones, R. J., and Hegarty, M. P. (1984). *Aust. J. Agric. Res.* **35,** 317–325.
Jones, R. J., and Lowry, J. B. (1984). *Experientia* **40,** 1435–1436.
Jones, R. J., and Megarrity, R. G. (1983). *Aust. J. Agric. Res.* **34,** 781–790.
Jones, R. J., and Megarrity, R. G. (1986). *Aust. Vet. J.* **63,** 259–262.
Jones, D. I. H., and Walters, R. J. K. (1969). *In* "Grass and Forage Breeding" (L. Phillips and R. Hughes, eds.), Occas. Symp. 5, pp. 37–45. Br. Grassl. Soc., Hurley, Maidenhead, England.
Jones, D. I. H., and Walters, R. J. K. (1975). *J. Sci. Food Agric.* **26,** 1436–1437.
Jones, R. J., and Winter, W. H. (1982). *Leucaena Res. Rep.* **3,** 2.
Jones, L. H. P., Milne, A. A., and Wadham, S. M. (1963). *Plant Soil* **18,** 358–371.
Jones, R. J., Griffiths Davies, J., and Waite, R. B. (1967a). *Aust. J. Exp. Agric. Anim. Husb.* **7,** 57–65.
Jones, D. I. H., Miles, D. G., and Sinclair, K. B. (1967b). *Br. J. Nutr.* **21,** 391–397.
Jones, R. J., Blunt, C. G., and Nurnberg, B. I. (1978). *Aust. Vet. J.* **54,** 387–392.
Jones, G. M., Larsen, R. E., and Lanning, N. M. (1980). *J. Dairy Sci.* **63,** 579–586.
Jones, A. L., Goetsch, A. L., Stokes, S. R., and Colberg, M. (1988). *J. Anim. Sci.* **66,** 194–203.
Joshi, D. C. (1973). *Acta Agric. Scand.* **23,** 5–10.
Jouany, J. P., and Thivend, P. (1986). *Anim. Feed Sci. Technol.* **15,** 215–229.
Joyce, J. P., and Brunswick, L. F. C. (1975). *N.Z. J. Exp. Agric.* **3,** 299–304.
Joyce, J. P., and Newth, R. P. (1967). *Proc. N.Z. Soc. Anim. Prod.* **27,** 166–180.
Joyce, J. P., and Rattray, P. V. (1970a). *N.Z. J. Agric. Res.* **13,** 792–799.
Joyce, J. P., and Rattray, P. V. (1970b). *N.Z. J. Agric. Res.* **13,** 800–807.
Judson, G. J., and Obst, J. M. (1975). *In* "Trace Elements in Soil–Plant–Animal Systems" (D. I. D. Nicholas and A. R. Egan, eds.), pp. 385–405. Academic Press, New York.
Judson, G. J., McFarlane, J. D., Riley, M. J., Milne, M. L., and Horne, A. C. (1981). *In* "Trace Element Metabolism in Man and Animals" (J. M. Howell, J. M. Gawthorne, and C. L. White, eds.), p. 35. Aust. Acad. Sci., Canberra, Australia.
Julien, W. E., Conrad, H. R., Jones, J. E., and Moxon, A. L. (1976). *J. Dairy Sci.* **59,** 1954–1959.
Kabaija, E., and Smith, O. B. (1988). *Anim. Feed Sci. Technol.* **20,** 171–176.
Kabata-Pendias, A., and Pendias, H. (1984). "Trace Elements in Soils and Plants." CRC Press, Boca Raton, Florida.
Kaiser, A. G. (1975). *Trop. Grassl.* **9,** 191–198.
Kaiser, C. J., Matches, A. G., Martz, F. A., and Mott, G. O. (1974). *Proc. Int. Grassl. Congr., 12th* **3,** 225–236.
Kaiser, A. G., Osbourn, D. F., England, P., and Dhanoa, M. S. (1982a). *Anim. Prod.* **34,** 179–190.
Kaiser, A. G., Osbourn, D. F., and England, P. (1982b). *J. Agric. Sci.* **95,** 357–369.
Kane, E. A., and Jacobson, W. C. (1954). *J. Dairy Sci.* **37,** 672.
Kane, E. A., and Moore, L. A. (1959). *J. Dairy Sci.* **42,** 936.

Kane, E. A., and Moore, L. A. (1961). *J. Dairy Sci.* **44,** 1457–1464.
Kappel, L. C., Morgan, E. B., Kilgore, L., Ingraham, R. H., and Babcock, D. K. (1985). *J. Dairy Sci.* **68,** 1822–1827.
Kassem, M. M., Thomas, P. C., Chamberlain, D. G., and Robertson, S. (1987). *Grass Forage Sci.* **42,** 175–183.
Kay, R. N. B. (1960). *J. Physiol. (London)* **150,** 515–537.
Kellaway R. C., and Leibholz, J. (1983). *World Anim. Rev.* **48,** 33–37.
Kellaway, R. C., Sitorus, P., and Leibholz, J. M. L. (1978). *Res. Vet. Sci.* **24,** 352–357.
Kemp, A. (1960). *Neth. J. Agric. Sci.* **8,** 281–304.
Kemp, A. (1964). *Neth. J. Agric. Sci.* **12,** 263–280.
Kemp, A. (1968). *Z. Tierphysiol., Tierernaehr. Futtermittelkd.* **23,** 267–278.
Kemp, A. (1971). *Proc. Colloq. Potash Inst., 1st* pp. 1–14.
Kemp, A., and Geurink, J. H. (1966). *Tijdschr. Diergeneeskd.* **91,** 580–613.
Kemp, A., and Hart, M. L. (1957). *Neth. J. Agric. Sci.* **5,** 4–17.
Kemp, A., and Hartmans, J. (1968). *Mineralstoffversorgung Tiergesundheit* **8,** 5–15.
Kemp, A., Deijs, W. B., Hemkes, O. J., and Van Es, A. J. H. (1960). "Conference on Hypomagnesaemia (London)," pp. 23–32. Br. Vet. Assoc., London.
Kemp, A., Deijs, W. B., Hemkes, O. K., and Van Es, A. J. H. (1961). *Neth. J. Agric. Sci.* **9,** 134–149.
Kemp, A., Deijs, W. B., and Kluvers, E. (1966). *Neth. J. Agric. Sci.* **14,** 290–295.
Kempton, T. J. (1982). *World Rev. Anim. Prod.* **18,** 7–14.
Kennedy, G. S. (1968). *Aust. J. Biol. Sci.* **21,** 529–538.
Kennedy, P. M. (1985). *Br. J. Nutr.* **53,** 159–173.
Kennedy, P. M., and Milligan, L. P. (1978). *Br. J. Nutr.* **39,** 105–117.
Kennedy, W. K., Carter, A. H., and Lancaster, R. J. (1959). *N.Z. J. Agric. Res.* **2,** 627–638.
Kennedy, P. M., Christopherson, R. J., and Milligan, L. P. (1982). *Br. J. Nutr.* **47,** 521–535.
Kerridge, P. C., and McLean, R. W. (1988). *Proc. Aust. Soc. Anim. Prod.* **17,** 426.
Keys, J. E., and Van Soest, P. J. (1970). *J. Dairy Sci.* **53,** 1502–1508.
Kiatoko, M., McDowell, L. R., Fick, K. R., and Fonseca, H. (1978). *J. Dairy Sci.* **61,** 324–330.
Kibon, A., and Holmes, W. (1987). *J. Agric. Sci.* **109,** 293–301.
Kick, C. H., Gerlaugh, P., Schalk, A. F., and Silver, E. A. (1937). *J. Agric. Res.* **55,** 587–597.
Kim, C. W., and Evans, J. L. (1975). *J. Dairy Sci.* **58,** 750.
Kivimae, A. (1959). *Acta Agric. Scand., Suppl.* **5.**
Kivimae, A. (1960). *Proc. Int. Grassl. Congr., 8th* pp. 466–470.
Kivimae, A. (1965). *Lantbrukshoegsk. Medd., Ser. A* **37.**
Kivimae, A. (1966). *Proc. Int. Grassl. Congr., 10th* pp. 389–393.
Kleese, R. A., Rasmusson, D. C., and Smith, L. H. (1968). *Crop Sci.* **8,** 591–593.
Kleiber, M., Goss, H., and Guilbert, H. R. (1936). *J. Nutr.* **12,** 121–153.
Klock, M. A., Schank, S. C., and Moore, J. E. (1975). *Agron. J.* **67,** 672–675.
Klosterman, E. W., Johnson, R. R., Moxon, A. L., and Ricketts, G. (1960). *Ohio, Agric. Exp. Stn., Mimeo Ser.* **121.**
Knapp, W. R., Holt, D. A., and Lechtenberg, V. L. (1975). *Agron. J.* **67,** 766–769.
Knight, R., and Yates, N. G. (1968). *Aust. J. Agric. Res.* **19,** 373–380.
Knights, G. I., O'Rourke, P. K., and Hopkins, P. S. (1979). *Aust. J. Exp. Agric. Anim. Husb.* **19,** 19–22.

Knott, P., Algar, B., Zervas, G., and Telfer, S. B. (1985). *In* "Trace Elements in Man and Animals" (C. F. Mills, I. Bremner, and J. K. Chesters, eds.), pp. 708–714. Commonw. Agric. Bur., Farnham Royal, England.

Knox, J. H., and Watkins, W. E. (1942). *N.M., Agric. Exp. Stn., Bull.* **287.**

Komisarczuk, S., Merry, R. J., McAllan, A. B., Smith, R. H., and Durand, M. (1984). *Can. J. Anim. Sci., Suppl.* **64,** 35–36.

Kotb, A. R., and Luckey, T. D. (1972). *Nutr. Abstr. Rev.* **42,** 813–845.

Kretschmer, A. E. (1964). *Proc.—Soil Crop Sci. Soc. Fla.* **24,** 167–176.

Kretschmer, A. E., Lazar, V. A., and Beeson, K. C. (1954). *Proc.—Soil Crop Sci. Soc. Fla.* **14,** 53–58.

Kromann, R. P., and Meyer, J. H. (1966). *J. Anim. Sci.* **25,** 1096–1101.

Kubota, J. (1968). *Soil Sci.* **106,** 122–130.

Kubota, J. (1975). *J. Range Manage.* **28,** 252–256.

Kubota, J., Lemon, E. R., and Allaway, W. H. (1963). *Soil Sci. Soc. Am. Proc.* **27,** 679–683.

Kubota, J., Allaway, W. H., Carter, D. L., Cary, E. E., and Lazar, V. A. (1967). *J. Agric. Food Chem.* **15,** 448–453.

Kuc, J., and Nelson, O. E. (1964). *Arch. Biochem. Biophys.* **105,** 103–113.

Kuchel, R. E., and Buckley, R. A. (1969). *Aust. J. Agric. Res.* **20,** 1099–1107.

Kuchel, R. D., and Godwin, K. O. (1976). *Proc. Aust. Soc. Anim. Prod.* **11,** 389–392.

Laby, R. H. (1973). Aust. Patent Appl. 555556/73.

Laby, R. H. (1980). *Proc. Aust. Soc. Anim. Prod.* **13,** 6–10.

Laby, R. H., Graham, C. A., Edwards, S. R., and Kautzner, B. (1984). *Can. J. Anim. Sci.* **64,** 337–338.

Laetsch, W. M. (1974). *Annu. Rev. Plant Physiol.* **25,** 27–52.

Laforest, J. P., Seoane, J. R., Dupuis, G., Phillip, L., and Filpot, P. M. (1986). *Can. J. Anim. Sci.* **66,** 117–127.

Lake, R. P., Clanton, D. C., and Karn, J. F. (1973). *J. Anim. Sci.* **36,** 1202.

Lake, R. P., Clanton, D. C., and Karn, J. F. (1974). *J. Anim. Sci.* **38,** 1291–1297.

Lakin, A. L. (1978). *In* "Developments in Food Analysis Techniques—1," pp. 43–74. Appl. Sci., Barking, Essex, England.

Lakke-Gowda, H. S., Kahar, N. D., and Ayyar, N. K. (1955). *Indian J. Med. Res.* **43,** 603–608.

Lamand, M. (1978). *Ann. Rech. Vet.* **9,** 495–500.

Lamand, M., Amboulou, D., Rayssiguier, Y., Tressol, J. C., Lab, C., and Bellanger, A. (1977). *Ann. Rech. Vet.* **8,** 303–306.

Lamand, M., Lab, C., and Tressol, J. C. (1980). *Ann. Rech. Vet.* **11,** 147–150.

Lamb, C. S., and Eadie, J. (1979). *J. Agric. Sci.* **92,** 235–241.

Lambourne, L. J. (1957). *J. Agric. Sci.* **48,** 415–425.

Lambourne, L. J., and Reardon, T. F. (1962). *Nature (London)* **196,** 961–962.

Lancaster, R. J. (1943). *N.Z. J. Sci. Technol., Sect. A.* **25,** 137–151.

Lancaster, R. J. (1947). *Proc. N.Z. Soc. Anim. Prod.* **7,** 125–127.

Lancaster, R. J. (1949) *Nature (London),* **163,** 330–331.

Lancaster, R. J. (1975). *N.Z. J. Exp. Agric.* **3,** 199–202.

Lancaster, R. J., and Bartrum, M. P. (1954). *N.Z. J. Sci. Technol., Sect. A.* **35,** 489–496.

Lancaster, R. J., and Wilson, R. K. (1975). *N.Z. J. Exp. Agric.* **3,** 203–206.

Lancaster, R. J., Brunswick, L. F. C., and Wilson, R. K. (1977). *N.Z. J. Exp. Agric.* **5,** 107–112.

Langille, J. E., and Calder, F. W. (1968). *Can. J. Plant Sci.* **48,** 626–628.

Langlands, J. P. (1966). *Anim. Prod.* **8,** 253–259.
Langlands, J. P. (1969). *Aust. J. Agric. Res.* **20,** 919–924.
Langlands, J. P. (1971). *Aust. J. Exp. Agric. Anim. Husb.* **11,** 9–13.
Langlands, J. P., Corbett, J. L., McDonald, I., and Reid, G. W. (1963). *Br. J. Nutr.* **17,** 219–226.
Langlands, J. P., Donald, G. E., Bowles, J. E., and Smith, A. J. (1980a). *Aust. J. Agric. Res.* **31,** 357–368.
Langlands, J. P., Bowles, J. E., Donald, G. E., and Smith, A. J. (1980b). *Aust. J. Exp. Agric. Anim. Husb.* **20,** 552–555.
Langlands, J. P., Bowles, J. E., Donald, G. E., Smith, A. J., and Paull, D. R. (1981). *Aust. J. Agric. Res.* **32,** 479–486.
Langlands, J. P., Bowles, J. E., Donald, G. E., and Smith, A. J. (1986). *Aust. J. Agric. Res.* **37,** 189–200.
Lannek, N., and Lindberg, P. (1975). *Adv. Vet. Sci. Comp. Med.* **19,** 127–164.
Laredo, M. A. (1974). Ph.D. thesis. Univ. of Queensland, Queensland, Australia.
Laredo, M. A., and Minson, D. J. (1973). *Aust. J. Agric. Res.* **24,** 875–888.
Laredo, M. A., and Minson, D. J. (1975a). *J. Br. Grassl. Soc.* **30,** 73–77.
Laredo, M. A., and Minson, D. J. (1975b). *Br. J. Nutr.* **33,** 159–170.
Laredo, M. A., and Minson, D. J. (1975c). *Aust. J. Exp. Agric. Anim. Husb.* **15,** 203–206.
Large, R. V., and Spedding, C. R. W. (1966). *J. Agric. Sci.* **67,** 41–52.
Lassiter, J. W., and Morton, J. D. (1968). *J. Anim. Sci.* **27,** 776–779.
Latteur, J. P. (1962). "Cobalt Deficiencies and Sub-deficiencies in Ruminants". Cent. Inf. Cobalt, Brussels. [Cited by Whitehead, D. C. (1966). *Mimeo Publ.—Commonw. Bur. Pastures Field Crops* **1/1966.**]
Laycock, K. A., Hazelwood, G. P., and Miller, E. L. (1985). *Proc. Nutr. Soc.* **44,** 54A.
Leaver, J. D. (1987). *Proc. N.Z. Soc. Anim. Prod.* **47,** 7–12.
Leche, T. F. (1977a). *Papua New Guinea Agric. J.* **28,** 11–17.
Leche, T. F. (1977b). *Papua New Guinea Agric. J.* **28,** 19–25.
Lechtenberg, V. L., Muller, L. D., Bauman, L. F., Rhykerd, C. L., and Barnes, R. F. (1972). *Agron. J.* **64,** 657–660.
Ledger, H. P., Rogerson, A., and Freeman, G. H. (1970). *Anim. Prod.* **12,** 425–431.
Le Du, Y. L. P., and Penning, P. D. (1982). *In* "Herbage Intake Handbook" (J. D. Leaver, ed.), pp. 37–75. Br. Grassl. Soc., Hurley, Maidenhead, England.
Le Du, Y. L. P., Combellas, J., Hodgson, J., and Baker, R. D. (1979). *Grass Forage Sci.* **34,** 249–260.
Lee, H. J. (1950). *Aust. Vet. J.* **26,** 152–159.
Lee, H. J. (1951). *J. Agric. South Aust.* **54,** 475–490.
Lee, H. J. (1975). *In* "Trace Elements in Soil–Plant–Animal Systems" (D. I. D. Nicholas and A. R. Egan, eds.), pp. 39–54. Academic Press, New York.
Lee, H. J., Kuchel, R. E., Good, B. F., and Trowbridge, R. F. (1957). *Aust. J. Agric. Res.* **8,** 494–501.
Lee, G. J., Hennessy, D. W., Williamson, P. J., Nolan, J. V., Kempton, T. J., and Leng, R. A. (1985). *Aust. J. Agric. Res.* **36,** 729–741.
Legg, S. P., and Sears, L. (1960). *Nature (London)* **186,** 1061–1062.
Lehr, J. J., Grashuis, J., and Van Koetsveld, E. E. (1963). *Neth. J. Agric. Sci.* **11,** 23–37.
Leibholz, J. (1981). *J. Agric. Sci.* **96,** 487–488.
Leigh, J. H., and Mulham, W. E. (1966a). *Aust. J. Exp. Agric. Anim. Husb.* **6,** 460–467.
Leigh, J. H., and Mulham, W. E. (1966b). *Aust. J. Exp. Agric. Anim. Husb.* **6,** 468–474.
Leonard, R. O., and Burns, R. H. (1955). *J. Anim. Sci.* **14,** 446–457.

Lessard, J. R., Hidiroglou, M., Carson, R. B., and Wauthy, J. M. (1970). *Can. J. Plant Sci.* **50,** 685–691.
Lewis, A. H. (1941). *Emp. J. Exp. Agric.* **9,** 43–49.
Lewis, C. E., Lowrey, R. W., Monson, W. G., and Knox, F. E. (1975). *J. Anim. Sci.* **41,** 208–212.
Leyva, V., Henderson, A. E., and Sykes, A. R. (1982). *J. Agric. Sci.* **99,** 249–259.
Lindberg, J. E., and Knutsson, P. G. (1981). *Agric. Environ.* **6,** 171–182.
Lindberg, P. E., and Lannek, N. (1970). *In* "Trace Element Metabolism in Animals" (C. F. Mills, ed.), pp. 421–426. Livingstone, Edinburgh, Scotland.
Lindberg, J. E., and Varvikko, T. (1982). *Swed. J. Agric. Res.* **12,** 163–171.
Lindberg, E. (1983). *Swed. J. Agric. Res.* **13,** 229–233.
Lindsay, J. R., Hogan, J. P., and Donnelly, J. B. (1980). *Aust. J. Agric. Res.* **31,** 589–600.
Lindsay, J. A., Mason, G. W. J., and Toleman, M. A. (1982). *Proc. Aust. Soc. Anim. Prod.* **14,** 67–68.
Lindsay, J. A., Kidd, J. F., Cervoni, M. D., Dodemaide, W. R., and Mulder, J. C. (1988). *Proc. Aust. Soc. Anim. Prod.* **17,** 430.
Linehan, P. A. (1952). *Proc. Int. Grassl. Congr., 6th* pp. 1328–1333.
Lines, E. W. (1935). *J. Counc. Sci. Ind. Res. (Aust.)* **8,** 117–119.
Lippke, H., Ellis, W. C., and Jacobs, B. F. (1986). *J. Dairy Sci.* **69,** 403–412.
Little, D. A. (1968). *Proc. Aust. Soc. Anim. Prod.* **7,** 376–380.
Little, D. A. (1972). *Aust. Vet. J.* **48,** 668–670.
Little, D. A. (1975). *Aust. J. Exp. Agric. Anim. Husb.* **15,** 437–439.
Little, D. A. (1980). *Res. Vet. Sci.* **28,** 259–260.
Little, D. A. (1983). Ph.D. thesis. Univ. of Queensland, Queensland, Australia.
Little, D. A., and McMeniman, N. P. (1973). *Aust. J. Exp. Agric. Anim. Husb.* **13,** 229–233.
Little, D. A., and Ratcliff, D. (1979). *Res. Vet. Sci.* **27,** 239–241.
Little, D. A. and Shaw, N. H. (1979). *Aust. J. Exp. Agric. Anim. Husb.* **19,** 645–651.
Little, D. A., Robinson, P. J., Playne, M. J., and Haydock, K. P. (1971). *Aust. Vet. J.* **47,** 153–156.
Little, D. A., McLean, R. W., and Winter, W. H. (1977). *J. Agric. Sci.* **88,** 533–538.
Little, W., Sansom, B. F., Manston, R., and Allen, W. M. (1978a). *Anim. Prod.* **27,** 79–87.
Little, D. A., Siemon, N. J., and Moodie, E. W. (1978b). *Aust. J. Exp. Agric. Anim. Husb.* **18,** 514–519.
Little, W., Vagg, M. J., Collis, K. A., Shaw, S. R., and Gleed, P. T. (1979). *Res. Vet. Sci.* **26,** 193–197.
Little, D. A., McIvor, J. G., and McLean, R. W. (1984). *In* "The Biology and Agronomy of Stylosanthes" (H. M. Stace and L. A. Edye, eds.), pp. 381–403. Academic Press, Orlando, Florida.
Littledike, E. T., and Cox, P. S. (1979). *In* "Grass Tetany" (V. V. Rendig and D. L. Grunes, eds.), pp. 1–50. Am. Soc. Agron., Madison, Wisconsin.
Littlejohn, A. I., and Lewis, G. (1960). *Vet. Rec.* **72,** 1137–1144.
Lloyd, L. E., Crampton, E. W., Donefer, E., and Beacom, S. E. (1960). *J. Anim. Sci.* **19,** 859–866.
Lloyd, L. E., Jeffers, H. F. M., Donefer, E., and Crampton, E. W. (1961). *J. Anim. Sci.* **20,** 468–473.
Lofgreen, G. P., and Kleiber, M. (1953). *J. Anim. Sci.* **12,** 366–371.
Lofgreen, G. P., and Kleiber, M. (1954). *J. Anim. Sci.* **13,** 258–264.
Lomba, F., Chauvaux, G., Fumiere, I., Bienfet, V., Paquay, R., De Baere, R., and Lousse, A. (1970). *Z. Tierphysiol., Tierernaehr. Futtermittelkd.* **27,** 9–18.

Long, M. I. E., and Marshall, B. (1973). *Trop. Agric. (Trinidad)* **50,** 121–128.
Long, M. I. E., Thorton, D. D., Ndyanabo, W. K., Marshal, B., and Ssekaalo, H. (1970). *Trop. Agric. (Trinidad)* **47,** 37–50.
Lonsdale, C. R., Thomas, C., and Haines, M. J. (1977). *J. Br. Grassl. Soc.* **32,** 171–176.
Loosli, J. K., and Harris, L. E. (1945). *J. Anim. Sci.* **4,** 435–437.
Loosmore, R. M., and Allcroft, R. (1951). *Vet. Rec.* **63,** 414–416.
Loper, G. M., and Smith, D. (1961). *Wis., Agric. Exp. Stn., Res. Rep.* **8.**
Losada, H., Aranda, E., Alderete, R., and Ruiz, J. (1979). *Trop. Anim. Prod.* **4,** 51–54.
Love, K. J., Egan, J. K., and McIntyre, J. S. (1978). *Proc. Aust. Soc. Anim. Prod.* **12,** 269.
Lovelace, D. A., Holt, E. C., Ellis, W. C., and Bashaw, E. C. (1972). *Agron. J.* **64,** 453–456.
Lowrey, R. S., Burton, G. W., Johnson, J. C., Marchant, W. H., and McCormick, W. C. (1968). *Res. Bull.—Ga. Agric. Exp. Stn.* **55.**
Lowther, W. L., Manley, T. R., and Barry, T. N. (1987). *N.Z. J. Agric. Res.* **30,** 23–25.
Lueker, C. E., and Lofgreen, G. P. (1961). *J. Nutr.* **74,** 233–238.
Lutz, J. A., Genter, C. F., and Hawkins, G. W. (1972). *Agron. J.* **64,** 583–585.
MacLusky, D. S. (1955). *Proc. Br. Soc. Anim. Prod.* pp. 45–51.
MacPherson, A. (1981). *In* "Trace Element Metabolism in Man and Animals" (J. M. Howell, J. M. Gawthorne, and C. L. White, eds.), pp. 175–178. Aust. Acad. Sci., Canberra, Australia.
MacPherson, A. (1983). *In* "Trace Elements in Animal Production and Veterinary Practice" (N. F. Suttle, R. G. Gunn, W. M. Allen, K. A. Linklater, and G. Weiner, eds.), Occas. Publ. 7, pp. 140–141. Br. Soc. Anim. Prod., London.
MacPherson, A. (1984). *Vet. Rec.* **115,** 354–355.
MacPherson, A., Moon, F. E., and Voss, R. C. (1973). *Br. Vet. J.* **129,** 414–426.
MacPherson, A., Voss, R. C., and Dixon, J. (1978). *In* "Trace Element Metabolism in Man and Animals" (M. Kirchgessner, ed.), pp. 497–498. Arbeitskreis Tiernaehrungsforsch., Freising, West Germany.
MacPherson, A., Voss, R. C., and Dixon, J. (1979). *Anim. Prod.* **29,** 91–99.
MacPherson, A., Gray, D., Mitchell, G. B. B., and Taylor, C. N. (1987). *Br. Vet. J.* **143,** 348–353.
MacRae, J. C., and Lobley, G. E. (1982). *Livest. Prod. Sci.* **9,** 447–456.
MacRae, J. C., and Ulyatt, M. J. (1974). *J. Agric. Sci.* **82,** 309–319.
MacRae, J. C., Ulyatt, M. J., Pearce, P. D., and Hendtlass, J. (1972). *Br. J. Nutr.* **27,** 39–50.
MacRae, J. C., Milne, J. A., Wilson, S., and Spence, A. M. (1979). *Br. J. Nutr.* **42,** 525–534.
MacRae, J. C., Smith, J. S., Dewey, P. J. S., Brewer, A. C., Brown, D. S., and Walker, A. (1985). *Br. J. Nutr.* **54,** 197–209.
MAFF (1975). *Tech. Bull.—Minist. Agric., Fish. Food* (G.B.) **33.**
Magill, B. F. (1960). Ph.D. thesis. Oregon State Uni., Corvallis, Oregon.
Mahmoud, O. M., Bakeit, A. O., and Elsamani, F. (1985). *In* "Trace Elements in Man and Animals" (C. F. Mills, I. Bremner, and J. K. Chesters, eds.). pp. 749–752. Commonw. Agric. Bur., Farnham Royal, England.
Makkar, H. P. S., Lall, D., and Negi, S. S. (1988). *Anim. Feed Sci. Technol.* **20,** 1–12.
Malan, A. I., Green, H. H., and Du, Toit, P. J. (1928). *J. Agric. Sci.* **18,** 376–383.
Mangan, J. L. (1982). *In* "Forage Protein in Ruminant Animal Production" (D. J. Thomson, D. E. Beever, and R. G. Gunn, eds.), Occas. Publ. 6, pp. 25–40. Br. Soc. Anim. Prod., London.
Mangan, J. L., Vetter, R. L., Jordan, D. J., and Wright, P. C. (1976). *Proc. Nutr. Soc.* **35,** 95A–97A.

Mannetje, L. (1974). *Proc. Int. Grassl. Congr., 12th* **3,** 299–304.
Mannetje, L., Ajit, S. S., and Murugaiah, M. (1976). *Mardi Res. Bull.* **4,** 90–98.
Manston, R., and Rowlands, C. J. (1973). *J. Dairy Res.* **40,** 85–92.
Manston, R., and Vagg, M. J. (1970). *J. Agric. Sci.* **74,** 161–167.
Margan, D. E., Graham, N. M., and Searle, T. W. (1985). *Aust. J. Exp. Agric.* **25,** 783–790.
Margan, D. E., Graham, N. M., Minson, D. J., and Searle, T. W. (1989). *Aust. J. Agric. Res.* **28,** 729–736.
Markley, R. A., Cason, J. L., and Baumgardt, B. R. (1959). *J. Dairy Sci.* **42,** 144–152.
Marston, H. R. (1935). *J. Counc. Sci. Ind. Res. (Aust.)* **8,** 111–116.
Marston, H. R. (1952). *Physiol. Rev.* **32,** 66–121.
Marston, H. R. (1959). *Med. J. Aust.* **2,** 105–113.
Marston, H. R., Lee, H. J., and McDonald, I. W. (1948a). *J. Agric. Sci.* **38,** 216–221.
Marston, H. R., Lee, H. J., and McDonald, I. W. (1948b). *J. Agric. Sci.* **38,** 222–228.
Marten, G. C. (1985). *U.S., Dep. Agric., Agric. Handb.* **643,** 45–48.
Marten, G. C., Barnes, R. F., Simons, A. B., and Wooding, F. J. (1973). *Agron. J.* **65,** 199–201.
Martens, H., and Rayssiguier, Y. (1980). *In* "Digestive Physiology and Metabolism in Ruminants" (Y. Ruckebusch and P. Thivend, eds.), pp. 447–466. AVI, Westport, Connecticut.
Martens, H., Kubel, O. W., Gabel, G., and Honig, H. (1987). *J. Agric. Sci.* **108,** 237–243.
Martin, J. E., Arrington, L. R., Moore, J. E., Ammerman, C. B., Davis, G. K., and Shirley, R. L. (1964). *J. Nutr.* **83,** 60–64.
Martinez, A., and Church, D. C. (1970). *J. Anim. Sci.* **31,** 982–990.
Martz, F. A., Noller, C. H., and Hill, D. L. (1960). *J. Dairy Sci.* **43,** 868.
Marum, P., Hovin, A. W., Marten, G. C., and Shenk, J. S. (1979). *Crop Sci.* **19,** 355–360.
Masaoka, Y., and Takano, N. (1980). *Nippon Sochi Gakkaishi* **26,** 179–184.
Mason, R. W. (1976). *Br. Vet. J.* **132,** 374–379.
Mason, R. W., and Laby, R. (1978). *Aust. J. Exp. Agric. Anim. Husb.* **18,** 653–657.
Mason, R. W., and Wilkinson, J. S. (1973). *Aust. Vet. J.* **49,** 44–49.
Masters, D. G. (1981). *In* "Trace Element Metabolism in Man and Animals"(J. M. Howell, J. M. Gawthorne, and C. L. White, eds.), pp. 331–333. Aust. Acad. Sci., Canberra, Australia.
Masters, D. G., and Fels, H. E. (1980). *Biol. Trace Elem. Res.* **2,** 281–290.
Masters, D. G., and Moir, R. J. (1980). *Aust. J. Exp. Agric. Anim. Husb.* **20,** 547–551.
Masters, D. G., and Somers, M. (1980). *Aust. J. Exp. Agric. Anim. Husb.* **20,** 20–24.
Masuda, Y. (1977). *J. Fac. Agric., Kyushu Univ.* **21,** 17–24.
Masuda, Y., Yano, M., Takahash, J., and Goto, I. (1977). Kyushu *Daigaku Nogakubu Gakugei Zasshi* **32,** 87–92.
Matches, A. G., Wedin, W. F., Marten, G. C., Smith, D., and Baumgardt, B. R. (1970). *Res. Rep.—Univ. Wis., Coll. Agric. Life Sci., Res. Div.* **73.**
Mathers, J. C., and Miller, E. L. (1981). *Br. J. Nutr.* **45,** 587–604.
Mathews, M. G., and McManus, W. R. (1976). *J. Agric. Sci.* **87,** 485–488.
Mathison, G. W., Milligan, L. P., and Weisenburger, R. D. (1986). *Agric. For. Bull. (Spec. Issue, 65th Annu. Feeders' Day Rep.)* pp. 55–57.
Maunsell, P. W., and Simpson, J. E. V. (1944). *N.Z. J. Sci. Technol., Sect. A* **26,** 142–145.
Mayes, R. W., and Lamb, C. S. (1982). *In* "Forage Protein in Ruminant Animal Production" (D. J. Thomson, D. E. Beever, R. G. Gunn, eds.), Occas. Publ. 6, pp. 145–150. Br. Soc. Anim. Prod., London.
Mayes, R. W., Lamb, C. S., and Colgrove, P. M. (1986). *J. Agric. Sci.* **107,** 161–170.
Mayland, H. F., and Grunes, D. L. (1979). *ASA Spec. Publ.* **35,** 123–175.
Mayland, H. F., Rosenau, R. C., and Florence, A. R. (1980). *J. Anim. Sci.* **51,** 966–974.

Mayne, C. S., Newberry, R. D., and Woodcock, S. C. F. (1988). *Grass Forage Sci.* **43,** 137–150.
McAleese, D. M., and Forbes, R. M. (1959). *Nature (London)* **184,** 2025.
McAleese, D. M., Bell, M. C., and Forbes, R. M. (1961). *J. Nutr.* **74,** 505–514.
McAnally, R. A. (1942). *Biochem. J.* **36,** 392–399.
McCarrick, R. B. (1963). *Ir. J. Agric. Res.* **2,** 49–60.
McCarrick, R. B. (1966). *Proc. Int. Grassl. Congr., 10th* pp. 575–580.
McCarrick, R. B., and Wilson, R. K. (1966). *J. Br. Grassl. Soc.* **21,** 195–199.
McClymont, G. L., Wynne, K. N., Briggs, P. K., and Franklin, M. C. (1957). *Aust. J. Agric. Res.* **8,** 83–90.
McCullough, T. A. (1972). *J. Br. Grassl. Soc.* **27,** 115–118.
McCullough, T. A. (1976). *J. Br. Grassl. Soc.* **31,** 105–109.
McCollum, F. T., and Galyean, M. L. (1985). *J. Anim. Sci.* **60,** 570–577.
McConaghy, S., McAllister, J. S. V., Todd, J. R., Rankin, J. E. F., and Kerr, J. (1963). *J. Agric. Sci.* **60,** 313–328.
McDonald, I. W. (1942). *Aust. Vet. J.* **18,** 107–115.
McDonald, I. W. (1968a). *Aust. Vet. J.* **44,** 145–150.
McDonald, I. W. (1968b). *Nutr. Abstr. Rev.* **38,** 381–400.
McDonald, P. (1982). *In* "Forage Protein in Ruminant Animal Production" (D. J. Thomson, D. E. Beever, and R. G. Gunn, eds.), Occas. Publ. 6, pp. 41–49. Br. Soc. Anim. Prod, London.
McDonald, P., and Whittenbury, R. (1973). *In* "Chemistry and Biochemistry of Herbage" (G. W. Butler and R. W. Bailey, eds.), Vol. 3, pp. 33–60. Academic Press, New York.
McDonald, R. C., and Wilson, K. R. (1980). *N.Z. J. Exp. Agric.* **8,** 105–109.
McDonald, P., Edwards, R. A., and Greenhalgh, J. F. D. (1988). "Animal Nutrition," 4th ed. Longman, Harlow, England.
McDowell, L. R. (1985). *In* "Nutrition of Grazing Ruminants in Warm Climates" (L. R. McDowell, ed.), pp. 189–315, 383–407. Academic Press, Orlando, Florida.
McDowell, L. R., Conrad, J. H., Thomas, J. E., Harris, L. E., and Fick, K. R. (1977). *Trop. Anim. Prod.* **2,** 273–279.
McDowell, L. R., Conrad, J. H., Ellis, G. L., and Loosli, J. K. (1983). "Minerals for Grazing Ruminants in Tropical Regions," Univ. Fla. Bull. Univ. of Florida, Gainesville.
McDowell, L. R., Conrad, J. H., and Ellis, G. L. (1984). *In* "Symposium on Herbivore Nutrition in the Subtropics and Tropics" (F. M. C. Gilchrist and R. I. Mackie, eds.), pp. 61–88. Science Press, Craighall, South Africa.
McGowan, A. C. (1983). *Proc. N.Z. Soc. Anim. Prod.* **43,** 135–136.
McIlmoyle, W. A., and Murdoch, J. C. (1977). *Anim. Prod.* **24,** 227–135.
McIlmoyle, W. A., and Murdoch, J. C. (1979). *Amim. Prod.* **28,** 223–229.
McIntosh, S., Crooks, P., and Simpson, K. (1973). *J. Agric. Sci.* **81,** 507–511.
McIvor, J. G. (1979). *Trop. Grassl.* **13,** 92–97.
McKenzie, R. A., and Schultz, K. (1983). *J. Agric. Sci.* **100,** 249–250.
McLachlan, B. P., and Ternouth, J. H. (1985). *Proc. Nutr. Soc. Aust.* **10,** 148.
McLaren, C. E., and Doyle, P. T. (1986). *Proc. Aust. Soc. Anim. Prod.* **16,** 275–278.
McLaren, R. G., Purves, D., MacKenzie, E. J., and MacKenzie, C. G. (1979). *J. Agric. Sci.* **93,** 509–511.
McLean, J. W., Thomson, G. G., and Claxton, J. H. (1959). *N.Z. Vet. J.* **7,** 47–52.
McLean, J. W., Thomson, G. G., Iversen, C. E., Jagusch, K. T., and Lawson, B. M. (1962). *Proc. N.Z. Grassl. Assoc.* **24,** 57–70.
McLean, R. W., Winter, W. H., Mott, J. J., and Little, D. A. (1981). *J. Agric. Sci.* **96,** 247–250.
McLeod, M. N. (1969). *J. Aust. Inst. Agric. Sci.* **35,** 122–124.

McLeod, M. N. (1973). *Aust. J. Exp. Agric. Anim. Husb.* **13,** 245–250.
McLeod, M. N. (1974). *Nutr. Abstr. Rev.* **44,** 803–815.
McLeod, M. N., and Minson, D. J. (1969a). *J. Br. Grassl. Soc.* **24,** 244–249.
McLeod, M. N., and Minson, D. J. (1969b). *J. Br. Grassl. Soc.* **24,** 296–298.
McLeod, M. N., and Minson, D. J. (1972). *J. Br. Grassl. Soc.* **27,** 23–27.
McLeod, M. N., and Minson, D. J. (1974). *J. Sci. Food Agric.* **25,** 907–911.
McLeod, M. N., and Minson, D. J. (1976). *Anim. Feed Sci. Technol.* **1,** 61–72.
McLeod, M. N., and Minson, D. J. (1978). *Anim. Feed Sci. Technol.* **3,** 277–287.
McLeod, M. N., and Minson, D. J. (1982). *Anim. Feed Sci. Technol.* **7,** 83–92.
McLeod, M. N., and Minson, D. J. (1988). *J. Anim. Sci.* **18,** 976–982.
McLeod, D. S., Wilkins, R. J., and Raymond, W. F. (1970). *J. Agric. Sci.* **75,** 311–319.
McManus, W. R., Dudzinski, M. L., and Arnold, G. W. (1967). *J. Agric. Sci.* **69,** 263–270.
McManus, W. R., Manta, L., McFarlane, J. D., and Gray, A. C. (1972a). *J. Agric. Sci.* **79,** 27–40.
McManus, W. R., Manta, L., McFarlane, J. D., and Gray, A. C. (1972b). *J. Agric. Sci.* **79,** 55–66.
McManus, W. R., Robinson, V. N. E., and Grout, L. L. (1977). *Aust. J. Agric. Res.* **28,** 651–662.
McMeniman, N. P. (1973). *Aust. Vet. J.* **49,** 150–152.
McMeniman, N. P. (1976). *Aust. J. Exp. Agric. Anim. Husb.* **16,** 818–822.
McMeniman, N. P., and Armstrong, D. G. (1977). *Anim. Feed Sci. Technol.* **2,** 255–266.
McMeniman, N. P., and Little, D. A. (1974). *Aust. J. Exp. Agric. Anim. Husb.* **14,** 316–321.
McMurray, C. H., and McEldowney, P. K. (1977). *Br. Vet. J.* **133,** 535–542.
McMurray, C. H., Rice, D. A., and Kennedy, S. (1983). *In* "Trace Elements in Animal Production and Veterinary Practice" (N. F. Suttle, R. G. Gunn, W. M. Allen, K. A. Linklater, and G. Wiener, eds.), Occas. Publ. 7, pp. 61–73. Br. Soc. Anim. Prod., London.
McNaught, K. J. (1948). *N.Z. J. Sci. Technol., Sect. A* **30,** 26–43.
McNaught, K. J. (1959). *N.Z. J. Agric.* **99,** 442.
McNaught, K. J. (1970). *Proc. Int. Grassl. Congr., 11th* pp. 334–338.
McNaught, K. J., and Karlovsky, J. (1964). *N.Z. J. Agric. Res.* **7,** 386–404.
McNaught, K. J., and Paul, C. W. (1939). *N.Z. J. Sci. Technol., Sect. B* **21,** 95–101.
McNaught, K. J., Dorofaeff, F. D., and Karlovsky, J. (1968). *N.Z. J. Agric. Res.* **11,** 533–550.
McNaught, K. J., Dorofaeff, F. D., Ludecke, T. E., and Cottier, K. (1973). *N.Z. J. Exp. Agric.* **1,** 329–347.
McQueen, I. P. M. (1984). *Proc. N.Z. Soc. Anim. Prod.* **44,**, 159–161.
McQueen, R., and Van Soest, P. J. (1975). *J. Dairy Sci.* **58,** 1482–1491.
Mehrez, A. Z., and Orskov, E. R. (1977). *J. Agric. Sci.* **88,** 645–650.
Meijs, J. A. C. (1981). Versl. Landbouwkd. Onderz. **909.**
Meijs, J. A. C. (1986). *Grass Forage Sci.* **41,** 229–235.
Meijs, J. A. C., and Hoekstra, J. A. (1984). *Grass Forage Sci.* **39,** 59–66.
Meijs, J. A. C., Walters, R. J. K., and Keen, A. (1982). *In* "Herbage Intake Handbook" (J. D. Leaver, ed.), pp. 11–36. Br. Grassl. Soc. Hurley, Maidenhead, England.
Meissner, H. H., Franck, F., and Hofmeyr, H. S. (1973). *S. Afr. J. Anim. Sci.* **3,** 51–52.
Meissner, H. H., Pienaar, J. P., Liebenberg, L. H. P., and Roux, C. Z. (1979) *Ann. Rech. Vet.* **10,** 219–222.
Mellin, T. N., Poulton, B. R., and Anderson, M. J. (1962). *J. Anim. Sci.* **21,** 123–126.
Merchen, N. R., Firkins, J. L., and Berger, L. L. (1986). *J. Anim. Sci.* **62,** 216–225.
Mertens, D. R. (1977). *Fed. Proc., Fed. Am. Soc. Exp. Biol.* **36,** 187–192.

Mertens, D. R. (1987). *J. Anim. Sci.* **64,** 1548–1558.
Metherell, A. K., Owens, J. L., and Mackintosh, C. G. (1987). *Proc. Anim. Sci. Congr. Asian–Australas. Assoc. Anim. Prod. Soc., 4th* p. 426.
Metson, A. J. (1974). *N.Z. J. Exp. Agric.* **2,** 277–319.
Metson, A. J., and Saunders, W. M. H. (1978a). *N.Z. J. Agric. Res.* **21,** 341–353.
Metson, A. J., and Saunders, W. M. H. (1978b). *N.Z. J. Agric. Res.* **21,** 355–364.
Metson, A. J., Gibson, E. J., Hunt, J. L., and Saunders, W. M. H. (1979). *N.Z. J. Agric. Res.* **22,** 309–318.
Meyer, H., and Busse, F. W. (1975). *Zentralbl. Veterinaermed., Reihe A* **22,** 864–876.
Meyer, J. H., Weir, W. C., Jones, L. G., and Hull, J. L. (1957). *J. Anim. Sci.* **16,** 623–632.
Meyer, J. H., Gaskill, R. L., Stoewsand, G. S., and Weir, W. C. (1959a). *J. Anim. Sci.* **18,** 336–346.
Meyer, J. H., Weir, W. C., Dobie, J. B., and Hull, J. L. (1959b). *J. Anim. Sci.* **18,** 976–982.
Meyer, J. H., Weir, W. C., Jones, L. G., and Hull, J. L. (1960). *J. Anim. Sci.* **19,** 283–294.
Miaki, T. (1970). *Nippon Chikusan Gakkai Ho* **41,** 459–464.
Michell, P. J. (1973a). *Aust. J. Exp. Agric. Anim. Husb.* **13,** 158–164.
Michell, P. J. (1973b). *Aust. J. Exp. Agric. Anim. Husb.* **13,** 165–170.
Miles, D. G., Walters, R. J. K., and Evans, E. M. (1969). *Anim. Prod.* **11,** 19–28.
Milford, R. (1960a). *Aust. J. Agric. Res.* **11,** 138–148.
Milford, R. (1960b). *Proc. Int. Grassl. Congr., 8th* pp. 474–479.
Milford, R. (1967). *Aust. J. Exp. Anim. Husb.* **7,** 540–545.
Milford, R., and Minson, D. J. (1965a). *J. Br. Grassl. Soc.* **20,** 177–179.
Milford, R., and Minson, D. J. (1965b). *Br. J. Nutr.* **19,** 373–382.
Milford, R., and Minson, D. J. (1965c). *Proc. Int. Grassl. Congr., 9th* pp. 814–822.
Milford, R., and Minson, D. J. (1966). *J. Br. Grassl. Soc.* **21,** 7–13.
Milford, R., and Minson, D. J. (1968a). *Aust. J. Exp. Anim. Husb.* **8,** 409–412.
Milford, R., and Minson, D. J. (1968b). *Aust. J. Exp. Agric. Anim. Husb.* **8,** 413–418.
Millar, K. R. (1983). *In* "The Mineral Requirements of Grazing Ruminants" (N. D. Grace, ed.), Occas. Publ. 9, pp. 38–47. N.Z. Soc. Anim. Prod., Wellington, New Zealand.
Millar, K. R., and Lorentz, P. P. (1979). *N.Z. Vet. J.* **27,** 90–92.
Millar, K. R., and Penrose, M. E. (1980). *N.Z. Vet. J.* **28,** 97–99.
Millard, P., Gordon, A. H., Richardson, A. J., and Chesson, A. (1987). *J. Sci. Food Agric.* **40,** 305–314.
Miller, W. J. (1969). *Am. J. Clin. Nutr.* **22,** 1323–1331.
Miller, W. J. (1970). *J. Dairy Sci.* **53,** 1123–1135.
Miller, W. J., Clifton, C. M., and Cameron, N. W. (1963). *J. Dairy Sci.* **46,** 715–719.
Miller, W. J., Adams, W. E., Nussbaumer, R., McCreery, R. A., and Perkins, H. F. (1964a). *Agron. J.* **56,** 198–201.
Miller, W. J., Pitts, W. J., Clifton, C. M., and Schmittle, S. C. (1964b). *J. Dairy Sci.* **47,** 556–558.
Miller, W. J., Clifton, C. M., Brooks, O. L., and Beaty, E. R. (1965a). *J. Dairy Sci.* **48,** 209–212.
Miller, W. J., Clifton. C. M., Fowler, P. R., and Perkins, H. F. (1965b). *J. Dairy Sci.* **48,** 450–453.
Miller, W. J., Powell, G. W., and Hiers, J. M. (1966). *J. Dairy Sci.* **49,** 1012–1013.
Miller, W. J., Blackmon, D. M., and Pate, F. M. (1970). *In* "Trace Element Metabolism in Animals" (C. F. Mills, ed.), pp. 231–237. Livingstone, Edinburgh, Scotland.
Mills, C. F. (1978). *Annu. Rep. Stud. Anim. Nutr. Allied Sci. (Rowett Res. Inst.) 34,* 105–115.
Mills, C. F. (1987). *J. Anim. Sci.* **65,** 1702–1711.

Mills, C. F., and Dalgarno, A. C. (1967). *Proc. Nutr. Soc.* **26,** 19.
Mills, C. F., and Dalgarno, A. C. (1972). *Nature (London)* **239,** 171–173.
Mills, C. F., and Dalgarno, A. C., Williams, R. B., and Quarterman, J. (1967). *Br. J. Nutr.* **21,**751–768.
Milne, J. (1953). *Vet. Rec.* **65,** 430.
Milne, J. A. (1977). *J. Br. Grassl. Soc.* **32,** 141–147.
Milne, J. A., and Bagley, L. (1976). *J. Agric. Sci.* **87,** 599–604.
Milne, J. A., and Campling, R. C. (1972). *J. Agric. Sci.* **78,** 79–86.
Milne, J. A., and Mayes, R. W. (1985). "The Hill Farming Research Organisation Biennial Report, 1984–85" pp. 115–119. Edinburgh, Scotland.
Milne, J. A., Bagley, L., and Grant, S. A. (1979). *Grass Forage Sci.* **34,** 45–53.
Milne, J. A., Maxwell, T. J., and Souter, W. (1981). *Anim. Prod.* **32,** 185–196.
Milne, J. A., Hodgson, J., Thompson, R., Souter, W., and Barthram, G. T. (1982). *Grass Forage Sci.* **37,** 209–218.
Miltimore, J. E., and Mason, J. L. (1971). *Can. J. Anim. Sci.* **51,** 193–200.
Miltimore, J. E., Mason, J. L., and Ashby, D. L. (1970). *Can. J. Anim. Sci.* **50,** 293–300.
Miltimore, J. E., Kalnin, C. M., and Clapp, J. B. (1978). *Can. J. Anim. Sci.* **58,** 525–529.
Milton, J. T. B., and Ternouth, J. H. (1978). *Proc. Aust. Soc. Anim. Prod.* **12,** 125.
Milton, J. T. B., and Ternouth, J. H. (1985). *Aust. J. Agric. Res.* **36,** 647–654.
Minson, D. J. (1963). *J. Br. Grassl. Soc.* **18,** 39–44.
Minson, D. J. (1966). *J. Br. Grassl. Soc.* **21,** 123–126.
Minson, D. J. (1967). *Br. J. Nutr.* **21,** 587–597.
Minson, D. J. (1971a). *Aust. J. Exp. Agric. Anim. Husb.* **11,** 18–25.
Minson, D. J. (1971b). *Aust. J. Agric. Res.* **22,** 589–598.
Minson, D. J. (1972). *Aust. J. Exp. Agric. Anim. Husb.* **12,** 21–26.
Minson, D. J. (1973). *Aust. J. Exp. Agric. Anim. Husb.* **13,** 153–157.
Minson, D. J. (1975). *Forage Res.* **1,** 1–10.
Minson, D. J. (1981). *J. Agric. Sci.* **96,** 239–242.
Minson, D. J. (1982). *Nutr. Abstr. Rev.* **52B,** 591–615.
Minson, D. J. (1984). *Aust. J. Exp. Agric. Anim. Husb.* **24,** 494–500.
Minson, D. J., and Bray, R. A. (1985). *Aust. J. Exp. Agric.* **25,** 306–310.
Minson, D. J., and Bray, R. A. (1986). *Grass Forage Sci.* **41,** 47–52.
Minson, D. J., and Brown, S. A. (1959). *Grassl. Res. Inst. (Hurley), Annu. Rep. 1957–1958* pp. 99–103.
Minson, D. J., and Hacker, J. B. (1986). *Aust. J. Exp. Agric.* **26,** 551–556.
Minson, D. J., and Haydock, K. P. (1971). *Aust. J. Exp. Agric. Anim. Husb.* **11,** 181–185.
Minson, D. J., and Kemp, C. D. (1961). *J. Br. Grassl. Soc.* **16,** 76–79.
Minson, D. J., and Lancaster, R. J. (1963). *N.Z. J. Agric. Res.* **6,** 140–146.
Minson, D. J., and McDonald, C. K. (1987). *Trop. Grassl.* **21,** 116–122.
Minson, D. J., and McLeod, M. N. (1970). *Proc. Int. Grassl. Congr., 11th* pp. 719–722.
Minson, D. J., and Milford, R. (1966). *Aust. J. Agric. Res.* **17,** 411–423.
Minson, D. J., and Milford, R. (1967a). *J. Br. Grassl. Soc.* **22,** 170–175.
Minson, D. J., and Milford, R. (1967b). *Aust. J. Exp. Agric. Anim. Husb.* **7,** 546–551.
Minson, D. J., and Milford, R. (1968a). *Aust. J. Exp. Agric. Anim. Husb.* **8,** 270–276.
Minson, D. J., and Milford, R. (1968b). *J. Agric. Sci.* **71,** 381–382.
Minson, D. J., and Norton, B. W. (1982). *Proc. Aust. Soc. Anim. Prod.* **14,** 357–360.
Minson, D. J., and Pigden, W. J. (1961). *J. Anim. Sci.* **20,** 962.
Minson, D. J., and Raymond, W. F. (1958). *Grassl. Res. Inst. (Hurley), Annu. Rep. 1956–1957* pp. 92–96.
Minson, D. J., and Ternouth, J. H. (1971). *Br. J. Nutr.* **26,** 31–39.

Minson, D. J., and Wilson, J. R. (1980). *J. Aust. Inst. Agric. Sci.* **46,** 247–249.
Minson, D. J., Tayler, J. C., Alder, F. E., Raymond, W. F., Rudman, J. E., Line, C., and Head, M. J. (1960a). *J. Br. Grassl. Soc.* **15,** 86–88.
Minson, D. J., Raymond, W. F., and Harris, C. E. (1960b). *J. Br. Grassl. Soc.* **15,** 174–180.
Minson, D. J., Harris, C. E., Raymond, W. F., and Milford, R. (1964). *J. Br. Grassl. Soc.* **19,** 298–305.
Minson, D. J., Butler, K. L., Grummitt, N., and Law, D. P. (1983). *Anim. Feed Sci. Technol.* **9,** 221–237.
Mislevy, P., and Everett, P. H. (1981). *Agron. J.* **73,** 601–604.
Mitchell, R. L. (1945). *Soil Sci.* **60,** 63–70.
Mitchell, R. L., Scott, R. O., Stewart, A. B., and Stewart, J. (1941). *Nature (London)* **148,** 725–726.
Mitchell, R. L., Reith, J. W. S., and Johnston, I. M. (1957a). *J. Sci. Food Agric.* **8,** S51–S59.
Mitchell, R. L., Reith, J. W. S., and Johnston, I. M. (1957b). *In* "Plant Analysis and Fertilizer Problems," pp. 248–259. Institut de Recherches pour les Huiles de Palme et oleagineux, Paris.
Mitchell, R. L., McLaren, J. B., and Fribourg, H. A. (1986). *Agron. J.* **78,** 675–680.
Moir, K. W. (1960a). *Queensl. J. Agric. Sci.* **17,** 361–371.
Moir, K. W. (1960b). *Queensl. J. Agric. Sci.* **17,** 373–383.
Moisey, F. R., and Leaver, J. D. (1982). *Anim. Prod.* **34,** 399.
Molloy, L. F., Metson, A. J., and Collie, T. W. (1973). *N.Z. J. Agric. Res.* **16,** 457–462.
Monson, W. G., and Reid, J. T. (1968). *Agron. J.* **60,** 610–612.
Monson, W. G., Lowrey, R. S., and Forbes, I. (1969). *Agron. J.* **61,** 587–589.
Montalvo-Hernandez, M. I., Garcia-Ciudad, A., and Garcia-Criado, B. (1984). *In* "The Impact of Climate on Grass Production and Quality" (H. Riley and A. O. Skjelvag, eds.), pp. 407–411. Nor. State Agric. Res. St., As, Norway.
Moon, F. E. (1954). *J. Agric. Sci.* **44,** 140–151.
Moore, L. A., Thomas, J. W., and Sykes, J. F. (1960). *Proc. Int. Grassl. Congr., 8th* pp. 701–704.
Moore, J. E., Ruelke, O. C., Rios, C. E., and Franke, D. E. (1970). *Proc.—Soil Crop Sci. Soc. Fla.* **30,** 211–221.
Moore, W. F., Fontenot, J. P., and Webb, K. E. (1972). *J. Anim. Sci.* **35,** 1046–1053.
Moore, K. J., Lechtenberg, V. L., Hendrix, K. S., and Hertel, J. M. (1983). *Proc. Int. Grassl. Congr., 14th* pp. 626–628.
Moore, K. J., Lechtenberg, V. L., and Hendrix, K. S. (1985). *Agron. J.* **77,** 67–71.
Moore, K. J., Lemenager, R. P., Lechtenberg, V. L., and Hendrix, K. S., and Risk, J. E. (1986). *J. Anim. Sci.* **62,** 235–243.
Moran, J. B., Norton, B. W., and Nolan, J. V. (1979). *Aust. J. Agric. Res.* **30,** 333–340.
Morgan, D. E. (1974). *Agric. Dev. Advis. Serv. (Sci. Arm), Annu. Rep.* pp. 94–103.
Morgan, C. A., Edwards, R. A., and McDonald, P. (1980). *J. Agric. Sci.* **94,** 287–298.
Morris, J. G. (1958). *Queensl. J. Agric. Sci.* **15,** 161–180.
Morris, J. G. (1980). *J. Anim. Sci.* 50, 145–152.
Morris, J. G., and Gartner, R. J. W. (1971). *Br. J. Nutr.* **25,** 191–205.
Morris, J. G., and Murphy, G. W. (1972). *J. Agric. Sci.* 78, 105–108.
Morris, J. G., and Peterson, R. G. (1975). *J. Nutr.* **105,** 595–598.
Moseley, G. (1974). *Br. J. Nutr.* **32,** 317–326.
Moseley, G., and Jones, D. I. H. (1974). *J. Agric. Sci.* **83,** 37–42.
Moseley, G., and Jones, D. I. H. (1984). *Br. J. Nutr.* **52,** 381–390.
Mowat, D. N., Fulkerson, R. S., Tossell, W. E., and Winch, J. E. (1965a). *Can. J. Plant Sci.* **45,** 321–331.

Mowat, D. N., Christie, B. R., and Winch, J. E. (1965b). *Can. J. Plant Sci.* **45,** 503–507.
Mowat, D. N., Fulkerson, R. S., Tossell, W. E., and Winch, J. E. (1965c). Proc. Int. Grassl. Congr., 9th pp. 803–806.
Mowat, D. N., Fulkerson, R. S., and Gamble, E. E. (1967). *Can. J. Plant Sci.* **47,** 423–426.
Mudd, A. J. (1970). *J. Agric. Sci.* **74,** 11–21.
Mulholland, J. G., Coombe, J. B., and McManus, W. R. (1974). *Aust. J. Exp. Agric. Anim. Husb.* **14,** 449–453.
Muller, L. D., Lechtenberg, V. L.., Bauman, L. F., Barnes, R. F., and Rhykerd, C. L. (1972). *J. Anim. Sci.* **35,** 883–889.
Munns, D. N., Johnson, C. M., and Jacobson, L. (1963). *Plant Soil* **29,** 115–126.
Murdoch, J. C. (1960). *J. Br. Grassl. Soc.* **15,** 70–73.
Murdoch, J. C. (1965). *J. Br. Grassl. Soc.* **20,** 54–58.
Murdock, F. R., Hodgson, A. S., and Harris, J. R. (1961). *J. Dairy Sci.* **44,** 1943–1945.
Murphy, G. M., and Connell, J. A. (1970). *Aust. Vet. J.* **46,** 595–598.
Murphy, G. M., and Plasto, A. W. (1973). *Aust. J. Exp. Agric. Anim. Husb.* **13,** 369–374.
Murphy, G. M., Dimmock, C. K., Kennedy, T. P., O'Bryan, M. S., Plasto, A. W., Powell, E. E., Twist, J. O., Wright, G. S., and Gartner, R. J. W. (1981). *In* "Trace Element Metabolism in Man and Animals" (J. M. Howell, J. M. Gawthorne, and C. L. White, eds.), pp. 183–186. Aust. Acad. Sci., Canberra, Australia.
Murphy, M. R., Baldwin, R. L., and Ulyatt, M. J. (1986). *J. Anim. Sci.* **62,** 1412–1422.
Murray, C. A., Romyn, A. E., Haylett, D. G., and Ericksen, F. (1936). *Rhod. Agric. J.* **33,** 422–441.
Murray, R. M., Teleni, E., and Playne, M. J. (1976). *Proc. Aust. Soc. Anim. Prod.* **11,** 9P.
Muth, O. H. (1963). *J. Am. Vet. Med. Assoc.* **142,** 272–277.
Muth, O. H., Oldfield, J. E., Remmert, L. F., and Schubert, J. R. (1958). *Science* **128,** 1090.
Muth, O. H., Oldfield, J. E., Schubert, J. R., and Remmert, L. F. (1959). *Am. J. Vet. Res.* **20,** 231–234.
Mwakatundu, A. G. K., and Owen, E. (1974). *East Afr. Agric. For. J.* **40,** 1–10.
Myers, B. J., and Ross, D. A. (1959). *N.Z. Agric. Res.* 2, 552–574.
Nagy, J. G., and Tengerdy, R. P. (1967). *Appl. Microbiol.* **15,** 819–821.
Nagy, J. G., and Tengerdy, R. P. (1968). *App. Microbiol.* **16,** 441–444.
Nagy, J. G., Steinhoff, H. F., and Ward, G. M. (1964). *J. Wildl. Manage.* **28,** 785–790.
Nakui, T., Kushibiki, H., Iwasaki, K., and Hayakawa, M. (1980). *Hokkaido Nogyo Shikenjo Kenkyu Hokoku* **126,** 149–162.
Neal, H. D. St. C., Thomas, C., and Cobby, J. M. (1984). *J. Agric. Sci.* **103,** 1–10.
Neathery, M. W. (1972). *J. Anim. Sci.* **34,** 1075–1084.
Nelson, A. B., and Waller, G. R. (1962). *J. Anim. Sci.* 21, 387.
Newbery, T. R., and Radcliffe, J. C. (1974). *Proc. Aust. Soc. Anim. Prod.* **10,** 387–390.
Newbery, T. R., and Radcliffe, J. C. (1975). *J. Aust. Inst. Agric. Sci.* **41,** 217–218.
Newman, D. M. R. (1972). *J. Aust. Inst. Agric. Sci.* **38,** 212–213.
Newton, H. P., and Toth, S. J. (1951). *Soil Sci.* **71,** 175–179.
Newton, J. E., and Young, N. E. (1974). *Anim. Prod.* **18,** 191–199.
Newton, G. L., Fontenot, J. P., Tucker, R. E., and Polan, C. E. (1972). *J. Anim. Sci.* **35,** 440–445.
Nicol, D. C., and Smith, L. D. (1981). *Aust. J. Exp. Agric. Anim. Husb.* **21,** 27–31.
Nicholson, J. W. G., Belanger, G., and Burgess, P. L. (1986). *Can. J. Anim. Sci.* **66,** 431–439.
Nielsen, F. H. (1984). *Annu. Rev. Nutr.* **4,** 21–41.
Nocek, J. E. (1985). *J. Anim. Sci.* **60,** 1347–1358.
Nocek, J. E., Russell, J. B., and Fallon, J. B. (1988). *Agron. J.* **80,** 525–532.
Nolan, J. V., and Leng, R. A. (1972). *Br. J. Nutr.* **27,** 177–194.

Norman, M. J. T. (1960). *Land Res. Reg. Surv., CSIRO Technical Pap.* **12.**
Norman, M. J. T. (1962). *Aust. J. Exp. Agric. Anim. Husb.* **2,** 27–34.
Norman, M. J. T. (1963). *Aust. J. Exp. Agric. Anim. Husb.* **3,** 119–124.
Norris, K. H., Barnes, R. F., Moore, J. E., and Shenk, J. S. (1976). *J. Anim. Sci.* **43,** 889–897.
Norrish, K. (1975). *In* "Trace Elements in Soil–Plant–Animal Systems" (D. I. D. Nicholas and A. R. Egan, eds.), pp. 55–81. Academic Press, New York.
Norton, B. W., and Hales, J. W. (1976). *Proc. Aust. Soc. Anim. Prod.* **11,** 393–396.
NRC (1978). "Nutrient Requirements of Domestic Animals. 3. Nutrient Requirements of Dairy Cattle." Natl. Res. Counc., Natl. Acad. Sci., Washington, D.C.
NRC (1984). "Nutrient Requirements of Beef Cattle." Natl. Res. Counc., Natl. Acad. Sci., Washington, D.C.
Oakes, A. J. (1966). *Agron. J.* **58,** 75–77.
Oakes, A. J., and Skov, O. (1962). *Agron J.* **54,** 176–178.
O'Connor, M. B., Foskett, H. R., and Smith, A. (1981). *Proc. N.Z. Soc. Anim. Prod.* **41,** 82–87.
O'Connor, M. B., Tower, N. R., Feyter, C., Edmeades, D. C., and MacMillan, K. L. (1987). *Proc. Anim. Sci. Congr. Asia–Australas. Assoc. Anim. Prod. Soc., 4th* p. 424.
O'Dell, B. L. (1981). *In* "Trace Element Metabolism in Man and Animals" (J. M. Howell, J. M. Gawthorne, and C. L. White, eds.), pp. 319–326. Aust. Acad. Sci., Canberra, Australia.
O'Donovan, P. B. (1984). *Nutr. Abstr. Rev.* **54B,** 389–410.
O'Donovan, P. B., Barnes, R. F., Plumlee, M. P., Mott, G. O., and Packett, L. V. (1967). *J. Anim. Sci.* **26,** 1144–1152.
O'Donovan, P. B., Conway, A., and O'Shea, J. (1972). *J. Agric. Sci.* **78,** 87–95.
Ogwang, B. H., and Mugerwa, J. S. (1976). *East Afr. Agric. J.* **41,** 213–242.
Oh, H. K., Baumgardt, B. R., and Scholl, J. M. (1966). *J. Diary Sci.* **49,** 850–855.
Oh, S. H., Pope, A. L., and Hoekstra, W. G. (1976). *J. Anim. Sci.* **42,** 984–992.
Ohyama, Y. (1960). *Nippon Chikusan Gakkai Ho* **31,** 55–62.
Oji, U. I., Mowat, D. N., and Winch, J. E. (1979). *Can. J. Anim. Sci.* 59, 813–816.
Oldfield, J. E., Muth, O. H., and Schubert, J. R. (1960). *Proc. Soc. Exp. Biol. Med.* **103,** 799–800.
Oliver, W. M. (1975). *J. Anim. Sci.* **41,** 999–1001.
Ololade, B. G., Mowat, D. N., and Winch, J. E. (1970). *Can. J. Anim. Sci.* **50,** 657–662.
Oltjen, R. R., Burns, W. C., and Ammerman, C. B. (1974). *J. Anim. Sci.* **38,** 975–983.
Olubajo, F. C., Van Soest, P. J., and Oyenuga, V. A. (1974). *J. Anim. Sci.* **38,** 149–153.
O'Moore, L. B. (1952). *Vet. Rec.* **64,** 475–480.
O'Moore, L. B. (1960a). *SCI Monogr.* **9,** 146–155.
O'Moore, L. B. (1960b). *SCI Monogr.* **9,** 155–158.
O'Moore, L. B., and Smyth, P. J. (1958). *Vet. Rec.* **70,** 773–774.
Oram, R. N., Clements, R. J., and McWilliam, J. R. (1974). *Aust. J. Agric. Res.* **25,** 265–274.
Orr, R. J., and Treacher, T. T. (1984). *J. Anim. Prod.* **39,** 89–98.
Orskov, E. R., and Fraser, C. (1975). *Br. J. Nutr.* **34,** 493–500.
Orskov, E. R., Grubb, D. A., Smith, J. S., Webster, A. J. F., and Corrigall, W. (1979). *Br. J. Nutr.* **41,** 541–551.
Orskov, E. R., Tait, C. A. G., Reid, G. W., and Flachowski, G. (1988a). *Anim. Prod.* **46,** 23–27.
Orskov, E. R., Reid, G. W., and Kay, M. (1988b). *Anim. Prod.* **46,** 29–34.

Osbourn, D. F., and Siddons, R. C. (1980). *Ann. Zootech.* **29,** 325–336.
Osbourn, D. F., Thomson, D. J., and Terry, R. A. (1966). *Proc. Int. Grassl. Congr., 10th* pp. 363–366.
Osbourn, D. F., Terry, R. A., Outen, G. E., and Cammell, S. B. (1974). *Proc. Int. Grassl. Congr., 12th* pp. 374–380.
Osbourn, D. F., Beever, D. E., and Thomson, D. J. (1976). *Proc. Nutr. Soc.* **35,** 191–200.
Osbourn, D. F., Terry, R. A., Spooner, M. C., and Tetlow, R. M. (1981). *Anim. Feed Sci. Technol.* **6,** 387–403.
Osuji, P. O. (1974). *J. Range Manage.* **27,** 437–443.
Osuji, P. O., Gordon, J. G., and Webster, A. J. F. (1975). *Br. J. Nutr.* **34,** 59–71.
Ota, Y. (1960). *Jpn. J. Vet. Res.* **8,** 161–172.
Ott, E. A., Smith, W. H., Stob, M., and Beeson, W. M. (1964). *J. Nutr.* **82,** 41–50.
Ott, E. A., Smith, W. H., Stob, M., Parker, H. E., and Beeson, W. M. (1965). *J. Anim. Sci.* **24,** 735–741.
Ott, E. A., Smith, W. H., Harrington, R. B., and Beeson, W. M. (1966a). J. Anim. Sci. **25,** 414–418.
Ott, E. A., Smith, W. H., Harrington, R. B., and Beeson, W. M. (1966b). *J. Anim. Sci.* **25,** 419–423.
Otto, J. S. (1938). *Onderstepoort J. Vet. Sci. Anim. Ind.* **10,** 281–364.
Overend, M. A., and Armstrong, D. G. (1982). *In* "Forage Protein in Ruminant Animal Production" (D. J. Thomson, D. E. Beever, and R. G. Gunn, eds.), Occas. Publ. 6, pp. 162–163. Br. Soc. Anim. Prod., London.
Owen, M. A. (1964). *East Afr. Agric. For. J.* **29,** 322–325.
Owen, E., and Nwadukwe, B. S. (1980). *Anim. Prod.* **30,** 489.
Owen, J. B., Lee, R. F., Lerman, P. M., and Miller, E. L. (1980). *J. Agric. Sci.* **94,** 637–644.
Ozanne, P. G., and Howes, K. M. W. (1971). *Aust. J. Agric. Res.* **22,** 941–950.
Ozanne, P. G., Purser, D. B., Howes, K. M. W., and Southey, I. (1976). *Aust. J. Exp. Agric. Anim. Husb.* **16,** 353–360.
Pain, B. F., and Broom, D. M. (1978). *Anim. Prod.* **26,** 75–83.
Paquay, R., De-Baere, R., Lousse, A., Lomba, F., Chauvaux, G., Fumiere, I., and Bienfet, V. (1970). *Z. Tierphysiol., Tierernaehr. Futtermittelkd.* **26,** 332–339.
Parr, W. H., and Allcroft, R. (1957). *Vet. Rec.* **69,** 1041–1047.
Paterson, D. D. (1933). *J. Agric. Sci.* **23,** 615–641.
Paterson, R., and Crichton, C. (1960). *J. Br. Grassl. Soc.* **15,** 100–105.
Patil, B. D., and Jones, D. I. H. (1970). *Proc. Int. Grassl. Congr., 11th* pp. 726–730.
Paulson, G. D., Broderick, G. A., Baumann, C. A., and Pope, A. L. (1968). *J. Anim. Sci.* **27,** 195–202.
Paynter, D. I. (1979). *Aust. J. Agric. Res.* **30,** 695–702.
Paynter, D. I., Anderson, J. W., and McDonald, J. W. (1979). *Aust. J. Agric. Res.* **30,** 703–710.
Peaker, M., and Linzell, J. L. (1973). *J. Endocrinol.* **58,** 139–140.
Pearce, G. R., ed. (1983). "The Utilisation of Fibrous Agricultural Residue." Aust. Gov. Publ. Serv., Canberra, Australia.
Pearce, G. R., and Moir, R. J. (1964). *Aust. J. Agric. Res.* **15,** 635–644.
Pelletier, G., and Donefer, E. (1973). *Can. J. Anim. Sci.* **53,** 257–263.
Pendlum, L. C., Boling, K. A., Bush, L. P., and Buckner, R. C. (1980). *J. Anim. Sci.* **51,** 704–711.
Penning, P. D. (1983). *Grass Forage Sci.* **38,** 89–96.

Penning, P. D., and Hooper, G. E. (1985). *Grass Forage Sci.* **40,** 79–84.

Penning, P. D., and Johnson, R. H. (1983a). *J. Agric. Sci.* **100,** 127–131.

Penning, P. D., and Johnson, R. H. (1983b). *J. Agric. Sci.* **100,** 133–138.

Penning, P. D., and Treacher, T. T. (1981). *Anim. Prod.* **32,** 374–375.

Penning, P. D., Steel, G. L., and Johnson, R. H. (1984). *Grass Forage Sci.* **39,** 345–351.

Penning, P. D., Hooper, G. E., and Treacher, T. T. (1986). *Grass Forage Sci.* **41,** 199–208.

Perdok, H. B., and Leng, R. A. (1987). *Anim. Feed. Sci. Technol.* **17,** 121–143.

Perdomo, J. T., Shirley, R. L., and Chicco, C. F. (1977). *J. Anim. Sci.* **45,** 1114–1119.

Perry, T. W., Beeson, W. M., Smith, W. H., and Mohler, M. T. (1968). *J. Anim. Sci.* **27,** 1674–1677.

Peter, D. W., Mann, A. W., and Hunter, R. A. (1981). *In* "Trace Element Metabolism in Animals and Man" (J. M. Howell, J. M. Gawthorne, and C. L. White, eds.), pp. 199–202. Aust. Acad. Sci., Canberra, Australia.

Peterson, P. J., and Spedding, D. J. (1963). *N.Z. J. Agric. Res.* **6,** 13–23.

Phillip, L. E., Buchanan-Smith, J. G., and Grovum, W. L. (1981). *J. Agric. Sci.* **96,** 429–438.

Phillippo, M., Humphries, W. R., and Garthwaite, P. H. (1987a). *J. Agric. Sci.* **109,** 315–320.

Phillippo, M., Humphries, W. R., Atkinson, T., Henderson, G. D., and Garthwaite, P. H. (1987b). *J. Agric. Sci.* **109,** 321–336.

Phillips, T. G., and Loughlin, M. E. (1949). *J. Agric. Res.* **78,** 389–395.

Phipps, R. H., and Weller, R. F. (1979). *J. Agric. Sci.* **92,** 471–483.

Pienaar, J. P., Hofmeyr, H. S., and Whiteford, G. L. (1983). *S. Afr. J. Anim. Sci.* **13,** 45–46.

Pigden, W. J., and Bell, J. M. (1955). *J. Anim. Sci.* **14,** 1239–1240.

Pigden, W. J., Pritchard, G. I., and Heaney, D. P. (1966). *Proc. Int. Grassl. Congr. 10th* pp. 397–400.

Piper, C. S. (1931). *J. Agric. Sci.* **21,** 762–779.

Piper, C. S. (1942). *J. Agric. Sci.* **32,** 143–177.

Piper, C. S., and Walkley, A. (1943). *J. Counc. Sci. Ind. Res. (Aust.)* **16,** 217–234.

Pitman, W. D., Vietor, D. M., and Holt, E. C. (1981). *CropSci.* **21,** 951–953.

Playne, M. J. (1969a). *Aust. J. Exp. Agric. Anim. Husb.* **9,** 192–195.

Playne, M. J. (1969b). *Aust. J. Exp. Agric. Anim. Husb.* **9,** 393–399.

Playne, M. J. (1970a). *Aust. J. Exp. Agric. Anim. Husb.* **10,** 32–35.

Playne, M. J. (1970b). *Proc. Aust. Soc. Anim. Prod.* **8,** 511–516.

Playne, M. J. (1972a). *Aust. J. Exp. Agric. Anim. Husb.* **12,** 373–377.

Playne, M. J. (1972b). *Aust. J. Exp. Agric. Anim. Husb.* **12,** 378–384.

Playne, M. J. (1974). *Proc. Aust. Soc. Anim. Prod.* **10,** 111.

Playne, M. J. (1976). *Rev. Rural Sci.* **3,** 155–164.

Playne, M. J. (1978). *Anim. Feed Sci. Technol.* **3,** 41–49.

Playne, M. J., and Haydock, K. P. (1972). *Aust. J. Exp. Agric. Anim. Husb.* **12,** 365–372.

Playne, M. J., and Kennedy, P. M. (1976). *J. Agric. Sci.* **86,** 367–372.

Playne, M. J., McLeod, M. N., and Dekker, R. F. H. (1972). *J. Sci. Food Agric.* **23,** 925–932.

Playne, M. J., Echevarria, M. G., and Megarrity, R. G. (1978). *J. Sci. Food. Agric.* **29,** 520–526.

Pless, C. D., Fontenot, J. P., and Webb, K. E. (1975). *J. Anim. Sci.* **40,** 198.

Plucknett, D. L., and Fox, R. L. (1965). *Proc. Int. Grassl. Congr., 9th* pp. 1525–1529.

Poe, J. H., Greene, L. W., Schelling, G. T., Byers, F. M., and Ellis, W. C. (1985). *J. Anim. Sci.* **60,** 578–582.

Poland, J. S., and Schnabel, J. A. (1980). *Trop. Agric. (Trinidad)* **57,** 259–264.

Pons, W. A., Stansbury, M. F., and Hoffpauir, C. L. (1953). *J. Assoc. Off. Agric. Chem.* **36,** 492–504.

Poole, D. B. R., and Connolly, J. F. (1967). *Ir. J. Agric. Res.* **6,** 281–284.

Poole, D. B. R., Moore, L., Finch, T. F., Gardiner, M. J., and Fleming, G. A. (1974a). *Ir. J. Agric. Res.* **13,** 119–122.

Poole, D. B. R., Rogers, P. A. M., and MacCarthy, D. D. (1974b). *In* "Trace Element Metabolism in Animals" (W. G. Hoekstra, J. W. Suttie, H. E. Ganther, and W. Mertz, eds.), pp. 618–620. Butterworths, London.

Poos, M. I., Hanson, T. L., and Klopfenstein, T. J. (1979). *J. Anim. Sci.* **48,** 1516–1524.

Pope, A. L., Moir, R. J., Somers, M., Underwood, E. J., and White, C. L. (1979). *J. Nutr.* **109,** 1448–1455.

Poppi, D. P. (1979). Ph.D. thesis. Univ. of Queensland, Queensland, Australia.

Poppi, D. P., Norton, B. W., Minson, D. J., and Hendricksen, R. E. (1980). *J. Agric. Sci.* **94,** 275–280.

Poppi, D. P., Minson, D. J., and Ternouth, J. H. (1981a). *Aust. J. Agric. Res.* **32,** 99–108.

Poppi, D. P., Minson, D. J., and Ternouth, J. H. (1981b). *Aust. J. Agric. Res.* **32,** 109–121.

Poppi, D. P., Minson, D. J., and Ternouth, J. H. (1981c). *Aust. J. Agric. Res.* **32,** 123–137.

Poppi, D. P., Hendricksen, R. E., and Minson, D. J. (1985). *J. Agric. Sci.* **105,** 9–14.

Potter, E. L., and Sparrow, S. (1973). *Fiji Agric. J.* **35,** 27–30.

Potter, E. L., Cooley, C. O., Raun, A. P., Richardson, L. F., and Rathmacher, R. P. (1974). *Proc., Annu. Meet.—Am. Soc. Anim. Sci., West. Sect.* **25,** 343–345.

Potter, E. L., Cooley, C. O., Richardson, L. F., Raun, A. P., and Rathmacher, R. P. (1976). *J. Anim. Sci.* **43,** 665–669.

Potter, B. J., Mano, M. T., Belling, G. B., Rogers, P. F., Martin, D. M., and Hetzel, B. S. (1981). *In* "Trace Element Metabolism in Man and Animals" (J. M. Howell, J. M. Gawthorne, and C. L. White, eds.), pp. 313–316. Aust. Acad. Sci., Canberra, Australia.

Potter, E. L., Muller, R. D., Wray, M. I., Carroll, L. H., and Meyer, R. M. (1986). *J. Anim. Sci.* **62,** 583–592.

Poulton, B. R., and Woelfel, C. G. (1963). *J. Dairy Sci.* **46,** 46–49.

Poulton, B. R., MacDonald, G. J., and Vander Noot, G. W. (1957). *J. Anim. Sci.* **16,** 462–466.

Preston, T. R., and Leng, R. A. (1987). "Matching Ruminant Production Systems with Available Resources in the Tropics and Sub-tropics." Penambul, Armidale, Australia.

Price, N. O., and Hardison, W. A. (1963). *Va., Agric. Exp. Stn., Tech. Bull.* **165.** [Cited by Whitehead, D. C. (1966). *Mimeo. Publ.—Commonw. Bur. Pastures Field Crops* **1/1966.**]

Price, J., and Humphries, W. R. (1980). *J. Agric. Sci.* **95,** 135–139.

Prins, R. A., Cline-Theil, W. C., and Van't Klooster, A. T. (1981). *Agric. Environ.* **6,** 183–194.

Pritchard, G. I., Pigden, W. J., and Minson, D. J. (1962). *Can. J. Anim. Sci.* **42,** 215–217.

Pritchard, G. I., Folkins, L. P., and Pigden, W. J. (1963). *Can. J. Plant Sci.* **43,** 79–87.

Pritchard, G. I., Pigden, W. J., and Folkins, L. P. (1964). *Can. J. Plant Sci.* **44,** 318–324.

Probasco, G. E., and Bjugstad, A. J. (1980). *J. Range Manage.* **33,** 244–246.

Pryor, W. J. (1959). *Aust. Vet. J.* **35,** 366–369.

Pullman, A. L., and Allden, W. G. (1971). *Aust. J. Agric. Res.* **22,** 401–413.

Quarterman, J., Mills, C. F., and Dalgarno, A. C. (1966). *Proc. Nutr. Soc.* **25,** 23.

Quesenberry, K. H., Sleper, D. A., and Cornell, J. A. (1978). *Crop Sci.* **18,** 847–850.

Quin, J. I., Van der Wath, J. G., and Myburgh, S. (1938). *Onderstepoort J. Vet. Sci. Anim. Ind.* **11,** 341–360.

Raab, L., Cafantaris, B., Jilg, T., and Menke, K. H. (1983). *Br. J. Nutr.* **50,** 569–582.
Rahman, H., McDonald, P., and Simpson, K. (1960). *J. Sci. Food Agric.* **11,** 422–428.
Rao, K. V., Van Soest, P. J., and Fick, G. M. (1987). *Crop Sci.* **27,** 601–603.
Rasmusson, D. C., Hester, A. J., Fick, G. N., and Bryne, I. (1971). *Crop Sci.* **11,** 623–626.
Rattray, P. V., and Joyce, J. P. (1969). *Proc. N.Z. Soc. Anim. Prod.* **29,** 102–113.
Rattray, P. V., and Joyce, J. P. (1974). *N.Z. J. Agric. Res.* **17,** 401–406.
Raun, A., Cheng, E., and Burroughs, W. (1956). *J. Agric. Food Chem.* **4,** 869–871.
Raymond, W. F. (1948). *Nature (London)* **161,** 937–938.
Raymond, W. F., and Harris, C. E. (1957). *J. Br. Grassl. Soc.* **12,** 166–170.
Raymond, W. F., and Minson, D. J. (1955). *J. Br. Grassl. Soc.* **10,** 282–296.
Raymond, W. F., and Spedding, C. R. W. (1965). *Proc.—Fert. Soc.* **88,** 5–34.
Raymond, W. F., Eyles, D. E., and Caukwell, V. G. (1949). *J. Br. Grassl. Soc.* **4,** 111–114.
Raymond, W. F., Harris, C. E., and Harker, V. G. (1953a). *J. Br. Grassl. Soc.* **8,** 301–314.
Raymond, W. F., Harris, C. E., and Harker, V. G. (1953b). *J. Br. Grassl. Soc.* **8,** 315–320.
Raymond, W. F., Kemp, C. D., Kemp, A. W., and Harris, C. E. (1954). *J. Br. Grassl. Soc.* **9,** 69–82.
Raymond, W. F., Minson, D. J., and Harris, C. E. (1956). *Proc. Int. Grassl. Congr., 7th* pp. 123–133.
Raymond, W. F., Minson, D. J., and Harris, C. E. (1959). *J. Br. Grassl. Soc.* **14,** 75–77.
Read, M. V. P., Engels, E. A. N., and Smith, W. A. (1986a). *S. Afr. J. Anim. Sci.* **16,** 1–6.
Read, M. V. P., Engels, E. A. N., and Smith, W. A. (1986b). *S. Afr. J. Anim. Sci.* **16,** 7–12.
Read, M. V. P., Engels, E. A. N., and Smith, W. A. (1986c). *S. Afr. J. Anim. Sci.* **16,** 13–17.
Read, M. V. P., Engels, E. A. N., and Smith, W. A. (1986d). *S. Afr. J. Anim. Sci.* **16,** 18–22.
Rearte, D. H., Kesler, E. M., and Hargrove, G. L. (1986). *J. Diary Sci.* **69,** 1366–1373.
Reay, P. F., and Marsh, B. (1976). *N.Z. J. Agric. Res.* **19,** 469–472.
Reddy, G. D., Alston, A. M., and Tiller, K. G. (1981a). *Aust. J. Exp. Agric. Anim. Husb.* **21,** 491–497.
Reddy, G. D., Alston, A. M., and Tiller, K. G. (1981b). *Aust. J. Exp. Agric. Anim. Husb.* **21,** 498–505.
Redshaw, E. S., Mathison, G. W., Milligan, L. P., and Weisenburger, R. D. (1986). *Can. J. Anim. Sci.* **66,** 103–115.
Reed, J. B. H., Smith, S. D., Doxey, D. L., Forbes, A. B., Finlay, R. S., Geering, I. W., and Wright, J. D. (1974a). *Trop. Anim. Health Prod.* **6,** 23–29.
Reed, J. B. H., Smith, S. D., Forbes, A. B., and Doxey, D. L. (1974b). *Trop. Anim. Health Prod.* **6,** 31–36.
Reed, J. B. H., Smith, S. D., Forbes, A. B., Finlay, R. S., Geering, I. W., Wright, J. D., and Doxey, D. L. (1974c). *Trop. Anim. Health Prod.* **6,** 37–38.
Rees, M. C., and Little, D. A. (1980). *J. Agric. Sci.* **94,** 483–485.
Rees, M. C., and Minson, D. J. (1976). *Br. J. Nutr.* **36,** 179–187.
Rees, M. C., and Minson, D. J. (1978). *Br. J. Nutr.* **39,** 5–11.
Rees, M. C., and Minson, D. J. (1982). *Aust. J. Agric. Res.* **33,** 629–636.
Rees, M. C., Minson, D. J., and Kerr, J. D. (1972). *Aust. J. Exp. Agric. Anim. Husb.* **12,** 553–560.
Rees, M. C., Minson, D. J., and Smith, F. W. (1974). *J. Agric. Sci.* **82,** 419–422.
Rees, M. C., Graham, N. M., and Searle, T. W. (1980). *Proc. Aust. Soc. Anim. Prod.* **13,** 466.
Reid, R. L., and Horvath, D. J. (1980). *Anim. Feed Sci. Technol.* **5,** 95–167.

Reid, R. L., and Jung, G. A. (1965a). *J. Anim. Sci.* **24,** 615–625.

Reid, R. L., and Jung, G. A. (1965b). *Proc. Int. Grassl. Congr., 9th* pp. 863–869.

Reid, R. L., Franklin, M. C., and Hallsworth, E. G. (1947). *Aust. Vet. J.* **23,** 136.

Reid, J. T., Woolfolk, P. G., Richard, C. R., Kaufmann, R. W., Loosli, J. K., Turk, K. L., Miller, J. I., and Blaser, R. E. (1950). *J. Diary Sci.* **33,** 60–71.

Reid, J. T., Woolfolk, P. G., Hardison, W. A., Martin, C. M., Brundage, A. L., and Kauffmann, R. W. (1952). *J. Nutr.* **46,** 255–269.

Reid, J. T., Kennedy, W. K., Turk, K. L., Slack, S. T., Trimberger, G. W., and Murphy, R. P. (1959a). *Agron. J.* **51,** 213–216.

Reid, J. T., Kennedy, W. K., Turk, K. L., Slack, S. T., Trimberger, G. W., and Murphy, R. P. (1959b). *J. Diary Sci.* **42,** 567–571.

Reid, C. S. W., Lyttleton, J. W., and Mangan, J. L. (1962). *N.Z. J. Agric. Res.* **5,** 237–248.

Reid, R. L., Clark, B., and Jung, G. A. (1964a). *Agron. J.* **56,** 537–542.

Reid, R. L., Jung, G. A., and Murray, S. (1964b). *J. Anim. Sci.* **23,** 700–710.

Reid, R. L., Jung, G. A., and Murray, S. J. (1966). *J. Anim. Sci.* **25,** 636–645.

Reid, R. L., Odhuba, E. K., and Jung, G. A. (1967a). *Agron. J.* **59,** 265–271.

Reid, R. L., Jung, G. A., and Kinsey, C. M. (1967b). *Agron. J.* **59,** 519–525.

Reid, R. L., Jung, G. A., and Post, A. J. (1969). *J. Anim. Sci.* **29,** 181.

Reid, R. L., Post, A. J., and Jung, G. A. (1970). *Bull.—W. Va., Agric. Exp. Stn.* **589T.**

Reid, R. L., Greenhalgh, J. F. D., and Aitken, J. N. (1972). *J. Agric. Sci.* **78,** 491–496.

Reid, R. L., Jung, G. A., Roemig, I. J., and Kocher, R. E. (1978a). *Agron. J.* **70,** 9–14.

Reid, R. L., Powell, K., Balasko, J. A., and McCormick, C. C. (1978b). *J. Anim. Sci.* **46,** 1493–1502.

Reid, R. L., Baker, B. S., and Vona, L. C. (1984). *J. Anim. Sci.* **59,** 1403–1410.

Reid, R. L., Jung, G. A., Stout, W. L., and Ranney, T. S. (1987). *J. Anim. Sci.* **64,** 1735–1742.

Reinach, N., and Louw, J. G. (1952). *Farming S. Afr.* **27,** 417–419.

Reinach, N., Louw, J. G., and Groenewald, J. W. (1952). *Onderstepoort J. Vet. Res.* **25,** 85–91.

Reis, P. J. (1969). *Aust. J. Biol. Sci.* **22,** 745–759.

Reith, J. W. S., Inkson, R. H. E., Holmes, W., MacLusky, D. S., Reid, D., Heddle, R. G., and Copeman, G. J. F. (1964). *J. Agric. Sci.* **63,** 209–219.

Rentsch, A. D. (1980). *Aust. Vet. J.* **56,** 458–459.

Reuter, D. J. (1975). *In* "Trace Elements in Soil–Plant–Animal Systems" (D. I. D. Nicholas and A. R., Egan, eds.), pp. 291–324. Academic Press, New York.

Rexen, F. P., and Knudsen, K. E. B. (1984). *In* "Straw and Other Fibrous By-Products as Feeds" (F. Sundstol and E. Owen, eds.), pp. 127–161. Elsevier, Amsterdam.

Rhodes, F. B., (1956). *Rhod. Agric. J.* **53,** 969–982.

Richards, C. R., and Reid, J. T. (1952). *J. Diary Sci.* **35,** 595–602.

Richards, I. R., and Wolton, K. M. (1975). *Anim. Prod.* **20,** 425–428.

Richards, C. R., Weaver, H. G., and Connolly, J. D. (1958). *J. Diary Sci.* **41,** 956–962.

Ritchie, N. S., and Fishwick, G. (1977). *J. Agric. Sci.* **88,** 71–73.

Rittenhouse, L. R., Clanton, D. C., and Streeter, C. L. (1970). *J. Anim. Sci.* **31,** 1215–1221.

Ritter, R. J., Boling, J. A., and Gray, N. (1984). *J. Anim. Sci.* **59,** 197–203.

Robb, T. W., Ely, D. G., Rieck, C. E., Thomas, R. J., Glenn, B. P., and Glenn, S. (1982). *J. Anim. Sci.* **54,** 155–163.

Robbins, C. T., Hanley, T. A., Hagerman, A. E., Hjeljord, O., Baker, D. L., Schwartz, C. C., and Mautz, W. W. (1987). *Ecology* **68,** 98–107.

Robertson, W. M. (1971). *Vet. Rec.* **89,** 5–12.

Robinson, R. R. (1942). *J. Am. Soc. Agron.* **34,** 933–939.

Robinson, M. F., and Thomson, C. D. (1983). *Nutr. Abstr. Rev., Ser. A* **53A,** 3–26.

Rodel, M. G. W. (1972). *Rhod. Agric. J.* **69,** 59–60.
Rodger, J. B. (1982). *J. Agric. Sci.* **99,** 199–206.
Rodrique, C. B., and Allen, N. N. (1960). *Can. J. Anim. Sci.* **40,** 23–29.
Rogers, G. L., and McLeay, L. M. (1977). *Proc. N.Z. Soc. Anim. Prod.* **37,** 46–49.
Rogers, P. A. M., and Poole, D. B. R. (1978). *In* "Trace Element Metabolism in Man and Animals" (M. Kirchgessner, ed.), pp. 481–485. Arbeitskreis Tiernaehrungsforsch., Freising, West Germany.
Rogers, G. L., Byrant, A. M., and McLeay, L. M. (1979). *N.Z. J. Agric. Res.* **22,** 533–541.
Rogers, G. L., Porter, R. H. D., Clarke, T., and Stewart, J. A. (1980). *Aust. J. Agric. Res.* **31,** 1147–1152.
Rojas, M. A., Dyer, I. A., and Cassatt, W. A. (1965). *J. Anim. Sci.* **24,** 664–667.
Rojas, D. O., McDowell, L. R., Moore, J. E., Martin, F. G., and Ocumpaugh, W. R. (1987). *Trop. Grassl.* **21,** 8–14.
Rollinson, D. H. L., and Bredon, R. M. (1960). *J. Agric. Sci.* **54,** 235–242.
Romberg, B., Pearce, G. R., and Tribe, D. E. (1969). *Aust. J. Exp. Agric. Anim. Husb.* **9,** 71–73.
Romero, F., Van Horn, H. H., Prine, G. M., and French, E. C. (1987). *J. Anim. Sci.* **65,** 786–796.
Ronning, M., and Dobie, J. B. (1967). *J. Dairy Sci.* **50,** 391–393.
Rook, J. A. F. (1964). *Proc. Nutr. Soc.* **23,** 71–80.
Rook, J. A. F., and Balch, C. C. (1958). *J. Agric. Sci.* **51,** 199–207.
Rook, J. A. F., and Balch, C. C. (1962). *J. Agric. Sci.* **59,** 103–108.
Rook, J. A. F., and Campling, R. C. (1962). *J. Agric. Sci.* **59,** 225–232.
Rook, J. A. F., and Storry, J. E. (1962). *Nutr. Abstr. Rev.* **32,** 1055–1077.
Rook, J. A. F., and Wood, M. (1960). *J. Sci. Food Agric.* **11,** 137–143.
Rook, J. A., Muller, L. D., and Shank, D. B. (1977). *J. Dairy Sci.* **60,** 1894–1904.
Rooke, J. A., Brookes, I. M., and Armstrong, D. G. (1982). *In* "Forage Protein in Ruminant Animal Production" (D. J. Thomson, D. E. Beever, and R. G. Gunn, eds.), Occas. Publ. 6, pp. 185–186. Br. Soc. Anim. Prod., London.
Rooke, J. A., Brett, P. A., Overend, M. A., and Armstrong, D. G., (1985). *Anim. Feed Sci. Technol.* **13,** 255–267.
Rooke, J. A., Maya, F. M., Arnold, J. A., and Armstrong, D. G. (1988). *Grass Forage Sci.* **43,** 87–95.
Rosenfeld, I., and Beath, O. A. (1964). "Selenium: Geobotany, Biochemistry, Toxicity, and Nutrition." Academic Press, New York.
Rosero, O. R., Tucker, R. E., Mitchell, G. E., and Schelling, G. T. (1980). *J. Anim. Sci.* **50,** 128–136.
Ross, J. G., Bullis, S. S., and Lin, K. C. (1970). *Crop Sci.* **10,** 672–673.
Ross, J. G., Thaden, R. T., and Tucker, W. L. (1975). *Crop Sci.* **15,** 303–306.
Rossiter, R. C., Curnow, D. H., and Underwood, E. J. (1948). *J. Aust. Inst. Agric. Sci.* **14,** 9–14.
Rotar, P. P. (1965). *Trop. Agric. (Trinidad)* **42,** 333–337.
Rowe, J. B. (1983). *In* "Feed Information and Animal Production" (G. E. Robards and R. G. Packam, eds.), pp. 273–277. Commonw. Agric. Bur., Farnham Royal, England.
Rowe, J. B. (1985). *Rev. Rural Sci.* **6,** 101–108.
Rowlands, G. J., and Pocock, R. M. (1976). *Vet. Rec.* **98,** 333–338.
Roy, J. H. B., Balch, C. C., Miller, E. L., Orskov, E. R., and Smith, R. H. (1977). *In* "Protein Metabolism and Nutrition" (S. Tamminga, ed.), European Association Animal Production Publ. 22, pp. 126–129. Cent. Agric. Publ. Doc., Wageningen, The Netherlands.

Rudert, C. P., and O'Donovan, W. M. (1974). *Rhod. J. Agric. Res.* **12,** 141–148.
Rudert, C. P., and Oliver, J. (1978). *Rhod. J. Agric. Res.* **16,** 23–29.
Ruiz, R., Cairo, J., Martinez, R. O., and Herrera, R. S. (1981). *Cuban J. Agric. Sci.* **15,** 133–144.
Rumball, W., Butler, G. W., and Jackman, R. H. (1972). *N.Z. J. Agric. Res.* **15,** 33–42.
Russell, J. B., and Duncan, D. L. (1956). "Minerals in Pasture: Deficiencies and Excesses in Relation to Animal Health." Commonw. Agric. Bur., Farnham Royal, England.
Russell, J. B., and Martin, S. A. (1984). *J. Anim. Sci.* **59,** 1329–1338.
Russell, J. B., and Sniffen, C. J. (1984). *J. Diary Sci.* **67,** 987–994.
Russell, J. B., Hurst, J. P., Jorgensen, N. A., and Barrington, G. P. (1978). *J. Anim. Sci.* **46,** 278–287.
Saenger, P. F., Lemenager, R. P., and Hendrix, K. S. (1982). *J. Anim. Sci.* **54,** 419–425.
Saibro, J. C., Hoveland, C. S., and Williams, J. C. (1978). *Agron. J.* **70,** 497–500.
Said, A. N. (1971). *East Afr. Agric. For. J.* **37,** 15–21.
Said, A. N., Wheeler, J. L., and Lindstad, P. (1977). *Forage Res.* **3,** 75–81.
Sarker, A. B., and Holmes, W. (1974). *J. Br. Grassl. Soc.* **29,** 141–143.
Satter, L. D., and Roffler, R. R. (1977). *Trop. Anim. Prod.* **2,** 238–259.
Saunders, W. M. H., and Metson, A. J. (1971). *N.Z. Agric. Res.* **14,** 307–328.
Scales, G. H. (1974). *Proc. N.Z. Soc. Anim. Prod.* **34,** 103–113.
Scales, G. H., Streeter, C. I., Denham, A. H., and Ward, G. M. (1974). *J. Anim. Sci.* **38,** 192–199.
Scaut, A. (1959). *Inst. Natl. Agron. Congo, Sci. Ser.* **81.**
Schank, S. C., Klock, M. A., and Moore, J. E. (1973). *Agron. J.* **65,** 256–258.
Schank, S. C., Day, J. M., and Delgado-De-Lucas, E. (1977). *Trop. Agric. (Trinidad)* **54,** 119–125.
Schlink, A. C., and Lindsay, J. A. (1988). *Proc. Aust. Soc. Anim. Prod.* **17,** 330–333.
Schneider, B. H. (1947). "Feeds of the World." West Virginia Univ. Morgantown, West Virginia.
Schneider, B. H., and Flatt, W. P. (1975). "The Evaluation of Feeds through Digestibility Experiments." Univ. of Georgia, Athens.
Schneider, K. M., Boston, R. C., and Leaver, D. D. (1982). *Aust. J. Agric. Res.* **33,** 827–842.
Schultz, I., Turner, M. A., and Cooke, J. G. (1979). *N.Z. J. Agric. Res.* **22,** 303–308.
Schwartz, C. C., Nagy, J. G., and Streeter, C. L. (1973). *J. Anim. Sci.* **37,** 821–826.
Scott, D. (1970). *Annu. Rep. Stud. Anim. Nutr. Allied Sci. (Rep. Rowett Inst.)* **26,** 98–107.
Scott, D., and Beastall, G. (1978). *Q. J. Exp. Physiol.* **63,** 147–156.
Scott, D., and Buchan, W. (1987). *Q. J. Exp. Physiol.* **72,** 331–338.
Scott, D., and Buchan, W. (1988). *J. Agric. Sci.* **110,** 411–413.
Seath, D. M., Lassiter, C. A., Davis, C. L., Rust, J. W., and Cole, M. (1956). *J. Dairy Sci.* **39,** 274–279.
Sen, K. M., and Mabey, G. L. (1965). *Proc. Int. Grassl. Congr., 9th* pp. 765–771.
Seoane, J. R., Cote, M., and Visser, S. A. (1982). *Can. J. Anim. Sci.* **62,** 473–480.
Setchell, B. P., Harris, A. N. A., Farleigh, E. A., and Clark, F. L. (1962). *Aust. Vet. J.* **38,** 62–65.
Sevilla, C. C., and Ternouth, J. H. (1980). *Proc. Aust. Soc. Anim. Prod.* **13,** 449.
Shaver, R. D., Erdman, R. A., O'Connor, A. M., and Vandersall, J. H. (1985). *J. Diary Sci.* **68,** 338–346.
Shaw, N. H. (1978). *Aust. J. Exp. Agric. Anim. Husb.* **18,** 800–807.
Sheaffer, C. C., and Marten, G. C. (1986). *Agron. J.* **78,** 75–79.
Shearer, D. A. (1961). *Can. J. Anim. Sci.* **41,** 197–204.

Sheehan, W. (1969). *Ir. J. Agric. Res.* **8,** 337–342.
Sheehy, E. J., O'Donovan, J., Day, W. R., and Curran, S. (1948). *Ir. Repub. Agric. Fish. J.* **45,** 5–28.
Shelton, D. C., and Reid, R. L. (1960). *Proc. Int. Grassl. Congr., 8th* pp. 524–528.
Shenk, J. S., Westerhaus, M. O., and Hoover, M. R. (1979). *J. Dairy Sci.* **62,** 807–812.
Shepperson, G. (1960). *Proc. Int. Grassl. Congr., 8th* pp. 704–708.
Sherrell, C. G. (1978). *N.Z. J. Exp. Agric.* **6,** 189–190.
Sherrell, C. G., and Rawnsley, J. S. (1982). *N.Z. J. Agric. Res.* **25,** 363–368.
Shimojo, M., and Goto, I. (1989). *Anim. Feed Sci. Technol.* **24,** 173–177.
Shirley, R. L. (1985). *In* "Nutrition of Grazing Ruminants in Warm Climates" (L. R. McDowell, ed.), pp. 37–57. Academic Press, Orlando, Florida.
Shirley, R. L., Easley, J. F., McCall, J. T., Davis, G. K., Kirk, W. G., and Hodges, E. M. (1968). *J. Anim. Sci.* **27,** 757–765.
Shrivastava, V. S., and Talapatra, S. K. (1962). *Indian J. Dairy Sci.* **15,** 154–160.
Sibbald, A. R., Maxwell, T. J., and Eadie, J. (1979). *Agric. Syst.* **4,** 119–134.
Siddons, R. C., and Paradine, J. (1981). *J. Sci. Food Agric.* **32,** 973–981.
Siddons, R. C., Evans, R. T., and Beever, D. E. (1979). *Br. J. Nutr.* **42,** 535–545.
Siddons, R. C., Beever, D. E., and Kaiser, A. G. (1982). *J. Sci. Food Agric.* **33,** 609–613.
Siebert, B. D., and Cameron, D. D. (1978). *Proc. Nutr. Soc. Aust.* **3,** 80.
Siebert, B. D., and Hunter, R. A. (1982). *In* "Nutrition Limits to Animal Production from Pastures" (J. B. Hacker, ed.), pp. 409–425. Commonw. Agric. Bur., Farnham Royal, England.
Siebert, B. D., and Kennedy, P. M. (1972). *Aust. J. Agric. Res.* **23,** 35–44.
Siebert, B. D., and Saunders, L. E. (1976). *Proc. Aust. Soc. Anim. Prod.* **11,** 8P.
Simesen, M. G. (1977). *Nord. Veterinaermed.* **29,** 284–286.
Simons, A. B., and Marten, G. C. (1971). *Agron. J.* **63,** 915–919.
Simpson, B. W. (1930). *N.Z. J. Agric.* **41,** 179–182.
Simpson, B. W. (1931). *N.Z. J. Agric.* **42,** 18–23.
Simpson, K. (1964). *Scott. Agric.* **43,** 197–201.
Sinclair, D. P., and Andrews, E. D. (1954). *N.Z. Vet. J.* **2,** 72–79.
Sinclair, D. P., and Andrews, E. D. (1958). *N.Z. Vet. J.* **6,** 87–95.
Sinclair, D. P., and Andrews, E. D. (1959). *N.Z. Vet. J.* **7,** 39–41.
Sinclair, D. P., and Andrews, E. D. (1961). *N.Z. Vet. J.* **9,** 96–100.
Sjollema, B. (1930). *Vet. Rec.* **10,** 425–430, 450–453.
Sjollema, B. (1932). *Nutr. Abstr. Rev.* **1,** 621–632.
Skerman, K. D. (1959). *Aust. Vet. J.* **35,** 369–373.
Skerman, K. D. (1962). *Proc. Aust. Soc. Anim. Prod.* **4,** 22–27.
Skerman, K. D., and O'Halloran, M. W. (1962). *Aust. Vet. J.* **38,** 98–102.
Skerman, K. D., Sutherland, A. K., O'Halloran, M. W., Bourke, J. M., and Munday, B. L. (1969). *Am. J. Vet. Res.* **20,** 977–984.
Slen, S. B., Demiruren, A. S., and Smith, A. D. (1961). *Can. J. Anim. Sci.* **41,** 263–265.
Sleper, D. A., and Mott, G. O. (1976). *Agron. J.* **68,** 993–995.
Sleper, D. A., Garner, G. B., Assay, K. H., Boland, R., and Pickett, E. E. (1977). *Crop Sci.* **17,** 433–438.
Sleper, D. A., Garner, G. B., Nelson, C. J., and Sevaugh, J. L. (1980). *Agron. J.* **72,** 720–722.
Smith, D. (1969). *Agron. J.* **61,** 470–473.
Smith, D. (1970a). *Proc. Grassl. Congr., 11th* pp. 510–514.
Smith, D. (1970b). *Agron. J.* **62,** 520–523.

Smith, D. (1970c). *J. Agric. Food Chem.* **18,** 652–656.
Smith, D. (1971). *Agron. J.* **63,** 497–500.
Smith, R. H. (1975). *In* "Digestion and Metabolism in the Ruminant" (I. W. McDonald and A. C. I. Warner, eds.), pp. 399–415. Univ. of New England, Armidale, Australia.
Smith, F. W. (1981). *J. Plant Nutr.* **3,** 813–826.
Smith, S. E., and Aines, P. D. (1959). *Bull.—N.Y. Agric. Exp. Stn. (Ithaca)* **938.**
Smith, B., and Coup, M. R. (1973). *N.Z. Vet. J.* **21,** 252–258.
Smith, G. S., and Edmeades, D. C. (1983). *N.Z. J. Agric. Res.* **26,** 223–225.
Smith, R. M., and Gawthorne, J. M. (1975). *In* "Trace Elements in Soil–Plant–Animal Systems" (D. I. D. Nicholas and A. R. Egan, eds.), pp. 243–258. Academic Press, New York.
Smith, S. E., and Loosli, J. K. (1957). *J. Diary Sci.* **40,** 1215–1227.
Smith, R. H., and McAllan, A. B. (1967). *Proc. Nutr. Soc.* **26,** xxxii.
Smith, G. S., and Middleton, K. R. (1978). *N.Z. J. Exp. Agric.* **6,** 217–225.
Smith, G. S., and Nelson, A. B. (1975). *J. Anim. Sci.* **41,** 891–898.
Smith, A. M., and Reid, J. T. (1955). *J. Diary Sci.* **35,** 515–524.
Smith, F. W., and Siregar, M. E. (1983). *In* "Sulphur in South East Asia and South Pacific Agriculture" (G. I. Blair and A. R. Till, eds.), pp. 76–86. Univ. of New England, Armidale, Australia.
Smith, R. J., and Thompson, J. M. (1978). *Proc. Aust. Soc. Anim. Prod.* **12,** 122.
Smith, R. J., and Thompson, J. M. (1980). *Aust. Vet. J.* **56,** 400–402.
Smith, S. E., Koch, B. A., and Turk, K. L. (1951). *J. Nutr.* **44,** 455–464.
Smith, E. F., Young, V. A., Anderson, K. L., Ruliffson, W. S., and Rogers, S. N. (1960). *J. Anim. Sci.* **19,** 388–391.
Smith, J. S., Wainman, F. W., and Dewey, P. J. S. (1976). *Proc. Nutr. Soc.* **35,** 97A.
Smith, R. M., Fraser, F. J., Russell, G. R., and Robertson, J. S. (1977a). *J. Comp. Pathol.* **89,** 119–128.
Smith, B. L., Embling, P. P., Towers, N. R., Wright, D. E., and Payne, E. (1977b). *N.Z. Vet J.* **25,** 124–127.
Smith, B. L., Coe, B. D., and Embling, P. P. (1978). *N.Z. Vet. J.* **26,** 314–315.
Smith, G. S., Young, P. W., and O'Connor, M. B. (1983). *Proc. N.Z. Grassl. Assoc.* **44,** 179–183.
Smith, B. L., Collier, A. J., Lawrence, R. J., and Towers, N. R. (1984). *N.Z. Vet. J.* **22,** 48–50.
Smyth, P. J., Conway, A., and Walsh, M. J. (1958). *Vet. Rec.* **70,** 846–849.
Snaydon, R. W. (1972). *Aust. J. Agric. Res.* **23,** 253–256.
Snook, L. C. (1955). *J. Dep. Agric., West. Aust.* **4,** 175–182.
Snook, L. C. (1962). *Aust. Vet. J.* **38,** 42–47.
Sollenberger, L. E., Moore, J. E., Quesenberry, K. H., and Beede, P. T. (1987). *Agron. J.* **79,** 1049–1054.
Somers, M., and Underwood, E. J. (1969). *Aust. J. Agric. Res.* **20,** 899–903.
Sotola, J. (1937). *J. Agric. Res.* **54,** 399–415.
Sotola, J. (1946). *J. Agric. Res.* **72,** 365–371.
Spahr, S. L., Kesler, E. M., Bratzler, J. W., and Washko, J. B. (1961). *J. Diary Sci.* 44, 503–510.
Spais, A. G., and Papasteriadis, A. A. (1974). *In* "Trace Element Metabolism in Animals" (W. G. Hoekstra, J. W. Suttie, H. E. Ganther, and W. Mertz, eds.), pp. 628–631. Butterworths, London.
Spears, J. W., Harvey, R. W., and Segerson, E. C. (1986). *J. Anim. Sci.* **63,** 586–594.

Spedding, C. R. W. (1954). *J. Comp. Pathol.* **64,** 5–14.

Spedding, C. R. W. (1955). *Ph.D. thesis.* Univ. of London.

Spedding, C. R. W., Large, R. V., and Kydd, D. D. (1966). *Proc. Int. Grassl. Congr., 10th* pp. 479–483.

Spencer, R. R., and Akin, D. E. (1980). *J. Anim. Sci.* **51,** 1189–1196.

Spencer, R. R., and Amos, H. E. (1977). *J. Anim. Sci.* 45, 126–131.

Spencer, R. R., Akin, D. E., and Rigsby, L. L. (1984). *Agron. J.* **76,** 819–824.

Stadtmore, D. L., and Reid, R. L. (1981). *J. Anim. Sci., Suppl.* **53,** 432–433.

Stallings, C. C., Donaldson, B. M., Thomas, J. W., and Rossman, E. C. (1982). *J. Diary Sci.* **65,** 1945–1949.

Stanton, D. J., and Kidson, E. B. (1939). *N.Z. J. Sci. Technol.* **21B,** 67–76.

Staten, H. I. W. (1949). *Bull.—Okla., Agric. Exp. Stn.* **T35.**

Statham, M., and Bray, A. C. (1975). *Aust. J. Agric. Res.* **26,** 751–768.

Statham, M., and Koen, T. B. (1982). *Aust. J. Exp. Agric. Anim. Husb.* **22,** 29–34.

Stehr, W., and Kirchgessner, M. (1976). *Anim. Feed Sci. Technol.* **1,** 53–60.

Stephenson, R. G. A., Edwards, J. C., and Hopkins, P. S. (1981). *Aust. J. Agric. Res.* **32,** 497–509.

Stewart, A. B., and Holmes, W. (1953). *J. Sci. Food Agric.* **4,** 401–408.

Stewart, J. W. B., and McConaghy, S. (1963). *J. Sci. Food Agric.* **14,** 613–621.

Stewart, J., and Reith, J. W. S. (1956). *J. Comp. Pathol.* **66,** 1–9.

Stewart, J., Mitchell, R. L., and Stewart, A. B. (1941). *Emp. J. Exp. Agric.* **9,** 145–152.

Stewart, J., Mitchell, R. L., and Stewart, A. B. (1942). *Emp. J. Exp. Agric.* **10,** 57–60.

Stewart, J., Mitchell, R. L., Stewart, A. B., and Young, H. M. (1946). *Emp. J. Exp. Agric.* **14,** 145–152.

Stillings, B. R., Bratzler, J. W., Marriot, L. F., and Miller, R. C. (1964). *J. Anim. Sci.* **23,** 1148–1154.

Stobbs, T. H. (1965). *Proc. Int. Grassl. Congr., 9th* pp. 939–942.

Stobbs, T. H. (1969). *East Afr. Agric. J.* **35,** 128–134.

Stobbs, T. H. (1973). *Aust. J. Agric. Res.* **24,** 809–819.

Stobbs, T. H. (1977). *Trop. Grassl.* **2,** 87–91.

Stobbs, T. H., and Cowper, L. J. (1972). *Trop. Grassl.* **6,** 107–112.

Stobbs, T. H., and Imrie, B. C. (1976). *Trop. Grassl.* **10,** 99–106.

Stobbs, T. H., Minson, D. J., and McLeod, M. N. (1977). *J. Agric. Sci.* **89,** 137–141.

Stockdale, C. R. (1985). *Grass Forage Sci.* **46,** 31–39.

Stockdale, C. R., and Beavis, G. W. (1988). *Proc. Aust. Soc. Anim. Prod.* **17,** 473.

Stratton, S. D., Sleper, D. A., and Matches, A. G. (1979). *Crop Sci.* **19,** 329–333.

Streeter, C. L. (1969). *J. Anim. Sci.* **29,** 757–768.

Strickland, R. W. (1970). *CSIRO, Div. Trop. Pastures, Annu. Rep. 1969–70* p. 67.

Strickland, R. W. (1974). *Aust. J. Exp. Agric. Anim. Husb.* **14,** 186–196.

Strickland, R. W., and Haydock, K. P. (1978). *Aust. J. Exp. Agric. Anim. Husb.* **18,** 817–824.

Sullivan, J. T. (1962). *Agron. J.* **54,** 511–515.

Sullivan, J. T. (1964). "Chemical Composition of Forages," Agric. Res. Serv. Publ., 34-62. U.S. Dep. Agric., Washington, D.C.

Sullivan, J. T., Phillips, T. G., Loughlin, M. E., and Sprague, V. G. (1956). *Agron. J.* **48,** 11–14.

Sundstol, F., and Coxworth, E. M. (1984). *In* "Straw and Other Fibrous By-products as Feed" (F. Sundstol and E. Owen, eds.), pp. 196–247. Elsevier, Amsterdam.

Sundstol, F., and Owen, E., eds. (1984). "Straw and Other Fibrous By-products as Feed." Elsevier, Amsterdam.

Sundstol, F., Coxworth, E., and Mowat, D. N. (1978). *World Anim. Rev.* **26,** 13–21.
Sutherland, A. K. (1956). *Proc. Aust. Soc. Anim. Prod.* **1,** 18–19.
Suttle, N. F. (1975). *Br. J. Nutr.* **34,** 411–420.
Suttle, N. F. (1979). *Proc. Nutr. Soc.* **38,** 135A.
Suttle, N. F. (1980). *Proc. Nutr. Soc.* **39,** 63A.
Suttle, N. F. (1981). *In* "Trace Element Metabolism in Man and Animals" (J. M. Howell, J. M. Gawthorne, and C. L. White, eds.), pp. 545–548. Aust. Acad. Sci., Canberra, Australia.
Suttle, N. F. (1983). *In* "Sheep Production" (I. W. Haresign, ed.), pp. 167–183. Butterworths, London.
Suttle, N. F. (1986). *Vet. Rec.* **119,** 519–522.
Suttle, N. F. (1987a). *Res. Vet. Sci.* **42,** 219–223.
Suttle, N. F. (1987b). *In* "The Nutrition of Herbivores" (J. B. Hacker and J. H. Ternouth, eds.), pp. 333–361. Academic Press, Orlando, Florida.
Suttle, N. F. (1988). *J. Comp. Pathol.* **99,** 241–258.
Suttle, N. F., and Angus, K. W. (1976). *J. Comp. Pathol.* **86,** 595–608.
Suttle, N. F., and Field, A. C. (1969). *Br. J. Nutr.* **23,** 81–90.
Suttle, N. F., and McMurray, C. H. (1983). *Res. Vet. Sci.* **35,** 47–52.
Suttle, N. F., Davies, H. L., and Field, A. C. (1982). *Br. J. Nutr.* **47,** 105–112.
Suttle, N. F., Abrahams, P., and Thornton, I. (1984). *J. Agric. Sci.* **103,** 81–86.
Swift, R. W., Cowan, R. L., Barron, G. P., Maddy, K. H., Grose, E. C., and Washko, J. B. (1952). *J. Anim. Sci.* **11,** 389–399.
Sykes, A. R., and Coop, R. L. (1976). *J. Agric. Sci.* **86,** 507–515.
Sykes, A. R., and Dingwall, R. A. (1976). *J. Agric. Sci.* **86,** 587–594.
Sykes, A. R., Field, A. C., and Slee, J. (1969). *Anim. Prod.* **11,** 91–99.
Tagari, H., and Ben-Ghedalia, D. (1977). *J. Agric. Sci.* **88,** 181–185.
Taji, K. (1967). *Nippon Chikusan Gakkai Ho* **38,** 537–546.
Tasker, J. B., Bewick, T. D., Clark, R. G., and Frazer, A. J. (1987). *Proc. Anim. Sci. Congr. Asian–Australas. Assoc. Anim. Prod. Soc., 4th* p. 428.
Tayler, J. C., and Deriaz, R. E. (1963). *J. Br. Grassl. Soc.* **18,** 29–38.
Tayler, J. C., and Rudman, J. E. (1965). Proc. Int. Grassl. Congr., 9th pp. 1639–1644.
Tayler, J. C., and Wilkinson, J. M. (1972). *Anim. Prod.* **14,** 85–96.
Taylor, A. J. (1941). *S. Afr., Dep. Agric. For., Bull.* **203.**
Tejada, R., McDowell, L. R., Martin, F. G., and Conrad, J. H. (1987). *Trop. Agric. (Trinidad)* **64,** 55–60.
Teleni, E. (1976). *Vet. Rev. Monogr.* **3,**
Teleni, E., Dean, H., and Murray, R. M. (1976). *Aust. Vet. J.* **52,** 529–533.
Tergas, L. E., and Blue, W. G. (1971). *Agron. J.* **63** 6–9.
Ternouth, J. H. (1967). *J. Aust. Inst. Agric. Sci.* **33,** 263–264.
Ternouth, J. H., and Sevilla, C. C. (1984). *Can. J. Anim. Sci., Suppl.* **64,** 221–222.
Ternouth, J. H., Poppi, D. P., and Minson, D. J. (1979). *Proc. Nutr. Soc. Aust.* **4,** 152.
Terry, R. A., and Tilley, J. M. A. (1964). *J. Br. Grassl. Soc.* **19,** 363–372.
Terry, R. A., Tilley, J. M. A., and Outen, G. E. (1969). *J. Sci. Food Agric.* **20,** 317–320.
Tetlow, R. M., and Wilkins, R. J. (1974). *Anim. Prod.* **19,** 193–200.
Tetlow, R. M., and Wilkins, R. J. (1977). *Anim. Prod.* **25,** 61–70.
Thaden, R. T., Ross, J. G., and Akyurek, A. (1975). *Crop Sci.* **15,** 375–378.
Theander, O., and Aman, P. (1984). *In* "Straw and Other Fibrous By-products as Feeds" (F. Sundstol and E. Owen, eds.), pp. 45–78. Elsevier, Amsterdam.
Theiler, A., Green, H. H., and Du Toit, P. J. (1924). *J. Dep. Agric., Union S. Afr.* **8,** 460–504.

Theiler, A., Green, H. H., and Du Toit, P. J. (1927). *J. Agric. Sci.* **7,** 291–314.
Theiler, A., Green, H. H., and Du Toit, P. J. (1928). *J. Agric. Sci.* **18,** 369–371.
Theiler, A., Du Toit, P. J., and Malan, A. I. (1937). *Onderstepoort J. Vet. Sci. Anim. Ind.* **8,** 375–414.
Thom, E. R., and Smith, G. S. (1980). *Proc. Agron. Soc. N.Z.* **10,** 23–26.
Thomas, C. (1977). *East Afr. Agric. For. J.* **42,** 328–336.
Thomas, C. (1978). *Trop. Agric. (Trinidad)* **55,** 325–327.
Thomas, B., and Trinder, N. (1947). *Emp. J. Exp. Agric.* **15,** 237–248.
Thomas, B., Thompson, A., Oyenuga, V. A., and Armstrong, R. H. (1952a). *Emp. J. Exp. Agric.* **20,** 10–22.
Thomas, W. E., Loosli, J. K., Williams, H. H., and Maynard, L. A. (1952b). *J. Nutr.* **43,** 515–523.
Thomas, J. W., Moore, L. A., Okamoto, M., and Sykes, J. F. (1961). *J. Dairy Sci.* **44,** 1471–1483.
Thomas, C., Njoroge, P. K., and Fenlon, J. S. (1980). *Trop. Agric. (Trinidad)* **57,** 75–82.
Thompson, R. H., and Todd, J. R. (1976). *Br. J. Nutr.* **36,** 299–304.
Thompson, J. K., and Warren, R. W. (1979). *Grass Forage Sci.* **34,** 83–88.
Thompson, J. K., MacDonald, D. C., and Warren, R. W. (1978). "Multiple Blood Analysis of Dairy Cows as a Management Aid." North Scotland Coll. Agric., Aberdeen, Scotland.
Thompson, N., Keiller, P. R., and Yates, C. W. (1989). *Grass Forage Sci.* **44,** 195–203.
Thomson, D. J. (1971). *J. Br. Grassl. Soc.* **26,** 149–155.
Thomson, D. J., and Cammell, S. B. (1979). *Br. J. Nutr.* **41,** 297–310.
Thomson, G. G., and Lawson, B. M. (1970). *N.Z. Vet. J.* **18,** 79–82.
Thomson, D. J., Beever, D. E., Harrison, D. G., Hill, I. W., and Osbourn, D. F. (1971). *Proc. Nutr. Soc.* **30,** 14A–15A.
Thomson, D. J., Beever, D. E., Coelho da Silva, J. F., and Armstrong, D. G. (1972). *Br. J. Nutr.* **28,** 31–41.
Thomson, D. J., Beever, D. E., Lonsdale, C. R., Haines, M. J., Cammell, S. B., and Austin, A. R. (1981). *Br. J. Nutr.* **46,** 193–207.
Thorley, C. M., Sharpe, M. E., and Bryant, M. P. (1968). *J. Dairy Sci.* **51,** 1811–1816.
Thornton, R. F., and Minson, D. J. (1973a). *Aust. J. Exp. Agric. Anim. Husb.* **13,** 537–543.
Thornton, R. F., and Minson, D. J. (1973b). *Aust. J. Agric. Res.* **24,** 889–898.
Thornton, I., and Webb, J. S. (1970). *In* "Trace Element Metabolism in Animals" (C. F. Mills, ed.), pp. 397–409. Livingstone, Edinburgh, Scotland.
Thornton, I., Atkinson, W. J., Webb, J. S., and Poole, D. B. R. (1966). *Ir. J. Agric. Res.* **5,** 280–283.
Thornton, I., Moon, R. N. B., and Webb, J. S. (1969). *Nature (London)* **221,** 457–458.
Thornton, I., Kershaw, G. F., and Davies, M. K. (1972). *J. Agric. Sci.* **78,** 165–171.
Thornton, I., Smith C., and Van Dorst, S. (1985). *In* "Trace Elements in Man and Animals" (C. F. Mills, I. Bremmer, and J. K. Chesters, eds.), pp. 853–855. Commonw. Agric. Bur., Farnham Royal, England.
Tilley, J. M. A., and Raymond, W. F. (1957). *Herb. Abstr.* **27,** 235–245.
Tilley, J. M. A., and Terry, R. A. (1963). *J. Br. Grassl. Soc.* **18,** 104–111.
Tilley, J. M. A., and Terry, R. A. (1969). *J. Br. Grassl. Soc.* **24,** 290–295.
Tilley, J. M. A., Deriaz, R. E., and Terry, R. A. (1960). *Proc. Int. Grassl. Congr., 8th* pp. 533–537.
Tillman, A. D., and Brethour, J. R. (1958). *J. Anim. Sci.* **17,** 782–786.
Tinnimit, P., and Thomas, J. W. (1976). *J. Anim. Sci.* **43,** 1058–1065.
Todd, J. R. (1961a). *J. Agric. Sci.* **56,** 411–415.
Todd, J. R. (1961b). *J. Agric. Sci.* **57,** 35–38.

Todd, J. R. (1962). *J. Agric. Sci.* **58,** 277–279.
Todd, J. R. (1965). *Br. Vet. J.* **121,** 371–380.
Todd, J. R. (1966). *Proc. Int. Grassl. Congr., 10th* pp. 178–180.
Todd, J. R. (1970). *In* "Trace Element Metabolism in Animals" (C. F. Mills, ed.), pp. 448–451. Livingstone, Edinburgh, Scotland.
Todd, J. R., and Morrison, N. E. (1964). *J. Br. Grassl. Soc.* **19,** 179–182.
Todd, J. R., Scally, W. C. P., and Ingram, J. M. (1966). *Vet. Rec.* **78,** 888–891.
Todd, J. R., Milne, A. A., and How, P. F. (1967). *Vet. Rec.* **81,** 653–656.
Tomas, F. M., and Potter, B. J. (1976). *Aust. J. Agric. Res.* **27,** 873–880.
Tomas, F. M., Moir, R. J., and Somers, M. (1967). *Aust. J. Agric. Res.* **18,** 635–645.
Topps, J. H., and Thompson, J. K. (1984). "Blood Characteristics and the Nutrition of Ruminants," Ref. Book 260. Minist. Agric., Fish. Food, London.
Topps, J. H., Reed, W. D. C., and Elliott, R. C. (1965). *J. Agric. Sci.* **64,** 397–402.
Torres, F., and Boelcke, C. (1978). *Anim. Prod.* **27,** 315–321.
Towers, N. R. (1977). *Proc. Nutr. Soc. N.Z.* **2,** 11–19.
Towers, N. R., and Grace, N. D. (1983). *In* "The Mineral Requirements of Grazing Ruminants" (N. D. Grace, ed.), Occas. Publ. 9, pp. 84–91. N.Z. Soc. Anim. Prod., Wellington, New Zealand.
Towers, N. R., Young, P. W., and Wright, D. E. (1981). *N.Z. Vet. J.* **29,** 113–114.
Towers, N. R., Young, P. W., and Smeaton, D. C. (1984). *Proc. N.Z. Soc. Anim. Prod.* **44,** 155–157.
Trigg, T. E., and Marsh, R. (1979). *Proc. N.Z. Soc. Anim. Prod.* **39,** 260–264.
Trinder, N., and Hall, R. J. (1972). *J. Sci. Food Agric.* **23,** 557–566.
Trinder, N., Hall, R. J., and Renton, C. P. (1973). *Vet. Rec.* **93,** 641–643.
Troelsen, J. E., and Bell, J. M. (1968). *Can. J. Anim. Sci.* **48,** 361–372.
Troelsen, J. E., and Bigsby, F. W. (1964). *J. Anim. Sci.* **23,** 1139–1142.
Troelsen, J. E., and Campbell, J. B. (1968). *Anim. Prod.* **10,** 289–296.
Troelsen, J. E., and Campbell, J. B. (1969). *J. Agric. Sci.* **73,** 145–154.
Troelsen, J. E., and Hanel, D. J. (1966). *Can. J. Anim. Sci.* **46,** 149–156.
Tudor, G. D., and Minson, D. J. (1982). *J. Agric. Sci.* **98,** 395–404.
Tudor, G. D., and Morris, J. G. (1971). *Aust. J. Exp. Agric. Anim. Husb.* **11,** 483–487.
Tuen, A. A., Wadsworth, J. C., and Murray, M. (1984). *Proc. Nutr. Soc. Aust.* **9,** 144–147.
Tullberg, J. N., and Minson, D. J. (1978). *J. Agric. Sci.* **91,** 557–561.
Tulloh, N. M., Hughes, J. W., and Newth, R. P. (1965). *N.Z. J. Agric. Res.* **8,** 636–651.
Turner, A. W., Kelley, R. B., and Dann, A. T. (1935). *J. Counc. Sci. Ind. Res. (Aust.)* **8,** 120–132.
Tyrrell, H. F., and Moe, P. W. (1975). *J. Dairy Sci.* **58,** 1151–1163.
Uden, P. (1988). *Anim. Feed Sci. Technol.* **19,** 145–157.
Ullrey, D. E., Brady, P. S., Whetter, P. A., Ku, P. K., and Magee, W. T. (1977). *J. Anim. Sci.* **45,** 559–565.
Ullrey, D. E., Light, M. R., Brady, P. S., Whetter, P. A., Tilton, J. E., Henneman, H. A., and Magee, W. R. (1978). *J. Anim. Sci.* **46,** 1515–1521.
Ulyatt, M. J. (1969). *Proc. N.Z. Soc. Anim. Prod.* **29,** 114–123.
Ulyatt, M. J. (1970). *Proc. Int. Grassl. Congr., 11th* pp. 709–713.
Ulyatt, M. J. (1981). *In* "Grazing Animals" (F. H. W. Morley, ed.), pp. 125–140. Elsevier, Amsterdam.
Ulyatt, J. M., MacRae, J. C., Clarke, R. T. J., and Pearce, P. D. (1975). *J. Agric. Sci.* **84,** 453–458.
Ulyatt, M. J., Lancashire, J. A., and Jones, W. T. (1976). *Proc. N.Z. Grassl. Assoc.* **38,** 107–118.

Ulyatt, J. M., Waghorn, G. C., John, A., Reid, C. S. W., and Monro, J. (1984). *J. Agric. Sci.* **102,** 645–657.
Ulyatt, M. J., Dellow, D. W., John, A., Reid, C. S. W., and Waghorn, G. C. (1986). *In* "Control of Digestion and Metabolism in Ruminants" (L. P. Milligan, W. L. Grovum, and A. Dobson, eds.), pp. 498–515. Prentice-Hall, Englewood Cliffs, New Jersey.
Underwood, E. J. (1971). "Trace Elements in Human and Animal Nutrition." Academic Press, New York.
Underwood, E. J. (1977). "Trace Elements in Human and Animal Nutrition," 4th ed. Academic Press, New York.
Underwood, E. J. (1981). "The Mineral Nutrition of Livestock," 2nd ed. Commonw. Agric. Bur., Farnham Royal, England.
Underwood, E. J., and Filmer, J. F. (1935). *Aust. Vet. J.* **11,** 84–92.
Underwood, E. J., and Harvey, R. J. (1938). *Aust. Vet. J.* **14,** 183–189.
Underwood, E. J., and Somers, M. (1969). *Aust. J. Agric. Res.* **20,** 889–897.
Ushida, K., Jouany, J. P., and Thivend, P. (1986). *Br. J. Nutr.* **56,** 407–419.
Utley, P. R., Hollis, D., Chapman, H. D., Monson, W. G., Marchant, W. H., and McCormick, W. C. (1974). *J. Anim. Sci.* **38,** 490–495.
Utley, P. R., Monson, W. G., Burton, G. W., Hellwig, R. E., and McCormick, W. C. (1978). *J. Anim. Sci.* **47,** 800–804.
Utley, P. R., Hellwig, R. E., McCormick, W. C., and Butler, J. L. (1982). *Can. J. Anim. Sci.* **62,** 499–505.
Vadiveloo, J., and Holmes, W. (1979a). *Anim. Prod.* **29,** 121–129.
Vadiveloo, J., and Holmes, W. (1979b). *J. Agric. Sci.* **93,** 553–562.
Valdivia-Rodriguez, R. J. (1979). *Diss. Abstr. B* **39,** 6.
Van de Riet, H., Verma, L. R., and Craig, W. M. (1988). *Appl. Eng. Agric.* **4,** 19–23.
Van der Kley, F. K. (1956). *Neth. J. Agric. Sci.* **4,** 197–204.
Van Dyne, G. M., and Torrell, D. T. (1964). *J. Range Manage.* **17,** 7–19.
Van Eenaeme, C., Istasse, L., Lambot, O., Bienfait, J. M., and Geilen, M. (1981). *Agric. Environ.* **6,** 161–170.
Van Heerden, A. J., Nel. J. W., and Mellet, P. (1974). *S. Afr. J. Anim. Sci.* **4,** 109–112.
Van Keulen, J., and Young, B. A. (1977). *J. Anim. Sci.* **44,** 282–287.
Van Schalkwyk, A., and Lombard, P. E. (1969). *Agroanimalia* **1,** 45–52.
Van Schalkwyk, A., Lombard, P. E., and Vorster, L. F. (1968). *S.-Afr. Tydskr. Landbouwet.* **11,** 113–121.
Van Soest, P. J. (1963). *J. Assoc. Off. Agric. Chem.* **46,** 829–835.
Van Soest, P. J. (1965a). *J. Dairy Sci.* **48,** 815.
Van Soest, P. J. (1965b). *J. Anim. Sci.* **24,** 834–843.
Van Soest, P. J. (1967). *J. Anim. Sci.* **26,** 119–128.
Van Soest, P. J., and Jones, L. H. P. (1968). *J. Dairy Sci.* **51,** 1644–1648.
Van Soest, P. J., Wine, R. H., and Moore, L. A. (1966). *Proc. Int. Grassl. Congr., 10th* pp. 438–441.
Van Wyk, H. P. D., Oosthuizen, S. A., and Basson, J. D. (1951). *Sci. Bull.—S. Afr., Dep. Agric. Tech. Serv.* **298.**
Ventura, M., Moore, J. E., Ruelke, O. C., and Franke, D. E. (1975). *J. Anim. Sci.* **40,** 769–774.
Vercoe, J. E., Pearce, G. R., and Tribe, D. E. (1962). *J. Agric. Sci.* **59,** 343–348.
Vérité, R., and Journet, M. (1970). *Ann. Zootech.* **19,** 225–268.
Vérité, R., Journet, M., and Jarrige, R. (1979). *Livest. Prod. Sci.* **6,** 349–368.
Vincente-Chandler, J., Silva, S., and Figarella, J. (1959a). *J. Agric. Univ. P.R.* **43,** 228–239.
Vincente-Chandler, J., Silva, S., and Figarella, J. (1959b). *J. Agric. Univ. P.R.* **43,** 240–248.
Vincente-Chandler, J., Figarellla, J., and Silva, S. (1961). *J. Agric. Univ. P.R.* **45,** 37–45.

Vincente-Chandler, J., Figarellla, J., and Silva, S. (1961). *J. Agric. Univ. P.R.* **45,** 37–45.
Vipperman, P. E., Preston, R. L., Kintner, L. D., and Pfander, W. H. (1969). *J. Nutr.* **97,** 449–462.
Virtanen, A. I. (1933). *Emp. J. Exp. Agric.* **1,** 143–155.
Voelker, H. H., Jorgensen, N. A., Mohanty, G. O., and Owens, M. J. (1969). *J. Dairy Sci.* **52,** 929.
Vogel, K. P., Haskins, F. A., and Gorz, H. J. (1981). *Crop Sci.* **21,** 39–41.
Vohnout, K., and Bateman, J. V. (1972). *J. Agric. Sci.* **78,** 413–416.
Voigt, P. W., Kneebone, W. R., McIlvain, E. H., Shoop, M. C., and Webster, J. E. (1970). *Agron. J.* **62,** 673–676.
Vose, P. B., and Breese, E. L. (1964). *Ann. Bot. (London)* **28,** 251–270.
Vough, L. R., and Marten, G. C. (1971). *Agron. J.* **63,** 40–42.
Wadsworth, J. C., and Cohen, R. D. H. (1976). *Rev. Rural Sci.* **3,** 143–153.
Waghorn, A. J., Jones, W. T., and Shelton, I. D. (1987). *Proc. N.Z. Soc. Anim. Prod.* **47,** 25–30.
Wagner, J. F., Brown, H., Bradley, N. W., Dinusson, W., Dunn, W., Elliston, N., Miyat, N., Mowrey, D., Moreman, J., Pendlum, L. C., Parrott, C., Richardson, L., Rush, I., and Woody, H. (1984). *J. Anim. Sci.* **58,** 1062–1067.
Wainman, F. W., and Blaxter, K. L. (1972). *J. Agric. Sci.* **79,** 435–445.
Wainman, F. W., Blaxter, K. L., and Smith, J. S. (1972). *J. Agric. Sci.* **78,** 441–447.
Waite, R., and Sastry, K. N. S. (1949). *Emp. J. Exp. Agric.* **17,** 179–187.
Waite, R., Holmes, W., and Boyd, J. (1952). *J. Agric. Sci.* **42,** 314–321.
Waite, R., Johnston, M. J., and Armstrong, D. G. (1964). *J. Agric. Sci.* **62,** 391–398.
Waldo, D. R. (1975). *J. Anim. Sci.* **41,** 424.
Waldo, D. R., and Goering, H. K. (1979). *J. Anim. Sci.* **49,** 1560–1568.
Waldo, D. R., Miller, R. M., Smith, L. W., Okamoto, M., and Moore, L. A. (1966). *Proc. Int. Grassl. Congr., 10th* pp. 570–574.
Walgenbach, R. P., Marten, G. C., and Blake, G. R. (1981). *Crop Sci.* **21,** 843–849.
Walker, C. A. (1957). *J. Agric. Sci.* **49,** 394–400.
Walker, W. R. (1980). *In* "CSIRO Symposium on the Importance of Copper in Biology and Medicine" (B. R. McAuslan, ed.), pp. 4–19. CSIRO, Canberra, Australia.
Walker, T. W., Edwards, G. H. A., Cavell, A. J., and Rose, T. H. (1953). *J. Br. Grassl. Soc.* **8,** 45–68.
Walker, D. J., Egan, A. R., Nader, C. J., Ulyatt, M. J., and Storer, G. B. (1975). *Aust. J. Agric. Res.* **26,** 699–708.
Walker, H. G., Kohler, G. O., and Garrett, W. N. (1982). *J. Anim. Sci.* **55,** 498–504.
Wallace, J. D., Sneva, F. A., Raleigh, R. J., and Rumburg, C. B. (1966). *J. Anim. Sci.* **25,** 598.
Walshe, M. J., and Conway, A. (1960). *Proc. Int. Grassl. Congr., 8th* pp. 548–553.
Walters, R. J. K. (1971). *J. Agric. Sci.* **76,** 243–252.
Walters, R. J. K. (1973). *Rep.—Welsh Plant Breed. Stn. (Aberystwyth, Wales)* p. 43.
Walters, R. J. K. (1974). *Vaextodling* **29,** 184–192.
Walters, R. J. K., and Evans, E. M. (1974). *Rep.—Welsh Plant Breed. Stn. (Aberystwyth, Wales)* p. 48.
Walters, R. J. K., and Evans, E. M. (1979). *Grass Forage Sci.* **34,** 37–44.
Ward, H. K. (1968). *Rhod. J. Agric. Res.* **6,** 93–101.
Ward, G. M., Boren, F. W., Smith, E. F., and Brethour, J. R. (1965). *J. Dairy Sci.* **48,** 811.
Ward, G., Harbers, L. H., and Blaha, J. J. (1979). *J. Dairy Sci.* **62,** 715–722.
Ward, R. G., Smith, G. S., Wallace, J. D., Urquhart, N. S., and Shenk, J. S. (1982). *J. Anim. Sci.* **54,** 399–402.
Warren, W. P., Martz, F. A., Asay, K. H., Hilderbrand, E. S., Payne, C. G., and Vogt, J. R. (1974). *J. Anim. Sci.* **39,** 93–96.

Watkins, B. E., Ullrey, D. E., Schmitt, S. M., and Nachreiner, R. F. (1981). *In* "Trace Element Metabolism in Man and Animals" (J. M. Howell, J. M. Gawthorne, and C. L. White, eds.), pp. 61–62. Aust. Acad. Sci., Canberra, Australia.
Watkinson, J. H. (1962). *Trans. Int. Soc. Soil Sci., Comm. IV and V* pp. 149–154.
Watkinson, J. H. (1983). *N.Z. Vet. J.* **31,** 78–85.
Watson, S. J., and Horton, E. A. (1936). *J. Agric. Sci.* **26,** 142–154.
Watson, C., and Norton, B. W. (1982). *Proc. Aust. Soc. Anim. Prod.* **14,** 467–470.
Watson, L. T., Ammerman, C. B., Feaster, J. P., and Roessler, C. E. (1973). *J. Anim. Sci.* **36,** 131–136.
Weast, R. C., Selby, S. M., and Hodgman, C. D., eds. (1965). "Handbook of Chemistry and Physics." Chem. Rubber Co., Cleveland, Ohio.
Weaver, D. E., Coppock, C. E., Lake, G. B., and Everett, R. W. (1978). *J. Dairy Sci.* **61** 1782–1788.
Webb, R. J., and Cmarik, G. F. (1958). *In* "Pelleted Feeds for Livestock and Poultry." Ill. Agric. Exp. Stn., Dixon Springs, Illinois.
Webb, R. J., Cmarik, G. F., and Cate, H. A. (1957). *J. Anim. Sci.* **16,** 1057.
Webb, J. S., Thornton, I., and Fletcher, K. (1968). *Nature (London)* **217,** 1010–1012.
Webb, K. E., Fontenot, J. P., and Wise, M. B. (1975). *J. Anim. Sci.* **40,** 760–768.
Webster, J. E., Hogan, J. W., and Elder, W. C. (1965). *Agron. J.* **57,** 323–325.
Weinmann, H. (1948). *Rhod. Agric. J.* **45,** 119–131.
Weinmann, H. (1950). *Rhod. Agric. J.* **47,** 435–454.
Weir, W. C., Rendig, V. V., and Ittner, N. R. (1958). *J. Anim. Sci.* **17,** 113–123.
Weir, W. C., Jones, L. G., and Meyer, J. H. (1960). *J. Anim. Sci.* **19,** 5–19.
Weiss, W. P., Conrad, H. R., Martin, C. M., Cross, R. F., and Shockey, W. L. (1968a). *J. Anim. Sci.* **63,** 525–532.
Weiss, W. P., Conrad, H. R., and Shockey, W. L. (1968b). *J. Dairy Sci.* **69,** 2658–2670.
Welch, J. G., Palmer, R. H., Gilman, B. E., and Bull, L. S. (1982). *J. Dairy Sci.* **65,** (Suppl. 1), 147.
Weller, R. F., Phipps, R. H., and Griffith, E. S. (1984). *J. Agric. Sci.* **103,** 223–227.
Wells, N. (1957). *N.Z. J. Sci. Technol., Sect. B* **38,** 884–902.
Wells, N. (1962). *"Available Sodium (Sodium in Sweet Vernal Grass) in Unfertilized Soil," Single Factor Maps 59 and 60.* Department Science Industrial Research, Wellington, New Zealand.
Wernli, C. G., and Wilkins, R. J. (1980a). *J. Agric. Sci.* **94,** 209–218.
Wernli, C. G., and Wilkins, R. J. (1980b). *J. Agric. Sci.* **94,** 219–227.
Westerman, D. T., and Robbins, C. W. (1974). *Agron. J.* **66,** 207–208.
Weston, R. H. (1967). *Aust. J. Agric. Res.* **18,** 983–1002.
Weston, R. H. (1970). *Aust. J. Exp. Agric. Anim. Husb.* **10,** 679–684.
Weston, R. H. (1985). *Proc. Nutr. Soc. Aust.* **10,** 55–62.
Weston, R. H., and Cantle, J. A. (1982). *Proc. Nutr. Soc. Aust.* **7,** 147.
Weston, R. H., and Hogan, J. P. (1967). *Aust. J. Agric. Res.* **18,** 789–801.
Weston, R. H., and Hogan, J. P. (1968a). *Aust. J. Agric. Res.* **19,** 567–576.
Weston, R. H., and Hogan, J. P. (1968b). *Aust. J. Agric. Res.* **19,** 963–979.
Weston, R. H., and Hogan, J. P. (1971). *Aust. J. Agric. Res.* **22,** 139–157.
Westra, R., and Christopherson, R. J. (1976). *Can. J. Anim. Sci.* **56,** 699–708.
Whanger, P. D., Weswig, P. H., Muth, O. H., and Oldfield, J. E. (1970). *Am. J. Vet Res.* **31,** 965–972.
Wheeler, J. L., Reardon, T. F., and Lambourne, L. J. (1963). *Aust. J. Agric. Res.* **14,** 364–372.

Wheeler, J. L., Said, A. N., and Lindstad, P. (1978). *Proc. Aust. Soc. Anim. Prod.* **12,** 128.
White, C. L., and Somers, M. (1978). *In* "Trace Element Metabolism in Man and Animals" (M. Kirchgessner, ed.), pp. 526–529. Arbeitskreis Tiernaehrungsforsch., Freising, West Germany.
Whitehead, D. C. (1966). *Grassl. Res. Inst. (Hurley), Tech. Rep.* **4.**
Whitehead, D. C. (1973). *J. Soil Sci.* **24,** 260–270.
Whitehead, D. C., and Jones, E. C. (1969). *J. Sci. Food Agric.* **20,** 584–591.
Whitehead, D. C., Jones, L. H. P., and Barnes, R. J. (1978). *J. Sci. Food Agric.* **29,** 1–11.
Whitehead, D. C., Barnes, R. J., and Jones, L. H. P. (1983). *J. Sci. Food Agric.* **34,** 901–909.
Whitlock, R. H., Kessler, M. J., and Tasker, J. B. (1975). *Cornell Vet.* **65,** 512–526.
Whitten, L. K. (1971). "Diseases of Domestic Animals in New Zealand," pp. 341–342. Edit. Serv., Wellington, New Zealand.
Wiedmeier, R. D., Clark, D. H., Arambel, M. J., and Lamb, R. C. (1986). *Nutr. Rep. Int.* **33,** 391–395.
Wiener, G., and Field, A. C. (1969). *J. Agric. Sci.* **73,** 275–278.
Wilkins, R. J. (1966). *J. Br. Grassl. Soc.* **21,** 65–69.
Wilkins, R. J. (1969). *J. Agric. Sci.* **73,** 57–64.
Wilkins, R. J., ed. (1976). "Green Crop Fractionation," Occas. Symp. 9. Br. Grassl. Soc., Hurley, Maidenhead, England.
Wilkins, R. J., Hutchinson, K. J., Wilson, R. F., and Harris, C. E. (1971). *J. Agric. Sci.* **77,** 531–537.
Wilkins, R. J., Lonsdale, C. R., Tetlow, R. M., and Forrest, T. J. (1972). *Anim. Prod.* **14,** 177–188.
Wilkins, J. F., Kilgour, R. J., Gleeson, A. C., Cox, R. J., Geddes, S. J., and Simpson, I. H. (1982). *Aust. J. Exp. Agric. Anim. Husb.* **22,** 24–28.
Wilkinson, J. M., and Phipps, R. J. (1979). *J. Agric. Sci.* **92,** 485–491.
Wilkinson, J. I. D., Appleby, W. G. G., Shaw, C. J., Lebas, G., and Pflug, R. (1980). *Anim. Prod.* **31,** 159–162.
Wilkinson, J. M., Le Du, Y. L. P., Cook, J. E., and Baker, R. D. (1982). *Grass Forage Sci.* **37,** 29–38.
Williams, A. J., and Miller, H. P. (1965). *Aust. J. Exp. Agric. Anim. Husb.* **5,** 385–389.
Williams, D. L., Whiteman, J. V., and Tillman, A. D. (1969). *J. Anim. Sci.* **28,** 807–812.
Williams, W. T., Haydock, K. P., Edye, L. A., and Ritson, J. B. (1971). *Aust. J. Agric. Res.* **22,** 979–991.
Williams, P. E. V., Innes, G. M., and Brewer, A. (1984). *Anim. Feed Sci. Technol.* **11,** 115–124.
Willoughby, W. M. (1958). *Proc. Aust. Soc. Anim. Prod.* **2,** 42–45.
Wilman, D. (1975). *J. Br. Grassl. Soc.* **30,** 141–147.
Wilman, D., and Owen, I. G. (1982). *J. Agric. Sci.* **99,** 577–586.
Wilman, D., Ojuederie, B. M., and Asare, E. O. (1976). *J. Br. Grassl. Soc.* **31,** 73–79.
Wilson, D. (1965). *J. Agric. Sci.* **65,** 285–292.
Wilson, A. D. (1966a). *Aust. J. Agric. Res.* **17,** 155–163.
Wilson, J. G. (1966b). *Vet. Rec.* **79,** 562–566.
Wilson, G. F. (1970). *Proc. N.Z. Soc. Anim. Prod.* **30,** 123–127.
Wilson, G. F. (1978). *N.Z. J. Exp. Agric.* **6,** 267–269.
Wilson, G. F. (1980). *Anim. Prod.* **31,** 153–157.
Wilson, G. F. (1981). *Proc. N.Z. Soc. Anim. Prod.* **41,** 53–60.
Wilson, J. R. (1982a). *In* "Nutritional Limits to Animal Production from Pastures" (J. B. Hacker, ed.), pp. 111–131. Commonw. Agric. Bur., Farnham Royal, England.

Wilson, G. F. (1982b). *In* "Dairy Production from Pasture," pp. 400–401. N.Z. Aust. Soc. Anim. Prod.
Wilson, J. R. (1983). *Aust. J. Agric. Res.* **34,** 377–390.
Wilson, J. R., and Ford, C. W. (1971). *Aust. J. Agric. Res.* **22,** 563–571.
Wilson, G. F., and Grace, N. D. (1978). *N.Z. J. Exp. Agric.* **6,** 267–269.
Wilson, J. R., and Mannetje, L. (1978). *Aust. J. Agric. Res.* **29,** 503–516.
Wilson, R. K., and McCarrick, R. B. (1966). *Proc. Int. Grassl. Congr., 10th* pp. 371–379.
Wilson, R. K., and McCarrick, R. B. (1967). *Ir. J. Agric. Res.* **6,** 267–279.
Wilson, J. R., and Minson, D. J. (1980). *Trop. Grassl.* **14,** 253–259.
Wilson, P. N., and Osbourn, D. F. (1960). *Biol. Rev.* **35,** 324–363.
Wilson, R. K., and Pigden, W. J.(1964). *Can. J. Anim. Sci.* **44,** 122–123.
Wilson, C. L., and Ritchie, N. S. (1981). *J. Sci. Food Agric.* **32,** 993–994.
Wilson, R. K., and Winter, K. A. (1984). *Ir. J. Agric. Res.* **23,** 97–98.
Wilson, J. R., and Wong, C. C. (1982). *Aust. J. Agric. Res.* **33,** 937–949.
Wilson, R. K., Flynn, A. V., and Conniffe, D. (1968). *Ir. J. Agric. Res.* **7,** 31–36.
Wilson, R. F., Terry, R. A., and Osbourn, D. F. (1969a). *J. Br. Grassl. Soc.* **24,** 119–122.
Wilson, G. F., Reid, C. S. W., Molloy, L. F., Metson, A. J., and Butler, G. W. (1969b). *N.Z. J. Agric. Res.* **12,** 467–488.
Wilson, A. D., Weir, W. C., and Torell, D. T. (1971). *J. Anim. Sci.* **32,** 1046–1050.
Wilson, T. R., Kromann, R. P., and Evans, D. W. (1978). *J. Anim. Sci.* **46,** 1351–1355.
Wilson, J. R., Brown, R. H., and Windham, W. R. (1983). *Crop Sci.* **23,** 141–146.
Wilson, J. R., Jones, P. N., and Minson, D. J. (1986). *Trop. Grassl.* **20,** 145–156.
Winch, J. E., and Major, H. (1981). *Can. J. Plant Sci.* **61,** 45–51.
Winks, L., and Laing, A. R. (1972). *Proc. Aust. Soc. Anim. Prod.* **9,** 253–257.
Winks, L., and O'Rourke, P. K. (1977). *J. Aust. Inst. Agric. Sci.* **43,** 76–77.
Winks, L., Lamberth, F. C., Moir, K. W., and Pepper, P. M. (1974). *Aust. J. Exp. Agric. Anim. Husb.* **14,** 146–154.
Winks, L., Laing, A. R., Wright, G. S., and Stokoe, J. (1976). *J. Aust. Inst. Agric. Sci.* **42,** 246–251.
Winks, L., Lambeth, F. C., and O'Rourke, P. K. (1977). *Aust. J. Exp. Agric. Anim. Husb.* **17,** 357–366.
Winks, L., Laing, A. R., O'Rourke, P. K., and Wright, G. S. (1979). *Aust. J. Exp. Agric. Anim. Husb.* **19,** 522–529.
Winter, K. A., and Collins, D. P. (1987). *Can. J. Plant Sci.* **67,** 445–449.
Winter, W. H., and Jones, R. K. (1977). *Trop. Grassl.* **11,** 247–255.
Winter, W. H., and McLean, R. W. (1988). *Proc. Aust. Soc. Anim. Prod.* **17,** 485.
Winter, W. H., Siebert, B. D., and Kuchel, R. E. (1977a). *Aust. J. Exp. Agric. Anim. Husb.* **17,** 10–15.
Winter, W. H., Edye, L. A., and Williams, W. T. (1977b). *Aust. J. Exp. Agric. Anim. Husb.* **17,** 187–196.
Wise, M. B., Ordoveza, A. L., and Barrick, E. R. (1963). *J. Nutr.* **79,** 79–84.
Witt, K. E., and Owens, F. N. (1983). *J. Anim. Sci.* **56,** 930–937.
Wodzicka-Tomaszewska, M. (1963). *N.Z. J. Agric. Res.* **6,** 440–447.
Wodzicka-Tomaszewska, M. (1966). *N.Z. J. Agric. Res.* **9,** 909–915.
Wohlt, J. E., Sniffen, C. J., and Hoover, W. H. (1973). *J. Dairy Sci.* **56,** 1052–1057.
Wong, E. (1973). *In* "Chemistry and Biochemistry of Herbage" (G. W. Butler and R. W. Bailey, eds.), Vol. 1, pp. 265–322. Academic Press, New York.
Woodman, H. E., Blunt, D. L., and Stewart, J. (1927). *J. Agric. Sci.* **17,** 209–263.
Woodman, H. E., Bee, J. W., and Griffith, G. (1930). *J. Agric. Sci.* **20,** 53–62.
Woodward, T. E. (1936). *J. Dairy Sci.* **19,** 347–357.

Woolliams, C., Suttle, N. F., Woolliams, J. A., Jones, D. G., and Wiener, G. (1986a). *Anim. Prod.* **43,** 293–301.
Woolliams, J. A., Woolliams, C., Suttle, N. F., Jones, D. G., and Weiner, G. (1968b). *Anim. Prod.* **43,** 303–317.
Wright, J. R., and Lawton, K. (1954). *Soil Sci.* **77,** 95–105.
Wright, D. E., Towers, N. R., and Sinclair, D. P. (1978). *N.Z. J. Agric. Res.* **21,** 215–221.
Wunsch, D. S. (1939). *Chem. Ind. (London)* **41,** 531–533.
Wylie, A. R. G., and Steen, R. W. J. (1988). *Grass Forage Sci.* **43,** 79–86.
Wylie, M. J., Fontenot, J. P., and Greene, L. W. (1985). *J. Anim. Sci.* **61,** 1219–1229.
Young, P. W. (1967). *Proc. Ruakura Farmers' Conf.* **19,** 70–78.
Young, P. W. (1975). *Proc. Ruakura Farmers' Conf.* **27,** 39–45.
Young, P. W., and Rys, G. (1977). *Proc. Ruakura Farmers' Conf.* **29,** 30–33.
Young, V. R., Richards, W. P. C., Lofgreen, G. P., and Luick, J. R. (1966a). *Br. J. Nutr.* **20,** 783–794.
Young, V. R., Lofgreen, G. P., and Luick, J. R. (1966b). *Br. J. Nutr.* **20,** 795–805.
Young, L. G., Jenkins, K. J., and Edmeades, D. M. (1977). *Can. J. Anim. Sci.* **57,** 793–799.
Young, P. W., O'Connor, M. B., and Feyter, C. (1979). *Proc. Ruakura Farmers' Conf.* **31,** 110–120.
Young, N. E., Newton, J. E., and Orr, R. J. (1980). *Grass Forage Sci.* **35,** 197–202.
Young, P. W., Rys, G., and O'Connor, M. B. (1981). *Proc. N.Z. Soc. Anim. Prod.* **41,** 61–67.
Younge, O. R., and Plucknett, D. L. (1965). *Proc. Int. Grassl. Congr., 9th* pp. 959–963.
Youssef, F. G. (1988). *Outlook Agric.* **17,** 104–111.
Yu, Y. (1978). *J. Anim. Sci.* **46,** 313–319.
Yungblut, D. H., Stone, J. B., MacLeod, G. K., Grieve, D. G., and Burnside, E. B. (1981). *Can. J. Anim. Sci.* **61,** 151–157.
Zadrazil, F. (1984). *In* "Straw and Other Fibrous By-products as Feed" (F. Sundstol and E. Owen, eds.), pp. 276–292. Elsevier, Amsterdam.
Zemmelink, G. (1980). *Versl. Landbouwkd. Onderz.* **896.**
Zemmelink, G., Haggar, R. J., and Davies, J. H. (1972). *Anim. Prod.* **15,** 85–88.
Zoby, J. L. F., and Holmes, W. (1983). *J. Agric. Sci.* **100,** 139–148.

Index

D

G

H

I

N

O

P